# Topics in Applied Physics
# Volume 80



## Available Online

Topics in Applied Physics is part of the Springer LINK service. For all customers with standing orders for Topics in Applied Physics we offer the full text in electronic form via LINK free of charge. Please contact your librarian who can receive a password for free access to the full articles by registration at:

http://link.springer.de/orders/index.htm

If you do not have a standing order you can nevertheless browse through the table of contents of the volumes and the abstracts of each article at:

http://link.springer.de/series/tap/

There you will also find more information about the series.

Springer-Verlag Berlin Heidelberg GmbH

# Topics in Applied Physics

Topics in Applied Physics is a well-established series of review books, each of which presents a comprehensive survey of a selected topic within the broad area of applied physics. Edited and written by leading research scientists in the field concerned, each volume contains review contributions covering the various aspects of the topic. Together these provide an overview of the state of the art in the respective field, extending from an introduction to the subject right up to the frontiers of contemporary research.

Topics in Applied Physics is addressed to all scientists at universities and in industry who wish to obtain an overview and to keep abreast of advances in applied physics. The series also provides easy but comprehensive access to the fields for newcomers starting research.

Contributions are specially commissioned. The Managing Editors are open to any suggestions for topics coming from the community of applied physicists no matter what the field and encourage prospective editors to approach them with ideas.

See also: http://www.springer.de/phys/books/tap/


**Managing Editors**

Dr. Claus E. Ascheron

Springer-Verlag Heidelberg
Topics in Applied Physics
Tiergartenstr. 17
69121 Heidelberg
Germany
Email: ascheron@springer.de

Dr. Hans J. Kölsch

Springer-Verlag Heidelberg
Topics in Applied Physics
Tiergartenstr. 17
69121 Heidelberg
Germany
Email: koelsch@springer.de

**Assistant Editor**

Dr. Werner Skolaut

Springer-Verlag Heidelberg
Topics in Applied Physics
Tiergartenstr. 17
69121 Heidelberg
Germany
Email: skolaut@springer.de

Mildred S. Dresselhaus Gene Dresselhaus
Phaedon Avouris (Eds.)

# Carbon Nanotubes

## Synthesis, Structure, Properties, and Applications

With 235 Figures

Springer

Prof. Mildred S. Dresselhaus
Department of Physics
Massachussetts Institute of Technology
77 Massachussetts Avenue
02139 Cambridge, MA
USA
milli@mgm.mit.edu

Dr. Gene Dresselhaus
Department of Physics
Massachussetts Institute of Technology
77 Massachussetts Avenue
02139 Cambridge, MA
USA
gene@mgm.mit.edu

Dr. Phaedon Avouris
T. J. Watson Research Center
IBM Research Division
10598 Yorktown Hights
New York
USA
avouris@us.ibm.com

Library of Congress Cataloging-in-Publication Data

Carbon nanotubes : synthesis, structure, properties, and applications / Mildred S. Dresselhaus, Gene Dresselhaus, Phaedon Avouris (eds.).
p. cm. -- (Topics in applied physics ; v. 80)
Includes bibliographical references and index.
ISBN 978-3-662-30789-2 ISBN 978-3-540-39947-6 (eBook)
DOI 10.1007/978-3-540-39947-6

1. Carbon. 2. Nanostructure materials. 3. Tubes. I. Dresselhaus, M. S. II. Dresselhaus, G. III. Avouris, Phaedon, 1945- IV. Series.

TA455.C3 C38 2000
620.1'93--dc21

00-048279

Physics and Astronomy Classification Scheme (PACS):
61.48.+c, 81.05.Tp, 72.80.Rj, 71.20.Tx, 78.30.Na

ISSN print edition: 0303-4216
ISSN electronic edition: 1437-0859
ISBN 978-3-662-30789-2



http://www.springer.de


Originally published by Springer-Verlag Berlin Heidelberg New York in 2001
Softcover reprint of the hardcover 1st edition 2001



Typesetting: DA-TEX Gerd Blumenstein, Leipzig
Cover design: *design & production* GmbH, Heidelberg

Printed on acid-free paper 57/3111 5

# Foreword

by R. E. Smalley, Chemistry Nobel Lauveate 1996

Since the discovery of the fullerenes in 1985 my research group and I have had the privilege of watching from a central location the worldwide scientific community at work in one of its most creative, penetrating, self-correcting, and divergent periods of the last century. In his recent book, "The Transparent Society", David Brin discusses the virtues of an open, information rich society in which individuals are free to knowledgeably criticize each other, to compete, to test themselves and their ideas in a free market place, and thereby help evolve a higher level of the social organism. He points out that modern science has long functioned in this mode, and argues that this open criticism and appeal to experiment has been the keystone of its success. This new volume, Carbon Nanotubes, is a wonderful example of this process at work.

Here you will find a summary of the current state of knowledge in this explosively growing field. You will see a level of creativity, breadth and depth of understanding that I feel confident is beyond the capability of any single human brain to achieve in a lifetime of thought and experiment. But many fine brains working independently in the open society of science have done it very well indeed, and in a very short time.

While the level of understanding contained in this volume is immense, it is clear to most of us working in this field that we have only just begun. The potential is vast. Here we have what is almost certainly the strongest, stiffest, toughest molecule that can ever be produced, the best possible molecular conductor of both heat and electricity. In one sense the carbon (fullerene) nanotube is a new man-made polymer to follow on from nylon, polypropylene, and Kevlar. In another, it is a new "graphite" fiber, but now with the ultimate possible strength. In yet another, it is a new species in organic chemistry, and potentially in molecular biology as well, a carbon molecule with the almost alien property of electrical conductivity, and super-steel strength.

Can it be produced in megatons?
Can it be spun into continuous fibers?
Can it grown in organized arrays or as a perfect single crystal?
Can it be sorted by diameter and chirality?
Can a single tube be cloned?
Can it be grown enzymatically?
Can it be assembled by the molecular machinery of living cells?
Can it be used to make nanoelectronic devices, nanomechanical memories, nano machines, ...?
Can it be used to wire a brain?

There is no way of telling at this point. Certainly for many researchers, the best, most exciting days of discovery in this field are still ahead. For the rest of us, it will be entertaining just to sit back and watch the worldwide organism of science at work. Hold on to your seats! Watch the future unfold.

Houston, Texas
January 2001

*Richard E. Smalley*

# Preface

Carbon nanotubes are unique nanostructures with remarkable electronic and mechanical properties, some stemming from the close relation between carbon nanotubes and graphite, and some from their one-dimensional aspects. Initially, carbon nanotubes aroused great interest in the research community because of their exotic electronic structure. As other intriguing properties have been discovered, such as their remarkable electronic transport properties, their unique Raman spectra, and their unusual mechanical properties, interest has grown in their potential use in nanometer-sized electronics and in a variety of other applications, as discussed in this volume.

An ideal nanotube can be considered as a hexagonal network of carbon atoms that has been rolled up to make a seamless hollow cylinder. These hollow cylinders can be tens of micrometers long, but with diameters as small as 0.7 nm, and with each end of the long cylinder "capped with half a fullerene molecule, i.e., 6 pentagons". Single-wall nanotubes, having a cylindrical shell with only one atom in thickness, can be considered as the fundamental structural unit. Such structural units form the building blocks of both multi-wall nanotubes, containing multiple coaxial cylinders of ever-increasing diameter about a common axis, and nanotube ropes, consisting of ordered arrays of single-wall nanotubes arranged on a triangular lattice.

The first reported observation of carbon nanotubes was by Iijima in 1991 for multi-wall nanotubes. It took, however, less than two years before single-wall carbon nanotubes were discovered experimentally by Iijima at the NEC Research Laboratory in Japan and by Bethune at the IBM Almaden Laboratory in California. These experimental discoveries and the theoretical work, which predicted many remarkable properties for carbon nanotubes, launched this field and propelled it forward. The field has been advancing at a breathtaking pace ever since with many unexpected discoveries. These exciting developments encouraged the editors to solicit articles for this book on the topic of carbon nanotubes while the field was in a highly active phase of development.

This book is organized to provide a snapshot of the present status of this rapidly moving field. After the introduction in Chap. 1, which provides some historical background and a brief summary of some basic subject matter and definitions, the connection between carbon nanotubes and other carbon materials is reviewed in Chap. 2. Recent developments in the synthesis and purification of single-wall and multi-wall carbon nanotubes are discussed in Chap. 3. This is followed in Chap. 4 by a review of our present limited understanding of the growth mechanism of single-wall and multi-wall carbon nanotubes. Chapter 5 demonstrates the generality of tubular nanostructures by discussing general principles for tubule growth, and providing the reader with numerous examples of inorganic nanotube formation. The unique electronic structure and properties of perfect and defective carbon nanotubes are reviewed from a theoretical standpoint in Chap. 6. The electrical properties, transport, and magneto-transport measurements on single-wall nanotubes and ropes, as well as simple device structures based on carbon nanotubes are presented in Chap. 7. Scanning tunneling microscopy is used to study that nanotube electronic structure and spectra. The use of nanotubes as atomic force microscope tips for ultra-high resolution and chemically sensitive imaging is also discussed in Chap. 8. The application of optical spectroscopy to nanotubes is presented in Chap. 9. In this chapter, the discussion of the optical properties focuses on the electronic structure, the phonon structure, and the coupling between electrons and phonons in observations of resonance Raman scattering and related phenomena. The contribution made by electron spectroscopies to the characterization of the electronic structure of the nanotubes is discussed in Chap. 10, in comparison with similar studies devoted to graphite and $C_{60}$. This is followed in Chap. 11 by a brief review of the phonon and thermal properties, with emphasis given to studies of the specific heat and the thermal conductivity, which are both sensitive to the low-dimensional aspects of carbon nanotubes. Chapter 12 discusses experiments and theory on the mechanical properties of carbon nanotubes. Linear elastic parameters, non-linear instabilities, yield strength, fracture and supra-molecular interactions are all reviewed. Chapter 13 discusses transport measurements, magnetotransport properties, electron spin resonance, and a variety of other exotic properties of multiwall nanotubes. The volume concludes in Chap. 14 with a brief review of the present early status of potential applications of carbon nanotubes.

Because of the relative simplicity of carbon nanotubes, we expect them to play an important role in the current rapid expansion of fundamental studies on nanostructures and their potential use in nanotechnology. This simplicity allows us to develop detailed theoretical models for predicting new phenomena connected with these tiny, one-dimensional systems, and then look for these phenomena experimentally. Likewise, new experimental effects, which have been discovered at an amazingly rapid rate, have provided stimulus for further theoretical developments, many of which are expected to be broadly applicable to nanostructures and nanotechnology research and development.

Cambridge, Massachusetts
Yorktown Heights, New York
January 2001

*Mildred S. Dresselhaus*
*Gene Dresselhaus*
*Phaedon Avouris*

# Contents

# Introduction to Carbon Materials Research

Mildred S. Dresselhaus[1] and Phaedon Avouris[2]

[1] Currently on leave from the
Department of Electrical Engineering and Computer Science
and Department of Physics
MIT Cambridge, Massachusetts 02139, USA
millie@mgm.mit.edu

[2] IBM Research Division, T. J. Watson Research Laboratory
Yorktown Heights, NY 10598, USA
avouris@us.ibm.com

**Abstract.** A brief historical review of carbon nanotube research is presented and some basic definitions relevant to the structure and properties of carbon nanotubes are provided.

Carbon nanotubes are unique nanostructures that can be considered conceptually as a prototype one-dimensional (1D) quantum wire. The fundamental building block of carbon nanotubes is the very long all-carbon cylindrical Single Wall Carbon Nanotube (SWNT), one atom in wall thickness and tens of atoms around the circumference (typical diameter ~1.4 nm). Initially, carbon nanotubes aroused great interest in the research community because of their exotic electronic properties, and this interest continues as other remarkable properties are discovered and promises for practical applications develop.

## 1 Historical Introduction

Very small diameter (less than 10 nm) carbon filaments were prepared in the 1970's and 1980's through the synthesis of vapor grown carbon fibers by the decomposition of hydrocarbons at high temperatures in the presence of transition metal catalyst particles of <10 nm diameter [1,2,3,4,5,6]. However, no detailed systematic studies of such very thin filaments were reported in these early years, and it was not until the observation of carbon nanotubes in 1991 by Iijima of the NEC Laboratory in Tsukuba, Japan (see Fig. 1) using High-Resolution Transmission Electron Microscopy (HRTEM) [7] that the carbon nanotube field was seriously launched. Independently, and at about the same time (1992), Russian workers also reported the discovery of carbon nanotubes and nanotube bundles, but generally having a much smaller length to diameter ratio [8,9].

A direct stimulus to the systematic study of carbon filaments of very small diameters came from the discovery of fullerenes by *Kroto*, *Smalley*, *Curl*, and coworkers at Rice University [10]. In fact, Smalley and others speculated publically in these early years that a single wall carbon nanotube might be a limiting case of a fullerene molecule. The connection between carbon

M. S. Dresselhaus, G. Dresselhaus, Ph. Avouris (Eds.): Carbon Nanotubes,
Topics Appl. Phys. **80**, 1–9 (2001)

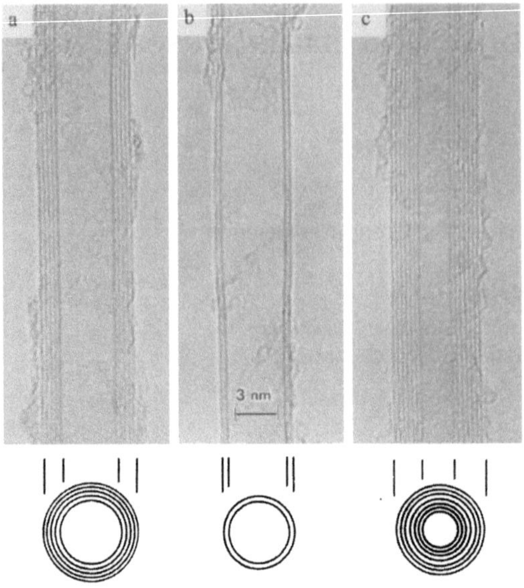


**Fig. 1.** The observation by TEM of multi-wall coaxial nanotubes with various inner and outer diameters, $d_\mathrm{i}$ and $d_\mathrm{o}$, and numbers of cylindrical shells $N$ reported by *Iijima* in 1991: (**a**) $N = 5$, $d_o$=67Å; (**b**) $N = 2$, $d_\mathrm{o}$=55Å; and (**c**) $N = 7$, $d_\mathrm{i}$=23Å, $d_\mathrm{o}$=65Å [7]

nanotubes and fullerenes was further promoted by the observation that the terminations of the carbon nanotubes were fullerene-like caps or hemispheres. It is curious that the smallest reported diameter for a carbon nanotube is the same as the diameter of the $C_{60}$ molecule, which is the smallest fullerene to follow the isolated pentagon rule. This rule requires that no two pentagons be adjacent to one another, thereby lowering the strain energy of the fullerene cage. While there is not, as yet, a definite answer to a provocative question raised by *Kubo* and directed to *Endo* in 1977, regarding the minimum size of a carbon fiber [11], this question was important for identifying carbon fibers with very small diameters as carbon nanotubes, the one-dimensional limit of a fullerene molecule. A recent report of a carbon nanotube which a diameter of 0.4 nm an a $C_{20}$ end cap may provide an answer to the question.

It was the *Iijima's* observation of the multiwall carbon nanotubes in Fig. 1 in 1991 [7] that heralded the entry of many scientists into the field of carbon nanotubes, stimulated at first by the remarkable 1D dimensional quantum effects predicted for their electronic properties, and subsequently by the promise that the remarkable structure and properties of carbon nanotubes might give rise to some unique applications. Whereas the initial experimental Iijima observation was for Multi-Wall Nanotubes (MWNTs), it was less than two years before Single-Wall Carbon Nanotubes (SWNTs) were discovered experimentally by Iijima and his group at the NEC Laboratory and by *Bethune* and coworkers at the IBM Almaden laboratory [12,13]. These findings were especially important because the single wall nanotubes are more fundamental, and had been the basis for a large body of theoretical studies and predictions that preceded the experimental observation of single wall carbon nanotubes. The most striking of these theoretical developments was the prediction that carbon nanotubes could be either semiconducting or metallic depending on their geometrical characteristics, namely their diameter and the orientation of their hexagons with respect to the nanotube axis (chiral angle) [14,15,16]. Though predicted in 1992, it was not until 1998 that these predictions regarding their remarkable electronic properties were corroborated experimentally [17,18].

A major breakthrough occurred in 1996 when *Smalley* and coworkers at Rice University [19] successfully synthesized bundles of aligned single wall carbon nanotubes, with a small diameter distribution, thereby making it possible to carry out many sensitive experiments relevant to 1D quantum physics, which could not previously be undertaken [19]. Of course, actual carbon nanotubes have finite length, contain defects, and interact with other nanotubes or with the substrate and these factors often complicate their behavior.

A great deal of progress has been made in characterizing carbon nanotubes and in understanding their unique properties since their 'discovery' in 1991. This progress is highlighted in the chapters of this volume.

## 2 Basic Background

To provide a framework for the presentations in the following chapters, we include in this introductory chapter some definitions about the structure and description of carbon nanotubes.

The structure of carbon nanotubes has been explored early on by high resolution Transmission Electron Microscopy (TEM) and Scanning Tunneling Microscopy (STM) techniques [20], yielding direct confirmation that the nanotubes are seamless cylinders derived from the honeycomb lattice representing a single atomic layer of crystalline graphite, called a graphene sheet, (Fig. 2a). The structure of a single-wall carbon nanotube is conveniently explained in terms of its 1D unit cell, defined by the vectors $\boldsymbol{C}_h$ and $\mathbf{T}$ in Fig. 2(a).

The circumference of any carbon nanotube is expressed in terms of the chiral vector $\boldsymbol{C}_h = n\hat{\mathbf{a}}_1 + m\hat{\mathbf{a}}_2$ which connects two crystallographically equivalent sites on a 2D graphene sheet (see Fig. 2a) [14]. The construction in Fig. 2a depends uniquely on the pair of integers $(n, m)$ which specify the chiral vector. Figure 2(a) shows the chiral angle $\theta$ between the chiral vector $\boldsymbol{C}_h$ and the "zigzag" direction ($\theta = 0$) and the unit vectors $\hat{\boldsymbol{a}}_1$ and $\hat{\boldsymbol{a}}_2$ of the hexagonal honeycomb lattice of the graphene sheet. Three distinct types of nanotube structures can be generated by rolling up the graphene sheet into a cylinder as described below and shown in Fig. 3. The zigzag and armchair nanotubes, respectively, correspond to chiral angles of $\theta = 0$ and 30°, and chiral nanotubes correspond to $0 < \theta < 30°$. The intersection of the vector $\overrightarrow{OB}$ (which is normal to $\boldsymbol{C}_h$) with the first lattice point determines the fun-

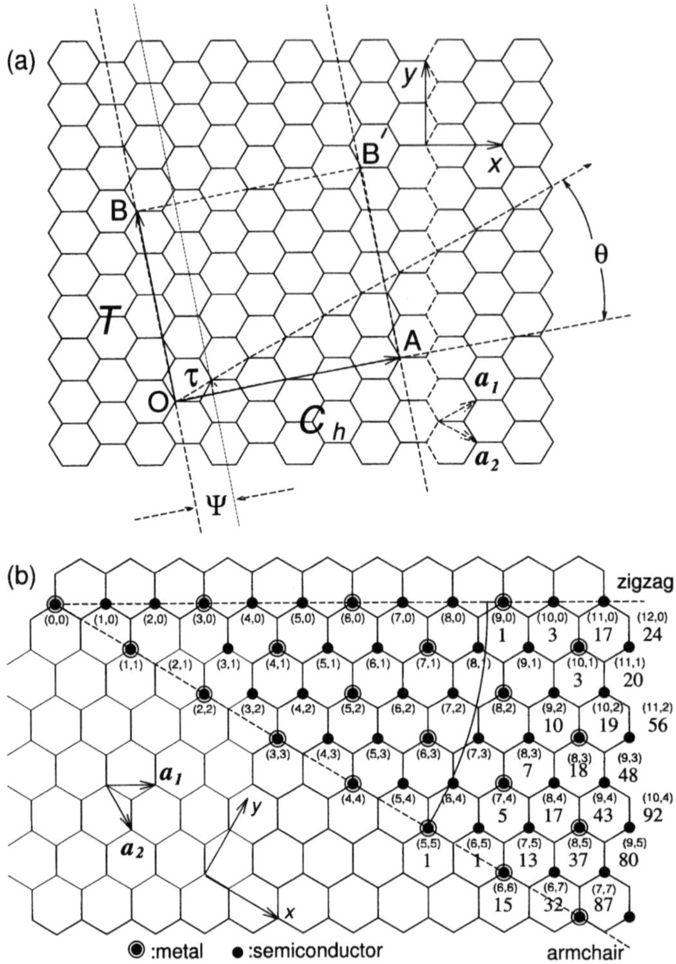


**Fig. 2.** (**a**) The chiral vector $\overrightarrow{OA}$ or $\boldsymbol{C}_h = n\hat{\boldsymbol{a}}_1 + m\hat{\boldsymbol{a}}_2$ is defined on the honeycomb lattice of carbon atoms by unit vectors $\hat{\boldsymbol{a}}_1$ and $\hat{\boldsymbol{a}}_2$ and the chiral angle $\theta$ with respect to the zigzag axis. Along the zigzag axis $\theta = 0°$. Also shown are the lattice vector $\overrightarrow{OB} = \boldsymbol{T}$ of the 1D nanotube unit cell and the rotation angle $\psi$ and the translation $\tau$ which constitute the basic symmetry operation $R = (\psi|\tau)$ for the carbon nanotube. The diagram is constructed for $(n, m) = (4, 2)$ [14]. (**b**) Possible vectors specified by the pairs of integers $(n, m)$ for general carbon nanotubes, including zigzag, armchair, and chiral nanotubes. Below each pair of integers $(n, m)$ is listed the number of distinct caps that can be joined continuously to the carbon nanotube denoted by $(n, m)$ [14]. The *encircled dots* denote metallic nanotubes while the *small dots* are for semiconducting nanotubes [21]

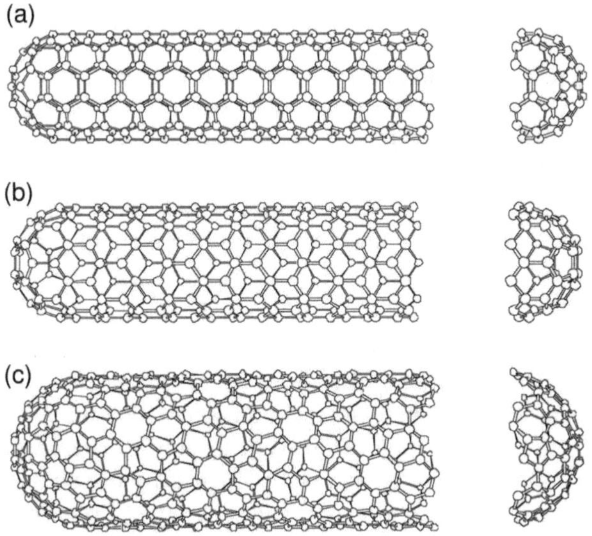


**Fig. 3.** Schematic models for single-wall carbon nanotubes with the nanotube axis normal to the chiral vector which, in turn, is along: (**a**) the $\theta = 30°$ direction [an "armchair" $(n, n)$ nanotube], (**b**) the $\theta = 0°$ direction [a "zigzag" $(n, 0)$ nanotube], and (**c**) a general $\theta$ direction, such as $\overrightarrow{OB}$ (see Fig. 2), with $0 < \theta < 30°$ [a "chiral" $(n, m)$ nanotube]. The actual nanotubes shown here correspond to $(n, m)$ values of: (**a**) (5, 5), (**b**) (9, 0), and (**c**) (10, 5). The nanotube axis for the (5,5) nanotube has 5-fold rotation symmetry, while that for the (9,0) nanotube has 3-fold rotation symmetry [22]

damental one-dimensional (1D) translation vector $\boldsymbol{T}$. The unit cell of the 1D lattice is the rectangle defined by the vectors $\boldsymbol{C}_h$ and $\boldsymbol{T}$ [Fig. 2(a)].

The cylinder connecting the two hemispherical caps of the carbon nanotube (see Fig. 3) is formed by superimposing the two ends of the vector $\boldsymbol{C}_h$ and the cylinder joint is made along the two lines $\overrightarrow{OB}$ and $\overrightarrow{AB'}$ in Fig. 2a. The lines $\overrightarrow{OB}$ and $\overrightarrow{AB'}$ are both perpendicular to the vector $\boldsymbol{C}_h$ at each end of $\boldsymbol{C}_h$ [14]. In the $(n, m)$ notation for $\boldsymbol{C}_h = n\hat{\boldsymbol{a}}_1 + m\hat{\boldsymbol{a}}_2$, the vectors $(n, 0)$ or $(0, m)$ denote zigzag nanotubes and the vectors $(n, n)$ denote armchair nanotubes. All other vectors $(n, m)$ correspond to chiral nanotubes [23]. The nanotube diameter $d_t$ is given by

$$d_t = \sqrt{3}a_{C-C}(m^2 + mn + n^2)^{1/2}/\pi = C_h/\pi \,, \tag{1}$$

where $C_h$ is the length of $\boldsymbol{C}_h$, $a_{C-C}$ is the C–C bond length (1.42 Å), The chiral angle $\theta$ is given by

$$\theta = \tan^{-1}[\sqrt{3}n/(2m + n)] \,. \tag{2}$$

From (2) it follows that $\theta = 30°$ for the $(n, n)$ armchair nanotube and that the $(n, 0)$ zigzag nanotube would have $\theta = 60°$. From Fig. 2a it follows that if we limit $\theta$ to be between $0 \leq \theta \leq 30°$, then by symmetry $\theta = 0°$ for a zigzag nanotube. Both armchair and zigzag nanotubes have a mirror plane and thus are considered as achiral. Differences in the nanotube diameter $d_t$ and chiral angle $\theta$ give rise to differences in the properties of the various carbon nanotubes. The symmetry vector $\boldsymbol{R} = (\psi|\tau)$ of the symmetry group for the nanotubes is indicated in Fig. 2a, where both the translation unit or pitch $\tau$ and the rotation angle $\psi$ are shown. The number of hexagons, $N$, per unit cell of a chiral nanotube, specified by integers $(n, m)$, is given by

$$N = 2(m^2 + n^2 + nm)/d_{\mathrm{R}}\,, \tag{3}$$

where $d_{\mathrm{R}} = d$ if $n - m$ is not a multiple of $3d$ or $d_R = 3d$, if $n - m$ is a multiple of $3d$ and d is defined as the largest common divisor of $(n, m)$. Each hexagon in the honeycomb lattice (Fig. 2a) contains two carbon atoms. The unit cell area of the carbon nanotube is $N$ times larger than that for a graphene layer and consequently the unit cell area for the nanotube in reciprocal space is correspondingly $1/N$ times smaller. Table 1 provides a summary of relations useful for describing the structure of single wall nanotubes [24]. Figure 2b indicates the nanotubes that are semiconducting and those that are metallic, and shows the number of distinct fullerene caps that can be used to close the ends of an $(n, m)$ nanotube, such that the fullerene cap satisfies the isolated pentagon rule.

The nanotube material produced by either the laser vaporization method or the carbon arc method appears in a Scanning Electron Microscope (SEM) image as a mat of carbon "bundles" or "ropes" 10–20 nm in diameter and up to 100μm or more in length. Under Transmission Electron Microscope (TEM) examination, an individual carbon nanotube bundle can be imaged and shown to consist primarily of an array of single-wall carbon nanotubes aligned along a common axis [19]. A collection of these intertwined bundles of single wall nanotubes is called a nanotube "rope" [26]. This picture is corroborated by X-ray diffraction measurements (which view many ropes at once) and transmission electron microscopy (which views a single bundle) [19]. Typical diameters of the single-wall nanotubes are ~1.4 nm, very close to the diameter of an ideal (10,10) nanotube. X-ray diffraction measurements showed that these single-wall nanotubes form a two-dimensional triangular lattice. For carbon nanotubes synthesized by the laser vaporization method, the lattice constant for this triangular lattice is ~1.7 nm, with an inter-tube separation of 0.32 nm at closest approach between adjacent nanotubes within a single bundle [19].

Multiwall carbon nanotubes (see Fig. 1) contain several coaxial cylinders, each cylinder being a single-wall carbon nanotube. Whereas multi-wall carbon nanotubes require no catalyst for their growth either by the laser vaporization or carbon arc methods, metal catalyst species such as the transition metals Fe, Co, or Ni are necessary for the growth of the single-wall nanotubes [19]. In most cases, two different catalyst species have been used to efficiently syn-

**Table 1.** Structural parameters for carbon nanotubes [24][a)]

| symbol | name | formula | value |
|---|---|---|---|
| $a$ | length of unit vector | $a = \sqrt{3}a_{\text{C-C}} = 2.49$ Å, | $a_{\text{C-C}} = 1.44$ Å |
| $\hat{\mathbf{a}}_1,\hat{\mathbf{a}}_2$ | unit vectors | $\left(\frac{\sqrt{3}}{2},\frac{1}{2}\right)a,\ \left(\frac{\sqrt{3}}{2},-\frac{1}{2}\right)a$ | $x, y$ coordinate |
| $\hat{\mathbf{b}}_1,\hat{\mathbf{b}}_2$ | reciprocal lattice vectors | $\left(\frac{1}{\sqrt{3}},1\right)\frac{2\pi}{a},\ \left(\frac{1}{\sqrt{3}},-1\right)\frac{2\pi}{a}$ | $x, y$ coordinate |
| $\mathbf{C}_h$ | chiral vector | $\mathbf{C}_h = n\mathbf{a}_1 + m\mathbf{a}_2 \equiv (n, m),$ | $(0 \leq \|m\| \leq n)$ |
| $L$ | length of $\mathbf{C}_h$ | $L = \|\mathbf{C}_h\| = a\sqrt{n^2+m^2+nm}$ | |
| $d_t$ | diameter | $d_t = L/\pi$ | |
| $\theta$ | chiral angle | $\sin\theta = \frac{\sqrt{3}m}{2\sqrt{n^2+m^2+nm}}$ | $0 \leq \|\theta\| \leq \frac{\pi}{6}$ |
| | | $\cos\theta = \frac{2n+m}{2\sqrt{n^2+m^2+nm}},$ | $\tan\theta = \frac{\sqrt{3}m}{2n+m}$ |
| $d$ | gcd($n$,$m$)[b)] | | |
| $d_R$ | gcd($2n+m$,$2m+n$)[b)] | $d_R = \begin{cases} d & \text{if } (n-m) \text{ is not multiple of } 3d \\ 3d & \text{if } (n-m) \text{ is multiple of } 3d \end{cases}$ | |
| $\mathbf{T}$ | translational vector | $\mathbf{T} = t_1\mathbf{a}_1 + t_2\mathbf{a}_2 \equiv (t_1, t_2)$ | $\gcd(t_1, t_2) = 1$[b)] |
| | | $t_1 = \frac{2m+n}{d_R},\ t_2 = -\frac{2n+m}{d_R}$ | |
| $T$ | length of $\mathbf{T}$ | $T = \|\mathbf{T}\| = \frac{\sqrt{3}L}{d_R}$ | |
| $N$ | Number of hexagons in the nanotube unit cell. | $N = \frac{2(n^2+m^2+nm)}{d_R}$ | |
| $\mathbf{R}$ | symmetry vector | $\mathbf{R} = p\hat{\mathbf{a}}_1 + q\hat{\mathbf{a}}_2 \equiv (p, q)$ <br> $t_1 q - t_2 p = 1,$ <br> $(0 < mp - nq \leq N)$ | $\gcd(p, q) = 1$[b)] |
| $\tau$ | pitch of $\mathbf{R}$ | $\tau = \frac{(mp-nq)T}{N} = \frac{MT}{N}$ | |
| $\psi$ | rotation angle of $\mathbf{R}$ | $\psi = \frac{2\pi}{N}$ | in radians |
| $M$ | number of $\mathbf{T}$ in $N\mathbf{R}$. | $N\mathbf{R} = \mathbf{C}_h + M\mathbf{T}$ | |

[a)] In this table $n$, $m$, $t_1$, $t_2$, $p$, $q$ are integers and $d$, $d_R$ $N$ and $M$ are integer functions of these integers.

[b)] $\gcd(n, m)$ denotes the greatest common divisor of the two integers $n$ and $m$.

thesize arrays of single wall carbon nanotubes by either the laser vaporization or carbon arc methods.

From an historical standpoint, the discovery in 1996 by the Rice group [19] of a synthetic method involving laser vaporization of a graphite target that leads to high quality single-wall nanotubes gave a great boost to the field. The technique involves the laser vaporization of a target composed of Co-Ni/graphite at 1200°C. Subsequently, an efficient carbon arc method for making single-wall nanotubes with a diameter distribution similar to that of the Rice group was developed at Montpellier in France [25]. These initial successes have triggered a worldwide effort to increase the yield and to narrow the diameter and chirality distribution of the nanotubes. The ability to synthesize nanotubes with a given chirality is particularly important in electronics applications since, as is discussed in several chapters in this volume, the electrical properties of the nanotubes depend on their chirality [see Fig. 2b]. Furthermore, for electronic applications, it is highly desirable to be able to grow a nanotube in a particular location and in a specified direction on a substrate. Progress in this direction was achieved recently by employing lithographic techniques to control the position of a metal catalyst particle and the use of the technique of Chemical Vapor-phase Deposition (CVD) to grow the nanotube at that location [27,28].

Along with the experimental efforts to improve the control and yield of the synthesis method, there are continuing theoretical efforts to interpret the experimental results and decipher the mechanism by which the nanotube structure is self-assembled from C atoms and small C clusters [29]. The various chapters of this book show that although the field is still very young, a tremendous amount of information has already been gathered on the structure and properties of both single- and multi-wall nanotubes and a multitude of applications have appeared on the horizon.

## References


1. M. Endo, *Mecanisme de croissance en phase vapeur de fibres de carbone (The growth mechanism of vapor-grown carbon fibers).* PhD thesis, University of Orleans, Orleans, France (1975) (in French)
2. M. Endo, PhD thesis, Nagoya University, Japan (1978) (in Japanese)
3. S. Iijima, J. Cryst. Growth **55**, 675–683 (1980)
4. A. Oberlin, M. Endo, T. Koyama, Carbon **14**, 133 (1976)
5. A. Oberlin, M. Endo, T. Koyama, J. Cryst. Growth **32**, 335–349 (1976)
6. Patrice Gadelle (unpublished)
7. S. Iijima, Nature (London) **354**, 56 (1991)
8. Z. Ya. Kosakovskaya, L. A. Chernozatonskiĭ, E. A. Fedorov, JETP Lett. (Pis'ma Zh. Eksp. Teor.) **56**, 26 (1992)
9. E. G. Gal'pern, I. V. Stankevich, A. L. Christyakov, L. A. Chernozatonskiĭ, JETP Lett. (Pis'ma Zh. Eksp. Teor.) **55**, 483 (1992)
10. H. W. Kroto, J. R. Heath, S. C. O'Brien, R. F. Curl, and R. E. Smalley, Nature (London) **318**, 162–163 (1985)

11. R. Kubo, 1977. (private communication to M. Endo at the Kaya Conference)
12. S. Iijima, T. Ichihashi, Nature (London) **363**, 603 (1993)
13. D. S. Bethune, C. H. Kiang, M. S. de Vries, G. Gorman, R. Savoy, J. Vazquez, R. Beyers, Nature (London) **363**, 605 (1993)
14. M. S. Dresselhaus, G. Dresselhaus, R. Saito, Phys. Rev. B **45**, 6234 (1992)
15. J. W. Mintmire, B. I. Dunlap, C. T. White, Phys. Rev. Lett. **68**, 631–634 (1992)
16. N. Hamada, S. Sawada, A. Oshiyama, Phys. Rev. Lett. **68**, 1579–1581 (1992)
17. J. W. G. Wildöer, L. C. Venema, A. G. Rinzler, R. E. Smalley, C. Dekker, Nature (London) **391**, 59–62 (1998)
18. T. W. Odom, J. L. Huang, P. Kim, C. M. Lieber, Nature (London) **391**, 62–64 (1998)
19. A. Thess, R. Lee, P. Nikolaev, H. Dai, P. Petit, J. Robert, C. Xu, Y. H. Lee, S. G. Kim, A. G. Rinzler, D. T. Colbert, G. E. Scuseria, D. Tománek, J. E. Fischer, R. E. Smalley, Science **273**, 483–487 (1996)
20. C. H. Olk, J. P. Heremans, J. Mater. Res. **9**, 259–262 (1994)
21. M. S. Dresselhaus, G. Dresselhaus, P. C. Eklund, *Science of Fullerenes and Carbon Nanotubes* (Academic, New York 1996)
22. M. S. Dresselhaus, G. Dresselhaus, R. Saito, Carbon **33**, 883–891 (1995)
23. R. Saito, M. Fujita, G. Dresselhaus, M. S. Dresselhaus, Appl. Phys. Lett. **60**, 2204–2206 (1992)
24. R. Saito, G. Dresselhaus, M. S. Dresselhaus, *Physical Properties of Carbon Nanotubes* (Imperial College Press, London 1998)
25. C. Journet, W. K. Maser, P. Bernier, A. Loiseau, M. Lamy de la Chapelle, S. Lefrant, P. Deniard, R. Lee, J. E. Fischer, Nature (London) **388**, 756–758 (1997)
26. A. Loiseau, N. Demoncy, O. Stéphan, C. Colliex, H. Pascard, pp. 1–16, in *Proceedings of the Nanotubs'99 Workshops,* East Lansing, MI, 1999, D. Tománek, R. J. Enbody (Eds.), Kluwer Academic/Plenum, New York (1999)
27. See also the chapter by H. Dai, in this volume
28. J. Liu, A. G. Rinzler, H. Dai, J. H. Hafner, R. K. Bradley, P. J. Boul, A. Lu, T. Iverson, K. Shelimov, C. B. Huffman, F. Rodriguex-Macia, D. T. Colbert, R. E. Smalley, Science **280**, 1253–1256 (1998)
29. See also the chapter by J. C. Charlier and S. Iijima, chapter 4 in this volume

# Relation of Carbon Nanotubes to Other Carbon Materials

Mildred S. Dresselhaus[1] and Morinobu Endo[2]

[1] Department of Electrical Engineering and Computer Science
and Department of Physics
MIT, Cambridge, MA 02139, USA
millie@mgm.mit.edu

[2] Faculty of Engineering
Department of Electrical and Electronic Engineering
Shinshu University, Nagano-shi, 380 Japan
endo@endomoribu.shinshu-u.ac.jp

**Abstract.** A review of the close connection between the structure and properties of carbon nanotubes and those of graphite and its related materials is presented in order to gain new insights into the exceptional properties of carbon nanotubes. The two dominant types of bonding ($sp^2$ and $sp^3$) that occur in carbon materials and carbon nanotubes are reviewed, along with the structure and properties of carbon materials closely related to carbon nanotubes, such as graphite, graphite whiskers, and carbon fibers. The analogy is made between the control of the properties of graphite through the intercalation of donor and acceptor species with the corresponding doping of carbon nanotubes.

Carbon nanotubes are strongly related to other forms of carbon, especially to crystalline 3D graphite, and to its constituent 2D layers (where an individual carbon layer in the honeycomb graphite lattice is called a graphene layer). In this chapter, several forms of carbon materials are reviewed, with particular reference to their relevance to carbon nanotubes. Their similarities and differences relative to carbon nanotubes with regard to structure and properties are emphasized.

The bonding between carbon atoms in the $sp^2$ and $sp^3$ configurations is discussed in Sect. 1. Connections are made between the nanotube curvature and the introduction of some $sp^3$ bonding to the $sp^2$ planar bonding of the graphene sheet. The unusual properties of carbon nanotubes are derived from the unusual properties of $sp^2$ graphite by imposing additional quantum confinement and topological constraints in the circumferential direction of the nanotube. The structure and properties of graphite are discussed in Sect. 2, because of their close connection to the structure and properties of carbon nanotubes, which is followed by reviews of graphite whiskers and carbon fibers in Sect. 3 and Sect. 4, respectively. Particular emphasis is given to the vapor grown carbon fibers because of their especially close connection to carbon nanotubes. The chapter concludes with brief reviews of liquid carbon

M. S. Dresselhaus, G. Dresselhaus, Ph. Avouris (Eds.): Carbon Nanotubes,
Topics Appl. Phys. **80**, 11–28 (2001)

and graphite intercalation compounds in Sect. 5 and Sect. 6, respectively, relating donor and acceptor nanotubes to intercalated graphite.

## 1 Bonding Between Carbon Atoms

Carbon-based materials, clusters, and molecules are unique in many ways. One distinction relates to the many possible configurations of the electronic states of a carbon atom, which is known as the hybridization of atomic orbitals and relates to the bonding of a carbon atom to its nearest neighbors.

Carbon is the sixth element of the periodic table and has the lowest atomic number of any element in column IV of the periodic table. Each carbon atom has six electrons which occupy $1s^2$, $2s^2$, and $2p^2$ atomic orbitals. The $1s^2$ orbital contains two strongly bound core electrons. Four more weakly bound electrons occupy the $2s^2 2p^2$ valence orbitals. In the crystalline phase, the valence electrons give rise to $2s$, $2p_x$, $2p_y$, and $2p_z$ orbitals which are important in forming covalent bonds in carbon materials. Since the energy difference between the upper $2p$ energy levels and the lower $2s$ level in carbon is small compared with the binding energy of the chemical bonds, the electronic wave functions for these four electrons can readily mix with each other, thereby changing the occupation of the $2s$ and three $2p$ atomic orbitals so as to enhance the binding energy of the C atom with its neighboring atoms. The general mixing of $2s$ and $2p$ atomic orbitals is called hybridization, whereas the mixing of a single $2s$ electron with one, two, or three $2p$ electrons is called $sp^n$ hybridization with $n = 1, 2, 3$ [1,2].

Thus three possible hybridizations occur in carbon: $sp$, $sp^2$ and $sp^3$, while other group IV elements such as Si and Ge exhibit primarily $sp^3$ hybridization. Carbon differs from Si and Ge insofar as carbon does not have inner atomic orbitals, except for the spherical $1s$ orbitals, and the absence of nearby inner orbitals facilitates hybridizations involving only valence $s$ and $p$ orbitals for carbon. The various bonding states are connected with certain structural arrangements, so that $sp$ bonding gives rise to chain structures, $sp^2$ bonding to planar structures and $sp^3$ bonding to tetrahedral structures.

The carbon phase diagram (see Fig. 1) guided the historical synthesis of diamond in 1960 [4], and has continued to inspire interest in new forms of carbon, as they are discovered [3]. Although we have learned much about carbon since that time, much ignorance remains about the possible phases of carbon. While $sp^2$ bonded graphite is the ground state phase of carbon under ambient conditions, at higher temperatures and pressures, $sp^3$ bonded cubic diamond is stable. Other regions of the phase diagram show stability ranges for hexagonal diamond, hexagonal carbynes [5,6,7], and liquid carbon [8]. It is believed that a variety of novel $\pi$-electron carbon bulk phases remain to be discovered and explored.

In addition to the bulk phases featured in the carbon phase diagram, much attention has recently focussed on small carbon clusters [9], since the

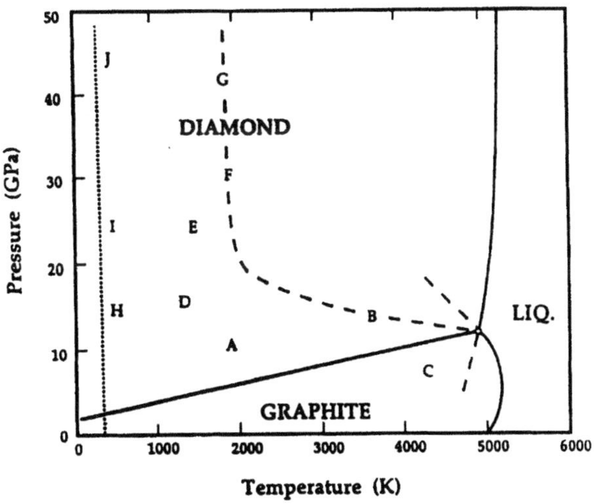


**Fig. 1.** A recent version of the phase diagram of carbon [3]. *Solid lines* represent equilibrium phase boundaries. A: commercial synthesis of diamond from graphite by catalysis; B: rapid solid phase graphite to diamond synthesis; C: fast transformation of diamond to graphite; D: hexagonal graphite to hexagonal diamond synthesis; E: shock compression graphite to hexagonal diamond synthesis; F: shock compression graphite to cubic-type diamond synthesis; B, F, G: graphite or hexagonal diamond to cubic diamond synthesis; H, I, J: compressed graphite acquires diamond-like properties, but reverts to graphite upon release of pressure

discovery of fullerenes in 1985 by *Kroto* et al. [10] and of carbon nanotubes in 1991 by *Iijima* [11]. The physical reason why these nanostructures form is that a graphene layer (defined as a single 2D layer of 3D graphite) of finite size has many edge atoms with dangling bonds,indexdangling bonds and these dangling bonds correspond to high energy states. Therefore the total energy of a small number of carbon atoms (30–100) is reduced by eliminating dangling bonds, even at the expense of increasing the strain energy, thereby promoting the formation of closed cage clusters such as fullerenes and carbon nanotubes.

The rolling of a single graphene layer, which is a hexagonal network of carbon atoms, to form a carbon nanotube is reviewed in this volume in the introductory chapter [12], and in the chapters by *Louie* [13] and *Saito/Kataura* [14], where the two indices $(n, m)$ that fully identify each carbon nanotube are specified [9,15]. Since nanotubes can be rolled from a graphene sheet in many ways [9,15], there are many possible orientations of the hexagons on the nanotubes, even though the basic shape of the carbon nanotube wall is a cylinder.

A carbon nanotube is a graphene sheet appropriately rolled into a cylinder of nanometer size diameter [13,14,15]. Therefore we can expect the planar $sp^2$ bonding that is characteristic of graphite to play a significant role in carbon nanotubes. The curvature of the nanotubes admixes a small amount of $sp^3$

bonding so that the force constants (bonding) in the circumferential direction are slightly weaker than along the nanotube axis. Since the single wall carbon nanotube is only one atom thick and has a small number of atoms around its circumference, only a few wave vectors are needed to describe the periodicity of the nanotubes. These constraints lead to quantum confinement of the wavefunctions in the radial and circumferential directions, with plane wave motion occurring only along the nanotube axis corresponding to a large number or closely spaced allowed wave vectors. Thus, although carbon nanotubes are closely related to a 2D graphene sheet, the tube curvature and the quantum confinement in the circumferential direction lead to a host of properties that are different from those of a graphene sheet. Because of the close relation between carbon nanotubes and graphite, we review briefly the structure and properties of graphite in this chapter. As explained in the chapter by *Louie* [13], $(n, m)$ carbon nanotubes can be either metallic ($n - m = 3q$, $q = 0, 1, 2, \ldots$) or semiconducting ($n - m = 3q \pm 1$, $q = 0, 1, 2, \ldots$), the individual constituents of multi-wall nanotubes or single-wall nanotube bundles can be metallic or semiconducting [13,15]. These remarkable electronic properties follow from the electronic structure of 2D graphite under the constraints of quantum confinement in the circumferential direction [13].

Actual carbon nanotube samples are usually found in one of two forms: (1) a Multi-Wall Carbon Nanotube (MWNT) consisting of a nested coaxial array of single-wall nanotube constituents [16], separated from one another by approximately 0.34 nm, the interlayer distance of graphite (see Sect. 2), and (2) a single wall nanotube rope, which is a nanocrystal consisting of ~10–100 Single-Wall Nanotubes (SWNTs), whose axes are aligned parallel to one another, and are arranged in a triangular lattice with a lattice constant that is approximately equal to $d_t + c_t$, where $d_t$ is the nanotube diameter and $c_t$ is approximately equal to the interlayer lattice constant of graphite.

## 2 Graphite

The ideal crystal structure of graphite (see Fig. 2) consists of layers in which the carbon atoms are arranged in an open honeycomb network containing two atoms per unit cell in each layer, labeled A and B. The stacking of the graphene layers is arranged, such that the A and A′ atoms on consecutive layers are on top of one another, but the B atoms in one plane are over the unoccupied centers of the adjacent layers, and similarly for the B′ atoms on the other plane [17]. This gives rise to two distinct planes, which are labeled by A and B. These distinct planes are stacked in the 'ABAB' Bernal stacking arrangement shown in Fig. 2, with a very small in-plane nearest-neighbor distance $a_{C-C}$ of 1.421 Å, an in-plane lattice constant $a_0$ of 2.462 Å, a $c$-axis lattice constant $c_0$ of 6.708 Å, and an interplanar distance of $c_0/2 = 3.354$ Å. This crystal structure is consistent with the $D^4_{6h}(P6_3/mmc)$ space group and has four carbon atoms per unit cell, as shown in Fig. 2.

(a) (b)

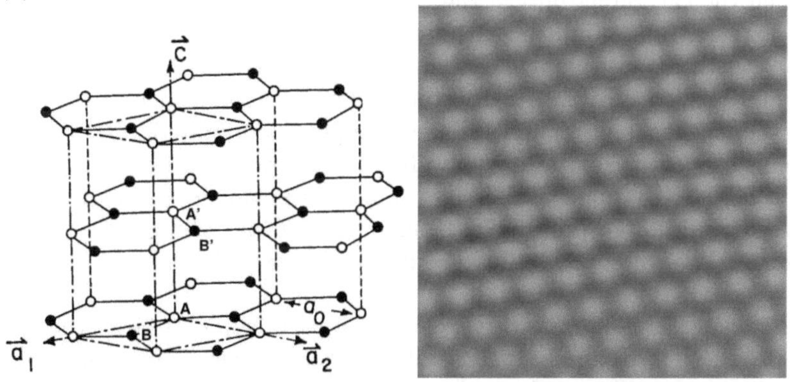


**Fig. 2.** (**a**) The crystal structure of hexagonal single crystal graphite, in which the two distinct planes of carbon hexagons called A and B planes are stacked in an ABAB... sequence with $P63/mmc$ symmetry. The notation for the A and B planes is not to be confused with the two distinct atoms A and B on a single graphene plane (note a rhombohedral phase of graphite with ABCABC... stacking also exists [17]). (**b**) An STM image showing the trigonal network of highly oriented pyrolytic graphite (HOPG) in which only one site of the carbon hexagonal network appears, as for example, the B site, denoted by black balls in (**a**)

Since the in-plane C-C bond is very strong and the nearest-neighbor spacing between carbon atoms in graphite is very small, the in-plane lattice constant is quite stable against external perturbations. The nearest neighbor spacing between carbon nanotubes is essentially the same as the interplanar spacing in graphite (∼3.4 Å). One consequence of the small value of $a_{\mathrm{C-C}}$ in graphite is that impurity species are unlikely to enter the covalently bonded in-plane lattice sites substitutionally (except for boron), but rather occupy some interstitial position between the graphene layer planes which are bonded by a weak van der Waals force. These arguments also apply to carbon nanotubes and explain why the substitutional doping of individual single wall carbon nanotubes with species other than boron is difficult. The weak interplanar bonding of graphite allows entire planes of dopant atoms or molecules to be intercalated between the carbon layers to form intercalation compounds. Also carbon nanotubes can adsorb dopant species on their external and internal surfaces and in interstitial sites between adjacent nanotubes, as is discussed in Sect. 6.

The graphene layers often do not stack perfectly and do not form the perfect graphite crystal structure with perfect Bernal 'ABAB' layer stacking. Instead, stacking faults are often formed (meaning departures from the ABAB stacking order). These stacking faults give rise to a small increase in the interlayer distance from the value 3.354 Å in 3D graphite until a value of about 3.440 Å is reached, at which interplanar distance, the stacking of the individual carbon layers become uncorrelated with essentially no site bonding between the carbon aatoms in the two layers. The resulting structure of these uncorrelated 2D graphene layers is called *turbostratic graphite* [1,18]. Because of the different diameters of adjacent cylinders of carbon atoms in a multiwall carbon nanotube [15,16], the structural arrangement of the adjacent carbon honeycomb cylinders is essentially uncorrelated with no site correlation between carbon atoms on adjacent nanotubes. The stacking arrangement of the nanotubes is therefore similar in behavior to the graphene sheets in turbostratic graphite. Thus, perfect nanotube cylinders at a large spatial separation from one another should be able to slide past one another easily.

Of significance to the properties expected for carbon nanotubes is the fact that the electronic structure of turbostratic graphite, a zero gap semiconductor, is qualitatively different from that of ideal graphite, a semimetal with a small band overlap (0.04 eV). The electronic structure of a 2D graphene sheet [15] is discussed elsewhere in this volume [14], where it is shown that the valence and conduction bands of a graphene sheet are degenerate by symmetry at the special point $K$ at the 2D Brillouin zone corner where the Fermi level in reciprocal space is located [19]. Metallic carbon nanotubes have an allowed wavevector at the $K$-point and therefore are effectively zero gap semiconductors like a 2D graphene sheet. However, semiconducting nanotubes do not have an allowed wavevector at the $K$ point (because of quantum confinement conditions in the circumferential direction) [14,15], thus resulting in an electronic band gap and semiconducting behavior, very different from that of a graphene sheet.

Several sources of crystalline graphite are available, but differ somewhat in their overall characteristics. Some discussion of this topic could be helpful to readers since experimentalists frequently use these types of graphite samples in making comparisons between the structure and properties of carbon nanotubes and $sp^2$ graphite.

Natural single-crystal graphite flakes are usually small in size (typically much less than 0.1 mm in thickness), and contain defects in the form of twinning planes and screw dislocations, and also contain chemical impurities such as Fe and other transition metals, which make these graphite samples less desirable for certain scientific studies and applications.

A synthetic single-crystal graphite, called "kish" graphite, is commonly used in scientific investigations. Kish graphite crystals form on the surface of high carbon content iron melts and are harvested as crystals from such high temperature solutions [20]. The kish graphite flakes are often larger than the natural graphite flakes, which makes kish graphite the material of choice when large single-crystal flakes are needed for scientific studies. However, these flakes may contain impurities.

The most commonly used high-quality graphitic material today is *Highly Oriented Pyrolytic Graphite* (HOPG), which is prepared by the pyrolysis of hydrocarbons at temperatures of above 2000°C and the resulting pyrolytic carbon is subsequently heat treated to higher temperatures to improve its crystalline order [21,22]. When stress annealed above 3300°C, the HOPG exhibits electronic, transport, thermal, and mechanical properties close to those of single-crystal graphite, showing a very high degree of $c$-axis alignment. For the high temperature, stress-annealed HOPG, the crystalline order extends to about 1 $\mu$m within the basal plane and to about 0.1 $\mu$m along the $c$-direction. This material is commonly used because of its good physical properties, high chemical purity and relatively large sample sizes. Thin-film graphite materials, especially those based on Kapton and Novax (polyimide) precursors, are also prepared by a pyrolysis/heat treatment method, and are often used.

## 3 Graphite Whiskers

A graphite whisker is a graphitic material formed by rolling a graphene sheet up into a *scroll* [23]. Except for the early work by *Bacon* [23], there is little literature about graphite whiskers. Graphite whiskers are formed in a dc discharge between carbon electrodes using 75–80 V and 70–76 A. In the arc apparatus, the diameter of the positive electrode is smaller than that of the negative electrode, and the discharge is carried out in an inert gas using a high gas pressure (92 atmospheres). As a result of this discharge, cylindrical boules with a hard shell were formed on the negative electrode. When these hard cylindrical boules were cracked open, scroll-like carbon whiskers up to ~3 cm long and 1–5 μm in diameter were found protruding from the fracture surfaces. The whiskers exhibited great crystalline perfection, high electrical conductivity, and high elastic modulus along the fiber axis. Since their discovery [23], graphite whiskers have provided the benchmark against which the performance of carbon fibers is measured. The growth of graphite whiskers by the arc method has many similarities to the growth of carbon nanotubes [24], especially MWNTs which do not require the use of a catalyst, except that graphite whiskers were grown at a higher gas pressure than is commonly used for nanotube growth. While MWNTs are generally found to be concentric cylinders of much smaller outer diameter, some reports have been given of scroll-like structures with outer diameters less than 100 nm [23].

## 4 Carbon Fibers

Carbon fibers represent an important class of graphite-related materials which are closely connected to carbon nanotubes, with regard to structure and properties. Despite the many precursors that can be used to synthesize carbon fibers, each having different cross-sectional morphologies (Fig. 3), the preferred orientation of the graphene planes is parallel to the fiber axis for all carbon fibers, thereby accounting for the high mechanical strength of carbon fibers [1]. Referring to the various morphologies in Fig. 3, the as-prepared vapor-grown fibers have an "onion skin" or "tree ring" morphology

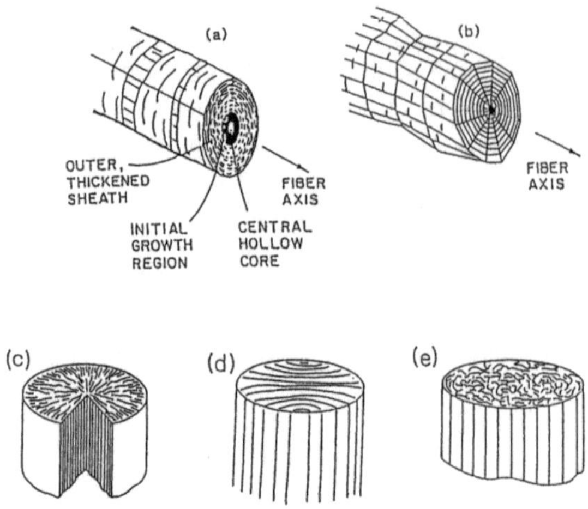


**Fig. 3.** Sketch illustrating the morphology of Vapor-Grown Carbon Fibers (VGCF): (**a**) as-deposited at 1100°C [1], (**b**) after heat treatment to 3000°C [1]. The morphologies for commercial mesophase-pitch fibers are shown in (**c**) for a "PAC-man" cross section with a radial arrangement of the straight graphene ribbons and a missing wedge and (**d**) for a PAN-AM cross-sectional arrangement of graphene planes. In (**e**) a PAN fiber is shown, with a circumferential arrangement of ribbons in the sheath region and a random structure in the core

and after heat treatment to about 2500°C bear a close resemblance to carbon nanotubes (Fig. 3a). After further heat treatment to about 3000°C, the outer regions of the vapor grown carbon fibers form facets (Fig. 3b), and become more like graphite because of the strong interplanar correlations resulting from the facets [1,25]. At the hollow core of a vapor grown carbon fiber is a multiwall (and also a single wall) carbon nanotube (MWNT), as shown in Fig. 4, where the MWNT is observed upon fracturing a vapor grown carbon fiber [1]. Of all carbon fibers, the faceted vapor grown carbon fibers (Fig. 3b) are closest to crystalline graphite in both crystal structure and properties.

The commercially available mesophase pitch-based fibers, are exploited for their extremely high bulk modulus and high thermal conductivity, while the commercial PAN (polyacrylonitrile) fibers are widely used for their high tensile strength [1]. The high modulus of the mesophase pitch fibers is related to the high degree of *c*-axis orientation of adjacent graphene layers, while the high strength of the PAN fibers is related to defects in the structure. These structural defects inhibit the slippage of adjacent graphene planes relative to each other, and inhibit the sword-in-sheath failure mode (Fig. 5) that dominates the rupture of a vapor grown carbon fiber [1]. Typical diameters for individual commercial carbon fibers are $\sim 7\mu m$, and they can be very long. These fibers are woven into bundles called tows and are then wound up as a continuous yarn on a spool. The remarkable high strength and modulus of carbon fibers (Fig. 6) are responsible for most of the commercial interest

(a)

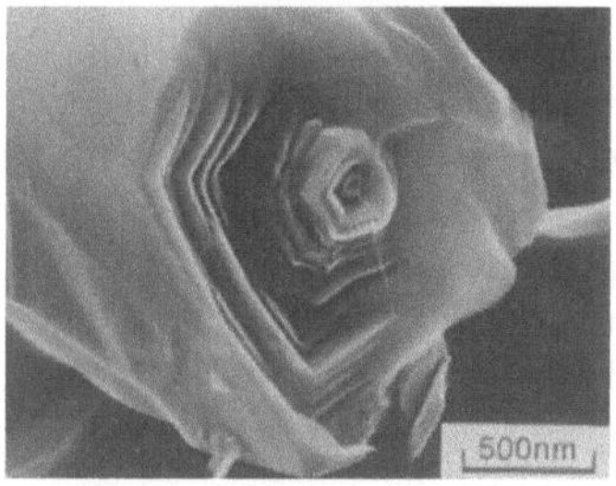


(b)

**Fig. 4.** (**a**) Carbon nanotube exposed on the breakage edge of a vapor grown carbon fiber as grown (**a**) and heat-treated at 3000°C (**b**). The sample is fractured by pulverization and the core diameter is ∼5 nm. (**b**) These photos suggest a structural discontinuity between the nanotube core of the fiber and the outer carbon layers deposited by chemical vapor deposition techniques. The photos show the strong mechanical properties of the nanotube core which maintain its form after breakage of the periphery [26]

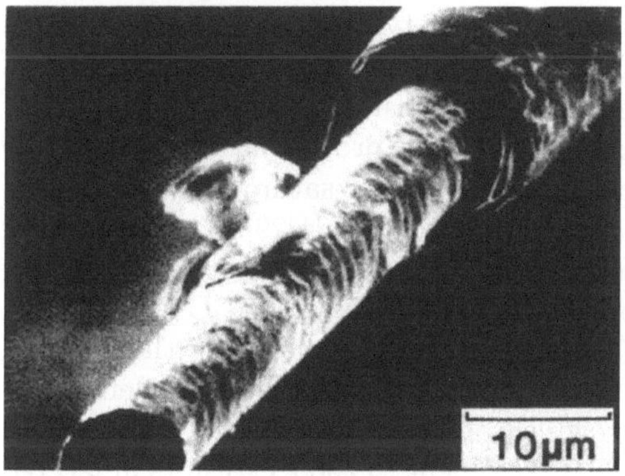


**Fig. 5.** The sword-in-sheath failure mode of heat treated vapor grown carbon fibers. Such failure modes are also observed in multiwall carbon nanotubes [1]

in these fibers and these superior mechanical properties (modulus and tensile strength) should be compared to steel, for which typical strengths and bulk modulus values are 1.4 and 207 GPa, respectively [1]. The excellent mechanical properties of carbon nanotubes are closely related to the excellent mechanical properties of carbon fibers, though notable differences in behavior are also found, such as the flexibility of single wall carbon nanotubes and the good mechanical properties of multiwall nanotubes (with only a few walls) under compression, in contrast with carbon fibers which fracture easily under compressive stress along the fiber axis.

Vapor-grown carbon fibers can be prepared over a wide range of diameters (from ∼10 nm to more than 100μm) and these fibers have central hollow cores. A distinction is made between vapor grown carbon fibers and fibers

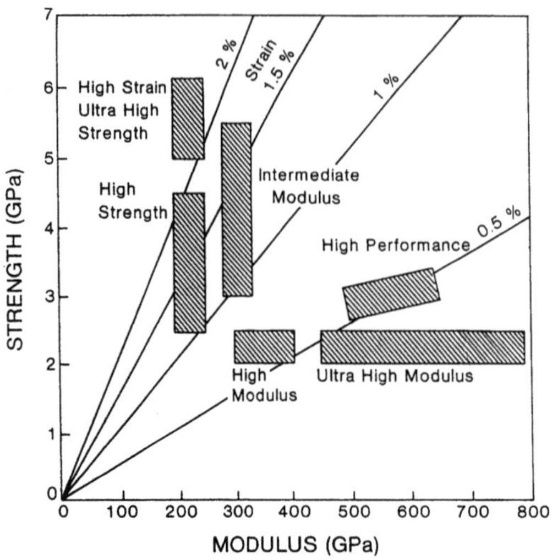


**Fig. 6.** The breaking strength of various types of carbon fibers plotted as a function of Young's modulus. Lines of constant strain can be used to estimate the breaking strains [1,27,28]

with diameters in the range 10–100 nm, which are called *nanofibers*, and exhibit properties intermediate between those of typical vapor grown carbon fibers, on the one hand, and MWNTs, on the other [24]. The preparation of vapor grown carbon fibers is based on the growth of a thin hollow tube of about 100 nm diameter (a nanofiber) by a catalytic process based on ultra-fine particles (∼10 nm diameter) which have been super-saturated with carbon from the pyrolysis of a hydrocarbon gas at ∼1050°C [1,29]. The thickening of the vapor-grown carbon fiber occurs through an epitaxial growth process, whereby the hydrocarbon gas is dehydrogenated at the ∼1050°C growth temperature, and the carbon deposit is adsorbed on the surface of the growing fiber. Subsequent heat treatment to ∼2500°C anneals the disordered carbon deposit and results in vapor grown carbon fibers with a tree ring coaxial cylinder morphology [29]. Further heat treatment to 2900°C results in faceted fibers (Fig. 3b) which exhibit structural and electronic properties very close to those of single crystal graphite [1,29]. If the growth process is stopped before the thickening step starts, MWNTs are obtained [30].

These vapor grown carbon fibers show $(h, k, l)$ X-ray diffraction lines indicative of the 3D graphite structure and a semimetallic band overlap and carrier density similar to 3D graphite. The infrared and Raman spectra are essentially the same as that of 3D graphite. Vapor-grown carbon fibers and nanofibers with micrometer and several tens of nanometer diameters, respectively, provide intermediate materials between conventional mesophase pitch-derived carbon fibers (see Fig. 3) and single wall carbon nanotubes. Since their smallest diameters (∼10 nm) are too large to observe significant quantum confinement effects, it would be difficult to observe band gaps in their electronic structure, or the radial breathing mode in their phonon spectra. Yet subtle differences are expected between nanofiber transport properties, electronic structure and Raman spectra relative to the corresponding phenomena in either MWNTs of graphitic vapor grown carbon fibers. At present, little is known in detail about the structure and properties of nanofibers, except that their properties are intermediate between those of vapor grown carbon fibers and MWNTs.

## 4.1 History of Carbon Fibers in Relation to Carbon Nanotubes

We provide here a brief review of the history of carbon fibers, the macroscopic analog of carbon nanotubes, since carbon nanotubes have become the focus of recent developments in carbon fibers.

The early history of carbon fibers was stimulated by needs for materials with special properties, both in the 19th century and more recently after World War II. The first carbon fiber was prepared by Thomas A. Edison to provide a filament for an early model of an electric light bulb. Specially selected Japanese Kyoto bamboo filaments were used to wind a spiral coil that was then pyrolyzed to produce a coiled carbon resistor, which could be heated ohmically to provide a satisfactory filament for use in an early model of an incandescent light bulb [31]. Following this initial pioneering work by Edison, further research on carbon filaments proceeded more slowly, since carbon filaments were soon replaced by a more sturdy tungsten filament in the electric light bulb. Nevertheless research on carbon fibers and filaments proceeded steadily over a long time frame, through the work of *Schützenberger* and *Schützenberger* (1890) [32], *Pelabon* [33], and others. Their efforts were mostly directed toward the study of vapor grown carbon filaments, showing filament growth from the thermal decomposition of hydrocarbons.

The second applications-driven stimulus to carbon fiber research came in the 1950's from the needs of the space and aircraft industry for strong, stiff light-weight fibers that could be used for building lightweight composite materials with superior mechanical properties. This stimulation led to great advances in the preparation of continuous carbon fibers based on polymer precursors, including rayon, polyacrylonitrile (PAN) and later mesophase pitch. The late 1950's and 1960's was a period of intense activity at the Union Carbide Corporation, the Aerospace Corporation and many other laboratories worldwide. This stimulation also led to the growth of a carbon whisker [23] (see Sect. 3), which has become a benchmark for the discussion of the mechanical and elastic properties of carbon fibers. The growth of carbon whiskers was also inspired by the successful growth of single crystal whisker filaments at that time for many metals such as iron, non-metals such as Si, and oxides such as $Al_2O_3$, and by theoretical studies [34], showing superior mechanical properties for whisker structures [35]. Parallel efforts to develop new bulk synthetic carbon materials with properties approaching single crystal graphite led to the development of highly oriented pyrolytic graphite (HOPG) in 1962 by *Ubbelohde* and co-workers [36,37], and HOPG has since been used as one of the benchmarks for the characterization of carbon fibers.

While intense effort continued toward perfecting synthetic filamentary carbon materials, and great progress was indeed made in the early 1960's, it was soon realized that long term effort would be needed to reduce fiber defects and to enhance structures resistive to crack propagation. New research directions were introduced because of the difficulty in improving the structure and microstructure of polymer-based carbon fibers for high strength and high modulus applications, and in developing graphitizable carbons for ultra-high modulus fibers. Because of the desire to synthesize more crystalline filamentous carbons under more controlled conditions, synthesis of carbon fibers by a catalytic Chemical Vapor Deposition (CVD) process proceeded, laying the scientific basis for the mechanism and thermodynamics for the vapor phase growth of carbon fibers in the 1960's and early 1970's [1]. In parallel to these scientific studies, other research studies focused on control of the process for the synthesis of vapor grown carbon fiber [38,39,40,41], leading to the more recent commercialization of vapor grown carbon fibers in the 1990's for various applications. Concurrently, polymer-based carbon fiber research has continued worldwide, mostly in industry, with emphasis on greater control of processing steps to achieve carbon fibers with ever-increasing modulus and strength, and on fibers with special characteristics, such as very high thermal conductivity, while decreasing costs of the commercial products.

As research on vapor grown carbon fibers on the micrometer scale proceeded, the growth of very small diameter filaments less than 10 nm (Fig. 7), was occasionally observed and reported [42,43], but no detailed systematic studies of such thin filaments were carried out. An example of a very thin vapor grown nanofiber along with a multiwall nanotube is shown in the bright field TEM image of Fig. 7 [42,43,44,45].

Reports of thin filaments below 10 nm inspired Kubo [46] to ask whether there was a minimum dimension for such filaments. Early work [42,43] on vapor grown carbon fibers, obtained by thickening filaments such as the fiber denoted by VGCF (Vapor Grown Carbon Fiber) in Fig. 7, showed very sharp lattice fringe images for the inner-most cylinders. Whereas the outermost layers of the fiber have properties associated with vapor grown carbon fibers,

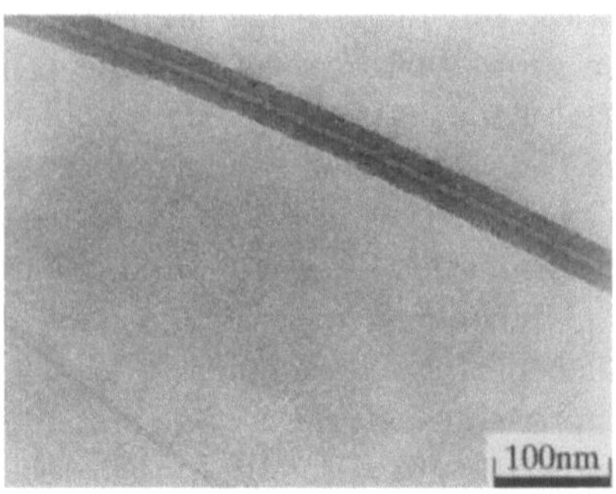


**Fig. 7.** High-resolution TEM micrograph showing carbon a vapor grown carbon nanofiber (VGCF) with an diameter less than 10 nm and a nanotube [42,43,44,45]

there may be a continuum of behavior of the tree rings as a function of diameter, with the innermost tree rings perhaps behaving like carbon nanotubes.

Direct stimulus to study carbon filaments of very small diameters more systematically [11] came from the discovery of fullerenes by *Kroto* and *Smalley* [10]. In December 1990 at a carbon-carbon composites workshop, papers were given on the status of fullerene research by *Smalley* [47], the discovery of a new synthesis method for the efficient production of fullerenes by *Huffman* [48], and a review of carbon fiber research by *Dresselhaus* [49]. Discussions at the workshop stimulated Smalley to speculate about the existence of carbon nanotubes of dimensions comparable to $C_{60}$. These conjectures were later followed up in August 1991 by discussions at a fullerene workshop in Philadelphia [50] on the symmetry proposed for a hypothetical single-wall carbon nanotubes capped at either end by fullerene hemispheres, with suggestions on how zone folding could be used to examine the electron and phonon dispersion relations of such structures. However, the real breakthrough on carbon nanotube research came with Iijima's report of experimental observation of carbon nanotubes using transmission electron microscopy [11]. It was this work which bridged the gap between experimental observation and the theoretical framework of carbon nanotubes in relation to fullerenes and as theoretical examples of 1D systems. Since the pioneering work of *Iijima* [11], the study of carbon nanotubes has progressed rapidly.

## 5 Liquid Carbon

Liquid carbon refers to the liquid phase of carbon resulting from the melting of pure carbon in a solid phase (graphite, diamond, carbon fibers or a variety of other carbons). The phase diagram for carbon shows that liquid carbon is stable at atmospheric pressure only at very high temperatures (the melting point of graphite $T_{\mathrm{m}} \sim 4450\,\mathrm{K}$) [4]. Since carbon has the highest melting point of any elemental solid, to avoid contamination of the melt, the crucible in which the carbon is melted must itself be made of carbon, and sufficient heat must be focused on the sample volume to produce the necessary temperature rise to achieve melting [8,51]. Liquid carbon has been produced in the laboratory by the laser melting of graphite, exploiting the poor interplanar thermal conductivity of the graphite [8], and by resistive heating [52], the technique used to establish the metallic nature of liquid carbon.

Although diamond and graphite may have different melting temperatures, it is believed that the same liquid carbon is obtained upon melting either solid phase. It is likely that the melting of carbon nanotubes also forms liquid carbon. Since the vaporization temperature for carbon ($\sim$4700 K) is only slightly higher than the melting point ($\sim$4450 K), the vapor pressure over liquid carbon is high. The high vapor pressure and the large carbon–carbon bonding energy make it energetically favorable for carbon clusters rather than independent atoms to be emitted from a molten carbon surface [53]. Energetic considerations suggest that some of the species emitted from a molten carbon surface have masses comparable to those of fullerenes [51]. The emission of fullerenes from liquid carbon is consistent with the graphite laser ablation studies of *Kroto, Smalley,* and co-workers [10].

Resistivity measurements on a variety of vapor grown carbon fibers provided important information about liquid carbon. In these experiments fibers were heated resistively by applying a single 28μs current pulse with currents up to 20 A [52]. The temperature of the fiber as a function of time was determined from the energy supplied in the pulse and the measured heat capacity for bulk graphite [54] over the temperature range up to the melting point, assuming that all the power dissipated in the current pulse was converted into thermal energy in the fiber. The results in Fig. 8 for well-graphitized fibers (heat treatment temperatures $T_{HT}$ = 2300°C, 2800°C) show an approximately linear temperature dependence for the resistivity $\rho(T)$ up to the melting temperature, where $\rho(T)$ drops by nearly one order of magnitude, consistent with metallic conduction in liquid carbon (Fig. 8). This figure further shows that both pregraphitic vapor grown carbon fibers ($T_{HT}$ = 1700°C, 2100°C), which are turbostratic and have small structural coherence lengths, and well graphitized fibers, all show the same behavior in the liquid phase, although their measured $\rho(T)$ functional forms in the solid phase are very different. Measurements of the resistance of a carbon nanotube through the melting transition have not yet been carried out.

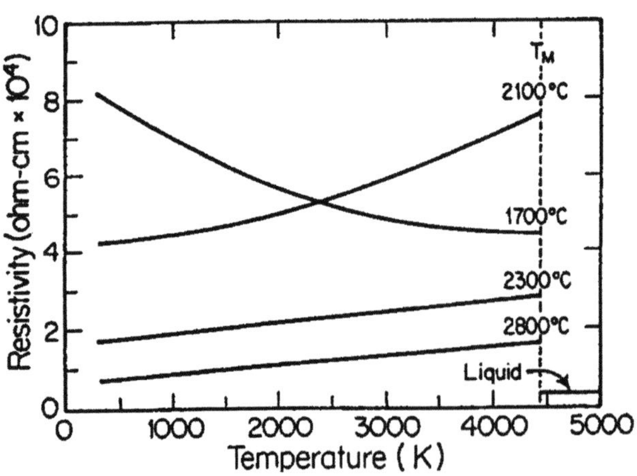


**Fig. 8.** The electrical resistivity vs temperature for vapor grown carbon fibers with various heat treatment temperatures ($T_{HT}$ = 1700, 2100, 2300, 2800°C). The sharp decrease in $\rho(T)$ above ~4000 K is identified with the melting of the carbon fibers. The measured electrical resistivity for liquid carbon is shown [52]

## 6 Graphite Intercalation Compounds

Because of the weak van der Waals interlayer forces associated with the $sp^2$ bonding in graphite, graphite intercalation compounds (GICs) may be formed by the insertion of layers of guest species between the layers of the graphite host material [56,57], as shown schematically in Fig. 9. The guest species may be either atomic or molecular. In the so-called donor GICs, electrons are transferred from the donor intercalate species (such as a layer of the alkali metal potassium) into the graphite layers, thereby raising the Fermi level $E_F$ in the graphitic electronic states, and increasing the mobile electron concentration by two or three orders of magnitude, while leaving the intercalate layer positively charged with low mobility carriers. Conversely, for acceptor GICs, electrons are transferred to the intercalate species (which is usually molecular) from the graphite layers, thereby lowering the Fermi level $E_F$ in the graphitic electronic states and creating an equal number of positively charged hole states in the graphitic $\pi$-band. Thus, electrical conduction in GICs (whether they are donors or acceptors) occurs predominantly in the graphene layers and as a result of the charge transfer between the intercalate and host layers. The electrical conductivity between adjacent graphene layers is very poor. Among the GICs, Li-based GICs are widely commercialized in

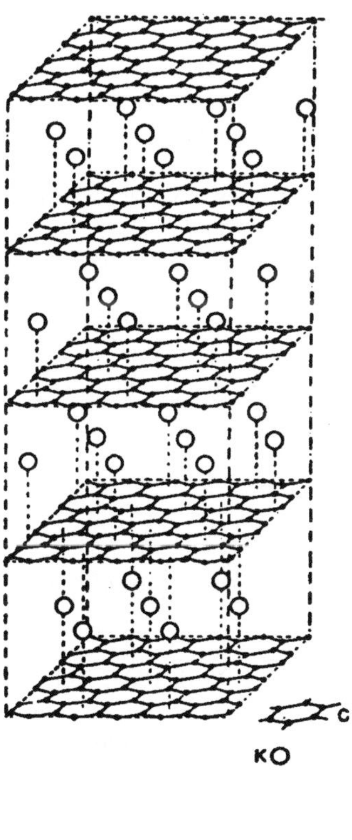


**Fig. 9.** Schematic model for a graphite intercalation compound showing the stacking of graphite layers (networks of hexagons on a sheet) and of intercalate (e.g., potassium) layers (networks of large hollow balls). For this stage 1 compound, each carbon layer is separated by an intercalate layer [55]

Li-ion secondary batteries for cell phones, personal computers, and electric vehicle batteries [58].

Carbon nanotubes also provide a host material for intercalation of donors (e.g., alkali metals) or acceptors (e.g., halogens such as bromine and iodine). In the case of nanotubes, the guest species is believed to decorate the exterior of the single wall nanotubes, and also to enter the hollow cores of the nanotubes. A similar charge transfer process is observed for alkali metals and halogens in SWNTs, based on measuring the downshifts and upshifts in the characteristic Raman $G$-band mode frequencies, similar to the standard characterization techniques previously developed for GICs [9,56]. Also, the distances of the dopants to the nearest-neighbor carbon atoms are similar for doped carbon nanotubes and for GICs [9,56]. Practical applications of intercalated carbon nanotubes are also expected in analogy with Li-GICs, especially for super high capacity batteries.

## References

1. M. S. Dresselhaus, G. Dresselhaus, K. Sugihara, I. L. Spain, H. A. Goldberg, *Graphite Fibers and Filaments*, *Springer Ser. Mater. Sci.*, Vol. 5 (Springer, Berlin, Heidelberg 1988)
2. B. T. Kelly, *Physics of Graphite* (Applied Science, London 1981)
3. F. P. Bundy, W. A. Bassett, M. S. Weathers, R. J. Hemley, H. K. Mao, A. F. Goncharov, Carbon **34**, 141–153 (1996)
4. F. P. Bundy. *Solid State Physics under Pressure: Recent Advance with Anvil Devices*, ed. by S. Minomura, (Reidel, Dordrecht 1985) p. 1
5. A. G. Whittaker, E. J. Watts, R. S. Lewis, E. Anders, Science **209**, 1512 (1980)
6. A. G. Whittaker, P. L. Kintner, Carbon **23**, 255 (1985)
7. V. I. Kasatochkin, V. V. Korshak, Y. P. Kudryavtsev, A. M. Sladkov, I. E. Sterenberg, Carbon **11**, 70 (1973)
8. M. S. Dresselhaus, J. Steinbeck, Tanso **132**, 44–56 (1988). (Journal of the Japanese Carbon Society)
9. M. S. Dresselhaus, G. Dresselhaus, P. C. Eklund, *Science of Fullerenes and Carbon Nanotubes* (Academic, New York 1996)
10. H. W. Kroto, J. R. Heath, S. C. O'Brien, R. F. Curl, R. E. Smalley, Nature (London) **318**, 162–163 (1985)
11. S. Iijima, Nature (London) **354**, 56 (1991)
12. M. S. Dresselhaus, P. Avouris, chapter 1 in this volume
13. S. G. Louie, chapter 6 in this volume
14. R. Saito, H. Kataura, chapter 9 in this volume
15. R. Saito, G. Dresselhaus, M. S. Dresselhaus, *Physical Properties of Carbon Nanotubes* (Imperial College Press, London 1998)
16. L. Forró, C. Schönenberger, chapter 13 in this volume
17. R. W. G. Wyckoff, *Crystal Structures*, (Interscience) New York 1964, Vol. 1
18. J. Maire, J. Méring, *Proceedings of the First Conference of the Society of Chemical and Industrial Conference on Carbon and Graphite* (London, 1958) p. 204
19. P. R. Wallace, Phys. Rev. **71**, 622 (1947)

20. S. B. Austerman, *Chemistry and Physics of Carbon*, Vol. 7, P. L. Walker, Jr. (Ed.) (Dekker, New York 1968) p. 137
21. A. W. Moore, *Chemistry and Physics of Carbon*, Vol. 11, P. L. Walker, Jr., P. A. Thrower (Eds.), (Dekker, New York 1973) p.69
22. A. W. Moore, *Chemistry and Physics of Carbon*, Vol. 17, P. L. Walker, Jr., P. A. Thrower (Eds.), (Dekker, New York 1981) p. 233
23. R. Bacon, J. Appl. Phys. **31**, 283–290 (1960)
24. J. C. Charlier , *Carbon Nanotubes*, M. S. Dresselhaus, G. Dresselhaus, P. Avouris (Eds.), (Springer, Berlin, 2000. Springer Series in Solid-State Sciences)
25. T. C. Chieu, G. Timp, M. S. Dresselhaus, M. Endo, A. W. Moore, Phys. Rev. B **27**, 3686 (1983)
26. M. Endo and M. S. Dresselhaus, Science Spectra (2000) (in press)
27. M. Endo, A. Katoh, T. Sugiura, M. Shiraishi, *Extended Abstracts of the 18th Biennial Conference on Carbon*, (Worcester Polytechnic Institute, 1987) p. 151
28. M. Endo, T. Momose, H. Touhara, N. Watanabe, J. Power Sources **20**, 99 (1987)
29. M. Endo, CHEMTECH **18** 568 (1988) (Sept.)
30. M. Endo, K. Takeuchi, K. Kobori, K. Takahashi, H. Kroto, A. Sarkar, Carbon **33**, 873 (1995)
31. T. A. Edison, US Patent 470,925 (1892) (issued March 15, 1892)
32. P. Schützenberger, L. Schützenberger, Compt. Rendue **111**, 774 (1890)
33. C. H. Pelabon, Compt. Rendue **137**, 706 (1905)
34. C. Herring, J. K. Galt, Phys. Rev. **85**, 1060 (1952)
35. A. P. Levitt, *Whisker Technology* (Wiley-Interscience, New York 1970)
36. A. W. Moore, A. R. Ubbelohde, D. A. Young, Brit. J. Appl. Phys. **13**, 393 (1962)
37. L. C. F. Blackman, A. R. Ubbelohde, Proc. Roy. Soc. (London) **A266**, 20 (1962)
38. T. Koyama, Carbon **10**, 757 (1972)
39. M. Endo, T. Koyama, Y. Hishiyama, Jap. J. Appl. Phys. **15**, 2073–2076 (1976)
40. G. G. Tibbetts, Appl. Phys. Lett. **42**, 666 (1983)
41. G. G. Tibbetts, J. Cryst. Growth **66**, 632 (1984)
42. M. Endo, *Mecanisme de croissance en phase vapeur de fibres de carbone (The growth mechanism of vapor-grown carbon fibers)*, PhD thesis, University of Orleans, Orleans, France, (1975) (in French)
43. M. Endo, PhD thesis, Nagoya University, Japan, (1978) (in Japanese)
44. A. Oberlin, M. Endo, T. Koyama, Carbon **14**, 133 (1976)
45. A. Oberlin, M. Endo, T. Koyama, J. Cryst. Growth **32**, 335–349 (1976)
46. R. Kubo, (private communication to M. Endo at the Kaya Conference) (1977)
47. R. E. Smalley, DoD Workshop in Washington, DC (Dec. 1990)
48. D. R. Huffman, DoD Workshop in Washington, DC (Dec. 1990)
49. M. S. Dresselhaus, DoD Workshop in Washington, DC (Dec. 1990)
50. M. S. Dresselhaus, G. Dresselhaus, P. C. Eklund, University of Pennsylvania Workshop (August 1991)
51. J. Steinbeck, G. Dresselhaus, M. S. Dresselhaus, Int. J. Thermophys. **11**, 789 (1990)
52. J. Heremans, C. H. Olk, G. L. Eesley, J. Steinbeck, G. Dresselhaus, Phys. Rev. Lett. **60**, 452 (1988)

53. J. S. Speck, J. Steinbeck, G. Braunstein, M. S. Dresselhaus, T. Venkatesan, in *Beam-Solid Interactions and Phase Transformations, MRS Symp. Proc.*, Vol. **51**, 263, H. Kurz, G. L. Olson, J. M. Poate (Eds.) (Materials Research Society Press, Pittsburgh PA, 1986)
54. F. P. Bundy, H. M. Strong, R. H. Wentdorf, Jr. *Chemistry and Physics of Carbon*, Vol. 10, P. L. Walker, Jr., P. A. Thrower (Eds.), (Dekker, New York 1973) p. 213
55. W. Rüdorff, E. Shultze, Z. Anorg. allg. Chem. **277**, 156 (1954)
56. M. S. Dresselhaus, G. Dresselhaus, Adv. Phys. **30**, 139–326 (1981)
57. H. Zabel, S. A. Solin (Eds.) *Graphite Intercalation Compounds I: Structure and Dynamics*, (Springer, Berlin, Heidelberg 1990)
58. M. Endo, C. Kim, T. Karaki, Y. Nishimura, M. J. Matthews, S. D. M. Brown, M. S. Dresselhaus, Carbon **37**, 561–568 (1999)

# Nanotube Growth and Characterization

Hongjie Dai

Department of Chemistry, Stanford University
Stanford, CA 94305-5080, USA
hdai@chem.stanford.edu

**Abstract.** This chapter presents a review of various growth methods for carbon nanotubes. Recent advances in nanotube growth by chemical vapor deposition (CVD) approaches are summarized. CVD methods are promising for producing high quality nanotube materials at large scales. Moreover, controlled CVD growth strategies on catalytically patterned substrates can yield ordered nanotube architectures and integrated devices that are useful for fundamental characterizations and potential applications of nanotube molecular wires.

In 1991, Iijima of the NEC Laboratory in Japan reported the first observation of multi-walled carbon nanotubes (MWNT) in carbon-soot made by arc-discharge [1]. About two years later, he made the observation of Single-Walled NanoTubes (SWNTs) [2]. The past decade witnessed significant research efforts in efficient and high-yield nanotube growth methods. The success in nanotube growth has led to the wide availability of nanotube materials, and is a main catalyst behind the recent progress in basis physics studies and applications of nanotubes.

The electrical and mechanical properties of carbon nanotubes have captured the attention of researchers worldwide. Understanding these properties and exploring their potential applications have been a main driving force for this area. Besides the unique and useful structural properties, a nanotube has high Young's modulus and tensile strength. A SWNT can behave as a well-defined metallic, semiconducting or semi-metallic wire depending on two key structural parameters, chirality and diameter [3]. Nanotubes are ideal systems for studying the physics in one-dimensional solids. Theoretical and experimental work have focused on the relationship between nanotube atomic structures and electronic structures, electron-electron and electron-phonon interaction effects [4]. Extensive effort has been taken to investigate the mechanical properties of nanotubes including their Young's modulus, tensile strength, failure processes and mechanisms. Also, an intriguing fundamental question has been how mechanical deformation in a nanotube affects its electrical properties. In recent years, progress in addressing these basic problems has generated significant excitement in the area of nanoscale science and technology.

Nanotubes can be utilized individually or as an ensemble to build functional device prototypes, as has been demonstrated by many research groups. Ensembles of nanotubes have been used for field emission based flat-panel

M. S. Dresselhaus, G. Dresselhaus, Ph. Avouris (Eds.): Carbon Nanotubes,
Topics Appl. Phys. **80**, 29–53 (2001)

display, composite materials with improved mechanical properties and electromechanical actuators. Bulk quantities of nanotubes have also been suggested to be useful as high-capacity hydrogen storage media. Individual nanotubes have been used for field emission sources, tips for scanning probe microscopy and nano-tweezers. Nanotubes also have significant potential as the central elements of nano-electronic devices including field effect transistors, single-electron transistors and rectifying diodes.

The full potential of nanotubes for applications will be realized until the growth of nanotubes can be optimized and well controlled. Real-world applications of nanotubes require either large quantities of bulk materials or device integration in scaled-up fashions. For applications such as composites and hydrogen storage, it is desired to obtain high quality nanotubes at the kilogram or ton level using growth methods that are simple, efficient and inexpensive. For devices such as nanotube based electronics, scale-up will unavoidably rely on self-assembly or controlled growth strategies on surfaces combined with microfabrication techniques. Significant work has been carried out in recent years to tackle these issues. Nevertheless, many challenges remain in the nanotube growth area. First, an efficient growth approach to structurally perfect nanotubes at large scales is still lacking. Second, growing defect-free nanotubes continuously to macroscopic lengths has been difficult. Third, control over nanotube growth on surfaces should be gained in order to obtain large-scale ordered nanowire structures. Finally, there is a seemingly formidable task of controlling the chirality of SWNTs by any existing growth method.

This chapter summarizes the progress made in recent years in carbon nanotube growth by various methods including arc-discharge, laser ablation and chemical vapor deposition. The growth of nanotube materials by Chemical Vapor Deposition (CVD) in bulk and on substrates will be focused on. We will show that CVD growth methods are highly promising for scale-up of defect-free nanotube materials, and enable controlled nanotube growth on surfaces. Catalytic patterning combined with CVD growth represents a novel approach to ordered nanowire structures that can be addressed and utilized.

## 1 Nanotube Growth Methods

Arc-discharge and laser ablation methods for the growth of nanotubes have been actively pursued in the past ten years. Both methods involve the condensation of carbon atoms generated from evaporation of solid carbon sources. The temperatures involved in these methods are close to the melting temperature of graphite, 3000–4000°C.

### 1.1 Arc-Discharge and Laser Ablation

In arc-discharge, carbon atoms are evaporated by plasma of helium gas ignited by high currents passed through opposing carbon anode and cathode

(Fig. 1a). Arc-discharge has been developed into an excellent method for producing both high quality multi-walled nanotubes and single-walled nanotubes. MWNTs can be obtained by controlling the growth conditions such as the pressure of inert gas in the discharge chamber and the arcing current. In 1992, a breakthrough in MWNT growth by arc-discharge was first made by Ebbesen and Ajayan who achieved growth and purification of high quality MWNTs at the gram level [5]. The synthesized MWNTs have lengths on the order of ten microns and diameters in the range of 5-30 nm. The nanotubes are typically bound together by strong van der Waals interactions and form tight bundles. MWNTs produced by arc-discharge are very straight, indicative of their high crystallinity. For as grown materials, there are few defects such as pentagons or heptagons existing on the sidewalls of the nanotubes. The by-product of the arc-discharge growth process are multi-layered graphitic particles in polyhedron shapes. Purification of MWNTs can be achieved by heating the as grown material in an oxygen environment to oxidize away the graphitic particles [5]. The polyhedron graphitic particles exhibit higher oxidation rate than MWNTs; nevertheless, the oxidation purification process also removes an appreciable amount of nanotubes.

For the growth of single-walled tubes, a metal catalyst is needed in the arc-discharge system. The first success in producing substantial amounts of SWNTs by arc-discharge was achieved by Bethune and coworkers in 1993 [6]. They used a carbon anode containing a small percentage of cobalt catalyst in the discharge experiment, and found abundant SWNTs generated in the soot material. The growth of high quality SWNTs at the 1–10 g scale was achieved by *Smalley* and coworkers using a laser ablation (laser oven) method (Fig. 1b) [7]. The method utilized intense laser pulses to ablate a carbon target containing 0.5 atomic percent of nickel and cobalt. The target was placed in a tube-furnace heated to 1200°C. During laser ablation, a flow of inert gas was passed through the growth chamber to carry the grown nanotubes downstream to be collected on a cold finger. The produced SWNTs are mostly in the form of ropes consisting of tens of individual nanotubes close-packed into hexagonal crystals via van der Waals interactions (Fig. 2). The optimization of SWNT growth by arc-discharge was achieved by *Journet* and coworkers using a carbon anode containing 1.0 atomic percentage of yttrium and 4.2 at. % of nickel as catalyst [8].

In SWNT growth by arc-discharge and laser ablation, typical by-products include fullerenes, graphitic polyhedrons with enclosed metal particles, and amorphous carbon in the form of particles or overcoating on the sidewalls of nanotubes. A purification process for SWNT materials has been developed by *Smalley* and coworkers [9] and is now widely used by many researchers. The method involves refluxing the as-grown SWNTs in a nitric acid solution for an extended period of time, oxidizing away amorphous carbon species and removing some of the metal catalyst species.

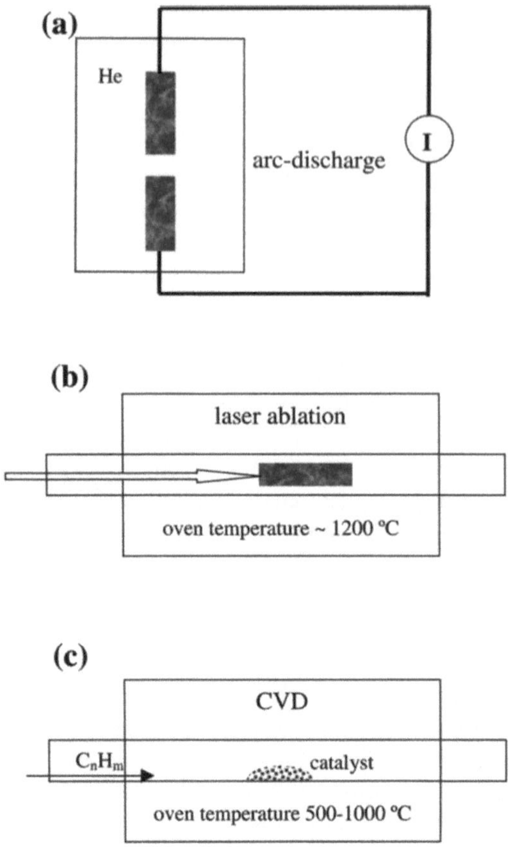


**Fig. 1 a–c.** Schematic experimental setups for nanotube growth methods

The success in producing high quality SWNT materials by laser-ablation and arc-discharge has led to wide availability of samples useful for studying fundamental physics in low dimensional materials and exploring their applications.

## 1.2 Chemical Vapor Deposition

Chemical vapor deposition (CVD) methods have been successful in making carbon fiber, filament and nanotube materials since more than 10–20 years ago [10,11,12,13,14,15,16,17,18,19].

### 1.2.1 General Approach and Mechanism

A schematic experimental setup for CVD growth is depicted in Fig. 1c. The growth process involves heating a catalyst material to high temperatures in a tube furnace and flowing a hydrocarbon gas through the tube reactor for a period of time. Materials grown over the catalyst are collected upon

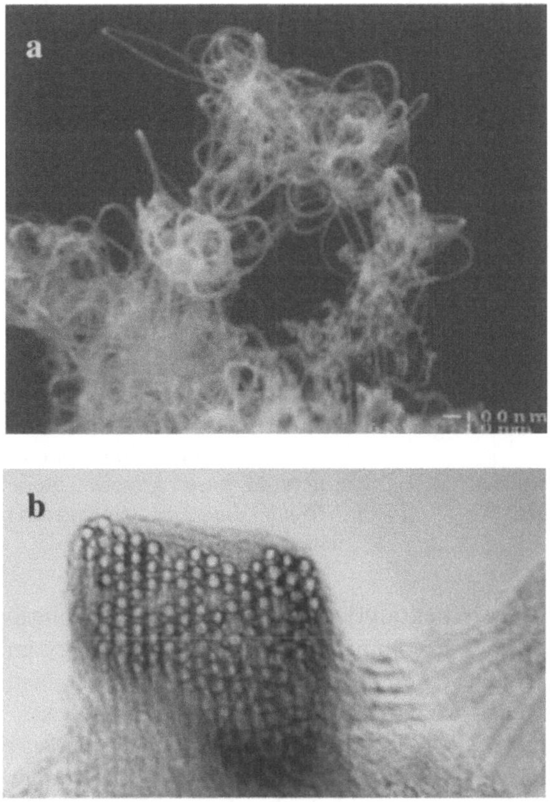


**Fig. 2 a,b.** Single-walled nanotubes grown by laser ablation (courtesy of R. Smalley)

cooling the system to room temperature. The key parameters in nanotube CVD growth are the hydrocarbons, catalysts and growth temperature. The active catalytic species are typically transition-metal nanoparticles formed on a support material such as alumina. The general nanotube growth mechanism (Fig. 3) in a CVD process involves the dissociation of hydrocarbon molecules catalyzed by the transition metal, and dissolution and saturation of carbon atoms in the metal nanoparticle. The precipitation of carbon from the saturated metal particle leads to the formation of tubular carbon solids in $sp^2$ structure. Tubule formation is favored over other forms of carbon such as graphitic sheets with open edges. This is because a tube contains no dangling bonds and therefore is in a low energy form. For MWNT growth, most of the CVD methods employ ethylene or acetylene as the carbon feedstock and the growth temperature is typically in the range of 550–750°C. Iron, nickel or cobalt nanoparticles are often used as catalyst. The rationale for choosing these metals as catalyst for CVD growth of nanotubes lies in the phase diagrams for the metals and carbon. At high temperatures, carbon has finite solubility in these metals, which leads to the formation of metal-carbon solutions and therefore the aforementioned growth mechanism. Noticeably, iron,

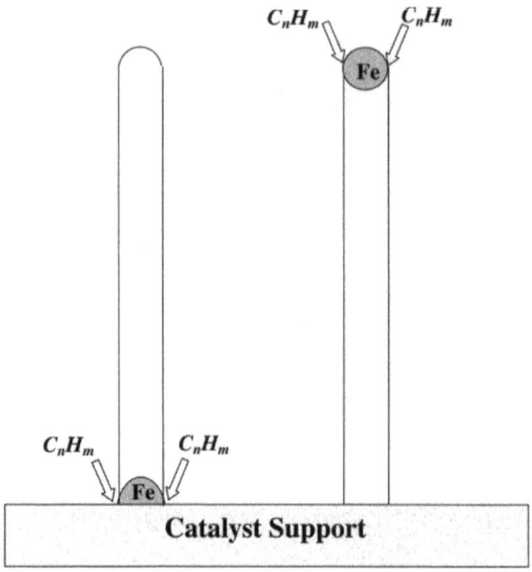


**Fig. 3.** Two general growth modes of nanotube in chemical vapor deposition. *Left diagram:* base growth mode. *Right diagram:* tip growth mode

cobalt and nickel are also the favored catalytic metals used in laser ablation and arc-discharge. This simple fact may hint that the laser, discharge and CVD growth methods may share a common nanotube growth mechanism, although very different approaches are used to provide carbon feedstock.

A major pitfall for CVD grown MWNTs has been the high defect densities in their structures. The defective nature of CVD grown MWNTs remains to be thoroughly understood, but is most likely be due to the relatively low growth temperature, which does not provide sufficient thermal energy to anneal nanotubes into perfectly crystalline structures. Growing perfect MWNTs by CVD remains a challenge to this day.

### 1.2.2 Single-Walled Nanotube Growth and Optimization

For a long time, arc-discharge and laser-ablation have been the principal methods for obtaining nearly perfect single-walled nanotube materials. There are several issues concerning these approaches. First, both methods rely on evaporating carbon atoms from solid carbon sources at $\geq 3000°C$, which is not efficient and limits the scale-up of SWNTs. Secondly, the nanotubes synthesized by the evaporation methods are in tangled forms that are difficult to purify, manipulate and assemble for building addressable nanotube structures.

Recently, growth of single-walled carbon nanotubes with structural perfection was enabled by CVD methods. For an example, we found that by using methane as carbon feedstock, reaction temperatures in the range of 850–1000°C, suitable catalyst materials and flow conditions one can grow high quality SWNT materials by a simple CVD process [20,21,22,23]. High temperature is necessary to form SWNTs that have small diameters and thus high strain energies, and allow for nearly-defect free crystalline nanotube structures. Among all hydrocarbon molecules, methane is the most stable at high temperatures against self-decomposition. Therefore, catalytic decomposition of methane by the transition-metal catalyst particles can be the dominant process in SWNT growth. The choice of carbon feedstock is thus one of the key elements to the growth of high quality SWNTs containing no defects and amorphous carbon over-coating. Another CVD approach to SWNTs was reported by Smalley and coworkers who used ethylene as carbon feedstock and growth temperature around 800°C [24]. In this case, low partial-pressure ethylene was employed in order to reduce amorphous carbon formation due to self-pyrolysis/dissociation of ethylene at the high growth temperature.

Gaining an understanding of the chemistry involved in the catalyst and nanotube growth process is critical to enable materials scale-up by CVD [22]. The choice of many of the parameters in CVD requires to be rationalized in order to optimize the materials growth. Within the methane CVD approach for SWNT growth, we have found that the chemical and textural properties of the catalyst materials dictate the yield and quality of SWNTs. This understanding has allowed optimization of the catalyst material and thus the synthesis of bulk quantities of high yield and quality SWNTs [22]. We have developed a catalyst consisting of Fe/Mo bimetallic species supported on a sol-gel derived alumina-silica multicomponent material. The catalyst exhibits a surface are of approximately 200 $m^2/g$ and mesopore volume of 0.8 mL/g. Shown in Fig. 4 are Transmission Electron Microscopy (TEM) and Scanning Electron Microscopy (SEM) images of SWNTs synthesized with bulk amounts of this catalyst under a typical methane CVD growth conditions for 15 min (methane flow rate = 1000 mL/min through a 1 inch quartz tube reactor heated to 900°C). The data illustrates remarkable abundance of individual and bundled SWNTs. Evident from the TEM image is that the nanotubes are free of amorphous carbon coating throughout their lengths. The diameters of the SWNTs are dispersed in the range of 0.7–3 nm with a peak at 1.7 nm. Weight gain studies found that the yield of nanotubes is up to 45 wt.% (1 gram of catalyst yields 0.45 gram of SWNT).

Catalyst optimization is based on the finding that a good catalyst material for SWNT synthesis should exhibit strong metal-support interactions, possess a high surface area and large pore volume. Moreover, these textural characteristics should remain intact at high temperatures without being sintered [22]. Also, it is found that alumina materials are generally far superior catalyst supports than silica. The strong metal-support interactions allow high metal dispersion and thus a high density of catalytic sites. The interactions prevent metal-species from aggregating and forming unwanted large particles that could yield to graphitic particles or defective multi-walled tube structures. High surface area and large pore volume of the catalyst support facilitate high-yield SWNT growth, owing to high densities of catalytic sites

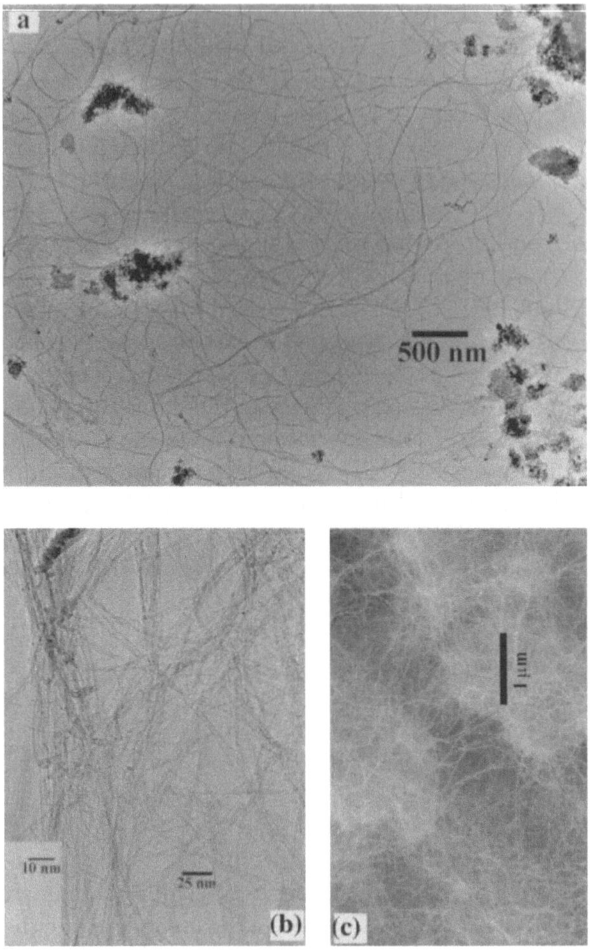


**Fig. 4.** Bulk SWNT materials grown by chemical vapor deposition of methane. (**a**) A low magnification TEM image. (**b**) A high magnification TEM image. (**c**) An SEM image of the as-grown material

made possible by the former and rapid diffusion and efficient supply of carbon feedstock to the catalytic sites by the latter.

Mass-spectral study of the effluent of the methane CVD system has been carried out in order to investigate the molecular species involved in the nanotube growth process [25]. Under the typical high temperature CVD growth condition, mass-spectral data (Fig. 1.2.2) reveals that the effluent consists of mostly methane, with small concentrations of $H_2$, $C_2$ and $C_3$ hydrocarbon species also detected. However, measurements made with the methane source at room temperature also reveals similar concentrations of $H_2$ and $C_2$–$C_3$ species as in the effluent of the 900°C CVD system. This suggests that the $H_2$ and $C_2$–$C_3$ species detected in the CVD effluent are due to impurities in the methane source being used. Methane in fact undergoes negligible self-pyrolysis under typical SWNT growth conditions. Otherwise, one would

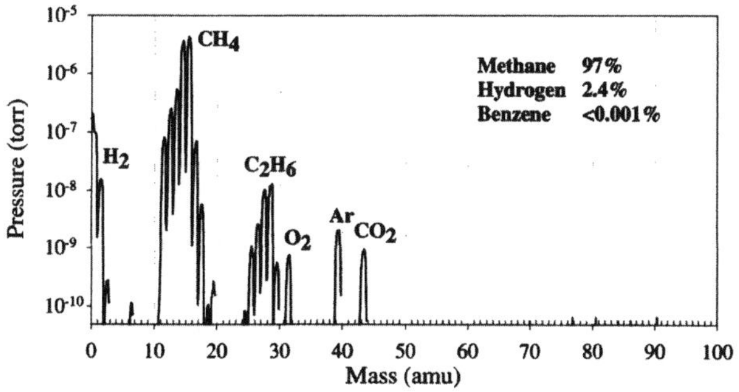


**Fig. 5.** Mass spectrum recorded with the effluent of the methane CVD system at 900°C

observe appreciable amounts of $H_2$ and higher hydrocarbons due to methane decomposition and reactions between the decomposed species. This result is consistent with the observation that the SWNTs produced by methane CVD under suitable conditions are free of amorphous carbon over-coating.

The methane CVD approach is promising for enabling scale-up of defect-free nanotube materials to the kilogram or even ton level. A challenge ist wheter it is possible to enable 1 g of catalyst producing 10, 100 g or even more SWNTs. To address this question, one needs to rationally design and create new types of catalyst materials with exceptional catalytic activities, large number of obtain active catalytic sites for nanotube nucleation with a given amount of catalyst, and learn how to grow nanotubes continuously into macroscopic lengths.

A significant progress was made recently by *Liu* and coworkers in obtaining a highly active catalyst for methane CVD growth of SWNTs [26]. Liu used sol-gel synthesis and supercritical drying to produce a Fe/Mo catalyst supported on alumina aerogel. The catalyst exhibits an ultra-high surface area ($\sim$ 540 m$^2$/g) and large mesopore volume ($\sim$ 1.4 mL/g), as a result of supercritical drying in preparing the catalyst. Under supercritical conditions, capillary forces that tend to collapse pore structures are absent as liquid and gas phases are indistinguishable under high pressure. Using the aerogel catalyst, Liu and coworkers were able to obtain $\sim$ 200% yield (1 g of catalyst yielding 2 g of SWNTs) of high quality nanotubes by methane CVD. Evidently, this is a substantial improvement over previous results, and is an excellent demonstration that understanding and optimization of the catalyst can lead to scale-up of perfect SWNT materials by CVD.

The growth of bulk amounts of SWNT materials by methane CVD has been pursued by several groups. *Rao* and coworkers used a catalyst based on mixed oxide spinels to growth SWNTs [27]. The authors found that good quality and yield of nanotubes were obtainable with FeCo alloy nanoparticles. Colomer and coworkers recently reported the growth of bulk quantities of SWNTs by CVD of methane using a cobalt catalyst supported on magnesium oxide [28]. They also found that the produced SWNTs can be separated from the support material by acidic treatment to yield a product with about 70–80% of SWNTs.

### 1.2.3 Growth Mode of Single-Walled Nanotubes in CVD

The states of nanotube ends often contain rich information about nanotube growth mechanisms [22]. High resolution TEM imaging of the SWNTs synthesized by the methane CVD method frequently observed closed tube ends that are free of encapsulated metal particles as shown in Fig. 6. The opposite ends were typically found embedded in the catalyst support particles when imaged along the lengths of the nanotubes. These observations suggest that SWNTs grow in the methane CVD process predominantly via the base-growth process as depicted in figure 3 [10,14,16,22,29]. The first step of the CVD reaction involves the absorption and decomposition of methane molecules on the surface of transition-metal catalytic nanoparticles on the support surface. Subsequently, carbon atoms dissolve and diffuse into the nanoparticle interior to form a metal-carbon solid state solution. Nanotube growth occurs when supersaturation leads to carbon precipitation into a crystalline tubular form. The size of the metal catalyst nanoparticle generally dictates the diameter of the synthesized nanotube. In the base-growth mode, the nanotube lengthens with a closed-end, while the catalyst particle remains on the support surface. Carbon feedstock is thus supplied from the 'base' where the nanotube interfaces with the anchored metal catalyst. Base-growth operates when strong metal-support interactions exist so that the metal species remain pinned on the support surface. In contrast, in the tip-growth mechanism, the nanotube lengthening involves the catalyst particle lifted off from the support and carried along at the tube end. The carried-along particle is responsible for supplying carbon feedstock needed for the tube growth. This mode operates when the metal-support interaction is weak [22].

In the methane CVD method, we have found that enhancing metal-support interactions leads to significant improvement in the performance of the catalyst material in producing high yield SWNTs [22]. This is due to the increased catalytic sites that favor base-mode nanotube growth. On the

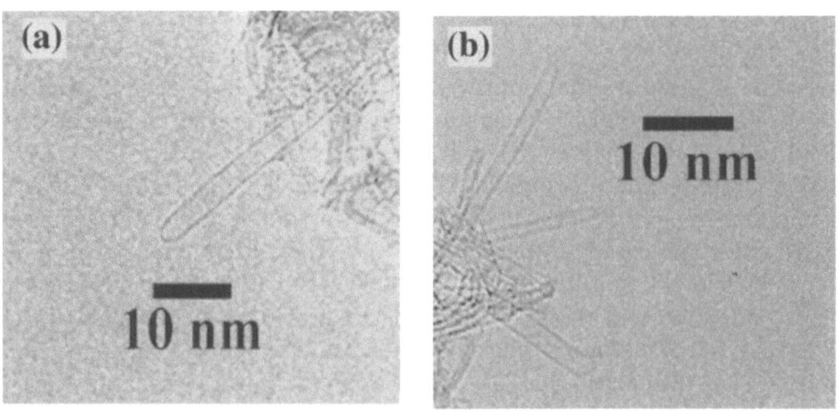


**Fig. 6 a,b.** TEM images of the ends of SWNTs grown by CVD

other hand, catalysts with weak metal-support interactions lead to aggregation of metal species and reduced nanotube yield and purity. Large metal particles due to the aggregation often lead to the growth of multi-layered graphitic particles or defective multi-walled tube structures. Metal-support interactions are highly dependent on the type of support materials and the type of metal precursor being used in preparing the catalyst [22].

### 1.3 Gas Phase Catalytic Growth

It has also been demonstrated that catalytic growth of SWNTs can be grown by reacting hydrocarbons or carbon monoxide with catalyst particles generated in-situ. *Cheng* and coworkers reported a method that employs benzene as the carbon feedstock, hydrogen as the carrier gas, and ferrocene as the catalyst precursor for SWNT growth [30]. In this method, ferrocene is vaporized and carried into a reaction tube by benzene and hydrogen gases. The reaction tube is heated at 1100–1200°C. The vaporized ferrocene decomposes in the reactor, which leads to the formation of iron particles that can catalyze the growth of SWNTs. With this approach however, amorphous carbon generation could be a problem, as benzene pyrolysis is expected to be significant at 1200°C. *Smalley* and coworkers has developed a gas phase catalytic process to grow bulk quantities of SWNTs [31]. The carbon feedstock is carbon monoxide (CO) and the growth temperature is in the range of 800–1200°C. Catalytic particles for SWNT growth are generated in-situ by thermal decomposition of iron pentacarbonyl in a reactor heated to the high growth temperatures. Carbon monoxide provides the carbon feedstock for the growth of nanotubes off the iron catalyst particles. CO is a very stable molecule and does not produce unwanted amorphous carbonaceous material at high temperatures. However, this also indicates that CO is not an efficient carbon source for nanotube growth. To enhance the CO carbon feedstock, Smalley and coworkers have used high pressures of CO (up to 10 atm) to significantly speed up the disproportionation of CO molecules into carbon, and thus enhance the growth of SWNTs. The SWNTs produced this way are as small as 0.7 nm in diameter, the same as that of a $C_{60}$ molecule. The authors have also found that the yield of SWNTs can be increased by introducing a small concentration of methane into their CO high pressure reactor at 1000–1100°C growth temperatures. Methane provides a more efficient carbon source than CO and does not undergo appreciable pyrolysis under these conditions. The high pressure CO catalytic growth approach is promising for bulk production of single-walled carbon nanotubes.

## 2 Controlled Nanotube Growth by Chemical Vapor Deposition

Recent interest in CVD nanotube growth is also due to the idea that aligned and ordered nanotube structures can be grown on surfaces with control that is not possible with arc-discharge or laser ablation techniques [23,32].

## 2.1 Aligned Multi-Walled Nanotube Structures

Methods that have been developed to obtain aligned multi-walled nanotubes include CVD growth of nanotubes in the pores of mesoporous silica, an approach developed by *Xie*'s group at the Chinese Academy of Science [33,34]. The catalyst used in this case is iron oxide particles created in the pores of silica, the carbon feedstock is 9% acetylene in nitrogen at an overall 180 torr pressure, and the growth temperature is 600°C. Remarkably, nanotubes with lengths up to millimeters are made (Fig. 7a) [34]. *Ren* has grown relatively large-diameter MWNTs forming oriented 'forests' (Fig. 7b) on glass substrates using a plasma assisted CVD method with nickel as the catalyst and acetylene as the carbon feedstock around 660°C [35].

Our group has been devising growth strategies for ordered multi-walled and single-walled nanotube architectures by CVD on catalytically patterned substrates [23,32]. We have found that multi-walled nanotubes can self-assemble into aligned structures as they grow, and the driving force for self-alignment is the Van der Waals interactions between nanotubes [36]. The growth approach involves catalyst patterning and rational design of the substrate to enhance catalyst-substrate interactions and control the catalyst particle

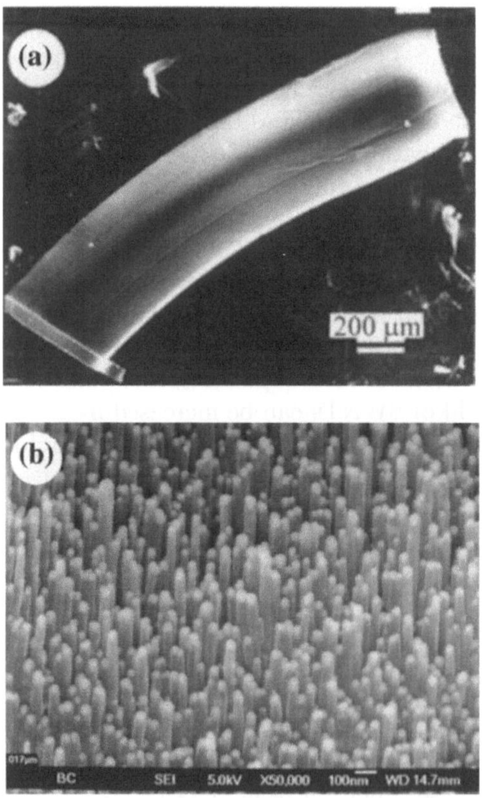


**Fig. 7.** Aligned multi-walled nanotubes grown by CVD methods. (**a**) An ultra-long aligned nanotube bundle (courtesy of S. Xie). (**b**) An oriented MWNT forest grown on glass substrate (courtesy of Z. Ren)

size. Porous silicon is found to be an ideal substrate for this approach and can be obtained by electrochemical etching of n-type silicon wafers in hydrofluoric acid/methanol solutions. The resulting substrate consisted of a thin nanoporous layer (pore size $\sim 3\,\mathrm{nm}$) on top of a macroporous layer (with submicron pores). Squared catalyst patterns on the porous silicon substrate are obtained by evaporating a 5 nm thick iron film through a shadow mask containing square openings. CVD growth with the substrate is carried out at 700°C under an ethylene flow of 1000 mL/min for 15 to 60 min. Figure 8 shows SEM images of regularly positioned arrays of nanotube towers grown from patterned iron squares on a porous silicon substrates. The nanotube towers exhibit very sharp edges and corners with no nanotubes branching away from the blocks. The high resolution SEM image (Fig. 8c) reveals that the MWNTs (Fig. 8c inset) within each block are well aligned along the direction perpendicular to the substrate surface. The length of the nanotubes and thus the height of the towers can be controlled in the range of 10–240μm by varying the CVD reaction time. The width of the towers is controlled by the size of the openings in the shallow mask. The smallest self-oriented nanotube towers synthesized by this method are 2μm × 2μm.

The mechanism of nanotube self-orientation involves the nanotube base-growth mode [36]. Since the nanoporous layer on the porous silicon substrate serves as an excellent catalyst support, the iron catalyst nanoparticles formed on the nanoporous layer interact strongly with the substrate and remain pinned on the surface. During CVD growth, the outermost walls of nanotubes interact with their neighbors via van der Waals forces to form a rigid bundle, which allows the nanotubes to grow perpendicular to the substrate (Fig. 8d). The porous silicon substrates exhibit important advantages over plain silicon substrates in the synthesis of self-aligned nanotubes. Growth on substrates containing both porous silicon and plain silicon portions finds that nanotubes grow at a higher rate (in terms of length/min) on porous silicon than on plain silicon. This suggests that ethylene molecules can permeate through the macroporous silicon layer (Fig. 8d) and thus efficiently feed the growth of nanotubes within the towers. The nanotubes grown on porous silicon substrates have diameters in a relatively narrow range since catalyst nanoparticles with a narrow size distribution are formed on the porous supporting surface, and the metal-support interactions prevent the catalytic metal particles from sintering at elevated temperatures during CVD.

## 2.2 Directed Growth of Single-Walled Nanotubes

Ordered, single-walled nanotube structures can be directly grown by methane CVD on catalytically patterned substrates. A method has been devised to grow suspended SWNT networks with directionality on substrates containing lithographically patterned silicon pillars [25,37]. Contact printing is used to transfer catalyst materials onto the tops of pillars selectively. CVD of

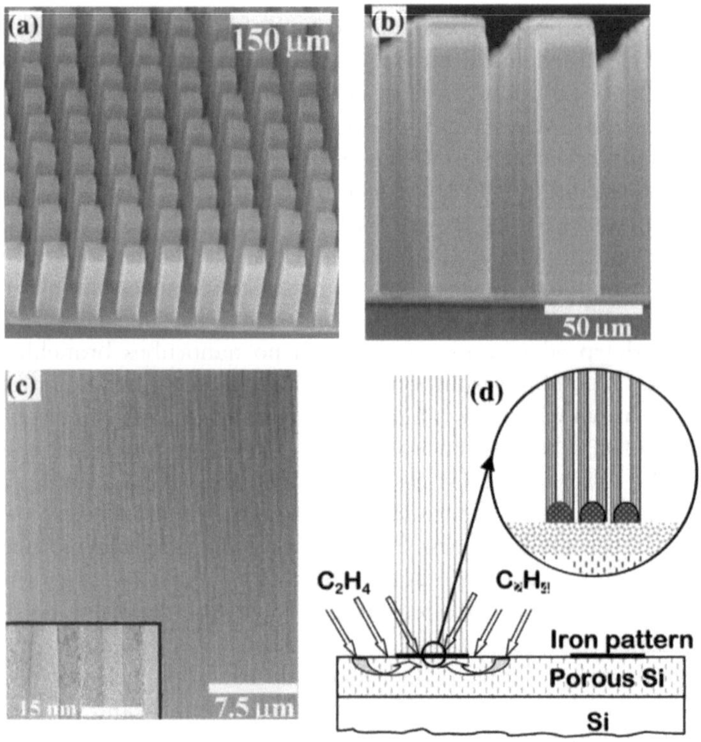


**Fig. 8.** Self-oriented MWNT arrays grown by CVD on a catalytically patterned porous silicon substrate. (**a**) SEM image of tower structures consisted of aligned nanotubes. (**b**) SEM image of the side view of the towers. (**c**) A high magnification SEM image showing aligned nanotubes in a tower. Inset: TEM image showing two MWNTs bundling together. (**d**) Schematic diagram of the growth process

methane using the substrates leads to suspended SWNTs forming nearly ordered networks with the nanotube orientations directed by the pattern of the pillars (Fig. 9).

The growth approach starts with developing a liquid-phase catalyst precursor material that has the advantage over solid-state supported catalysts in allowing the formation of uniform catalyst layers and for large-scale catalytic patterning on surfaces [25,37]. The precursor material consists of a triblock copolymer, aluminum, iron and molybdenum chlorides in mixed ethanol and butanol solvents. The aluminum chloride provides an oxide framework when oxidized by hydrolysis and calcination in air. The triblock copolymer directs the structure of the oxide framework and leads to a porous catalyst structure upon calcination. The iron chloride can lead to catalytic particles needed for the growth of SWNTs in methane CVD. The catalyst precursor material is first spun into a thin film on a poly-dimethyl siloxane (PDMS) stamp, followed by contact-printing to transfer the catalyst precursor selectively onto

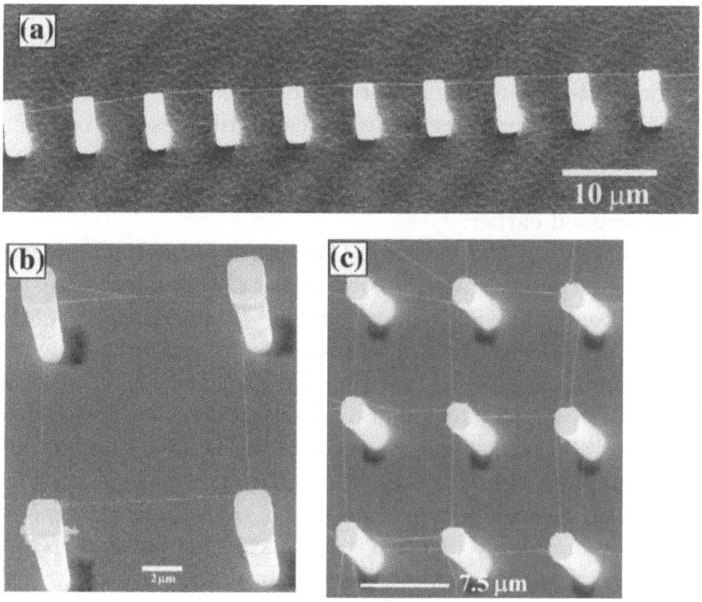


**Fig. 9.** Directed growth of suspended SWNT. (**a**) A nanotube power-line structure. (**b**) A square of nanotubes. (**c**) An extensive network of suspended SWNTs

the tops of pillars pre-fabricated on a silicon substrate. The stamped substrate is calcined and then used in CVD growth.

Remarkably, the SWNTs grown from the pillar tops tend to direct from pillar to pillar. Well-directed SWNT bridges are obtained in an area of the substrate containing isolated rows of pillars as shown in Fig. 9a, where suspended tubes forming a power-line like structure can be seen. In an area containing towers in a square configuration, a square of suspended nanotube bridges is obtained (Fig. 9b). Suspended SWNTs networks extending a large area of the substrate are also formed (Fig. 9c). Clearly, the directions of the SWNTs are determined by the pattern of the elevated structures on the substrate. Very few nanotubes are seen to extend from pillars and rest on the bottom surface of the substrate.

The self-directed growth can be understood by considering the SWNT growth process on the designed substrates [25]. Nanotubes are nucleated only on the tower-tops since the catalytic stamping method does not place any catalyst materials on the substrate below. As the SWNTs lengthen, the methane flow keeps the nanotubes floating and waving in the 'wind' since the flow velocity near the bottom surface is substantially lower than that at the level of the tower-tops. This prevents the SWNTs from being caught by the bottom surface. The nearby towers on the other hand provide fixation points for the growing tubes. If the waving SWNTs contact adjacent towers, the tube-tower van der Waals interactions will catch the nanotubes and hold them aloft.

Growing defect-free nanotubes continuously to macroscopic lengths has been a challenging task. In general, continuous nanotube growth requires the catalytic sites for SWNTs remaining active indefinitely. Also, the carbon feedstock must be able to reach and feed into the catalytic sites in a continuous fashion. For CVD growth approaches, this means that catalyst materials with high surface areas and open pore structures can facilitate the growth of long nanotubes, as discussed earlier.

We have found that within the methane CVD approach, small concentrations (~0.03%) of benzene generated in-situ in the CVD system can lead to an appreciable increase in the yield of long nanotubes. Individual SWNTs with length up to 0.15 mm (150μm) can be obtained [25]. An SEM image of an approximately 100μm long tube is shown in Fig. 10. The SWNTs appear continuous and strung between many pillars. In-situ generation of benzene is accomplished by catalytic conversion of methane in the CVD system using bulk amounts of alumina supported Fe/Mo catalyst [25]. This leads to enhanced SWNT growth, and a likely explanation is that at high temperatures, benzene molecules are highly reactive compared to methane, therefore enhancing the efficiency of carbon-feedstock in nanotube growth. However, we find that when high concentrations of benzene are introduced into the CVD growth system, exceedingly low yields of SWNTs result with virtually no suspended tubes grown on the sample. This is because high concentrations of benzene undergo extensive pyrolysis under the high temperature CVD condition, which causes severe catalyst poisoning as amorphous carbon is deposited on the catalytic sites therefore preventing SWNT growth.

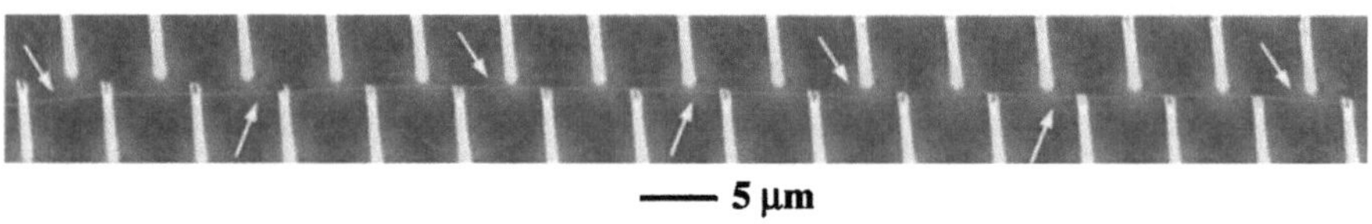


**Fig. 10.** SEM image of a CVD grown approximately 100μm long SWNT strung between silicon pillars

## 2.3 Growth of Isolated Nanotubes on Specific Surface Sites

A CVD growth strategy has been developed collaboratively by our group and *Quate* to grow individual single-walled nanotubes at specific sites on flat silicon oxide substrates [20]. The approach involves methane CVD on substrates containing catalyst islands patterned by electron beam lithography. 'Nanotube chips' with isolated SWNTs grown from the islands are obtained. Atomic Force Microscopy (AFM) images of SWNTs on such nanotube-chips are shown in Fig. 10, where the synthesized nanotubes extending from the catalyst islands are clearly observed. This growth approach readily leads to SWNTs originating from well controlled surface sites, and have enabled us to develop a controlled method to integrate nanotubes into addressable structures for the purpose of elucidating their fundamental properties and building devices with interesting electrical, electromechanical and chemical functions [23,38,39,40,41,42,43,44].

## 2.4 From Growth to Molecular-Wire Devices

Integrating individual nanotubes into addressable structures is important to the characterization of nanotubes. It is necessary to investigate individual tubes because the properties of nanotubes are highly sensitive to their structural parameters, including chirality and diameter. Currently, all of the growth methods yield inhomogeneous materials containing nanotubes with various chiralities. Therefore, measurements of ensembles of nanotubes can only reveal their bulk averaged properties.

### 2.4.1 Electrical Properties of Individual Nanotubes

Previous approaches to individual SWNT electrical devices include randomly depositing SWNTs from liquid suspensions onto pre-defined electrodes [4,45], or onto a flat substrate followed by locating nanotubes and patterning electrodes [46,47,48,49]. We have demonstrated that controlled nanotube growth strategies open up new routes to individually addressable nanotubes. The SWNTs grown from specific sites on substrates can be reliably contacted by electrodes (Figs. 11c, 12a) and characterized [38]. Metal electrodes are placed onto the two ends of a nanotube via lithography patterning and electron beam evaporation. Detailed procedure for contacting a SWNT can be found in [38].

The formation of low ohmic contacts with SWNTs is critical to elucidating their intrinsic electrical properties [40] and building devices with advanced characteristics. This is accomplished by our controlled approach of growing and contacting nanotubes. We found that titanium metal contacts give rise to the lowest contact resistance compared to other metals. Metallic SWNTs that are several microns long typically exhibit two-terminal resistance on the order of tens to hundreds of kilo-ohms. The lowest single-tube resistance measured with our individual metallic SWNT is $\sim 12\,\mathrm{k}\,\Omega$ (Fig. 12b). The low contact resistance achieved in our system can be attributed to several factors. The first is that our method allows the two metal electrodes to contact the two ends of a nanotube. Broken translational symmetry at the nanotube ends could be responsible for the strong electrical coupling between the tube and metal [50]. Secondly, titanium–carbon (carbide) bond formation at the metal–tube interface may have occurred during the electron-beam evaporation process. This is based on our result that titanium yields lower contact resistance than other metals, including aluminum and gold, and the fact that

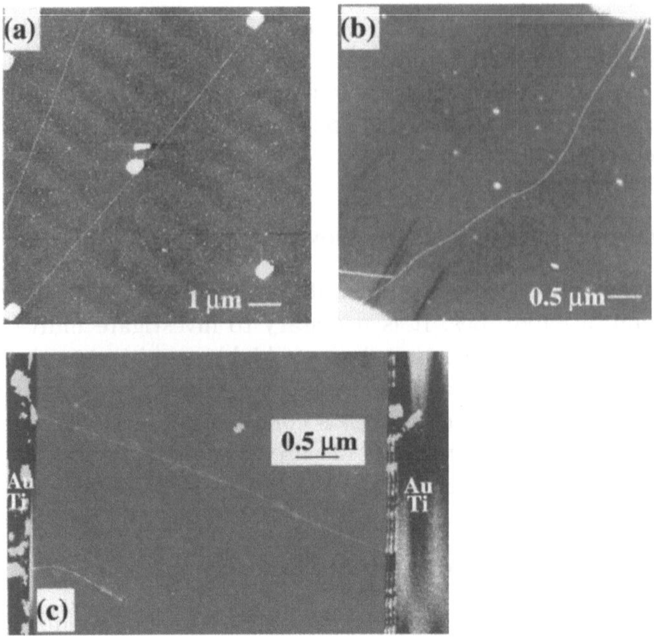


**Fig. 11.** Growth of single-walled nanotubes on controlled surface sites and device integration. (**a**) An atomic force microscopy image of SWNTs grown from patterned catalyst dots (while spots). (**b**) AFM image of a SWNT grown between two catalyst sites (white corners). (**c**) AFM image of a SWNT grown from a catalyst island and contacted by metal electrodes

aluminum and gold do not form strong bonding with carbon and stable carbide compounds.

For individual semiconducting SWNTs grown on surfaces, relatively low resistance devices on the order of hundreds of kilo-ohms can be made by our approach [40]. These nanotubes exhibit *p*-type transistor behavior at room temperature as their conductance can be dramatically changed by gate voltages (Fig. 12c). This property is consistent with the initial observation made with high resistance samples (several mega-ohms) by *Dekker* [51] and by *Avouris'* group [52]. The transconductance (ratio of current change over gate-voltage change) of our semiconducting tube samples can be up to ~200 nA/V [40] which is two orders of magnitude higher than that measured with earlier samples. The high transconductance is a direct result of the relatively low resistance of our semiconducting SWNT samples, as high currents can be transported through the system at relatively low bias voltages. This result should not be underestimated, given the importance of high transconductance and voltage gain to transistors. Nevertheless, future work is needed in order to create semiconducting SWNT devices with transconductance and voltage gains that match existing silicon devices.

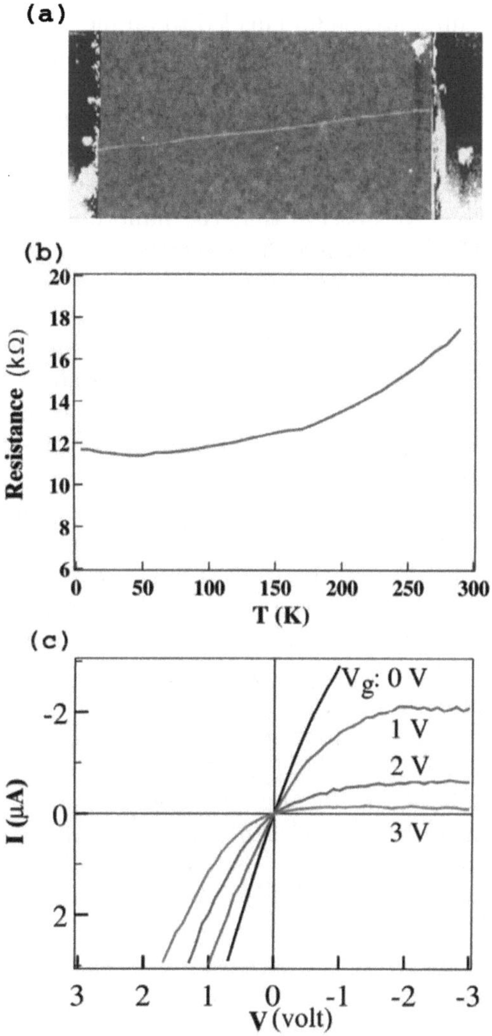


**Fig. 12.** Electrical properties of individual single-walled carbon nanotubes. (**a**) AFM image of an individual single-walled nanotube contacted by two metal electrode. The length of the nanotube between electrodes is $\sim 3\mu m$. (**b**) Temperature dependent resistance of a metallic SWNT. (**c**) Current ($I$) vs. voltage ($V$) characteristics of a semiconducting nanotube under various gate-voltages ($V_g$)

### 2.4.2 Nanotube Electromechanical Properties and Devices

Controlled growth of nanotubes on surfaces combined with microfabrication approaches allows the construction of novel nanotube devices for new types of studies. For instance, the question of how mechanical deformation affects the electrical properties of carbon nanotubes has been intriguing [53,54,55,56,57], and is important to potential applications of nanotubes as building blocks for nanoscale electro-mechanical devices. To address this question experimentally, it will be desired to obtains suspended nanotubes that can be manipulated mechanically and at the same time addressed electrically. To this end, we have grown individual SWNTs from patterned catalyst sites across pre-fabricated trenches on $SiO_2/Si$ substrates [43]. This led to an individual SWNTs partially suspended over the trenches (Fig. 13a). The nanotube is then contacted at the two ends by metal electrodes. The suspended part of the nanotube can be manipulated with an AFM tip, while the resistance of the sample is being monitored. For a metallic SWNT with $\sim$ 610 nm suspended length, an AFM tip is used to repeatedly push down the suspended nanotube and then retract (Fig. 13b). We observe that the nanotube conductance decreases each time the AFM tip pushes the nanotube down, but recovers as the tip retracts. The repeated pushing and retracting cause oscillations in the cantilever/nanotube deflection and sample conductance, with equal periodicity in the two oscillations (Fig. 13c). The full reversibility of the nanotube electrical conductance upon tip retraction suggests that the metal-tube contacts are not affected each time the tip deflects the suspended part of the nanotube. The observed change in sample conductance is entirely due to the mechanical deformation of the SWNT caused by the pushing tip. The reversibility also suggests that the suspended part of the SWNT is firmly fixed on the substrate next to the trench. The length of the SWNT resting on the $SiO_2$ surface ($\geq$ 1.5 μm on both sides of the trench) does not stretch or slide on the substrate when the suspended part is pushed. Anchoring of the nanotube on the $SiO_2$ substrate is due to strong tube-substrate van der Waals interactions.

The conductance is found to decrease by a factor of two at $\sim 5°$ bending angle (strain $\sim$ 0.3%), but decrease more dramatically by two orders of magnitude at a bending angle $\sim 14°$ (strain $\sim$ 3%, Fig. 13d) [44]. To understand these electromechanical characteristics, *Wu* and coworkers at the University of Louisville have carried out order-$N$ non-orthogonal tight-binding molecular-dynamics simulations of a tip deflecting a metallic (5,5) SWNT, with the tip modeled by a short and stiff (5,5) SWNT cap [44,58]. At relatively small bending angles, the nanotube is found to retain $sp^2$ bonding throughout its structure, but exhibits significant bond distortion for the atoms in the region near the tip. As tip-pushing and bending proceed, the nanotube structure progressively evolves and larger structural-changes occur in the nanotube region in the vicinity of the tip. At a 15° bending angle, the average number of bonds per atom in this region is found to increase to $\sim$ 3.6, suggesting the appearance of $sp^3$-bonded atoms (marked in red in Fig. 13e). This causes a significant decrease in the local $\pi$-electron density as revealed by electronic structure calculations. Since the $\pi$-electrons are delocalized and responsible for electrical conduction, a drastic reduction in the $\pi$-electron density is responsible for the significant decrease in conductance. Simulations find that the large local $sp^3$ deformation is highly energetic, and its appearance is entirely due to the forcing tip. The structure is found to fully reverse to $sp^2$ upon moving the tip away in the simulation.

Using the Landau-Buttiker formula, *Wu* and coworkers have calculated the conductance of the (5,5) tube vs. bending angle [58]. It is found that

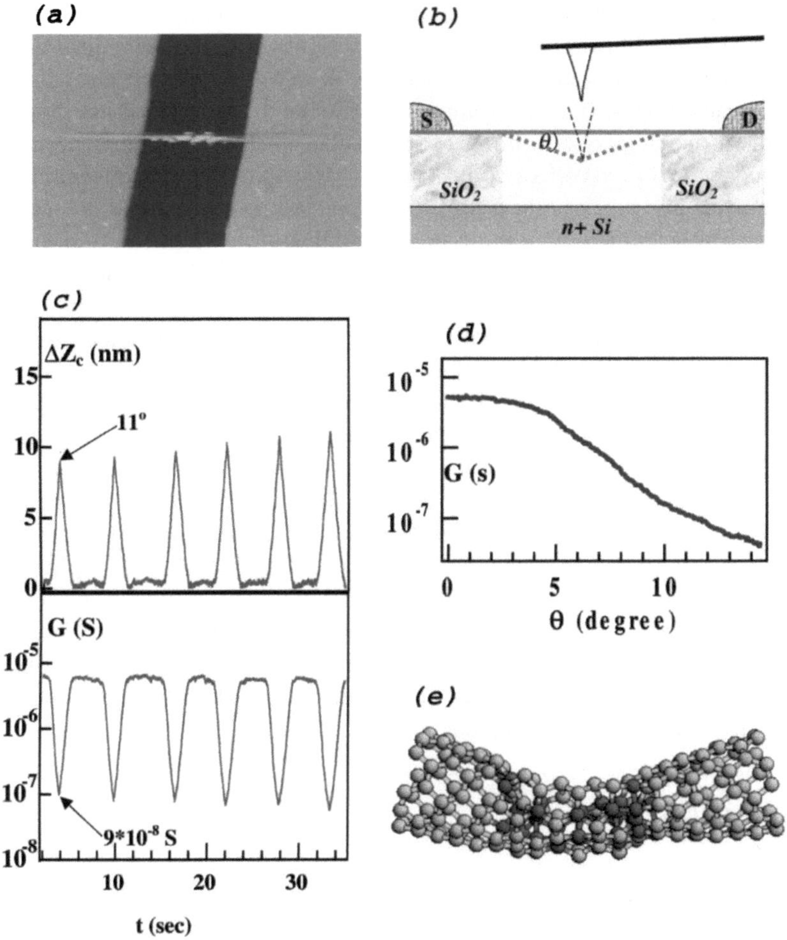


**Fig. 13.** Electromechanical characteristics of suspended nanotubes. (**a**) AFM image of a SWNT with a ~ 605 nm long suspension over a trench. The bright spots around the suspended tube part are caused by tube touching and sticking to the side of the pyramidal scanning tip. (**b**) A schematic view of the electromechanical measurement setup. (**c**) Cantilever deflection ($\Delta Z_c$, upper graph) and nanotube electrical conductance (G, lower graph) evolution during repeated cycles of pushing the suspended SWNT and retracting. (**d**) Electrical conductance (G) of the SWNT sample vs. bending angle ($\theta$). (**e**) Simulated atomic configurations of a (5,5) SWNT pushed to a 15 degrees bending angle by an AFM tip

at relatively small bending angles, the conductance of the SWNT exhibits appreciable decrease, but the decrease is relatively gradual. The conductance decrease is caused by the relatively large C–C bond distortions in the nanotube region near the tip. The decrease is gradual because the overall nanotube structure remains in the $sp^2$ state. At large bending angles, the nanotube conductance decreases dramatically, as $sp^3$ bonding sets into the nanotube structure. These results agree qualitatively with experimental data and thus provide a detailed rationale to the observed nanotube electromechanical characteristics. The combined experimental and theoretical study leads to an in-depth understanding of nanotube electromechanical properties, and suggests that SWNTs could serve as reversible electro-mechanical transducers that are potentially useful for nano-electro-mechanical devices.

The electro-mechanical characteristics are elucidated when the nanotube deformation is caused by a manipulating local probe. This should be distinguished from previous theoretical considerations of 'smoothly' bent nanotubes. Nanotube bending in most of the earlier investigations is modeled by holding the ends of a nanotube such that the nanotube is at a certain bending configuration [54,55,56]. In most of the cases studied, the nanotube remains in the $sp^2$ state with only small bond distortions throughout the structure. Therefore, the nanotube conductance has been found to be little changed ($<$ 10 fold) under bending angles up to 20°. Nevertheless, *Rochefort* et al. did find that at larger bending angles (e.g. $\theta$=45°), the electrical conductance of a metallic (6,6) SWNT is lowered up to 10-fold [55,56]. The physics studied in our case should be applicable to SWNTs containing large local deformations caused by other forces. For instance, if a highly kinked SWNT stabilized by van der Waals forces on a substrate develops $sp^3$ type of bonding characteristics at the kink, the electrical conductance should be significantly reduced compared to a straight tube.

## 3 Conclusions

We have presented an overview of various growth methods for multi-walled and single-walled carbon nanotubes. It is shown that chemical vapor deposition approaches are highly promising for producing large quantities of high quality nanotube materials at large scales. Controlling nanotube growth with CVD strategies has led to organized nanowires that can be readily integrated into addressable structures useful for fundamental characterization and potential applications. It can be envisioned that in a foreseeable future, controlled growth will yield nanotube architectures used as key components in next generations of electronic, chemical, mechanical and electromechanical devices.

It is fair to say that progress in nanotube research has been built upon the successes in materials synthesis. This trend shall continue. It is perhaps an ultimate goal for growth to gain control over the nanotube chirality and diameter, and be able to direct the growth of a semiconducting or metallic nanowire from and to any desired sites. Such control will require significant future effort, and once successful, is likely to bring about revolutionary opportunities in nanoscale science and technology.

**Acknowledgements**

The work carried out at Stanford were done by J. Kong, N. Franklin, T. Tombler, C. Zhou, R. Chen, A. Cassell, M. Chapline, E. Chan and T. Soh. We thank Professors C. Quate, S. Fan, S. Y. Wu, C. Marcus and Dr. J. Han for fruitful collaborations. This work was supported financially by National Science Foundation, DARPA/ONR, a Packard Fellowship, a Terman Fellowship, Semiconductor Research Corporation/Motorola Co., Semiconductor Research Corporation/Semetech., the National Nanofabrication Users Network at Stanford, Stanford Center for Materials Research, the Camile Henry-Dreyfus Foundation and the American Chemical Society.

## References

1. S. Iijima, Nature **354**, 56–58 (1991)
2. S. Iijima, T. Ichihashi, Nature **363**, 603–605 (1993)
3. M. S. Dresselhaus, G. Dresselhaus, P. C. Eklund, *Science of Fullerenes and Carbon Nanotubes* (Academic, San Diego 1996)
4. C. Dekker, Phys. Today **52**, 22–28 (1999)
5. T. W. Ebbesen, P. M. Ajayan, Nature **358**, 220–222 (1992)
6. D. S. Bethune, C. H. Kiang, M. DeVries, G. Gorman, R. Savoy, J. Vazquez, R. Beyers, Nature **363**, 605–607 (1993)
7. A. Thess, R. Lee, P. Nikolaev, H. J. Dai, P. Petit, J. Robert, C. H. Xu, Y. H. Lee, S. G. Kim, A. G. Rinzler, D. T. Colbert, G. E. Scuseria, D. Tomanek, J. E. Fischer, R. E. Smalley, Science **273**, 483–487 (1996)
8. C. Journet, W. K. Maser, P. Bernier, A. Loiseau, M. L. Delachapelle, S. Lefrant, P. Deniard, R. Lee, J. E. Fischer, Nature **388**, 756–758 (1997)
9. J. Liu, A. G. Rinzler, H. Dai, J. H. Hafner, R. K. Bradley, P. J. Boul, A. Lu, T. Iverson, K. Shelimov, C. B. Huffman, F. Rodriguez-Macias, Y.-S. Shon, T. R. Lee, D. T. Colbert, R. E. Smalley, Science **280**, 1253–1256 (1998)
10. G. G. Tibbetts, J Cryst. Growth **66**, 632–638 (1984)
11. G. G. Tibbetts, Carbon **27**, 745-747 (1989)
12. G. G. Tibbetts: Filaments and Composites, in *Carbon Fibers* (Kluwer Academic, Amsterdam 1990) pp. 73–94
13. G. G. Tibbetts, J. Cryst. Growth **73**, 431–438 (1985)
14. G. G. Tibbetts, M. G. Devour, E. J. Rodda, Carbon **25**, 367–375 (1987)
15. R. T. K. Baker, *Physics and Chemistry of Carbon*, Vol.14, P. Walker, P. Thrower, (Eds.) (Dekker, New York 1978) pp. 83–165
16. R. T. K. Baker, Carbon **27**, 315–323 (1989)
17. R. T. K. Baker, L. L. Murrell, (Eds.), *Novel Materials in Heterogeneous Catalysis* (Washington, DC 1990)
18. H. G. Tennent, Hyperion Catalysis International, Inc., US patent no. 4663230, USA, (1987)
19. C. E. Snyder, W. H. Mandeville, H. G. Tennent, L. K. Truesdale, Hyperion Catalysis International, US patent, (1989)
20. J. Kong, A. M. Cassell, H. Dai, Chem. Phys. Lett. **292**, 567-574 (1998)
21. J. Kong, H. Soh, A. Cassell, C. F. Quate, H. Dai, Nature **395**, 878–879 (1998)

22. A. Cassell, J. Raymakers, J. Kong, H. Dai, J. Phys. Chem. **103**, 6484–6492 (1999)
23. H. Dai, J. Kong, C. Zhou, N. Franklin, T. Tombler, A. Cassell, S. Fan, M. Chapline, J. Phys. Chem. **103**, 11246–11255 (1999)
24. J. Hafner, M. Bronikowski, B. Azamian, P. Nikolaev, D. Colbert, R. Smalley, Chem. Phys. Lett. **296**, 195–202 (1998)
25. N. Franklin, H. Dai, Adv. Mater. **12**, 890 (2000)
26. M. Su, B.Zheng, J. Liu, Chem. Phys. Lett. **322**, 321-326 (2000)
27. E. Flahaut, A. Govindaraj, A. Peigney, C. Laurent, C. N. Rao, Chem. Phys. Lett. **300**, 236–242 (1999)
28. J.-F. Colomer, C. Stephan, S. Lefrant, G. V. Tendeloo, I. Willems, Z. Kánya, A. Fonseca, C. Laurent, J. B.Nagy, Chem. Phys. Lett. **317**, 83–89 (2000)
29. S. Amelinckx, X. B. Zhang, D. Bernaerts, X. F. Zhang, V. Ivanov, J. B. Nagy, Science **265**, 635–639 (1994)
30. H. Cheng, F. Li, G. Su, H. Pan, M. Dresselhaus, Appl. Phys. Lett. **72**, 3282–3284 (1998)
31. P. Nikolaev, M. J. Bronikowski, R. K. Bradley, F. Rohmund, D. T. Colbert, K. A. Smith, R. E. Smalley, Chem. Phys. Lett. **313**, 91–97 (1999)
32. H. Dai, Phys. World (2000)
33. W. Z. Li, S. S. Xie, L. X. Qian, B. H. Chang, B. S. Zou, W. Y. Zhou, R. A. Zhao, G. Wang, Science **274**, 1701–1703 (1996)
34. Z. Pan, S. S.Xie, B. Chang, C. Wang, Nature **394**, 631–632 (1998)
35. Z. F. Ren, Z. P. Huang, J. W. Xu, J. H. Wang, Science **282**, 1105–1107 (1998)
36. S. Fan, M. Chapline, N. Franklin, T. Tombler, A. Cassell, H. Dai, Science **283**, 512–514 (1999)
37. A. Cassell, N. Franklin, T. Tombler, E. Chan, J. Han, H. Dai, J. Am. Chem. Soc. **121**, 7975-7976 (1999)
38. H. Soh, C. Quate, A. Morpurgo, C. Marcus, J. Kong, H. Dai, Appl. Phys. Lett. **75**, 627–629 (1999)
39. A. Morpurgo, J. Kong, C. Marcus, H. Dai, Science **286**, 263-265 (1999)
40. C. Zhou, J. Kong, H. Dai, Appl. Phys. Lett. **76**, 1597–1599 (1999)
41. J. Kong, N. Franklin, C. Zhou, M. Chapline, S. Peng, K. Cho, H. Dai, Science **287**, 622-625 (2000)
42. C. Zhou, J. Kong, H. Dai, Phys. Rev. Lett. **84**, 5604 (2000)
43. T. Tombler, C. Zhou, J. Kong, H. Dai, Appl. Phys. Lett. **76**, 2412–2414 (2000)
44. T. Tombler, C. Zhou, L. Alexeyev, J. Kong, H. Dai, L. Liu, C. Jayanthi, M. Tang, S. Y. Wu, Nature **405**, 769 (2000)
45. S. J. Tans, M. H. Devoret, H. Dai, A. Thess, R. E. Smalley, L. J. Geerligs, C. Dekker, Nature **386**, 474-477 (1997)
46. M. Bockrath, D. H. Cobden, P. L. McEuen, N. G. Chopra, A. Zettl, A. Thess, R. E. Smalley, Science **275**, 1922-1925 (1997)
47. D. H. Cobden, M. Bockrath, N. G. Chopra, A. Zettl, P. McEuen, A. Rinzler, R. E. Smalley, Phys. Rev. Lett. **81**, 681-684 (1998)
48. M. Bockrath, D. Cobden, J. Lu, A. Rinzler, R. Smalley, L. Balents, P. McEuen, Nature **397**, 598–601 (1999)
49. J. Nygard, D. H. Cobden, M. Bockrath, P. L. McEuen, P. E. Lindelof, Appl. Phys. A **69**, 297–304 (1999)
50. J. Tersoff, Appl. Phys. Lett. **74**, 2122–2124 (1999)
51. S. Tans, A. Verschueren, C. Dekker, Nature **393**, 49–52 (1998)

52. R. Martel, T. Schmidt, H. R. Shea, T. Hertel, Ph. Avouris, Appl. Phys. Lett. **73**, 2447–2449 (1998)
53. C. L. Kane, E. J. Mele, Phys. Rev. Lett. **78**, 1932–1935 (1997)
54. M. Nardelli, J. Bernholc, Phys. Rev. B **60**, R16338–16341 (1998)
55. A. Rochefort, D. Salahub, Ph. Avouris, Chem. Phys. Lett. **297**, 45–50 (1998)
56. A. Rochefort, F. Lesage, D. Salhub, Ph. Avouris, Phys. Rev. B **60**, 13824–13830 (1999)
57. S. Paulson, M. Falvo, N. Snider, A. Helser, T. Hudson, A. Seeger, R. Taylor, R. Superfine, S. Washburn, Appl. Phys. Lett. **75**, 2936–2938 (1999)
58. L. Liu, C. Jayanthi, M. Tang, S. Y. Wu, T. Tombler, C. Zhou, L. Alexeyev, J. Kong, H. Dai, Phys. Rev. Lett. **84**, 4950 (2000).

# Growth Mechanisms of Carbon Nanotubes

Jean-Christophe Charlier[1] and Sumio Iijima[2,3]

[1] Université Catholique de Louvain
Unité de Physico-Chimie et de Physique des Matériaux
Place Croix du Sud, 1, Bâtiment Boltzmann,
1348 Louvain-la-Neuve, Belgium
charlier@pcpm.ucl.ac.be

[2] NEC Corporation, R & D Group
34 Miyukigaoka, Tsukuba, Ibaraki 305, Japan

[3] Meijo University, Faculty of Science and Engineering
1-501, Shiogamaguchi Tenpaku, Nagoya, Aichi 4688502, Japan
iijimas@meijo-u.ac.jp

**Abstract.** This review considers the present state of understanding of the growth of carbon nanotubes based on TEM observations and numerical simulations. In the highlights of these experimental and theoretical approaches, various mechanisms, already proposed in the literature, for single and multi-shell nanotubes nucleation and growth are reviewed. A good understanding of nanotube growth at the atomic level is probably one of the main issues either to develop a nanotube mass production process or to control growth in order to obtain well-designed nanotube structures.

Nearly ten years after their discovery [1], carbon nanotubes (CNs) are still attracting much interest for their potential applications, which largely derives from their unusual structural and electronic properties. Since all these properties are directly related to the atomic structure of the tube, it is essential to understand what controls nanotube size, the number of shells, the helicity, and the structure during synthesis. A thorough understanding of the formation mechanisms for these nanotubular carbon systems is crucial to design procedures for controlling the growth conditions to obtain more practical structures, which might be directly available for nanotechnology.

In order to optimize the single- and multi-walled nanotube yield and quality, three main production methods have been used up to now for their synthesis. Multi-wall carbon nanotubes [1,2] typically grow on the cathode during an arc discharge between two graphitic electrodes (temperature ~3000 K). Single-wall carbon nanotubes were first observed in the arc discharge apparatus by co-evaporating iron [3] or cobalt [4] (as metal catalysts) in a methane atmosphere. The discovery of a *single-shell tube* stimulated intense research to find an efficient way to produce bundles of ordered single-wall nanotubes (called *ropes*).

The laser ablation technique uses two lasers to vaporize a graphite target mixed with a small amount of Co and/or Ni in order to condense the

M. S. Dresselhaus, G. Dresselhaus, Ph. Avouris (Eds.): Carbon Nanotubes,
Topics Appl. Phys. **80**, 55–80 (2001)

carbon into single-wall tubes. Through this technique [5], the growth conditions are well controlled and maintained over for a long time, leading to a more uniform vaporization. Another technique [6], based on the carbon-arc method, also provides similar arrays of single-wall nanotubes, produced from an ionized carbon plasma which is generated by joule heating during the discharge. Both techniques synthesize, with yields of more than 70–90 percent, single-wall nanotubes (~1.4 nm in diameter) self-organized into bundles. These ropes, which consist of up to a hundred parallel nanotubes packed in a perfect triangular lattice, are typically more than one-tenth of a millimeter long and look very promising for engineering applications. Another attractive synthesis technique, based on a vapor phase growth mechanism which utilizes the decomposition of hydrocarbons at a temperature of ~1500 K, has also been recently used to produce large quantities of rope-like bundles of high-purity single-wall nanotubes at a very low cost [7]. However, methods for large-scale production have not really been developed, probably because so little is understood about the synthesis process at the microscopic level.

In the following, we will address from both an experimental and a theoretical point of view, the growth of carbon nanotubes. One of our objectives will be to show that presently available simulation techniques (semi-empirical and *ab initio*) can provide quantitative understanding not only of the stability, but also of the dynamics of the growth of carbon nanotube systems. We will try to summarize the microscopic insight obtained from these theoretical simulations, which will allow us to isolate the essential physics and to propose good models for multi-shell and single-shell nanotube growth, and to analyze a possible consensus for certain models based on experimental data.

## 1 Experimental Facts for Growth Models

In spite of the enormous progress in synthesis techniques, the theoretical understanding of the growth of mutli-wall and single-wall nanotubes has lagged behind. Transition metal catalysts are necessary to produce single-wall nanotubes, but are not required for producing multi-wall nanotubes, suggesting different growth mechanisms. In each synthesis method, the selective formation of a single-wall tube is triggered by enriching the graphite source material with a single species or a mixture of transition metal catalysts (Co, Ni, Y,...). Experiments show that the width and the diameter distribution depend on the composition of the catalyst, the growth temperature, and various other growth conditions. Since the electric arc-discharge [6] and the laser-ablation [5] techniques lead to very similar materials, a unique mechanism for single-wall nanotube growth should be found. Such a mechanism should not strongly depend on the details of the experimental conditions, but more on the kinetics of carbon condensation in a non-equilibrium situation. The necessity of transition metal catalysts for the synthesis of single-wall tubes is obvious as they do not grow in the absence of catalysts. However, the precise role performed by the catalysts when assisting the growth process still remains highly controversial.

Years of study on the growth of catalyst-grown carbon fibers suggested that growth occurs via precipitation of dissolved carbon from a moving catalytic particle surface [8]. Growth terminates when the catalyst particle gets poisoned by impurities or after the formation of a stable metal carbide. The reason put forward for the tubular nature of carbon fibers is that it is energetically favorable for the newly-formed surface of the growing fiber to precipitate as low-energy basal planes of graphite rather than as high-energy prismatic planes. However, the curving of the graphite layers introduces an extra elastic term into the free-energy equation of nucleation and growth, leading to a lower limit ($\sim$10 nm) to the diameters of carbon fibers that can form from curved graphite layers [9]. This implies that to explain the growth of carbon nanotubes, where diameters can be much smaller than this threshold value, new mechanisms have to be considered. There are, however, some notable differences between multi-wall carbon nanotubes and carbon fibers. In the former case, no catalyst particle or any external agent seem to be involved during growth. In addition, the tips of the multishell nanotubes are frequently observed to be closed [10], in contrast with the open ends or metal-particle terminated ends of the catalyst-grown fibers (see Fig. 1). One important question arises, and that is whether the tube remains open or closed during growth.

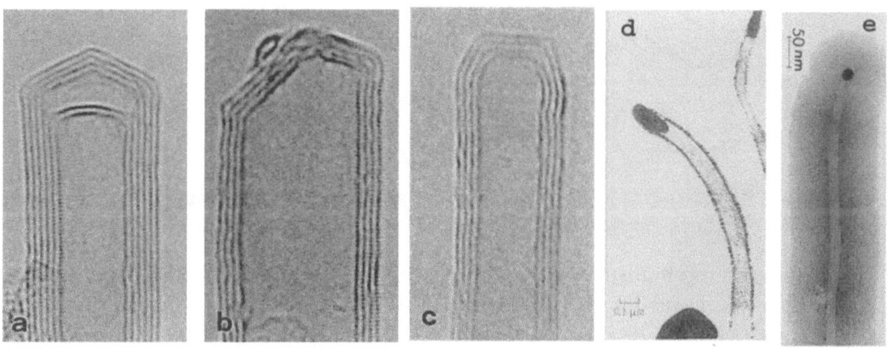


**Fig. 1.** Electron micrographs showing common multishell carbon nanotube caps (**a**,**b**,**c**) [11] and typical carbon filaments [9] (**d**) and fibers [8] (**e**), both showing the catalyst particles at their tips [12]

## 2 Open- or Close-Ended Growth for Multi-walled Nanotubes ?

Assuming first that the tube remains closed during growth, the longitudinal growth of the tube occurs by the continuous incorporation of small carbon clusters ($C_2$ dimers). This $C_2$ absorption process is assisted by the presence of pentagonal defects at the tube end, allowing bond-switching in order to reconstruct the cap topology [13,14]. Such a mechanism implies the use of the Stone-Wales mechanism to bring the pentagons of the tip at their canonical positions at each $C_2$ absorption. This model explains the growth of tubes at relatively low temperatures (~1100°C), and assumes that growth is nucleated at active sites of a vapor-grown carbon fiber of about 1000Å diameter. Within such a lower temperature regime, the closed tube approach is favorable compared to the open one, because any dangling bonds that might participate in an open tube growth mechanism would be unstable. However, many observations regarding the structure of the carbon tubes (see Sect. 1) produced by the arc method (temperatures reaching 4000°C) cannot be explained within such a model. For instance, the present model fails to explain multilayer tube growth and how the inside shells grow often to a different length compared with the outer ones [15]. In addition, at these high temperatures, nanotube growth and the graphitization of the thickening deposits occur simultaneously, so that all the coaxial nanotubes grow at once, suggesting that open nanotube growth may be favored.

In the second assumption, the nanotubes are open during the growth process and carbon atoms are added at their open ends [15,16]. If the nanotube has a random chirality, the adsorption of a $C_2$ dimer at the active dangling bond edge site will add one hexagon to the open end (Fig. 2).

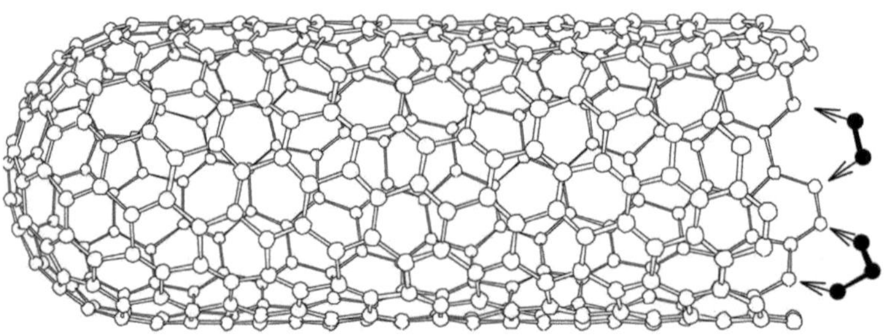

**Fig. 2.** Growth mechanism of a carbon nanotube (white ball-and-stick atomic structure) at an open end by the absorption of $C_2$ dimers and $C_3$ trimers (in *black*), respectively [14]

The sequential addition of $C_2$ dimers will result in the continuous growth of chiral nanotubes. However, for achiral edges, $C_3$ trimers are sometimes required in order to continue adding hexagons, and not forming pentagons. The introduction of pentagons leads to positive curvature which would start a capping of the nanotube and would terminate the growth (see Fig. 3). However, the introduction of a heptagon leads to changes in nanotube size and orientation (Fig. 3). Thus, the introduction of heptagon-pentagon pairs can produce a variety of tubular structures, as is frequently observed experimentally (Fig. 3).

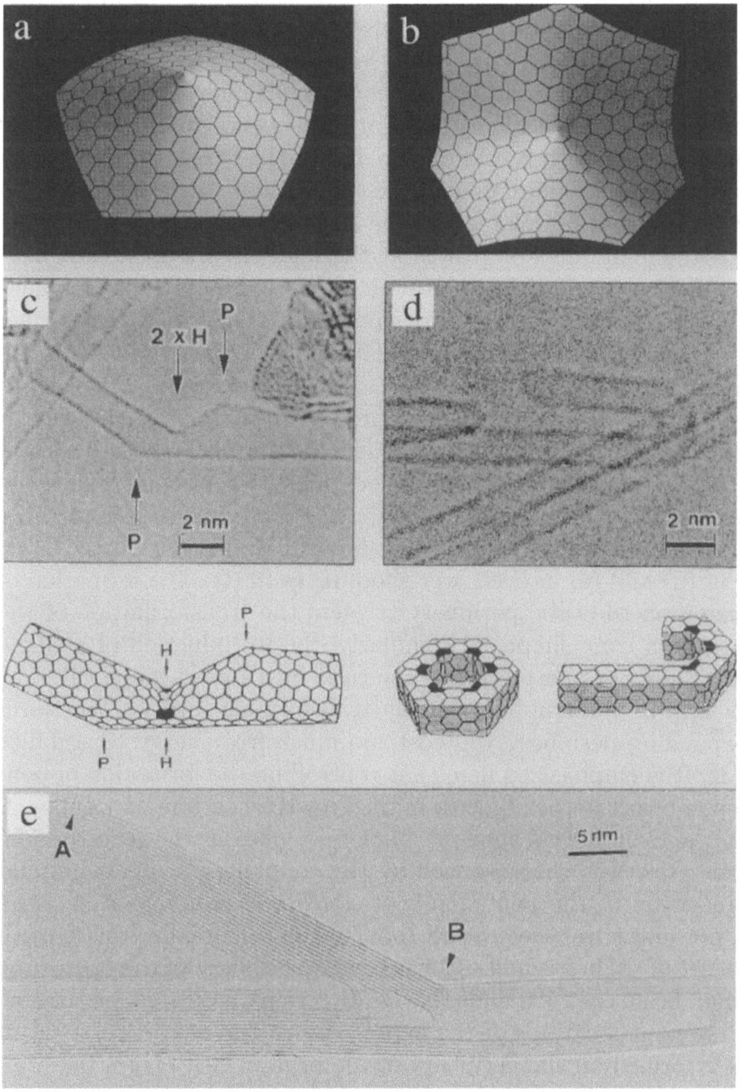


**Fig. 3.** (**a**) Model for a $+\pi/3$ disclination, pentagon ring which causes a positive curvature in a hexagonal network. (**b**) Model for a $-\pi/3$ disclination, heptagon ring which causes a negative curvature in a hexagonal network. (**c**) A deflecting single-shell nanotube. Letters P and H represent the locations of pentagons and heptagons, as illustrated in the model. (**d**) A "candy cane"-shaped nanotube and its possible model, which contains part of a torus structure. (**e**) An electron micrograph showing a bill-like termination of a multi-shell tube. A positive and a negative disclination are formed at the locations, indicated by A and B, probably due to the correlated presence of pentagons and heptagons, respectively, within the different concentric layers of the nanotube [11]

This model is thus a simple scenario where all the growing layers of a tube remain open during growth and grow in the axial direction by the addition of carbon clusters to the network at the open ends to form hexagonal rings [16]. Closure of the layer is caused by the nucleation of pentagonal rings due to local perturbations in growth conditions or due to the competition between different stable structures. Thickening of the tubes occurs by layer growth on already grown inner-layer templates and the large growth anisotropy results from the vastly different rates of growth at the high-energy open ends having dangling bonds in comparison to growth on the unreactive basal planes (Fig. 4a). Figure 4c is a summary of various possibilities of growth as revealed by the diversity of observed capping morphologies. The open-end tube is the starting form (nucleus) as represented in (a). A successive supply of hexagons on the tube periphery results in a longer tube as illustrated in (b). Enclosure of this tube can be completed by introducing six pentagons to form a polygonal cap (c). Open circles represent locations of pentagons. Once the tube is enclosed, there will no more growth on that tube. A second tube, however, can be nucleated on the first tube side-wall and eventually cover it, as illustrated in (d) and (e) or even over-shoot it, as in (f). The formation of a single pentagon on the tube periphery triggers the transformation of the cylindrical tube to a cone shape (g). Similarly, the introduction of a single heptagon into a tube periphery changes the tube into a cone shape (h). The latter growth may be interrupted soon by transforming into another form because an expanding periphery will cost too much free energy to stabilize dangling bonds. It is emphasized here that controlling the formation of pentagons and heptagons is a crucial factor in the growth of carbon nanotubes. A final branch in the variations of tube morphologies concerns the semi-toroidal tube ends. This growth is characterized by the coupling of a pentagon and a heptagon. Insertion of the pentagon-heptagon (5–7) pair into a hexagon network does not affect the sheet at all topologically. To realize this growth process, first a set of six heptagons is formed on a periphery of the open-tube (i). The circular brim then expands in the directions indicated by arrows. Solid circles represent locations of the heptagons. In the next step, a set of six pentagons is formed on the periphery of the brim, which makes the brim turn around by 180°, as illustrated in (j). An alternative structure is shown in (k), in which a slightly thicker tube is extended in the original tube direction, yielding a structure which has actually not yet been observed. Finally, it should be emphasized that an open-end tube can choose various passes or their combinations. One example is shown in (l), in which the first shell grows as a normal tube but the second tube follows a semi-toroidal tube end.

Although open-ended tubes are only occasionally spotted in the arc-grown deposit, these can be considered as quenched-growth structures, suggesting evidence for an open-ended growth model (Fig. 5). Terminated regions of such tubules are often decorated with thin granular objects which might be carbonaceous amorphous material. This observation suggests that dangling

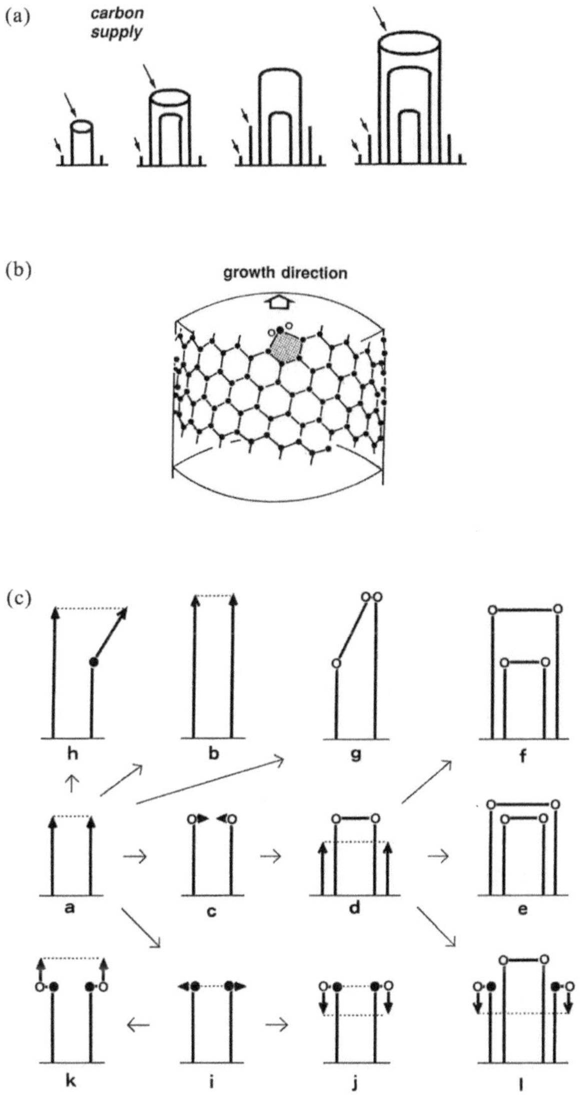


**Fig. 4.** Model for the open-end growth of the nanotube. (**a**) The tube ends are open while growing by accumulating carbon atoms at the tube peripheries in the carbon arc. Once the tube is closed, there will be no more growth on that tube but new tube shells start to grow on the side-walls. (**b**) Schematic representation of a kink-site on the tube end periphery. Supplying two carbon atoms (○) to it, the kink advances and thus the tube grows. But the supply of one carbon atom results in a pentagon which transforms the tube to a cone shape. (**c**) Evolution of carbon nanotube terminations based on the open-end tube growth. *Arrows* represent passes for the evolution. Arrow heads represents terminations of the tubes and also growth directions. *Open* and *solid circles* represents locations of pentagons and heptagons, respectively [17] (see text)

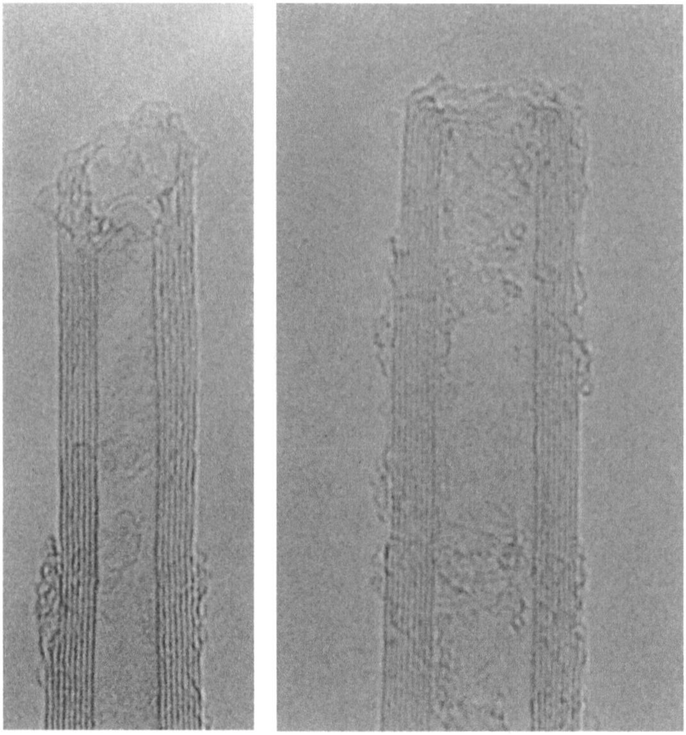

**Fig. 5.** Transmission electron micrographs showing open-ended multi-wall carbon nanotubes [17]

bonds of carbon atoms at the peripheries of the open tube ends could have been stabilized by a reorganization of the carbon atoms. A rare chance of seeing such open tubes also suggests that under normal growing conditions the tube ends close rapidly.

An electric field ($\sim 10^8$ V/ cm) was also suggested as being the cause for the stability of open ended nanotubes during the arc discharge [15]. Because of the high temperature of the particles in the plasma of the arc discharge, many of the species in the gas phase are expected to be charged, thereby screening the electrodes. Thus the potential energy drop associated with the electrodes is expected to occur over a distance of $\sim$ 1μm or less, causing very high electric fields. Later experiments and simulations confirmed that the electric field is in fact neither a necessary nor a sufficient condition for the growth of carbon nanotubes [18,19]. Electric fields at nanotube tips have been found to be inadequate in magnitude to stabilize the open ends of tubes, even in small diameter nanotubes (for larger tubes, the field effect drops drastically).

## 3 Macroscopic Model for Multi-walled Nanotube Growth

A model based on simple thermodynamic arguments and using physical parameters of the arc plasma has also been considered [20], in order to explain the formation of multishell tubes and spherical nanoparticles as well. Within this model, the growth of carbon structures on a microscopic scale is governed by the attachment probabilities of carbon atoms, ions and clusters of various sizes and shapes, controlled by a set of time- and space-dependent parameters in the arc plasma formed in the inter-electrode region. In a region close to the cathode surface, the growth of carbon structures occurs due to the competitive input of two groups of carbon species having different velocity distributions. An isotropic Maxwellian velocity distribution corresponding to a temperature of 4000 K is thought to cause the formation of spherical carbon nanoparticles. A directed ion current due to singly charged carbon ions accelerated in a region of potential drop leads to the formation of elongated structures, as a consequence of the creation of an axis of symmetry. The growth rates of nanotubes in the arc plasma can be estimated from the collision probabilities; i.e., the growth time for a nanotube 5 nm diameter and 1 mm long is approximately $10^{-3}$–$10^{-4}$ s.

## 4 "Lip–lip" Interaction Models for Multi-walled Nanotube Growth

Additional carbon atoms (spot-weld), bridging the dangling bonds between shells of a multilayered structure, have also been proposed for the stabilization of the open growing edge of multishell tubes [21]. For multiwall species, it is quite likely that the presence of the outer walls should stabilize the innermost wall, keeping it open for continued growth (Fig. 6). Static *tight-binding* calculations performed on multilayered structures where the growing edge is stabilized by bridging carbon adatoms, show that such a mechanism could prolong the lifetime of the open structure [21].

Quantum molecular dynamics simulations were also performed to understand the growth process of multi-walled carbon nanotubes [22,23]. Within such calculations, the topmost atoms (dangling bonds) of the inner and outer edges of a bilayer tube rapidly move towards each other, forming several bonds to bridge the gap between the adjacent edges, thus verifying the assumption that atomic bridges could keep the growing edge of a nanotube open without the need of "spot-weld" adatoms (Fig. 7). At ~3000 K (a typical experimental growth temperature), the "lip–lip" interactions stabilize the open-ended bilayer structure and inhibit the spontaneous dome closure of the inner tube as observed in the simulations of single-shell tubes. These calculations also show that this end geometry is highly active chemically, and easily

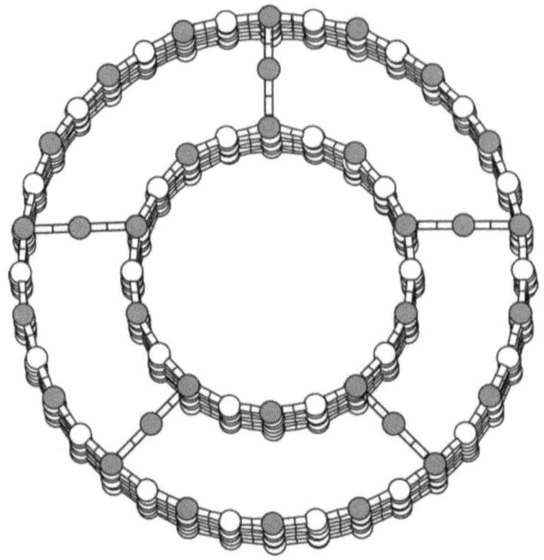

**Fig. 6.** Schematic ball-and-stick representation of the top of a multiwall nanotube with an open zigzag edge. Only two of many layers are shown for clarity. Three-coordinated carbon atoms are represented by white spheres, while low coordinated carbon atoms (dangling bonds and bridging atoms) are represented as light grey spheres on the top of the structure. Several "spot-weld" adatoms are shown occupying sites between doubly coordinated edge atoms of adjacent layers. The nanotube growth is enabled by these adatom spot-welds which stabilize the open configuration [21]

accommodates incoming carbon clusters, supporting a model of growth by chemisorption from the vapor phase.

In the "lip–lip" interaction model, the strong covalent bonds which connect the exposed edges of adjacent walls are also found to be highly favorable energetically within *ab initio* static calculations [24]. In the latter work, the open-ended growth is stabilized by the "lip–lip" interactions, involving rearrangement of the carbon bonds, leading to significant changes in the growing edge morphology. However, when classical three-body potentials are used, the role of the "lip–lip" interactions is relegated to facilitate tube closure by mediating the transfer of atoms between the inner and outer shells [25]. This example shows that very-accurate *ab initio* techniques are required to simulate the quantum effects in a "fluctuating dangling bond network", which is likely present at the growing open edge of a multiwall carbon nanotube.

Successful synthesis of multi-wall nanotubes raises the question why the growth of such tubular structures often prevail over their more stable spherical fullerene counterpart [24]. It is furthermore intriguing that these nanotubes are very long, largely defect-free, and (unless grown in the presence of a metal catalyst) always have multiple walls. The "lip–lip" model explains that the sustained growth of defect-free multi-wall carbon nanotubes

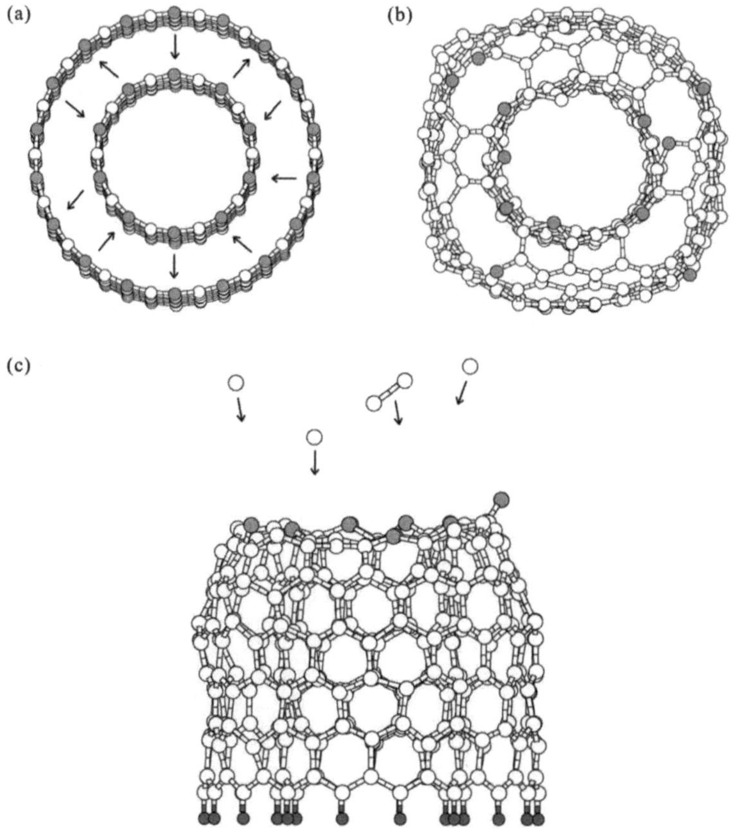


**Fig. 7.** Creation (**a**) and stabilization (**b**) of a double-walled (10,0)@(18,0) nanotube open edge by "lip–lip" interactions at ~3000 K. The notation (10,0)@(18,0) means that a (10,0) nanotube is contained within an (18,0) nanotube. The direct incorporation (**c**) of extra single carbon atoms and a dimer with thermal velocity into the fluctuating network of the growing edge of the nanotube is also illustrated [22,23]. The present system contains 336 carbon atoms (large white spheres) and 28 hydrogen atoms (small dark grey spheres) used to passivate the dangling bonds on one side of the cluster (bottom). The other low coordinated carbon atoms (dangling bonds) are represented as light grey spheres on the top of the structure

is closely linked to efficiently preventing the formation of pentagon defects which would cause a premature dome closure. The fluctuating dangling bond network present at the nanotube growing edge will also help topological defects to heal out, yielding tubes with low defect concentrations. With nonzero probability, two pentagon defects will eventually occur simultaneously at the growing edge of two adjacent walls, initiating a double-dome closure. As this probability is rather low, carbon nanotubes tend to grow long, reaching length to diameter ratios on the order of $10^3$–$10^4$.

Semi-toroidal end shapes for multi-wall nanotubes are sometimes observed experimentally [11,17]. The tube, shown in Fig. 8a does not have a simple double sheet structure, but rather consists of six semi-toroidal shells. The lattice images turn around at the end of the tube, so that an even number of lattice fringes is always observed. Another example is shown in Fig. 8b, in which some of the inner tubes are capped with a common carbon tip structure, but outer shells form semi-toroidal terminations. Such a semi-toroidal termination is extremely informative and supports the model of growth by "lip–lip" interactions for multi-wall nanotubes.

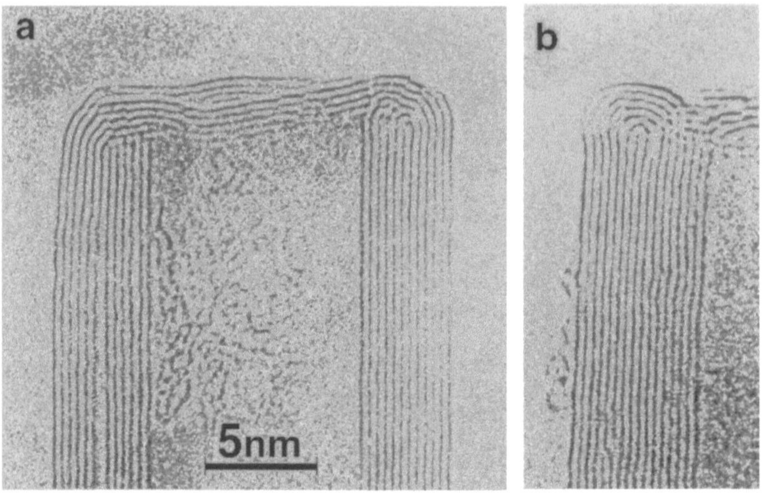


**Fig. 8.** Transmission electron micrographs of the semi-toroidal termination of multi-wall tubes, which consists of six graphitic shells (**a**). A similar semi-toroidal termination, where three inside tube shells are capped (**b**) [17]

## 5 Is Uncatalyzed Growth Possible for Single-Shell Nanotubes ?

The growth of single-shell nanotubes, which have a narrow diameter distribution (0.7–2 nm), differs from that of multishell tubes insofar as catalysts are necessary for their formation. This experimental fact is consequently an indirect proof of the existence of covalent lip–lip interactions which are postulated to be indispensable in a *pure carbon atmosphere* and imply that all nanotubes should have multiple walls. Single-wall tubes with unsaturated carbon dangling bonds at the growing edge are prone to be etched away in the aggressive atmosphere that is operative under typical synthesis conditions, which again explains the absence of single wall tubes in a pure carbon environment. However, the growth process of single-walled tubes differs also from the growth of conventional catalyst-grown carbon fibers in that no "observable" catalyst particle is seen at the tips of single-shell tubes, which appear to be closed by domes of half-fullerene hemispheres.

There have been several works based on classical, semi-empirical and quantum molecular dynamics simulations attempting to understand the growth process of single-shell tubes [22,23,26,27,28]. Most importantly, these studies have tried to look at the critical factors that determine the kinetics of open-ended tube growth, as well as studies that determine the relative stability of local-energy minimum structures that contain six-, five-, and seven-membered carbon rings in the lattice. Classical molecular dynamics simulations show that wide tubes which are initially open can continue to grow straight and maintain an all-hexagonal structure [27,28]. However, tubes narrower than a critical diameter, estimated to be about $\sim 3\,\mathrm{nm}$, readily nucleate curved, pentagonal structures that lead to tube closure with further addition of carbon atoms, thus inhibiting further growth. Continued carbon deposition on the top of a closed tube yields a highly disordered cap structure, where only a finite number of carbon atoms can be incorporated, implying that uncatalyzed defect-free growth cannot occur on single-shell tubes.

First-principles molecular dynamics simulations [22] show that the open end of single-walled nanotubes closes spontaneously, at experimental temperatures (2000 K–3000 K), into a graphitic dome with no residual dangling bonds (see Fig. 9). Similar self-closing processes should also occur for other nanotubes in the same diameter range, as is the case for most single walled nanotubes synthesized so far [3,5,6]. The reactivity of closed nanotube tips was also found to be considerably reduced compared to that of open end nanotubes. It is therefore unlikely that single-walled nanotubes could grow by sustained incorporation of C atoms on the closed hemifullerene-like tip. This is in agreement with the theoretical finding that C atoms are not incorporated into $C_{60}$ [30].

## 6 Catalytic Growth Mechanisms for Single-Shell Nanotubes

All these classical and quantum simulations, described above, may explain why single-shell nanotubes do not grow in the absence of transition metal catalysts. However, the role played by these metal atoms in determining the growth has been inaccessible to direct observation and remains controversial. The catalytic growth mechanism has not yet been clearly understood, but plausible suggestions include metal atoms initially decorating the dangling bonds of an open fullerene cluster, thus preventing it from closing. As more carbon atoms collide with this metal-decorated open-carbon cluster, they are inserted between the metal and the existing carbon atoms in the shell [10].

Planar polyyne rings have also been proposed to serve as nuclei for the formation of single-wall tubes, whose diameter would be related to the ring

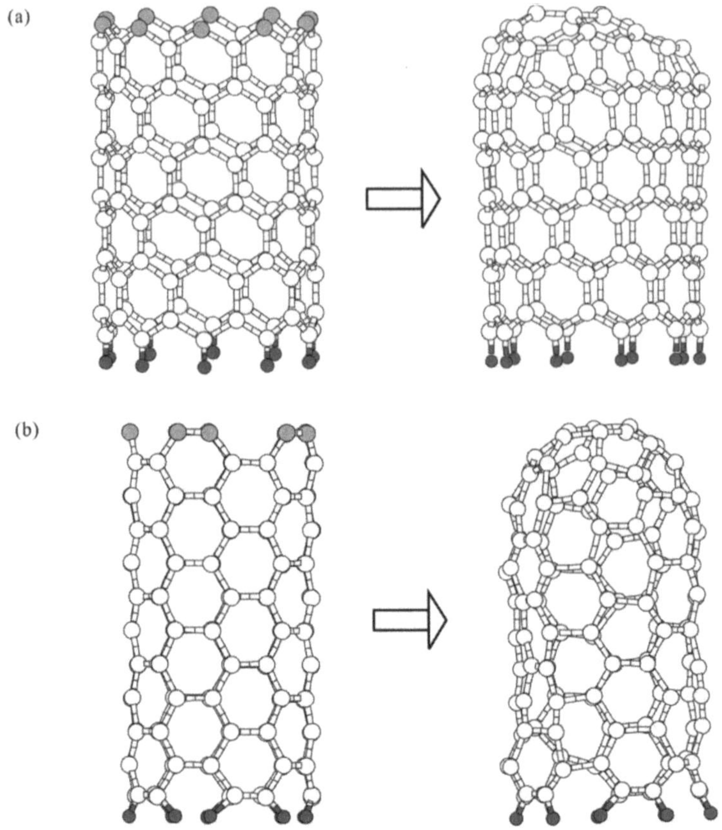


**Fig. 9.** Spontaneous closure of two single-walled nanotubes: (**a**) (10,0) zigzag tube and (**b**) (5,5) armchair tube. The notation of [29] is adopted. Both nanotubes have a fully reconstructed closed-end configuration with no residual dangling bonds [22,23]. The present systems contains 120 carbon atoms (*large white spheres*) and 10 hydrogen atoms (*small dark grey spheres*) used to passivate the dangling bonds on one side of the two clusters (*bottom*). The other low coordinated carbon atoms (dangling bonds) are represented as *light grey spheres* on the top of the structure

size [31]. In this model, the starting materials for producing nanotubes are monocyclic carbon rings, acting as nanotube precursors and $Co_mC_n$ species acting as catalysts. The composition and structure of the Co carbide cluster are undetermined, but the cluster should be able to bond to $C_n$ and/or to add the $C_n$ to the growing tube. The helical angle of the single-shell tube produced is determined by the ratio of *cis* to *trans* conformations of the first benzene ring belt during the growth initiation process. The present model may explain the formation of larger tubes upon the addition of S, Bi, or Pb to the Co catalyst [32], as these promoters modify the growth at the nucleation stage and may stabilize larger monocyclic rings, providing the nuclei necessary to build larger diameter tubes [7].

Since transition metals have the necessary high propensity for decorating the surface of fullerenes, a layer of Ni/Co atoms, adsorbed on the surface of $C_{60}$, also provides a possible catalyst template for the formation of single-walled nanotubes of a quite uniform diameter [33]. In this scenario, the metal coated fullerene acts as the original growth template inside the cylinder (Fig. 10). Once the growth has been initiated, nanotube propagation may occur without the particle. The optimum diameter of the nanotube product can thus be predicted to be the sum of the diameter of the fullerenes (0.7 nm) and two times the metal ring distances ($2 \times 0.3$–$0.35$ nm), which is close to the observed value of 1.4 nm [5].

Another possibility is that one or a few metal atoms sit at the open end of a precursor fullerene cluster [5], which will be determining the uniform diameter of the tubes (the optimum diameter being determined by the competition between the strain energy due to curvature of the graphene sheet and the dangling bond energy of the open edge). The role of the metal catalyst is to prevent carbon pentagons from forming by "scooting" around the growing edge (Fig. 11).

Static *ab initio* calculations have investigated this scooter model and have shown that a Co or Ni atom is strongly bound but still very mobile at the growing edge [34]. However, the metal atom locally inhibits the formation of pentagons that would initiate dome closure. In addition, in a concerted ex-

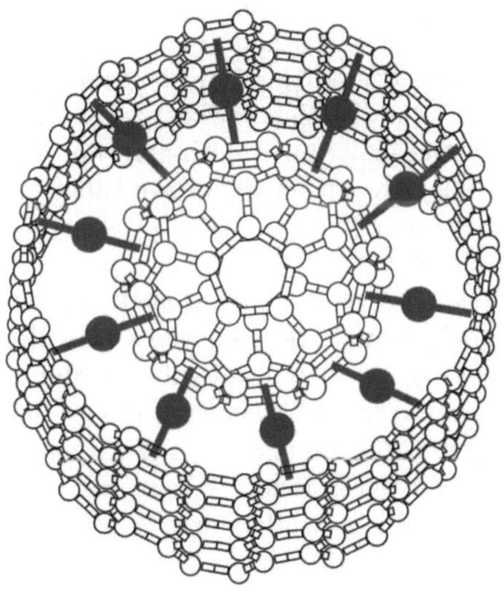

**Fig. 10.** Schematic ball-and-stick representation of a transition metal surface decorated fullerene ($C_{60}$) inside a open-ended (10,10) carbon (*white spheres*) nanotube. The Ni and Co atoms (*large dark spheres*) adsorbed on the $C_{60}$ surface are possible agents for the creation of single-walled nanotubes of uniform diameter [33]

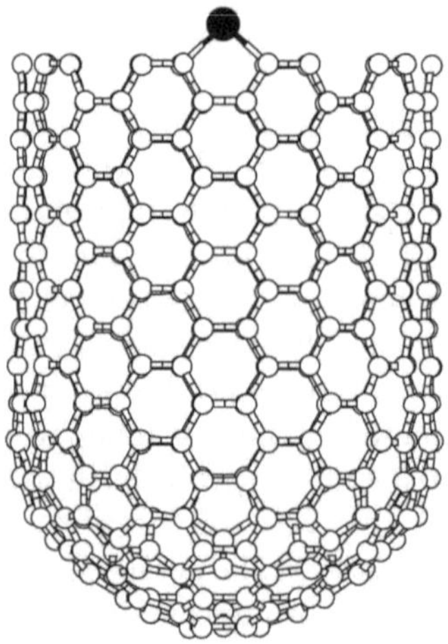

**Fig. 11.** View of a (10,10) armchair nanotube (*white ball-and-stick atomic structure*) with a Ni (or Co) atom (*large black sphere*) chemisorbed onto the open edge [5]. The metal catalyst keeps the tube open by "scooting" around the open edge, insuring that any pentagons or other high energy local structures are rearranged to hexagons. The tube shown has 310 C atoms

change mechanism, the metal catalyst assists incoming carbon atoms in the formation of carbon hexagons, thus increasing the tube length. With a non-vanishing concentration of metal atoms in the atmosphere, several catalyst atoms will eventually aggregate at the tube edge, where they will coalesce. The adsorption energy per metal atom is found to decrease with increasing size of the adsorbed cluster [34]. The ability of metal clusters to anneal defects is thus expected to decrease with their increasing size, since they will gradually become unreactive and less mobile. Eventually, when the size of the metal cluster reaches some critical size (related to the diameter of the nanotube), the adsorption energy of the cluster will decrease to such a level that it will peel off from the edge. In the absence of the catalyst at the tube edge, defects can no longer be annealed efficiently, thus initiating tube closure. This mechanism is consistent with the experimental observation (see Fig. 12) that no "observable" metal particles are left on the grown tubes [5]. This also suggests that too high a concentration of the metal catalyst will be detrimental to the formation of long nanotubes.

Although the scooter model was initially investigated using static *ab initio* calculations [34], first-principles molecular dynamics simulations were also performed [35] in order to study the growth of single-wall tubes within a scheme where crucial dynamical effects are explored by allowing the system to evolve free of constraints at the experimental temperature. Within such a simulation at 1500 K, the metal catalyst atom is found to help the open end of the single-shell tube close into a graphitic network which incorporates the catalyst atom (see Fig. 13). However, the cobalt-carbon chemical bonds are frequently breaking and reforming at experimental temperatures, providing the necessary pathway for carbon incorporation, leading to a closed-end catalytic growth mechanism.

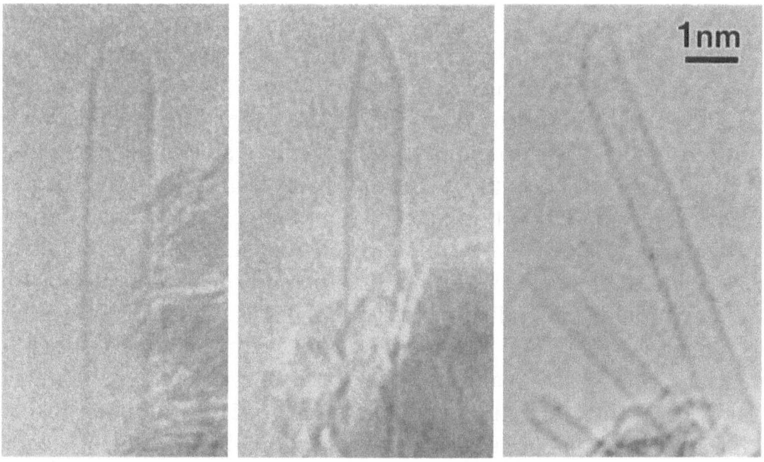


**Fig. 12.** Transmission electron micrographs of single-shell nanotube ends. No particles of metals are observed at the nanotube caps [17]

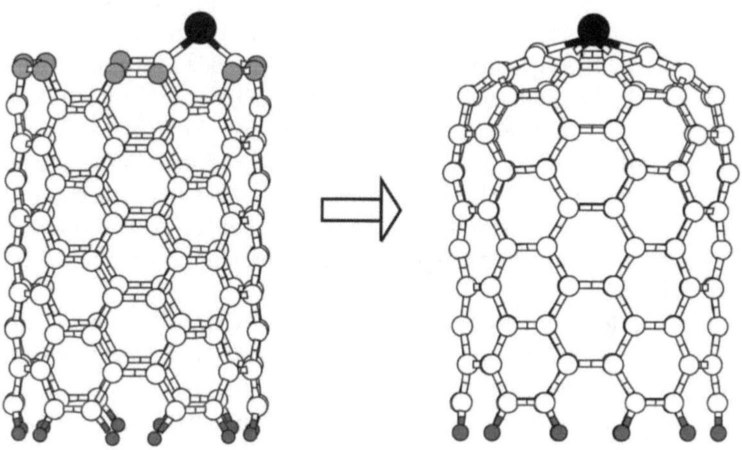

**Fig. 13.** Catalytic growth of a (6,6) armchair single-walled nanotube. The notation of [29] is adopted. The metal catalyst atom cannot prevent the formation of pentagons, leading to tube closure at experimental temperatures (1500 K) [35]. The present systems contains 1 cobalt atom (*large black sphere*), 120 carbon atoms (*white spheres*) and 12 hydrogen atoms (*small dark grey spheres*) used to passivate the dangling bonds on one side of the two clusters (*bottom*). The other low coordinated carbon atoms (dangling bonds) are represented as *light grey spheres* on the top of the structure (*left*)

This model, where the Co or Ni catalyst keeps a high degree of chemical activity on the nanotube growth edge, clearly differs from the uncatalyzed growth mechanism of a single-wall nanotube discussed above, which instantaneously closes into an chemically-inert carbon dome. The model, depicted in Fig. 13, supports the growth by chemisorption from the vapor phase, as proposed long ago for carbon filaments by *Baker* et al. [36], *Oberlin* et al. [8], and *Tibbetts* [9], which adopts the concepts of the vapor-liquid-solid (VLS) model introduced in the 1960s to explain the growth of silicon whiskers [37]. In the VLS model, growth occurs by precipitation from a super-saturated catalytic liquid droplet located at the top of the filament, into which carbon atoms are preferentially absorbed from the vapor phase. From the resulting super-saturated liquid, the solute continuously precipitates, generally in the form of faceted cylinders (VLS-silicon whiskers [37]) or tubular structures [9]. Tibbetts also demonstrates that it is energetically favorable for the fiber to precipitate with graphite basal planes parallel to the exterior planes, arranged around a hollow core, thus forming a wide diameter ($\sim 1\,\mu m$) multiwall tube.

Although the VLS model for catalyst-grown carbon fibers is a macroscopic model based on the fluid nature of the metal particle which helps to dissolve carbon from the vapor phase and to precipitate carbon on the fiber walls, the catalytic growth model for the single-wall nanotubes can be seen as its analogue, with the only difference being that the catalytic particle is reduced to one or a few metal atoms in the case of the single-wall nanotube growth. In terms of this analogy, the quantum aspects of a few metal atoms at the growing edge of the single-shell tube has to be taken into account. In the catalytic growth of single-shell nanotubes, it is no longer the "fluid nature" of the metal cluster (VLS model), but the chemical interactions between the Co or Ni $3d$ electrons and the $\pi$ carbon electrons, that makes possible a rapid incorporation of carbon atoms from the plasma. The cobalt $3d$ states increase the DOS near the Fermi level, thus enhancing the chemical reactivity of the Co-rich nanotube tip [35].

The catalytic particle is not frequently observed at the tube end [5], although some experiments [38] report the presence of small metal particles on the walls of nanotube bundles (Fig. 14).

Precisely because of the enhancement of the chemical activity due to the presence of metal catalyst particles at the experimental growth temperature, the Co–C bonds are frequently reopened and the excess of cobalt could be ejected from the nanotube tip. Moreover, single-layer tubes have also been produced by pre-formed catalytic particles [39], which are attached to the tube end and thus are closely correlated with the tube diameter size (from 1 to 5 nm), demonstrating the validity of the VLS model on a nanometer scale. When the metal particle is observable [39], the metal-catalytic growth of carbon nanotubes is thus believed to proceed (in analogy to the catalyst-grown carbon fibers) via the solvation of carbon vapor into tiny metal clusters, followed by the precipitation of excess carbon in the form of nanotubes.

The presence of any remaining metal catalyst atoms at the nanotube tip (even in very small – undetectable amounts) cannot be excluded. The presence of such ultra-small catalyst particles, which is certainly not easy to es-

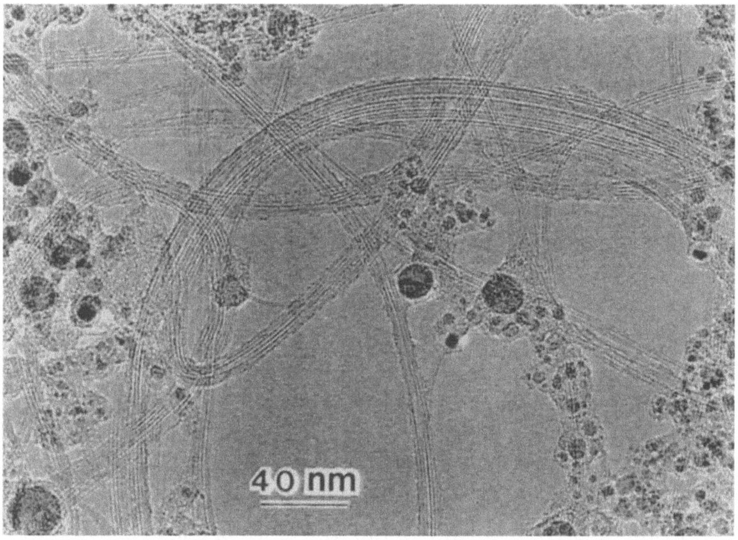


**Fig. 14.** A high-resolution electron micrograph showing raft-like bundles of single-walled carbon nanotubes. Most of the nanotubes form arch-like structures and they terminate at or originate from carbon blocks composed of mostly disordered cage-like fullerene molecules [38]. Some of the nanotube bundles are terminated in metal particles of a few nanometers in diameter. Smaller metal particles were also observed on the walls of the nanotube bundles [38]

tablish experimentally, should strongly influence the field emission properties of these single-shell tubes [40] and could explain some field emission patterns observed at the nanotube tip [41]. Magnetic susceptibility measurements and magnetic STM could be used to investigate the presence of metal catalysts at the nanotube tip, as such experimental techniques are sensitive to tiny amounts of magnetic transition metals. At this stage, a better experimental characterization method for the atomic structure of single-walled nanotube tips is required.

## 7 Catalytic Growth Mechanism for Nanotube Bundles

Both in the vaporization [5] and in the arc-discharge [6] catalytic techniques, nearly perfect single-wall nanotubes of uniform diameter ($\sim 1.4\,\text{nm}$) are produced in high yield (70–90%), and are arranged in bundles or ropes (2D triangular lattice) of a few tens of tubes with similar spacing between adjacent neighbors ($\sim 3.2\,\text{Å}$ [42]). Only a few isolated carbon single-shell tubes are observed. Additionally, the use of a second element beside the catalyst (bi-metal mixtures : Ni–Co, Ni–Y, ...) is also found to produce the greatest yield, sometimes 10–100 times that for single metals alone.

Actually no coherent explanation of the growth mechanism has clearly emerged yet to explain the formation of ropes of single-shell tubes. However, some crucial information can be extracted from TEM observations. Larger-diameter bundles consist mostly of an assembly of smaller bundles separated by twin-like boundaries [6], suggesting for their formation an alignment of the pre-formed smaller bundles due to van der Waals forces. In contrast, these pre-formed smaller bundles are probably the result of a correlated growth between single-wall nanotubes which constitute this rope.

Such a correlated growth could be explained by a modified "spot-weld" model proposed for the stabilization of the open growing edge of multishell tubes [21]. During the growth of ropes, bridging metal catalyst adatoms could connect the open edges of two neighboring single-wall nanotubes (see Fig. 15), preventing the spontaneous closure of each single-shell tube. The chemical activity of the open end would still be very important due to the presence of the metal catalyst as a surfactant, favoring the formation of long ropes.

The feed stock for the rope arrives at the growing end of the nanotube mostly by the diffusion of carbon clusters along the sides of the tubes and also from direct incorporation of carbon from the vapor phase. While only a few

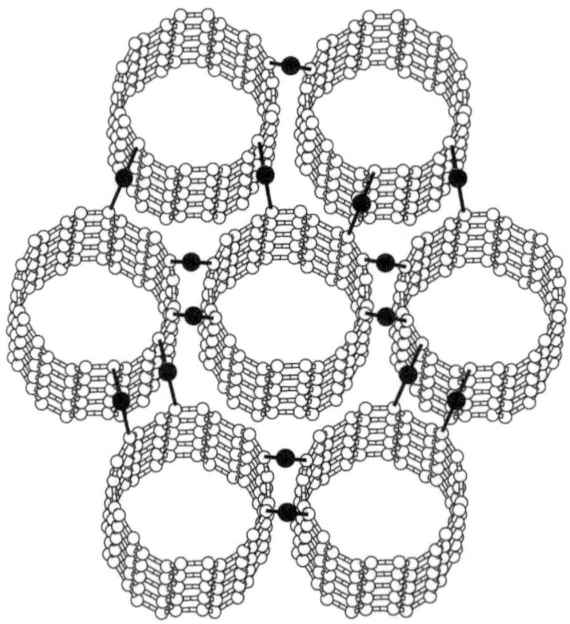

**Fig. 15.** Schematic ball-and-stick representation of a nanotube bundle composed of seven (6,6) armchair carbon nanotubes (*white spheres*). Several transition metal catalyst Ni and/or Co atoms (in *black*) are shown, occupying sites between the growing edge of adjacent single-shell nanotubes, thus stabilizing the open edge configuration of the nanotube bundle

catalytic bridges need to be involved in the growth of nearest-neighboring single-wall tubes, other metal atoms will soon congregate at the growing edge, forming other links or small metal clusters. The metal catalyst that exists at the live end of the growing rope, needs to be sufficiently efficient to incorporate the large amount of available carbon feed-stock in the perfect cylindrical network of the tubes. If the cluster is too large, it can be ejected from the growing edge of the rope, explaining its observation along the walls of nanotube bundles [38]. Some static and dynamical *ab initio* calculations are now being performed [35] to check the validity of the proposed mechanism.

## 8 Root Growth Mechanism for Single-Shell Nanotubes

In the catalytic growth experiments, two situations arise. In the first case, the metal particle, which is not frequently observed, is found to be reduced to one or a few metal atoms, as described above. In the second situation, carbon and metal atoms can condense and form alloy particles during the arc discharge. As these particles are cooled, carbon atoms, dissolved in the particle, segregate onto the surface, because the solubility of the surface decreases with decreasing temperature. As the system is cooled, soot is formed. The sizes of the soot particles are several tens of nm wide (much larger than single-shell nanotube diameters), and are identified by TEM observations as embedded metal clusters surrounded by a coating of a few graphitic layers. Some singularities at the surface structure or atomic compositions may catalyze the formation of single-wall tubes (as described by sea-urchin models [43,44]), thereby providing another mechanism for the growth of single-wall carbon nanotubes. After the formation of the tube nuclei, carbon is supplied from the core of the particle to the root of the tubes, which grow longer maintaining hollow capped tips. Addition of carbon atoms (or dimers) from the gas phase at the tube tips (opposite side) also probably helps the growth.

Many single-shell nanotubes are observed to coexist with catalytic particles and often appear to be sticking out of the particle surfaces, as shown in Fig. 16. One end of the tubule is thus free, the other one being embedded in the particle, which often has a size exceeding the nanotube diameter by well over one order of magnitude.

Classical molecular-dynamics simulations [45] reveal a possible atomistic picture for this root growth mode for single-wall tubes. According to the model, carbon atoms precipitate from the metal particle, migrate to the tube base, and are incorporated into the nanotube network, thereby leading to defect-free growth. The simulation consists of treating the outermost layer of a large metal particle, embedded in a few carbon layers, out of which a capped single-shell nanotube protrudes (Fig. 17).

The calculation shows that the addition of new hexagons at the tube base occurs through a sequence of processes involving a pair of "handles" on the opposite bonds of a heptagon. The heptagon defects are present due to the

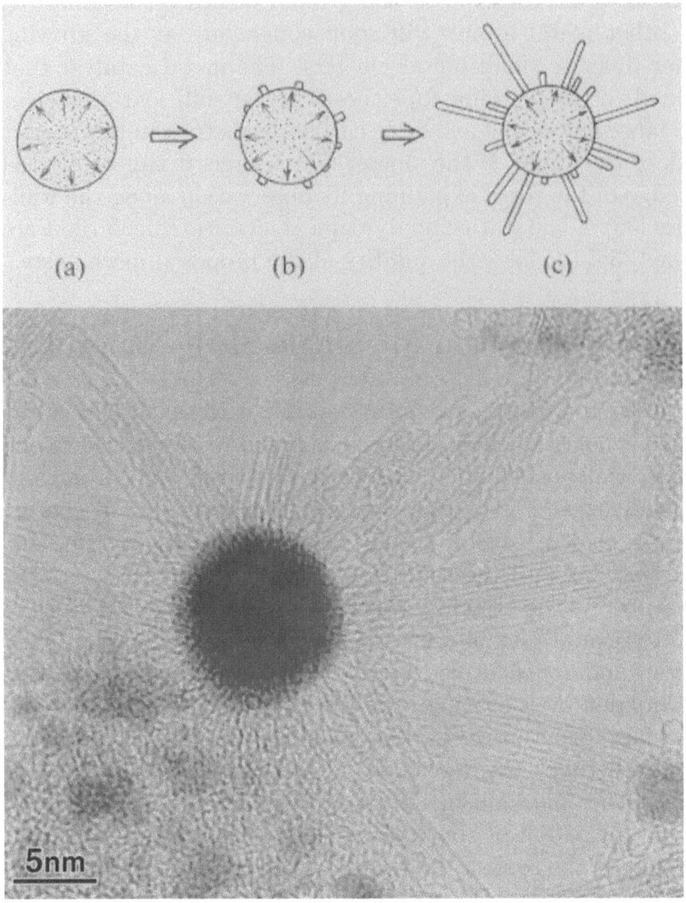


**Fig. 16.** Transmission electron micrograph of single-wall nanotubes growing radially from a Ni-carbide particle (*bottom*) [38]. The top inset illustrates the hypothetical growth process of single-shell tubes from a metal-carbon alloy particle: (**a**) segregation of carbon towards the surface, (**b**) nucleation of single-wall tubes on the particle surface, and (**c**) growth of these nanotubes [44]

positive curvature needed to create the carbon protrusion (Fig. 17). These "handles" are formed by adatoms between a pair of nearest-neighboring carbon atoms. Calculations show that the energy of a single handle is a minimum at the point of highest curvature, explaining why these handles are attracted to the tube base region, where the heptagons are located [45]. Figure 18 schematically displays the process of hexagon addition (without creating another defect) when an isolated heptagon or a 5–7 pair (pentagon and heptagon rings connected together) is introduced into the carbon network. An even number of handles leads to the creation of new hexagons at the tube stem, and thus to a defect-free root growth mechanism.

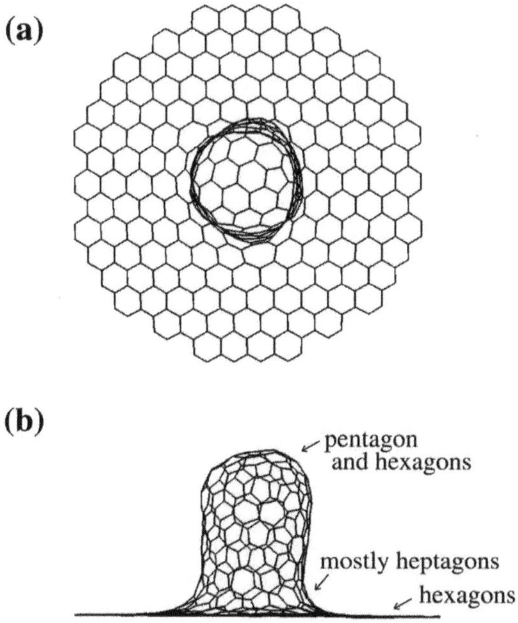


**Fig. 17.** Schematic wireframe representation of the top (**a**) and side (**b**) views of a (11,3) nanotube growing out of a flat all-hexagonal graphene sheet by a root mechanism involving the presence of heptagons at the tube base [45]

Providing an atomistic picture of nanotube nucleation is more difficult [45]. The chemical processes involved in the co-evaporation of carbon and metal in the arc discharge, the solvation of carbon into the metal particle, the condensation of metal-carbide particles into quasi-spherical droplets upon cooling, and the subsequent precipitation of carbon, are extremely complex processes. Simulating all these steps requires a microscopically detailed understanding of the process of arc discharge evaporation which is not presently available. However, the present model does show that protrusions with a diameter small compare to their height can lead to nanotube nucleation, while wider protrusions lead only to strained graphene sheets and no nanotube growth.

## 9 Conclusion

Since their discovery in 1991, great progress has been made in the understanding of carbon nanotube growth. There has been a constant fruitful interplay between theoretical calculations and experimental measurements which has enhanced our insight into the formation processes of these ultimate carbon fibers.

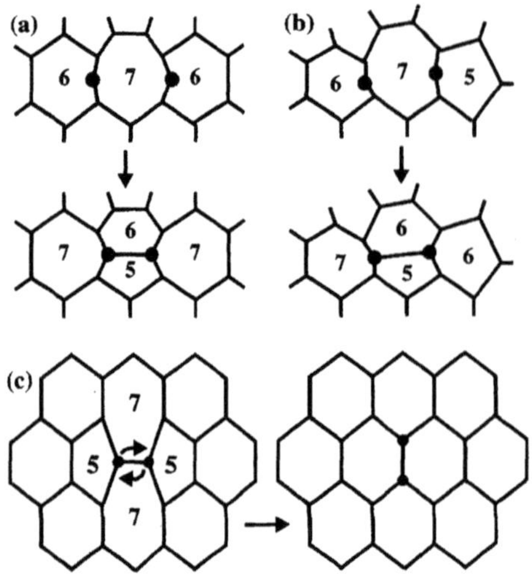


**Fig. 18.** The atomistics of the addition of a single hexagon at the nanotube base by bond formation between a pair of handle atoms (*black dots*) at the opposite sides of a heptagon [45]. (**a**) for an isolated heptagon, a 5–7 pair forms in addition to a hexagon; (**b**) for a 5–7–6 complex, only an additional hexagon forms. (**c**) Shows the annihilation of two adjacent 5–7 pairs into four hexagons by a Stone-Wales switch, but retaining the same number of carbon atoms [46]

Several mechanisms, as described above, have been proposed to account for the growth mechanisms of single-wall, multi-wall and ropes of carbon nanotubes with or without the presence of any catalyst. The key role played by the metal catalyst is crucial for understanding the growth of single-shell tubes at the microscopic level. However, the actual role of the metal or alloy is still very confusing, although some interesting models have been proposed. At the present stage, experimental observations of quenched-growth single-wall structures are required to validate one or another of these models. In-situ experimental studies have just started [47,48], and are very promising for a direct determination of important characteristics, such as the temperature gradient, time evolution of the matter aggregation after the initial vaporization of the different chemical species, ... Actually, experimental observations to date have relied only on the study of the soot after the synthesis has occurred.

Probably the most intriguing problem is to understand the microscopic mechanism and optimum conditions for the formation of well-designed single-wall nanotubes. Although it has been argued that the armchair nanotube structure is favored energetically [5], experimental conditions under which these tubules would be grown with good control are still not yet well known. Nonetheless, as more experimental data become available to correlate the atomic structure and the synthesis conditions, and more is known about the growth at the atomic level, it is hoped that controlled growth of single-walled carbon nanotubes with designed structures will be achieved soon. In addition, with further experimental confirmation of their unique properties, there will be a great incentive to develop industrial-scale production methods.

**Acknowledgments**

J.C.C. acknowledges the National Fund for Scientific Research (FNRS) of Belgium for financial support. This paper presents research results of the Belgian Program on Inter-university Attraction Poles initiated by the Belgian State-Prime Minister's Office-Science Policy Programming. This work is carried out within the framework of the EU Human Potential - Research Training Network project under contract N$^o$ HPRN-CT-2000-00128 and within the framework of the specific research and technological development European Union (EU) programme "Competitive and Sustainable Growth" under contract G5RD-CT-1999-00173. J.C.C. is indebted to X. Blase, A. De Vita and R. Car for their collaborative efforts and to the *Institut de Recherche Numérique en Physique des Matériaux* (IRRMA-EPFL, Lausanne, Switzerland) for hosting part of the research presented above.

## References

1. S. Iijima, Nature **354**, 56 (1991)
2. T. W. Ebbesen and P. M. Ajayan, Nature **358**, 220 (1992)
3. S. Iijima, T. Ichihashi, Nature **363**, 603 (1993)
4. D. S. Bethune, C. H. Kiang, M. S. de Vries, G. Gorman, R. Savoy, J. Vazquez, R. Beyers, Nature **363**, 605 (1993)
5. A. Thess, R. Lee, P. Nikolaev, H. Dai, P. Petit, J. Robert, C. Xu, Y. H. Lee, S. G. Kim, D. T. Colbert, G. Scuseria, D. Tomanek, J. E. Fischer, R. E. Smalley, Science **273**, 483 (1996)
6. C. Journet, W. K. Maser, P. Bernier, A. Loiseau, M. Lamy de la Chapelle, S. Lefrant, P. Deniard, R. Lee, J. E. Fischer, Nature **388**, 756 (1997)
7. H. M. Cheng, F. Li, G. Su, H. Y. Pan, L. L. He, X. Sun, M. S. Dresselhaus, Appl. Phys. Lett. **72**, 3282 (1998)
8. A. Oberlin, M. Endo, T. Koyama, J. Crystl. Growth **32**, 335 (1976)
9. G. G. Tibbetts, J. Crystl. Growth **66**, 632 (1984)
10. P. M. Ajayan, T. W. Ebbesen, Rep. Prog. Phys. **60**, 1025 (1997)
11. S. Iijima, MRS Bulletin **19**, 43 (1994)
12. M. S. Dresselhauss, G. Dresselhaus, K. Sugihara, I. L. Spain, H. A. Goldberg, *Graphite Fibers and Filaments*, Springer Ser. Mater. Sci. **5** (Springer, Berlin, Heidelberg 1988)
13. M. Endo, H. W. Kroto, J. Phys. Chem. **96**, 6941 (1992)
14. R. Saito, M. Fujita, G. Dresselhaus, M. S. Dresselhaus, Mater. Sci. Eng. B **19**, 185 (1993)
15. R. E. Smalley, Mater. Sci. Eng. B **19**, 1 (1993)
16. S. Iijima, T. Ichihashi, Y. Ando, Nature **356**, 776 (1992)
17. S. Iijima, Mater. Sci. Eng. B **19**, 172 (1993)

18. A. Maiti, C. J. Brabec, C. M. Roland, J. Bernholc, Phys. Rev. Lett. **73**, 2468 (1994)
19. L. Lou, P. Nordlander, R. E. Smalley, Phys. Rev. B **52**, 1429 (1995)
20. E. G. Gamaly, T. W. Ebbesen, Phys. Rev. B **52**, 2083 (1995)
21. T. Guo, P. Nikolaev, A. G. Rinzler, D. Tománek, D. T. Colbert, R. E. Smalley, J. Phys. Chem. **99**, 10694 (1995)
22. J.-C. Charlier, A. De Vita, X. Blase, R. Car, Science **275**, 646 (1997)
23. J.-C. Charlier, X. Blase, A. De Vita, R. Car, Appl.Phys. A **68**, 267 (1999)
24. Y.-K. Kwon, Y. H. Lee, S.-G. Kim, P. Jund, D. Tománek, R. E. Smalley, Phys. Rev. Lett. **79**, 2065 (1997)
25. M. Buongiorno Nardelli, C. Brabec, A. Maiti, C. Roland, J. Bernholc, Phys. Rev. Lett. **80**, 313 (1998)
26. D. H. Robertson, D. H. Brenner, J. W. Mintmire, Phys. Rev. B **45**, 12592 (1992)
27. A. Maiti, C. J. Brabec, C. M. Roland, J. Bernholc, Phys. Rev. B **52**, 14850 (1995)
28. C. J. Brabec, A. Maiti, C. Roland, J. Bernholc, Chem. Phys. Lett. **236**, 150 (1995)
29. N. Hamada, S.-I. Sawada, A. Oshiyama, Phys. Rev. Lett. **68**, 1579 (1992)
30. B. R. Eggen, M. I. Heggie, G. Jungnickel, C. D. Latham, R. Jones, P. R. Briddon, Science **272**, 87 (1996)
31. C. H. Kiang, W. A. Goddard, Phys. Rev. Lett. **76**, 2515 (1996)
32. C. H. Kiang, W. A. Goddard, R. Beyers, J. R. Salem, D. Bethune, J. Phys. Chem. Solids. **57**, 35 (1996)
33. P. R. Birkett, A. J. Cheetham, B. R. Eggen, J. P. Hare, H. W. Kroto, D. R. M. Walton, Chem. Phys. Lett. **281**, 111 (1997)
34. Y. H. Lee, S.-G. Kim, P. Jund, D. Tománek, Phys. Rev. Lett. **78**, 2393 (1997)
35. J.-C. Charlier, A. De Vita, X. Blase, R. Car, manuscript in preparation
36. R. T. K. Baker, M. A. Barber, P. S. Harris, F. S. Feates, R. J. Waite, *J. Catalysis* **26**, 51–62 (1972); R. T. K. Baker, Carbon **27**, 315 (1989)
37. R. S. Wagner, W. C. Ellis, Appl. Phys. Lett. **4**, 89 (1964)
38. L.-C. Qin, S. Iijima, *Chem. Phys. Lett.* **269**, 65 (1997)
39. H. Dai, A. G. Rinzler, P. Nikolaev, A. Thess, D. T. Colbert, R. E. Smalley, Chem. Phys. Lett. **260**, 471 (1996)
40. D. L. Carroll, P. Redlich, P. M. Ajayan, J.-C. Charlier, X. Blase, A. De Vita, R. Car, Phys. Rev. Lett. **78**, 2811 (1997)
41. J. M. Bonard, T. Stöckli, W. A. de Heer, A. Châtelain, J.-C. Charlier, X. Blase, A. De Vita, R. Car, J.-P. Salvetat, L. Forró, manuscript in preparation (2000)
42. J.-C. Charlier, X. Gonze, J.-P. Michenaud, Europhys. Lett. **29**, 43 (1995)
43. D. Zhou, S. Seraphin, S. Wang, Appl. Phys. Lett. **65**, 1593 (1994)
44. Y. Saito, M. Okuda, N. Fujimoto, T. Yoshikawa, M. Tomita, T. Hayashi, Jpn. J. Appl. Phys. **33**, L526 (1994)
45. A. Maiti, C. J. Brabec, J. Bernholc, Phys. Rev. B **55**, R6097 (1997)
46. A. J. Stone, D. J. Wales, Chem. Phys. Lett. **128**, 501 (1986)
47. S. Arapelli, C. D. Scott, Chem. Phys. Lett. **302**, 139 (1999)
48. F. Kokai, K. Takahashi, M. Yudasaka, S. Iijima, J. Phys. Chem. B **103**, 8686 (1999)

# Nanotubes from Inorganic Materials

Reshef Tenne[1] and Alex K. Zettl[2]

[1] Weizmann Institute of Science, Department of Materials and Interfaces
Rehovot 76100, Israel
reshef.tenne@weizmann.ac.il

[2] Department of Physics, University of California Berkeley
Berkeley, CA 94720, USA
azettl@physics.berkeley.edu

**Abstract.** The inorganic analogs of carbon fullerenes and nanotubes, like $MoS_2$ and BN, are reviewed. It is argued that nanoparticles of 2D layered compounds are inherently unstable in the planar configuration and prefer to form closed cage structures. The progress in the synthesis of these nanomaterials, and, in particular, the large-scale synthesis of BN, $WS_2$ and $V_2O_5$ nanotubes, are described. Some of the electronic, optical and mechanical properties of these nanostructures are reviewed. The red-shift of the energy gap with shrinking nanotube diameter is discussed as well as the suggestion that zigzag nanotubes exhibit a direct gap rather than an indirect gap, as is prevalent in many of the bulk 2D materials. Some potential applications of these nanomaterials are presented as well, most importantly the superior tribological properties of $WS_2$ and $MoS_2$ nested fullerene-like structures (onions).

Following the discovery of carbon fullerenes [1] and later on carbon nanotubes [2], it was recognized that polyhedral structures are the thermodynamically stable form of carbon under the constraint that the number of atoms is not allowed to grow beyond a certain limit. However, if one considers the stimulus for the formation of such nanostructures, it is realized that these kinds of perfectly organized nanostructures should not be limited to carbon, only. As shown in Fig. 1, the propensity of nanoparticles of graphite (Fig. 1a) to form hollow closed structures stems from the high energy of the dangling bonds at the periphery of the nanoparticles, a property which is also common to materials like $MoS_2$ Fig. (1b). It was therefore hypothesized [3,4] that the formation of closed polyhedra and nanotubes is a generic property of materials with anisotropic 2D layered structures. These kind of structures are often called Inorganic Fullerene-like (IF) structures. Thus Fig. 2 shows a multi-wall nanotube of $MoS_2$, which was observed in ultra-thin films of such compounds grown on quartz substrates. Numerous examples of the validity of this concept have been provided over the last few years. Phrased in different terms, a significant body of evidence suggests that the phase diagram of elements, which form layered compounds, include the new phase of hollow and closed nano-materials (nanostructures) within the phase diagram of the layered compound itself. Provided that the crystallites cannot grow beyond a certain size (less than say 0.2μm), this nanostructured phase would be the

M. S. Dresselhaus, G. Dresselhaus, Ph. Avouris (Eds.): Carbon Nanotubes,
Topics Appl. Phys. **80**, 81–112 (2001)

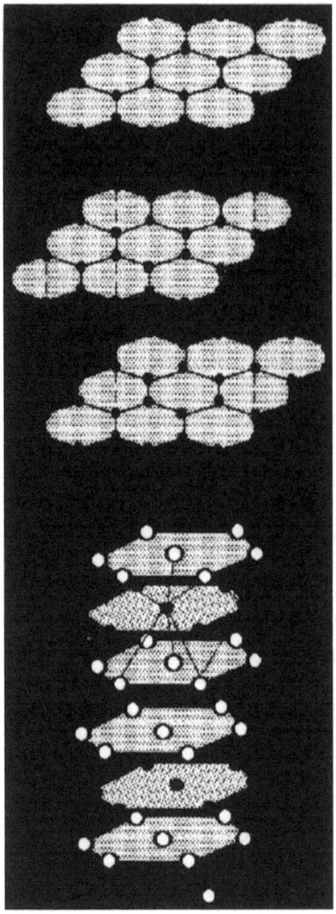

**Fig. 1.** Schematic drawings of graphite (*upper figure*) and $WS_2$ nanoclusters (*lower figure*) in a 2H polytype structure. Note that in both cases the surface energy, which destabilizes the planar topology of the nanocluster, is concentrated in the prismatic edges parallel to the $c$-axis ($\| c$) [3]

thermodynamically preferred phase. Nanotubular structures were produced also from 3D compounds, like $TiO_2$ [5], and there seems to be no limit to the kind of compound that can serve as a precursor for the formation of nanotubular structures. However, as will become clear from the discussion below, a clear distinction holds between nanotubular structures obtained from 3D and layered 2D compounds. Intuitively, a 3D compound cannot form a perfectly ordered, flawless nanotubular or polyhedral structure, since some of the bonds, particularly on the surface of the nanotube remain unsatisfied. On the other hand, 2D (layered) compounds form perfectly crystalline closed cage structures, by introducing elements of lower symmetry into the generically hexagonal tiling of the crystalline planes. For a recent review of the subject, the reader is referred to [6].

A number of different synthetic strategies have been developed, which yield large amounts of nanotubes of $V_2O_5$ [7], $WS_2$ [8,9], BN [10] and related

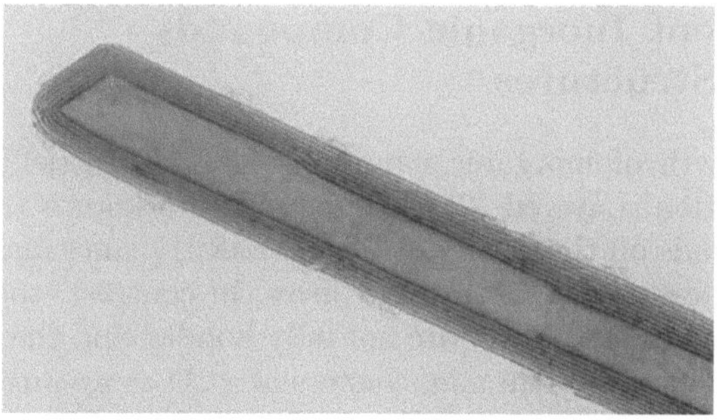

**Fig. 2.** TEM image of an edge of a multi-wall nanotube (MWNT) of $MoS_2$, 7 molecular layers thick. The distance between adjacent layers is 6.15 Å. The *c*-axis is always normal to the surface of the nanotube [4]

structures. Although the detailed growth mechanism of the inorganic nanotubes is not fully understood, some progress has been realized in unraveling the growth mechanism.

Structural aspects of these nanoparticles are also discussed in this review. It was found that different growth strategies lead to inorganic nanotubes with quite distinct structures, as discussed below.

In some cases (such as for BN [10]) methods have been found to fine-tune the synthesis procedure and, as for carbon-based systems, it is possible to generate inorganic nanoparticles and nanotubes of a specified and monodisperse size and with a uniform number of layers. However, in general such size and product selection has not yet been demonstrated for the majority of the inorganic fullerene and nanotube analogs which we here describe. It is believed that this reflects more an "early development stage" phenomenon rather than intrinsic fundamental synthesis barriers.

So far, the properties of inorganic nanotubes have been studied rather scantily. Optical measurements in the UV–vis range and Raman scattering have provided important clues regarding the electronic structure of these nanoparticles. Generically, semiconducting nanoparticles of 3D compounds exhibit a blue shift in the absorption and luminescence spectrum due to quantum size effects. In contrast, the bandgap of semiconducting nanotubes, like BN [11], shrinks with decreasing nanotube diameter, which is attributed to the strain in the folded structure.

The mechanical properties of inorganic fullerenes and nanotubes are expected to be unique and in some cases truly exceptional. The strong covalent $sp^2$ bonds of BN-based systems, for example, yield nanotubes with the highest Young's modulus of any known insulating fiber [12]. Very good tips for scanning probe microscopes have been prepared from $WS_2$ nanotubes [13]. The mechanical and chemical stability of these structures is attributed to their structural perfection and rigidity. The potential applications of inorganic nanotubes as conducting or non-conducting structural reinforcements, or tips for scanning probe microscopy for the study of soft tissues, rough surfaces, and for nano-lithography are further discussed in this review. Most importantly, these kinds of nanoparticles exhibit interesting tribological properties, also briefly discussed.

# 1 Categorizing Different Inorganic Compounds Forming Nanotubular Structures

The driving force for the growth of inorganic nanotubes has been briefly mentioned in the previous section. Layered (2D) compounds are known to have fully satisfied chemical bonds on their van der Waals (basal) planes and consequently their (0001) surfaces are generally very inert. In contrast, the atoms on the prismatic $(10\bar{1}0)$ and $(11\bar{2}0)$ faces are not fully bonded and they are therefore chemically very reactive. When nanoclusters of a 2D compound are formed, the prismatic edges are decorated by atoms with dangling bonds, which store enough chemical energy to destabilize the planar structure. One way to saturate these dangling bonds is through a reaction with the environment, e.g., reaction with ambient water or oxygen molecules. However, in the absence of reactive chemical species, an alternative mechanism for the annihilation of the peripheral dangling bonds may be provoked, leading to the formation of hollow closed nanoclusters. For this process to take place, sufficient thermal energy is required in order to overcome the activation barrier associated with the bending of the layers (elastic strain energy). In this case, completely seamless and stable hollow nanoparticles are obtained in the form of either polyhedral structures or elongated nanotubes.

However, this is not the sole mechanism which can lead to the formation of nanotubular or microtubular structures. One mechanism, which was already proposed by *Pauling* [14], involves 2D compounds with a non-symmetric unit cell along the *c*-axis, like that of kaolinite. The structure of this compound is made by the stacking of layers consisting of $SiO_2$ tetrahedra and $AlO_6$ octahedra, the latter having a larger *b* parameter. To compensate for this geometric mismatch, hollow whiskers are formed, in which the $AlO_6$ octahedra are on the outer perimeter and the $SiO_2$ tetrahedra are in the inner perimeter of the layer. In this geometry, all the chemical bonds are satisfied with relatively little strain. Consequently, the chemical and structural integrity of the compound is maintained.

Nanotubes and microtubes of a semi-crystalline nature can be formed by almost any compound, using a template growth mechanism. Amphiphilic molecules with a hydrophilic head group, like carboxylate or an -OH group, and a hydrophobic carbon-based chain are known to form very complex phase diagrams, when these molecules are mixed with, e.g., water and an aprotic (non-aqueous) solvent [15]. Structures with a tubular shape are typical for at least one of the phases in this diagram. This mode of packing can be exploited for the templated growth of inorganic nanotubes, by chemically attaching a metal atom to the hydrophilic part of the molecule. Once the tubular phase has been formed, the template for the tubular structure can be removed, e.g., by calcination. In this way, stable metal-oxide nanotubes, can be obtained from various oxide precursors [16]. Nonetheless, the crystallinity of these phases is far from being perfect, which is clearly reflected by their X-Ray Diffraction (XRD) and electron diffraction patterns. Thus, whereas a net of sharp diffraction spots is observed in the electron diffraction patterns of nanotubes from 2D (layered) compounds, the electron diffraction of nanotubes from 3D (isotropic) materials, appears as a set of diffuse diffraction rings, which alludes to their imperfect crystallinity. Also, the sharpness of the diffraction pattern in the latter case may vary from point to point, alluding to the variation in crystallinity of the different domains on the nanotube.

Interestingly, the stability of inorganic nanotubes and nanoparticles can be quite high. Figure 3 shows theoretical results for the strain energy needed to form a nanotube of a given diameter for both BN nanotubes and C nanotubes [17]. The closed circles represent the strain energy needed to form a BN tube relative to a sheet of hexagonal BN, while the open circles indicate the energy of a carbon nanotube relative to graphite. Clearly both organic and inorganic nanotubes are high energy metastable structures, but compared to their respective sheet materials, BN nanotubes are energetically even more favorable than carbon nanotubes.

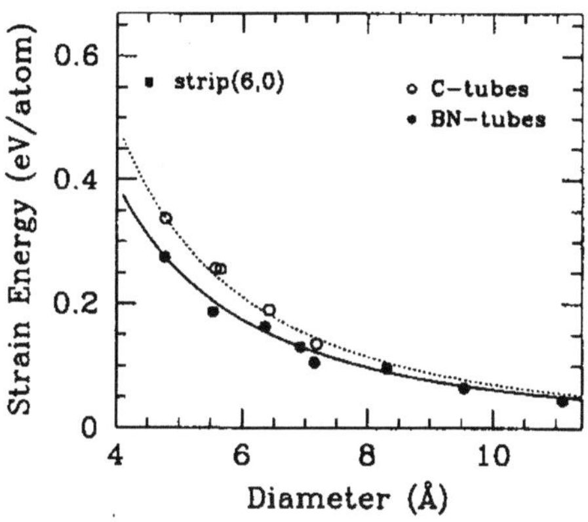


**Fig. 3.** Strain energy versus diameter for the formation of BN and carbon nanotubes relative to their sheet structures. Closed and open circles indicate the energy for BN and carbon nanotubes, respectively. (Courtesy of *X. Blase*) [17]

## 2 Synthesis of Inorganic Nanotubes

Recently, a few techniques for the synthesis of large amounts of $WS_2$ and $MoS_2$ multiwall inorganic nanotubes have been described [8,9,18,19,20,21,22,23]. Each of these techniques is very different from the others and produces nanotube material of somewhat different characteristics. This fact by itself indicates that the nanotubes of 2D metal-dichalcogenides are a genuine part of the phase-diagram of the respective constituents. It also suggests that, with slight changes, these or related techniques can be used for the synthesis of nanotubes from other inorganic layered compounds.

The inorganic nanotube series consisting of $B_xC_yN_z$ is particularly interesting, in that various stoichiometries, including BN, $BC_3$, and $BC_2N$ have been predicted [11] and experimentally realized [24,25]. For BN, single-wall nanotubes have been synthesized and techniques now exist for the mass-production of mono-disperse double-walled BN nanotubes, including crystalline “ropes” of double-walled BN nanotubes. Multiwall $B_xC_yN_z$ tubes are also easily produced.

One of the earliest synthesis methods [18,19,20,21,22] of inorganic nanotubes made use of the chemical vapor transport method, which is the standard growth technique for high quality single crystals of layered metal-dichalcogenide ($MX_2$) compounds. According to this method, a powder of $MX_2$ (or M and X in the 1:2 ratio) is placed on the hot side of an evacuated quartz ampoule, together with a transport agent, like bromine or iodine. A temperature gradient of 20–50°C is maintained. After a few days, a single crystal of the same compound grows on the cold side of the ampoule. Accidentally, microtubes and nanotubes of $MoS_2$ were found on the cold end of the ampoule. These preliminary studies were extended to other compounds, like $WS_2$ (see for example, Fig. 4), and the method was optimized with respect to the production of nanotubes rather than for bulk crystals. Since the synthesis duration of this process is rather long (over two weeks time), it would be rather difficult to optimize the present synthesis method for the production of bulk amounts of such nanotubes. Whereas nanotubes produced by chemical vapor transport were found to consist of the 2H polytype [19], microtubes of diameters exceeding 1 μm and lengths of a few hundreds of micrometers were found to prefer the 3R polytype [21] (see below).

An alternative method for the synthesis of $MoS_2$ nanotubes has also been reported [23]. This synthesis is based on a generic deposition strategy, which has been advanced mostly through the work of *Martin* [26]. Nonuniform electrochemical corrosion of aluminum foil in an acidic solution produces a dense pattern of cylindrical pores, which serve as a template for the deposition of nanofilaments from a variety of materials. Thermal decomposition of a $(NH_4)_2MoS_4$ precursor, which was deposited from solution at 450°C and the subsequent dissolution of the alumina membrane in a KOH solution led to the isolation of large amounts of $MoS_2$ nanotubes. However, due to the limited stability of the alumina membrane, the annealing temperatures of the nano-

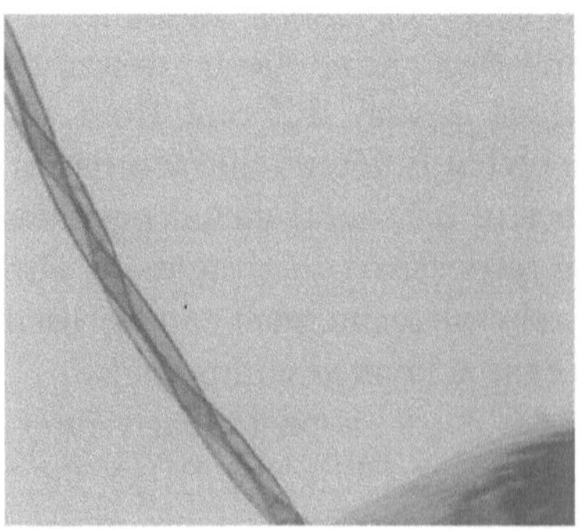

**Fig. 4.** Two entangled spiral $WS_2$ nanotubes obtained by the chemical vapor transport method [19]

tubes were relatively low, and consequently their structure was imperfect. In fact, the $MoS_2$ nanotubes appeared like bamboo-shaped hollow fibers [26].

The synthesis of a pure phase of $WS_2$ nanotubes, 2–10µm long and with diameters in the range of 20–30 nm has been recently reported [8,9]. A typical assemblage of such nanotubes is shown at two magnifications in Fig. 5. Here, short tungsten-oxide nano-whiskers (50–300 nm long) react with $H_2S$ under mild reducing conditions. The oxide precursors are by heating a tungsten filament prepared in the presence of water vapor. One can visualize the growth process of the encapsulated nanowhisker as follows. At the first instant of the reaction (Fig. 6a), the short oxide nanowhisker reacts with $H_2S$ and forms a protective tungsten disulfide monomolecular layer, which covers the entire surface of the growing nanowhisker, excluding its tip. This $WS_2$ monomolecular skin prohibits coalescence of the nanoparticle with neighbor-

**Fig. 5.** Scanning electron microscopy image (two magnifications) of a mat of $WS_2$ nanotubes obtained from a powder of short asymmetric oxide nanoparticles by a combined reduction/sulfidization process [9]

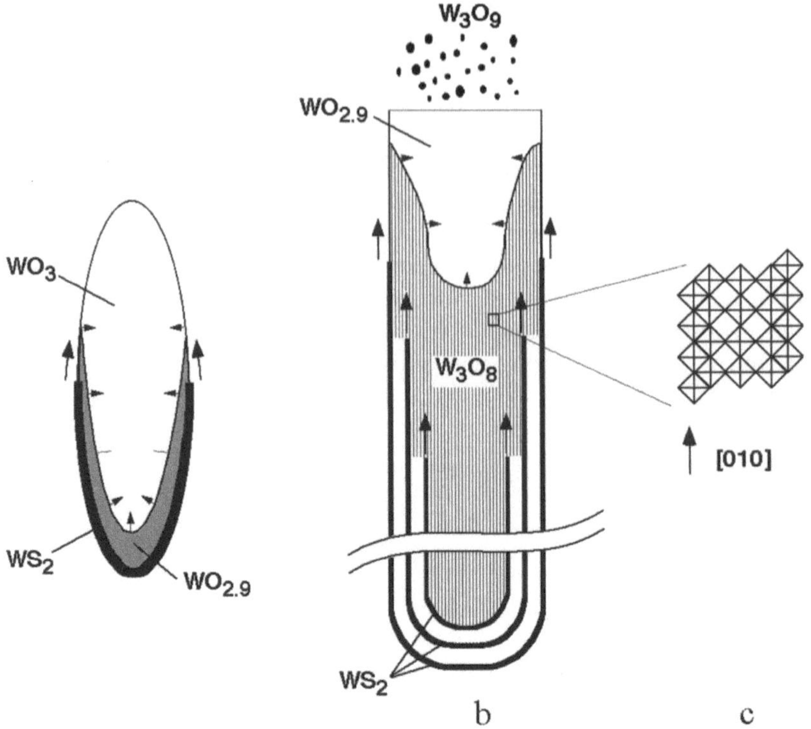


**Fig. 6.** Schematic representation of the growth model of the $WS_2$ nanotubes [9]

ing oxide nanoparticles, which therefore drastically slows their coarsening. Simultaneous condensation of $(WO_3)_n$ or $(WO_{3-x}{\cdot}H_2O)_n$ clusters on the uncovered (sulphur-free) nanowhisker tip, and their immediate reduction by hydrogen gas, lead to the lowering of the volatility of these clusters and therefore to the tip growth. This concerted mechanism leads to a fast growth of the sulfide-coated oxide nanowhisker. Once the oxide source is depleted, the vapor pressure of the tungsten oxide in the gas phase decreases and the rate of tip growth slows down. This leads to the termination of the growth process, since the rate of the sulfidization of the oxide skin continues at the same pace and the exposed whisker tip becomes coated with the protective sulfide skin. This process leads to the formation of oxide nanowhiskers 2–10μm long, coated with an atomic layer of tungsten sulfide. After the first layer of sulfide has been formed, in an almost instantaneous process, the conversion of the oxide core into tungsten sulfide is a rather slow diffusion-controlled process. The oxide nanowhisker growth is schematically illustrated in Fig. 6b. Note that during the gradual reduction of the oxide core, the CS planes in the oxide phase rearrange themselves and approach each other [27] until a stable reduced oxide phase-$W_3O_8$ is reached [28]. This phase provides a sufficiently open structure for the sulfidization to proceed until the entire oxide core is consumed and converted into a respective sulfide. Furthermore, the highly ordered nature of the reduced oxide provides a kind of a template for a virtually dislocation-free sulfide layer growth. Further reduction of the oxide core would bring the sulfidization reaction to a halt [29]. It was indeed shown that in the absence of the sulfide skin, the oxide nanowhisker is reduced rather swiftly to a pure tungsten nanorod. Therefore, the encapsulation of the oxide nanowhisker, which tames the reduction of the core, allows for the gradual conversion of this nanoparticle into a hollow $WS_2$ nanotube. The present model alludes to the highly synergistic nature between the reduction and sulfidization processes during the $WS_2$ nanotube growth and the conversion of the oxide core into multiwall tungsten sulfide nanotubes. It is important to emphasize that the diameter of the nanotube is determined by the precursor diameter in this process. The number of $WS_2$ layers increases with the duration of the reaction until the entire oxide is consumed. This permits very good control of the number of walls in the nanotube, but with the oxide core inside. The oxide whisker can be visualized as a template for the sulfide, which grows from outside inwards. This method does not lend itself to the synthesis of (hollow) single wall nanotubes, since long whiskers, a few nm thick, would not be stable.

The multiwall hollow $WS_2$ nanotubes, which are obtained at the end of the process are quite perfect in shape, which has a favorable effect on some of their physical and electronic properties. This strategy, i.e., the preparation of nanowhiskers from a precursor and their subsequent conversion into nanotubes, is likely to become a versatile vehicle for the synthesis of pure nanotube phases from other 2D layered compounds, as well.

Another strategy for the synthesis of inorganic nanotubes is through the sol-gel process [6,29,30,31,32]. Here, a metal organic compound is dissolved together with a template-forming species in alcohol, and a template structure for the growth of nanotubes is formed. The addition of small amounts of water leads to a slow hydrolysis of the organic-metal compound, i.e., the formation of metal-oxide sol, but the template structure is retained. Upon tuning of the pH, the sol transforms into a gel, which consists of a -M–O- polymer with longer chains. In the case of the layered compound $V_2O_5$ [29,30], a sol was first prepared by mixing vanadium (V) oxide tri-isoporpoxide with hexadecylamine in ethanol and aging the solution while stirring, which resulted in the hydrolysis of the vanadium oxide. Subsequent hydrothermal treatment at 180°C led to the formation of nanotubes with the formal composition $VO_{2.4}{\cdot}(C_{16}H_{33}NH_2)_{0.34}$. These nanotubes are crystalline, which is evident from their X-ray and electron diffraction patterns. Facile and (quite) reversible Li intercalation into such nanotubes has been demonstrated [31], which puts this material into the forefront of high energy density battery research.

Thus far, only metal oxide nanotubes have been synthesized by this process. Whereas crystalline nanotubes were obtained from 2D (layered) oxides, the 3D oxide compounds, like $SiO_2$ [32] resulted in semicrystalline or amorphous nanotubes only. In principle, this kind of process could be extended to the synthesis of nanotubes from chalcogenide and halide compounds in the future.

In another procedure, carbon nanotubes were used as templates for the deposition of $V_2O_5$ nanotubes; the template was subsequently removed by burning the sample in air at 650°C [33]. This strategy can be easily adopted for the synthesis of different oxide nanotubes, as shown below.

The heating of an ammonium thiomolybdate compound with the formula $(NH_4)_2Mo_3S_{13} \cdot xH_2O$ was shown to lead to its decomposition around 673 K and to the formation of a non-homogeneous $MoS_2$ phase, which also contains elongated $MoS_2$ particles (mackles) [34]. The mackles are ascribed to a topochemical conversion of the precursor particles, which form closed layers of $MoS_2$ on top of an amorphous core (possibly $a$-$MoS_3$).

Crystallization of amorphous $MoS_3$ ($a$-$MoS_3$) nanoparticle precursors by the application of microsecond long electrical pulses from the tip of a scanning tunneling microscope, led to the formation of composite nanoparticles with an IF-$MoS_2$ envelope and an $a$-$MoS_3$ core [35]. The $MoS_2$ shells were found to be quite perfect in shape and fully closed. This observation is indicative of the fast kinetics of the crystallization of fullerene-like structures. A self-propogating self-limiting mechanism, has been proposed [35]. According to this model, the process is maintained by the local heating due to the exothermic nature of the chemical reaction and of the crystallization process until the $MoS_2$ layers are completed and closed [35].

A novel room-temperature method for producing nested fullerene-like $MoS_2$ with an $a$-$MoS_3$ core using a sono-electrochemical probe has been described recently [36]. $MoS_2$ nanotubes also occur occasionally in this product. Ultrasonically-induced reactions are attributed to the effect of cavitation, whereupon very high temperatures and pressures are obtained inside imploding gas-bubbles in liquid solutions [37]. The combination of a sono-chemical probe with electrochemical deposition has been investigated for some time now [38,39]. Generally, the decomposition of the gas molecules in the bubble and the high cooling rates lead to the production of amorphous nanomaterials. It appears, however, that in the case of layered compounds, crystalline nanomaterials with structures related to fullerenes, are obtained by sonochemical reactions [36]. In this case, the collapsing bubble serves as an isolated reactor and there is a strong thermodynamic driving force in favor of forming the seamless (fullerene-like) structure, rather than the amorphous or plate-like nanoparticle [35,40]. In fact, there appears to exist a few similarities between the above two methods for the preparation of IF-$MoS_2$ [35,36]. First, both processes consist of two steps, where the $a$-$MoS_3$ nanoparticles are initially prepared and are subsequently crystallized by an electrical pulse [35] in one case, and by a sonochemical pulse in the other process [36]. A compelling factor in favor of the fast kinetics of fullerene-like nanoparticle formation is that, in both processes, the envelope is complete, while the core of the nanoparticles (>20 nm) remains amorphous. Since, the transformation of the amorphous core into a crystalline structure involves a slow out-diffusion of sulfur atoms, the core of the nanoparticles is unable to crystallize during the short (ns – μs) pulses. It is likely that the sonochemical formation of IF-$MoS_2$ nanoparticles can also be attributed to the self-propagating self-limiting process described above.

In a related study, scroll-like structures were prepared from the layered compound GaOOH by sonicating an aqueous solution of $GaCl_3$ [41]. This study shows again the preponderance of nanoparticles with a rolled-up structure from layered compounds. It has been pointed-out [42] that in the case of non-volatile compounds, the sonochemical reaction takes place on the interfacial layer between the liquid solution and the gas bubble. It is hard to envisage that a gas bubble of such an asymmetric shape, as the GaOOH scroll-like structure, is formed in an isotropic medium. Alternatively, one can envisage that a monomolecular layer of GaOOH is formed on the bubble's envelope, which rolls into a scroll-like shape once the bubble is collapsed. As indicated in [41], rolled-up scroll-like structures were obtained by sonication of $InCl_3$, $TlCl_3$ and $AlCl_3$, which demonstrates the generality of this process. Hydrolysis of this group of compounds ($MCl_3$) results in the formation of the layered compounds MOOH, which, upon crystallization, prefer the fullerene-like structures. Nevertheless, MOCl compounds with a layered structure are also known to exist and their formation during the sonication of $MCl_3$ solutions has not been convincingly excluded. More recently, the same group reported the formation of fullerene-like $Tl_2O$ nanoparticles by the sonochemical reaction of $TlCl_3$ in aqueous solution [43]. Some compounds with the formula $M_2O$, where M is a metal atom, possess the anti-$CdCl_2$ structure, with the anion layer sandwiched between two cation layers. Currently, the yield of the IF-$Tl_2O$ product is not very high (~10%), but purification of this phase by the selective heating of the sample to 300°C has been demonstrated. Furthermore, size and shape control of the fullerene-like particles is not easy in this case. Nonetheless, the fact that this is a room temperature process is rather promising, and future developments will hopefully permit better control of the reaction products. Perhaps most important, the versatility of the sonochemical technique is a clear-cut asset for nanoparticle synthesis. It is also to be noted that $Tl_2O$ is a rather unstable compound in bulk form, and the formation of a closed IF structure appears to render it a stable phase. $NiCl_2$ "onions" and nanotubes prepared by sublimation of a $NiCl_2$ powder at 950°C was reported [44]. These nanotubes have potentially interesting magnetic properties. However, their large scale synthesis has only met with partial success so far, mostly due to the hygroscopic nature of the precursor.

As discussed above briefly, semi-crystalline or amorphous nanotubes can be obtained from 3D compounds and metals, by depositing a precursor on a nanotube-template intermediately, and subsequent removing the template by, for example, calcination. Since a nanotube is a rolled-up structure of a 2D molecular sheet, there is no way that all the chemical bonds of a 3D compound will be fully satisfied on the nanotube inner and outer surface. Therefore, in this case, the nanotubes cannot form a fully crystalline structure and the nanotube surface is not going to be inert. Nonetheless, there are certain applications, like in catalysis, where such a high surface area pattern with reactive surface sites (i.e., unsaturated bonds) is highly desirable. The first report of $SiO_2$ nanotubes [45] came serendipitously during the synthesis of spherical silica particles by the hydrolysis of tetraethylorthosilicate in a mixture of water, ammonia, ethanol and tartaric acid. More recently, nanotubes of $SiO_2$ [46], $TiO_2$ [5,47], $Al_2O_3$ and $ZrO_2$ [5,48], etc. have been prepared by the self-assembly of molecular moieties on preprepared templates, like carbon nanotubes, elongated micelles, or other templates which instigate uniaxial growth. In fact, there is almost no limitation on the type of inorganic compound which can be 'molded' into this shape, using this strategy. We note in passing, that the tuneability of the carbon nanotube radii and the perfection of its structure could be important for their use as a template for the growth of inorganic nanotubes with a controlled radius. The formation of nanootubular structures can be rather important for the selective catalysis of certain reactions, where either the reaction precursor or the product must diffuse through the (inorganic) nanotube inner core.

The rational synthesis of peptide based nanotubes by the self-assembling of polypeptides into a supramolecular structure, was demonstrated. This self-organization leads to peptide nanotubes having channels 0.8 nm in diameter and a few hundred nm long [49]. The connectivity of the proteins in these nanotubes is provided by weak bonds, like hydrogen bonds. These structures benefit from the relative flexibility of the protein backbone, which does not exist in nanotubes of covalently bonded inorganic compounds.

We now discuss the synthesis of $B_xC_yN_z$ nanotubes and nanoparticles. The similarity between graphite and hexagonal BN suggests that some of the successful synthesis methods used for carbon nanotube production might be adapted to $B_xC_yN_z$ nanotube growth. This is indeed the case. A non-equilibrium plasma arc technique has been used to produce pure BN nanotubes [25]. To avoid the possibility of carbon contamination, no graphite components are used in this synthesis. The insulating nature of bulk BN prevents the use of a pure BN electrode. Instead, a pressed rod of hexagonal BN is inserted into a hollow tungsten electrode forming a compound anode. The cathode consists of a rapidly cooled pure copper electrode. During discharge, the environmental helium gas is maintained at 650 torr and a dc current between 50A and 140 A is applied to maintain a constant potential drop of 30 V between the electrodes. The arc temperature exceeds 3700 K.

Arcing the BN/W compound electrode results in a limited amount of dark gray soot deposit on the copper cathode, in contrast to the cohesive cylindrical boule which typically grows on the cathode upon graphite arcing. High resolution TEM studies of the soot reveal numerous multiwalled nanotubes. Figure 7 shows a representative TEM image of a BN nanotube thus produced. Electron Energy-Loss Spectroscopy (EELS) performed on individual nanotubes inside the TEM confirms the B–N 1:1 stoichiometry. Figure 7 shows that the end of the BN nanotube contains a dense particle, most likely tungsten or a tungsten-boron-nitrogen compound. In contrast to carbon nanotubes, where the capping is fullerene-like or involves pentagons and heptagons, BN tube closure by pentagon formation is suppressed in order to inhibit the formation of less favorable B–B bonds. The small metal cluster may thus simultaneously act as a growth catalyst and as an aid to capping the nanotube by relieving strain energy.

Different synthesis methods have been successfully employed to produce BN nanotubes. An arc-discharge method employing $HfB_2$ electrodes in a nitrogen atmosphere yields BN nanotubes with a wall number ranging from many to one [50]. In this synthesis method, the Hf is apparently not incorporated into the tube itself, but rather acts as a catalyzing agent. The source of nitrogen for tube growth is from the $N_2$ environmental gas. The ends of the BN nanotubes thus produced appear to be pure BN, with unusual geometries reflecting the bond frustration upon closure. Tantalum has also been used as the catalyzing agent in the synthesis of various nanoscale BN structures using arc-vaporization methods [51]. Pure BN nanotubes are produced, along with other nanoparticles, including onion-like spheres similar to those produced by a high intensity electron irradiation method [52]. In other studies, circumstantial evidence is found for the presence of $B_2N_2$ squares at the BN nanotube tips, as well as $B_3N_3$ hexagons in the main fabric of the nanotubes [51].

A most intriguing observation is that of $B_xC_yN_z$ nanotubes with segregated tube-wall stoichiometry [53]. Multiwall nanotubes containing pure-carbon walls adjacent to pure BN walls have been achieved, forming a sort of nanotube coaxial structure. In one specific tube studied carefully by EELS, the innermost three walls of the tube contained only carbon, the next six walls

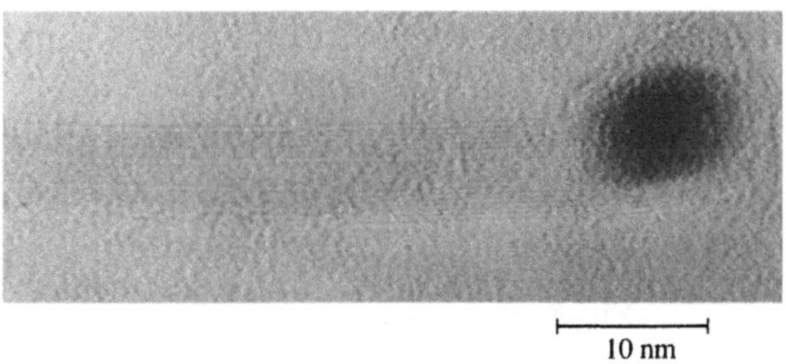


**Fig. 7.** TEM micrograph of the end of a BN nanotube showing termination by a metal particle. (courtesy of *N. G. Chopra* [25])

were comprised of BN, and the last five outermost walls were again pure carbon. The entire 14-walled composite nanotube was 12 nm in diameter. Similar layer segregation is obtained for onion-like coverings over nanoparticles (quite often the core nanoparticle is composed of the catalyst material). Related studies have also found $B_xC_yN_z$ nanotubes consisting of concentric cylinders of $BC_2N$ and pure carbon. A reactive laser ablation method has also been used to synthesize multi-element nanotubes containing BN [54]. The nanotubes contain a silicon carbide core followed by an amorphous silicon oxide intermediate layer; this composite nano-rod is then sheathed with BN and carbon nanotube layers, segregated in the radial direction. It has been speculated that the merging BN and carbon nanotube structures may be the basis for novel electronic device architectures.

The above-described methods for $B_xC_yN_z$ nanotube synthesis unfortunately result in relatively low product yields, and do not produce monodisperse materials. Recently, a method has been developed which produces virtually mono-disperse BN double-walled nanotubes in large quantity [10]. Multi-wall BN "nano-cocoons" with etchable cores have also been synthesized in large quantity [10] using this method. The synthesis employs nitrogen-free, boron-rich electrodes arced in a pure nitrogen gas environment. The electrodes incorporate 1 at.% each of nickel and cobalt. Arcing in a dynamic $N_2$ gas environment with a pressure near 380 Torr results in an abundance of gray web-like material inside the synthesis chamber; this material contains a high percentage of double-walled BN nanotubes, as shown in Fig. 8. The

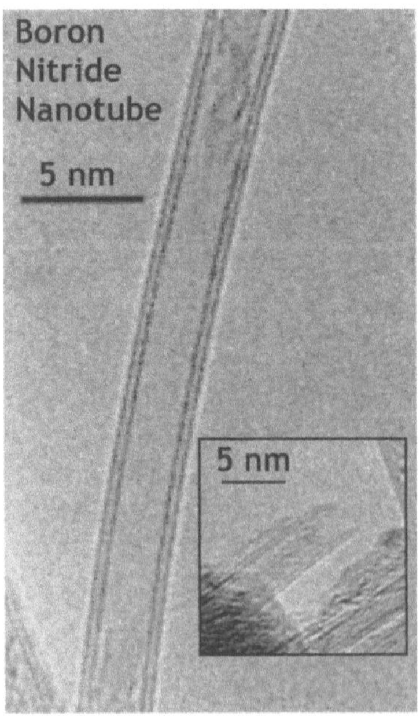


**Fig. 8.** High resolution TEM image of a BN nanotube with two layers. Inset: Image of a BN nanotube end. (courtesy of *J. Cumings* [10])

tubes have a narrow size distribution, centered on 2.7 nm (outer diameter) with a standard deviation of 0.4 nm. The double walled BN nanotubes often aggregate as "ropes", very similar to the crystalline rope-like structures previously observed for mono-disperse single-walled carbon nanotubes [55]. Figure 9 shows BN nano-cocoons before and after their cores have been etched out with acid [10]. The nano-cocoons can be formed with a relatively uniform size distribution.

The double-wall BN structure has been observed to form in the case of $HfB_2$ arced in nitrogen gas [50] and is claimed to be the dominant structure in laser ablation of BN with a Ni/Co catalyst carried out in a nitrogen carrier gas [56]. It has been proposed that lip–lip interactions [57] could stabilize the growth of open-ended multiwall carbon nanotubes [17]. It has been predicted that the B–N bond of a single-wall BN nanotube should naturally undergo out-of-plane buckling [58], producing a dipolar shell. Buckling (absent in carbon tubes) may thus favor double-walled BN nanotube growth.

Multi-walled nanotubes with different $B_xC_yN_z$ stoichiometries have also been produced in small quantity, including $BC_2N$ and $BC_3$ nanotubes [25,59]. Importantly, both $BC_2N$ and $BC_3$ are known to exist in bulk (i.e., layered sheet) form. The bulk materials can be synthesized [60] via the following chemical reactions:

$$CH_3CN + BCl_3 \longrightarrow BC_2N + 3HCl\ (>800\ ^\circ C)$$
$$2BCl_3 + C_6H_6 \longrightarrow 2BC_3 + 6HCl\ (\ 800\ ^\circ C).$$

Both bulk $BC_2N$ and $BC_3$ have bright metallic luster and resemble the lay-

a)

b)

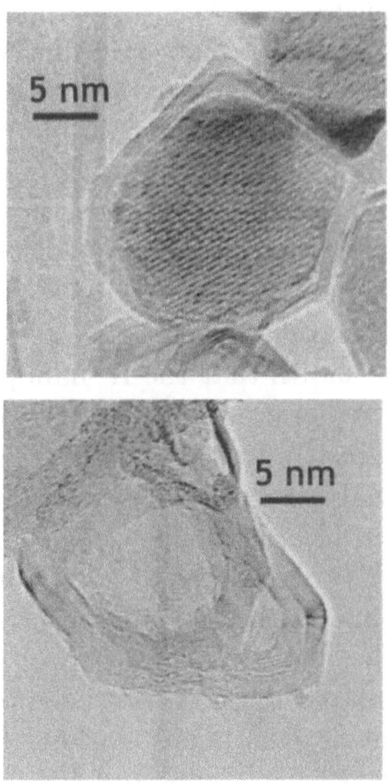


**Fig. 9.** (**a**) BN-coated boron nanocrystal. (**b**) Same as (**a**), after treating with nitric acid. The boron core has been chemically removed, leaving an empty BN nano-cocoon. (courtesy of *J. Cumings* [10])

ered structure of graphite. Resistivity measurements of the layered bulk compounds indicate that $BC_2N$ is semiconducting with an energy gap of about 0.03 eV and $BC_3$ is semi-metallic [60]. $BC_2N$ and $BC_3$ nanotubes have been produced by arc-synthesis methods using a compound anode formed by inserting a pressed BN rod inside a hollowed out graphite electrode. The compound anode is arced against a copper cathode in a 450 torr helium gas environment. The stoichiometry of the resulting multiwalled nanotubes is identified using EELS, but further experimental characterization is lacking. In the next section we discuss some of the theoretically predicted properties of $BC_2N$ and $BC_3$ nanotubes.

## 3 Thermodynamic and Topological Considerations

The thermodynamic stability of the fullerene-like materials is rather intricate and far from being fully understood. Such structures are not expected to be globally stable, but they are probably the stable phase of a layered compound, when the particles are not allowed to grow beyond, say a fraction of a micron. Therefore, there seems to exist a narrow window of conditions in the vicinity of the layered compound, itself, where nanophases of this kind exist. This idea is supported by a number of observations. For example, the W-S phase diagram provides a very convenient pathway for the synthesis of IF-$WS_2$. The compound $WS_3$, which is stable below 850°C under excess of sulfur, is amorphous. This compound will therefore lose sulfur atoms and crystallize into the compound $WS_2$, which has a layered structure, upon heating or when sulfur is denied from its environment. If nanoparticles of $WS_3$ are prepared and they are allowed to crystallize under the condition that no crystallite can grow beyond say 0.2μm, fullerene-like $WS_2$ ($MoS_2$) particles and nanotubes will become the favored phase. This principle serves as a principal guideline for the synthesis of bulk amounts of the IF-$WS_2$ phase [40] and $WS_2$ nanotubes in particular [8]. Unfortunately, in most cases, the situation is not as favorable, and more work is needed to clarify the existence zone of the IF phase in the phase diagram (in the vicinity of the layered compound).

Another very important implication of the formation of nanoparticles with IF structures is that in several cases it has been shown that the IF nanoparticles are stable, but the bulk form of the layered compound is either very difficult to synthesize or is totally unstable. The reason for this surprising observation is probably related to the fact that the IF structure is always closed and hence it does not expose reactive edges and interacts only very weakly with the ambient, which in many cases is hostile to the layered compound. For example, Na intercalated $MoS_2$ is unstable in a moistured ambient, since water is sucked between the layers and into the van der Waals gap of the platelet and exfoliates it. In contrast, Na intercalated IF-$MoS_2$ has been produced and was found to be stable in the ambient or even in suspensions [61]. Chalcogenides of the first row of transition metals, like $CrSe_2$ and $VS_2$, are not stable in the layered structure. However, Na intercalation endows extra stability to the layered structure, due to the charge transfer of electrons from the metal into the partially empty valence band of the host [62]. Thus, for example, $NaCrSe_2$ and $LiVS_2$ form a superlattice, in which the alkali metal layer and the transition metal layer alternate. The structure of this compound can be visualized akin to the layered structure $CrSe_2$, in which the octahedral sites in the van der Waals gap between adajacent layers are fully occupied by the Na (Li) atoms. Nevertheless, $VS_2$ nanoparticles with a fullerene-like structure, i.e., consisting of layered $VS_2$, were found to be stable [61]. The unexpected extra stability of this structure emanates from the closed seamless structure of the IF, which does not expose the chemically reactive sites to the hostile environment. This idea opens new avenues for the synthesis of layered compounds, which could not be previously obtained or could not be exposed to the ambient [43], and therefore could only be studied to a limited extent.

Many layered compounds come in more than one stacking polytype [63]. For example, the two most abundant polytypes of $MoS_2$ are the 2H and 3R. The 2H polytype is an abbreviation for the hexagonal structure consisting of two S–Mo–S layers in the unit cell (AbA···BaB···AbA···BaB, etc.). The 3R polytype has a rhombohedral unit cell of three repeating layers (AbA···BcB···CaC···AbA···BcB···CaC, etc.). In the case of $MoS_2$, the most common polytype is the 2H form, but the 3R polytype was found, for example, in thin $MoS_2$ films prepared by sputtering [64]. The nanotubes grown by the gas phase reaction between $MoO_3$ and $H_2S$ at 850°C were found to belong to the 2H polytype [4,65]. The same is true for $WS_2$ nanotubes obtained from $WO_3$ and $H_2S$ [8]. The appearance of the 3R polytype in such nanotubes can probably be associated with strain. For example, a "superlattice" of 2H and 3R polytypes was found to exist in $MoS_2$ nanotubes grown by chemical vapor transport [22]. Strain effects are invoked to explain the preference of the rhombohedral polytype in both $MoS_2$ and $WS_2$ microtubes grown in the same way [21]. These observations indicate that the growth kinetics of the nanotubes and of thin films influence the strain relief mechanism, and therefore different polytypes can be adopted by the nanotubes.

The trigonal prismatic structure of $MoS_2$ alludes to the possibility to form stable point defects consisting of a triangle or a rhombus [29]. In the past, evidence in support of the existence of "bucky-tetrahedra" [65] and "bucky-cubes" [66], which have four triangles and six rhombi in their corners, respectively, were found. However, the most compelling evidence in support of this idea was obtained in nanoparticles collected from the soot of laser ablated $MoS_2$ [67]. Sharp cusps and even a rectangular apex were noticed in $WS_2$ nanotubes, as well [8,9]. These features are probably a manifestation of the inherent stability of elements of symmetry lower than pentagons, such as triangles and rhombi, in the structure of $MoS_2$, etc. They are preferentially formed by the cooling of the hot plasma soot of the ablated targets and are located in the nanotube apex or corners of the octahedra. Point defects of this symmetry were not observed in carbon fullerenes, most likely because the $sp^2$ bonding of carbon atoms in graphite is not favorable for such topological elements. These examples and others illustrate the influence of the lattice structure of the layered compound on the detailed topology of the fullerene-like nanoparticle or of the nanotube cap obtained from such compounds.

The chemical composition of the IF phase deviates only very slightly, if at all, from the composition of the bulk layered compound. Deviations from stoichiometry can only occur in the cap of the nanotube. In fact, even the most modern analytical techniques, like scanning probe techniques and high (spatial) resolution electron energy loss spectroscopy are unable to resolve such a tiny change in the stoichiometry, like the excess or absence of a single Mo (W) or S (Se) atom in the nanotube cap.

The crystal structures of bulk graphite, BN, $BC_2N$, and $BC_3$ are quite similar to each other. They are all hexagonal layered structures, with ABAB packing being the most common arrangement of the layers. In the case of $BC_2N$, two different sheet configurations are possible, leading to two different nanotube isomers with the same $BC_2N$ stoichiometry [68]. Figure 10 shows examples of such isomers with similar diameters.

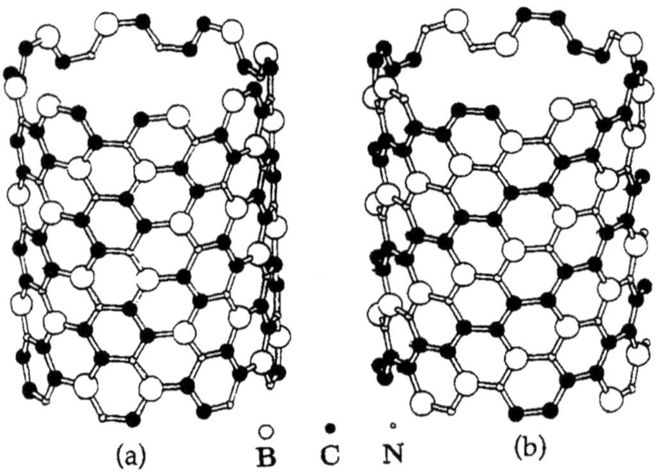


**Fig. 10.** Theoretically determined tubules of isomers of $BC_2N$. These are the (4,4) tubes, using the indexing protocol for carbon nanotubes. (courtesy of *Y. Miyamoto* [68])

## 4 Physical Properties

Early-on, a few groups used powerful theoretical tools to calculate the stability, band-structure and other physical properties of boron-nitride and boron carbo-nitride nanotubes [11,69,70,71]. A few striking conclusions emerged from these studies. First, it was found that B–B and N–N nearest neighbors do not lead to stable polyhedral structures. Instead, distinct B–N pairs of atoms were found to be thermodynamically preferred. This observation implies that $B_2N_2$ rectangles, rather than the 5-member rings found in carbon fullerenes and nanotubes, are required in order to stabilize the BN polyhedra and nanotubes. Experimental verification of this hypothesis has been obtained in the work of a few groups [51,72]. Secondly, in contrast to carbon nanotubes, which can be metallic or semiconducting depending on their chirality, all BN nanotubes were found to be semiconductors, independent of their chirality. Thirdly, whereas the smallest forbidden gap of the achiral $(n,0)$ nanotubes is a direct bandgap ($\Gamma$–$\Gamma$), an indirect bandgap ($\Delta$–$\Gamma$) is calculated for the chiral nanotubes $(n,m)$. Bulk BN material has an indirect bandgap of 5.8 eV. This is to be contrasted with carbon nanotubes, which are either metallic or semiconducting, depending on their $(n,m)$ values (see *Louie* [73]). The fourth point to be noted is that, in contrast with carbon nanotubes in which the band gap increases with decreasing diameter, the bandgap of inorganic nanotubes was found to decrease with decreasing diameter of the IF nanotubes. This effect is attributed to the strain, and also to zone folding in the closed nanotube. The strain increases with decreasing diameter ($D$) of the nanotube as $1/D^2$. It should also be noted that generically, the bandgap of semiconducting nanoparticles increases with decreasing particle diameter, which is attributed to the quantum size confinement of the electron wavefunction.

### 4.1 Band Structure Calculations

Figure 11 contrasts the band structures of BN in a sheet structure to that of BN in a nanotube structure. As mentioned above, BN nanotubes have a fairly robust bandgap, largely independent of the geometrical details of the nanotube. This uniformity suggests that BN nanotubes may present significant advantages over carbon nanotubes for specific applications. Details of the nanotube band structure show that the lowest lying conduction band state is a nearly free-electron like state which has a maximum charge density located about 0.2 nm interior to the tube wall. Thus, if BN tubes were injected with charge (say by modest doping), the resulting metallic tube would carry a cylinder of charge internally along its length.

Due to the greater complexity of $BC_2N$, the unit cell of the bulk material is "double" that of graphite, and there are two possible arrangements of the B,C, and N atoms in the sheet, as reflected in the nanotubes of Fig. 10. The Type A sheet on the left (Fig. 10) has inversion symmetry (as does graphite) while the Type B sheet on the right (Fig. 10) does not (similar to BN). Consequently, the predicted electronic properties of Types A and B $BC_2N$ nanotubes parallel the properties of carbon and BN nanotubes, respectively. Type

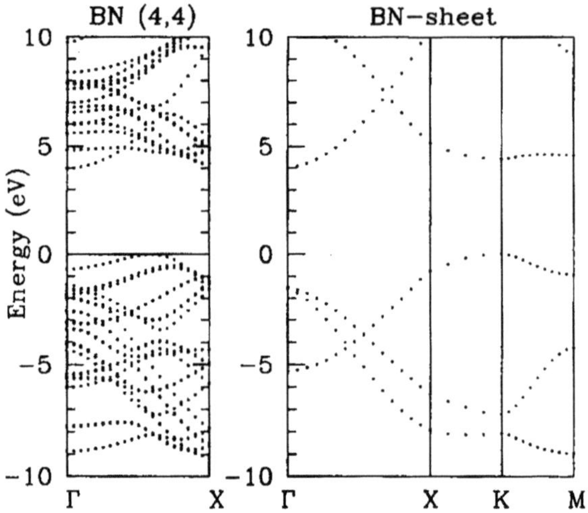


**Fig. 11.** Band structure of (**a**) BN sheet and (**b**) a BN (4,4) nanotube. (courtesy of *S.G. Louie* [17])

A $BC_2N$ nanotubes (Fig. 10a) range from semiconducting to metallic depending on diameter and chirality, while Type B $BC_2N$ nanotubes (Fig. 10b) are predicted to be semiconducting, independent of tube parameters. An interesting feature of Type B $BC_2N$ nanotubes is the arrangement of atoms in the tube wall fabric: a chain of potentially conducting carbon atoms alternating with a string of insulating BN. This resembles a solenoid, and doping a semiconducting Type B $BC_2N$ nanotube should result in a conducting tube where the electrical current spirals along the axis of the nanotube, forming a nano-coil.

The electrical behavior of $BC_3$ is rather complex, but the most significant result from the theoretical calculations is that concentric tubes of $BC_3$, or a close-packed array of mono-disperse single-walled $BC_3$ tubes, are metallic, while isolated single-walled $BC_3$ tubes are semiconducting [68].

Further work was carried out on nanotubes of the semiconducting layered compound GaSe [74]. In this compound, each atomic layer consists of a Ga–Ga dimer sandwiched between two outer selenium atoms in a hexagonal arrangement. This work indicated that some of the early observations made for BN and boro-carbonitride are not unique to these layered compounds, and are valid for a much wider group of structures. First, it was found that like the bulk material, GaSe nanotubes are semiconductors. Furthermore, the strain energy in the nanotube was shown to increase, and consequently the bandgap was found to shrink as the nanotube diameter becomes smaller. Recent work on $WS_2$ and similar nanotubes [75] confirmed these earlier results. While the lowest bandgap of the armchair $(n, n)$ nanotubes were found to be indirect, a direct transition was predicted for the zigzag $(n, 0)$ nanotubes. Additionally, a similar dependence of the strain energy and bandgap energy on the nanotube diameter was predicted for $WS_2$ nanotubes. These findings suggest a new mechanism for optical tuning through strain effects in the hollow nanocrystalline structures of layered compounds. The existence of a direct gap in zigzag nanotubes is rather important, since it suggests that such nanostructures may exhibit strong electroluminescence, which has never been observed for the bulk material.

The transport properties of inorganic nanotubes have not yet been reported. However, a wealth of information exists on the transport properties of the corresponding bulk quasi-2D materials, which is summarized in a few review articles [63,76].

## 4.2 Optical Studies in the UV and Visible

Measurements of the optical properties in this range of wavelengths can probe the fundamental electronic transitions in these nanostructures. Some of the aforementioned effects have in fact been experimentally revealed [77,78]. As mentioned above, the IF nanoparticles in this study were prepared by a careful sulfidization of oxide nanoparticles. Briefly, the reaction starts on the surface of the oxide nanoparticle and proceeds inwards, and hence the number of closed (fullerene-like) sulfide layers can be controlled quite accurately during the reaction. Also, the deeper the sulfide layer in the nanoparticle, the smaller is its radius and the larger is the strain in the nanostructure. Once available in sufficient quantities, the absorption spectra of thin films of the fullerene-like particles and nanotubes were measured at various temperatures (4–300 K). The excitonic nature of the absorption of the nanoparticles was established, which is a manifestation of the semiconductive nature of the material. Furthermore, a clear red shift in the exciton energy, which increased with the number of sulfide layers of the nanoparticles, was also observed (Fig. 12). The temperature dependence of the exciton energy was not very different from the behavior of the exciton in the bulk material. This observation indicates that the red shift in the exciton energy cannot be attributed to defects or dislocations in the IF material, but rather it is a genuine property of the inorganic fullerene-like and nanotube structures. In contrast to the previous observations, IF phases with less than 5 layers of sulfide revealed a clear blue shift in the excitonic transition energy, which was associated with the quantum size effect. Figure 13 summarizes this series of experiments and the two effects. The red shift of the exciton peak in the absorption measurements, due to strain in the bent layer on one hand, and the blue shift for the IF structures with very few layers and large diameter (minimum strain), on the other hand, can be discerned.

The $WS_2$ and $MoS_2$ nanotubes and the nested fullerene-like structures used for the experiments shown in Figs. 12 and 13 had relatively large diameters (>20 nm). Therefore, the strain energy is not particularly large in the first few closed layers of the sulfide, but the strain energy increases as

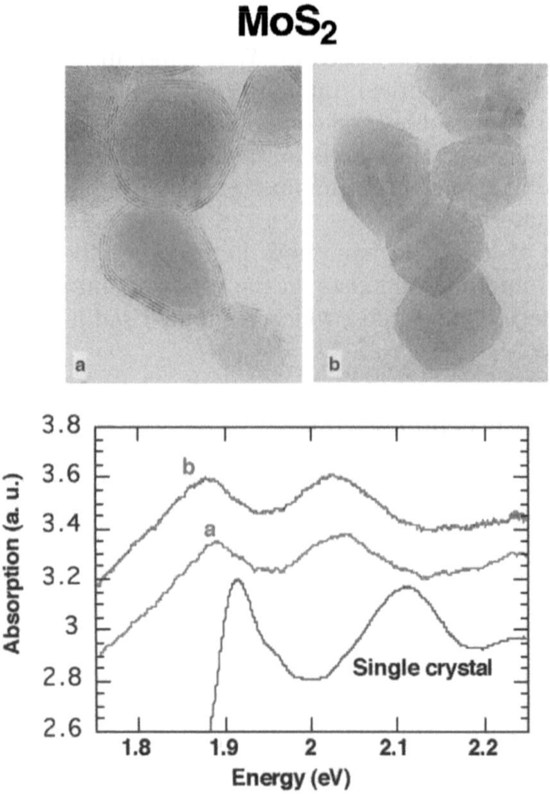


**Fig. 12.** Transmission electron microscopy (TEM) images and absorption spectra of crystalline and fullerene-like (IF) $MoS_2$ films. (**a**) TEM micrograph of a partially converted nanoparticles with 5 layers of $MoS_2$ and $MoO_2$ core. (**b**) TEM micrograph of a fully converted IF-$MoS_2$ nanoparticles. (**c**) Absorption spectra of various $MoS_2$ particles. *Curve-1* IF-$MoS_2$ nanoparticles with $MoO_2$ cores shown in (**a**); *Curve-2* the fully converted (sulfidized) IF-$MoS_2$ nanoparticles shown in (**b**). *Curve-3* single crystal $MoS_2$. Note the red shift of the excitonic peaks of the IF structure compared to those for the crystalline film. This shift increases as the number of closed $MoS_2$ layers increases at the expense of the oxide core and their radii shrink [77]

the oxide core is progressively converted into sulfide, i.e., closed sulfide layers of smaller and smaller diameter are formed. This unique experimental opportunity permitted a clear distinction to be made between the strain effect and the quantum size effect. In the early stages of the reaction, the strain is not very large and therefore the confinement of the exciton along the $c$-axis is evident from the blue shift in the exciton peak. The closed and therefore seamless nature of the $MS_2$ layer is analogous to an infinite crystal in the $a$–$b$ plane and hence quantum size effects in this plane can be ruled out. However, there is a clear confinement effect observable perpendicular to the $a$–$b$ plane, i.e., in the $c$-direction. The quantum size effect in layered compounds was

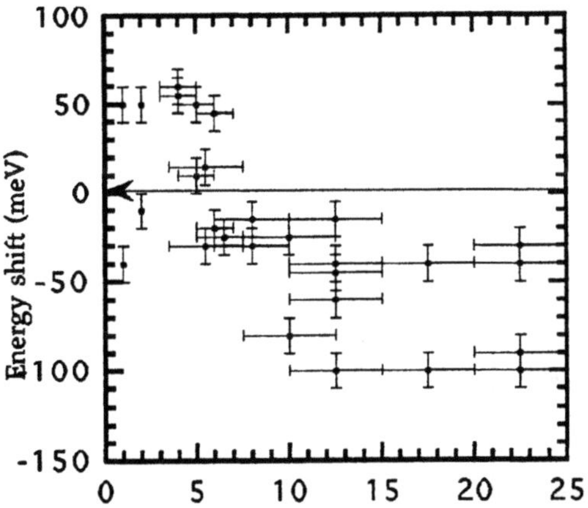


**Fig. 13.** The dependence of the A exciton shifts on the number of layers in the IF structure. The $x$ error bar represents the distribution of the number of layers determined with TEM for each sample. The $y$-axis error bar is $\pm 10\,\text{meV}$ [77]

studied in the past [79,80]. The energy shift due to this effect ($\Delta E_g$) can be expressed as:

$$\Delta E_g = \frac{h^2}{4\mu_{\|} L_z^2} \cdot = \frac{\pi^2 \hbar^2}{\mu_{\|} L_z^2} \,. \tag{1}$$

Here, $\mu_{\|}$ is the exciton effective mass parallel to the $c$-axis and $L_z$ is the (average) thickness of the $WS_2$ nested structure ($L_z = n \times 0.62\,\text{nm}$, where $n$ is the number of $WS_2$ layers) in the nanoparticle. In a previous study of ultra-thin films of 2H-$WSe_2$, $\Delta E_g$ of the A exciton was found to obey (1) over a limited thickness range. $\Delta E_g$ was found to depend linearly on $1/L_z^2$ for $L_z$ in the range of 4–7 nm and $\Delta E_g$ became asymptotically constant for $L_z > 8\,\text{nm}$ [79]. A similar trend is observed for IF-$WS_2$ and $MoS_2$, as shown in Fig. 14 [77]. Therefore, the quantum-size effect is indeed observed for IF structures with a very small number of $WS_2$ layers ($n < 5$). Note that in the current measurements, IF films 150 nm thick were used, but since each IF structure is isolated and the exciton cannot diffuse from one nanoparticle to another, the quantum size effect can be observed in this case. Note also, that due to the (residual) strain effect, the energy for both the A and B excitons is smaller than for their bulk counterparts. The corresponding red shift in the absorption spectrum has also been found for $MoS_2$ nanotubes [23].

These studies suggest a new kind of optical tuneability. Combined with the observation that zigzag inorganic nanotubes are predicted to exhibit direct optical transitions [75], new opportunities for optical device technology, e.g., $MoS_2$ nanotube based light-emitting diodes and lasers, could emerge from such studies in the future. The importance of strong light sources a few nm in size in future opto-electronic applications involving nanotechnology can be appreciated from the need to miniaturize current sub-micron light sources for lithography.

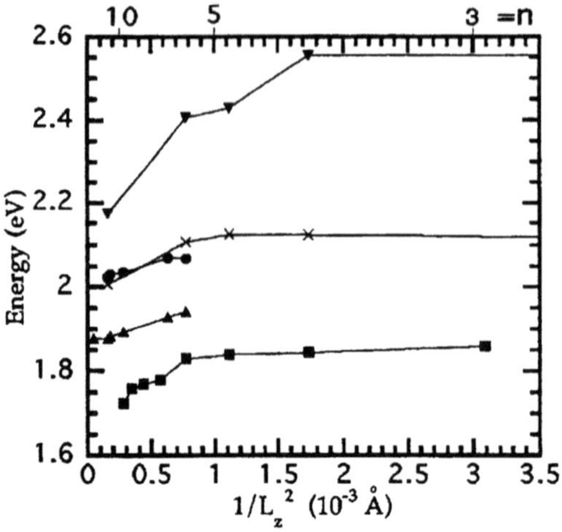


**Fig. 14.** Plot of the A and B exciton energies of IF-$MoS_2$ and IF-$WS_2$ vs $1/L_z^2$, where $L_z$ is the particle size and $n$ on the upper scale is the number of $MS_2$ layers. The ▲ and ● symbols represent the data for the A and B excitons of IF-$MoS_2$, respectively; the × and ▼ show the data for the A and B excitons of IF-$WS_2$ (25 K). The ■ represent the A exciton of 2H-$WS_2$ at 77 K [77,79]

## 4.3 Raman Spectroscopy

Raman and Resonance Raman (RR) measurements of fullerene-like particles of $MoS_2$ have been carried out recently (see Figs. 14, 15) [78,81]. Using 488 nm excitation from an Ar-ion laser light source, the two strongest Raman features in the Raman spectrum of the crystalline particles, at 383 and 408 $cm^{-1}$, which correspond to the $E_{2g}^1$ and $A_{1g}$ modes, respectively, (see Table 1), were also found to be dominant in IF-$MoS_2$ and in $MoS_2$ platelets of a very small size. A distinct broadening of these two features could be discerned as the size of the nanoparticles was reduced. In analogy to the models describing quantum confinement in electronic transitions, it was assumed that quantum confinement leads to contributions of modes from the edge of the Brillouin zone with a high density of phonon states to the Raman spectra. Taking account of the phonon dispersion curves near the zone edge and carrying out a lineshape analysis of the peaks led to the conclusion that the phonons are confined by coherent domains in IF nanoparticles of about 10 nm in size. Such domains could be associated with the faceting of the polyhedral IF structures.

RR spectra were obtained by using the 632.8 nm (1.96 eV) line of a He-Ne laser [81]. Figure 15 shows the RR spectra of a few $MoS_2$ samples. Table 1 lists the peak positions and the assignments of the various peaks for the room temperature spectra. A few second-order Raman transitions were also identified. The intensity of the 226 $cm^{-1}$ peak did not vary much by lowering the temperature, and therefore it cannot be assigned to a second-order transition. This peak was therefore attributed to a zone-boundary phonon, activated by

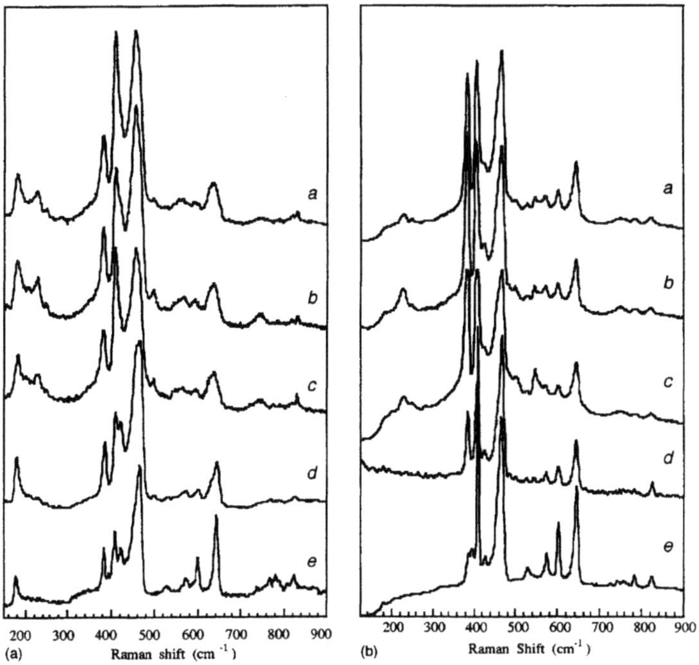


**Fig. 15a,b.** Resonance Raman (RR) spectra excited by the 632.8 nm (1.96 eV) laser line at room temperature (*left*) and 125 K (*right*), showing second-order Raman (SOR) bands for several $MoS_2$ nanoparticle samples: IF-$MoS_2$ 200 Å (*curve a*), IF-$MoS_2$ 800 Å (*curve b*), PL-$MoS_2$ 50 × 300 Å$^2$ (*curve c*), PL-$MoS_2$ 5000 Å (*curve d*), bulk 2H-$MoS_2$ (*curve e*), where IF denotes inorganic fullerene-like particles and PL denotes platelets [81]

the relaxation of the $q = 0$ selection rule in the nanoparticles. Lineshape analysis of the intense 460 cm$^{-1}$ mode revealed that it is a superposition of two peaks at 456 cm$^{-1}$ and 465 cm$^{-1}$. The lower frequency peak is assigned to a 2LA ($M$) process, while the higher energy peak is associated with the $A_{2u}$ mode, which is Raman inactive in crystalline $MoS_2$, but is activated by the strong resonance Raman effect in the nanoparticles.

## 4.4 Mechanical Properties

The mechanical properties of the inorganic nanotubes have only been investigated to a small extent. An elastic continuum model, which takes into account the energy of bending, the dislocation energy and the surface energy, was used as a first approximation to describe the mechanical properties [82]. A first-order phase transition from an evenly curved (quasi-spherical) particle into a polyhedral structure was predicted for nested fullerenes with shell thicknesses larger than about 1/10 of the nanotube radius. Indeed, during the synthesis of IF-$WS_2$ particles [83], it was observed that the nanoparticles were

**Table 1.** Raman peaks observed in the $MoS_2$ nanoparticle spectra at room temperature and the corresponding symmetry assignments. All peak positions are in $cm^{-1}$ [81]

| Bulk $MoS_2$ | PL-$MoS_2$ 5000 Å | PL-$MoS_2$ $50 \times 30$ Å$^2$ | IF-$MoS_2$ 800 Å | IF-$MoS_2$ 200 Å | Symmetry assignment |
|---|---|---|---|---|---|
| 177 | 179 | 180 | 180 | 179 | $A_{1g}$ $(M)$ – LA $(M)$ |
| – | – | 226 | 227 | 226 | LA $(M)$ |
| – | – | – | 248 | 248 | – |
| – | – | – | – | 283 | $E_{1g}$ $(\Gamma)$ |
| 382 | 384 | 381 | 378 | 378 | $E^1_{2g}$ $(\Gamma)$ |
| 407 | 409 | 408 | 407 | 406 | $A_{1g}$ $(\Gamma)$ |
| 421 | 419 | – | weak | weak | – |
| 465 | 460 | 455 | 452 | 452 | 2× LA $(M)$ |
| – | – | 498 | 495 | 496 | Edge phonon |
| 526 | 529 | – | – | – | $E_{1g}$ $(M)$ + LA $(M)$ |
| – | – | 545 | 545 | 543 | – |
| 572 | 572 | ~557 | 565 | 563 | 2× $E_{1g}$ $(\Gamma)$ |
| 599 | 601 | 595 | 591 | 593 | $E^1_{2g}$ $(M)$ + LA $(M)$ |
| 641 | 644 | 635 | 633 | 633 | $A_{1g}$ $(M)$ + LA $(M)$ |

transformed into a highly faceted structure, when the shell of the nanoparticles exceeded a few nm in thickness. Theoretical and experimental work is underway to elucidate the mechanical properties of inorganic nanotubes.

Along with unusual electronic and optical properties, inorganic nanotubes can also display dramatic mechanical properties [12]. Indeed, the axial Young's modulus of individual BN nanotubes has been measured using vibration reed techniques inside a TEM [12]. The elastic modulus is found to be of order 1 TPa, comparable to that of high quality carbon nanotubes. BN nanotubes thus have the highest elastic modulus of any known insulating fiber. Table 2 summarizes some of the predicted and measured properties of $B_xC_yN_z$ nanotubes. Less is known about other IF nanotubes. Clearly these fascinating materials deserve a great deal of further study.

**Table 2.** Summary of predicted and measured properties of $B_xC_yN_z$ nanotubes

| Type of Nanotube | Predicted Property | | | Experimental | |
|---|---|---|---|---|---|
| | Electrical[a] | $E_{gap}$ (eV) | $Y$ (TPa) | $Y$ (TPa) | SWNT |
| Carbon | SC or M | 0 to 1.5 | 1 to 7 | 1.35 | yes |
| BN | SC | 4 to 5.5 | 0.95 to 6.65 | 1.18 | yes |
| $BC_2N$ (I) | SC or M | – | – | – | no |
| $BC_2N$ (II) | SC | 1.28 | – | – | no |
| $BC_3$ | M | – | – | – | no |

[a] SemiConducting (SC) or Metallic (M)

# 5 Applications

The spherical shape of the fullerene-like nanoparticles and their inert sulfur-terminated surface suggest that $MoS_2$ particles could be used as a solid-lubricant additive in lubrication fluids, greases, and even in solid matrices. Applications of a pure IF-$MoS_2$ powder could be envisioned in high vacuum and microelectronics equipment, where organic residues with high vapor pressure can lead to severe contamination problems [84,85]. Since the $MoS_2$ layers are held together by weak van der Waals forces, they can provide easy shear between two close metal surfaces, which slide past each other. At the same time, the $MoS_2$ particles, which come in the form of platelets, serve as spacers, eliminating contact between the two metal surfaces and minimizing the metal wear. Therefore $MoS_2$ powder is used as a ubiquitous solid-lubricant in various systems, especially under heavy loads, where fluid lubricants cannot support the load and are squeezed out of the contact region between the two metal surfaces. Unfortunately, $MoS_2$ platelets tend to adhere to the metal surfaces through their reactive prismatic $(10\bar{1}0)$ edges, in which configuration they "glue" the two metal surfaces together rather than serve as a solid lubricant. During the mechanical action of the engine parts, abrasion and burnishing of the solid lubricant produces smaller and smaller platelets, increasing their surface area and consequently their tendency to stick to the metal surfaces through their reactive prismatic edges. Furthermore, the exposed prismatic edges are reactive sites, which facilitate chemical oxidation of the platelets. These phenomena adversely affect the tribological benefits of the solid lubricant. In contrast, the spherical IF-$MS_2$ nanoparticles are expected to behave like nano-ball bearings and upon mechanical stress they would slowly exfoliate or mechanically deform to a rugby-shape ball, but would not lose their tribological benefits, until they are completely gone, or oxidized. To test this hypothesis, various solid-fluid mixtures were prepared and tested under standard conditions [86]. The beneficial effect of IF powder as a solid lubricant additive has been thus confirmed through a long series of experiments [87].

The mechanism of the action of the IF nanoparticles as additives in lubrication fluids is more complicated than was initially thought. First, it is clear that the more spherical the nanoparticles and the fewest structural defects they include, the better is their performance as solid lubricant additives [88]. Three main mechanisms responsible for the onset of failure of the nanoparticles in tribological tests have been clearly identified. They include: exfoliation of the nanoparticles; deformation into a rugby ball shape, and explosion. The partially damaged nanoparticles are left with reactive edges, which can undergo further oxidation and can lead to a complete loss of their tribological action. Recent nanotribological experiments, using the surface force apparatus with the lubricant between two perpendicular mica surfaces, revealed that material transfer from the IF nanoparticles onto the mica surface is a major factor in reducing the friction between the two mica surfaces [87]. This experiment and many others, carried out over the last few years, suggest an important application for these nanoparticles both as an additive in lubrication fluids or greases, as well as in composites with metals, plastics, rubber, and ceramics.

Another important field where inorganic nanotubes can be useful is as tips in scanning probe microscopy [13]. Here applications in the inspection of microelectronics circuitry have been demonstrated and potential applications in nanolithography are being contemplated. A comparison between a $WS_2$ nanotube tip and a microfabricated Si tip indicates, that while the microfabricated conically-shaped Si tip is unable to probe the bottom of deep and narrow grooves, the slender and inert nanotube can go down and image the bottom of the groove faithfully [13]. This particular tip has been tested for a few months with no signs of deterioration[1], which is indicative of its resilience and passive surface. Although other kinds of tips have been in use in recent years for high resolution imaging using scanning probe microscopy, the present tips are rather stiff and inert, and consequently they are likely to serve in high resolution imaging of rough surfaces having features with large aspect ratios. Furthermore, inorganic nanotubes exhibit strong absorption of light in the visible part of the spectrum and their electrical conductivity can be varied over many orders of magnitude by doping and intercalation. This suggests numerous applications, in areas such as nanolithography, photocatalysis and others.

The shape of the $B_xC_yN_z$ IF nanoparticles (onion or cocoon like) again suggests important tribological applications (such as lubricants), and the inherently strong covalent bond of some of the IF nanotubes, such as those composed of $B_xC_yN_z$, suggest high-strength, high stiffness fiber applications.

## 6 Conclusions

Inorganic fullerene-like structures and inorganic nanotubes, in particular, are shown to be a generic structure of nanoparticles from inorganic layered (2D) compounds. Various synthetic approaches to obtain these structures are presented. In some cases, like IF-$WS_2$, IF-$MoS_2$, $V_2O_5$ and BN nanotubes, bulk synthetic methods are already available; however, size and shape control is still at its infancy. Study of these novel structures has led to the observation of a few interesting properties and some potential applications in tribology, high energy density batteries, and nanoelectronics.

[1] Recent test results at the Weizmann Institute

**Acknowledgments**

The work at the Weizmann Institute was supported by the following grants: "Krupp von Bohlen and Halbach" Stiftung (Germany), Israel Academy of Sciences ("Bikkura"), Israel Science Foundation, Israeli Ministry of Science ("Tashtiot").

## References

1. H. W. Kroto, J. R. Heath, S. C. O'Brien, R. F. Curl, R. E. Smalley, Nature (London) **318**, 162–163 (1985)
2. S. Iijima, Nature (London) **354**, 56–58 (1991)
3. R. Tenne, L. Margulis, M. Genut, G. Hodes, Nature (London) **360**, 444–445 (1992)
4. Y. Feldman, Wasserman E., D. J. Srolovitz, R. Tenne, Science **267**, 222–225 (1995)
5. B. C. Satishkumar, A. Govindaraj, E. M. Vogel, L. Basumallick, C. N. R. Rao, J. Mater. Res. **12**, 604–606 (1997)
6. W. Tremel, Angew. Chem. Intl. Ed. **38**, 2175–2179 (1999)
7. M. E. Spahr, P. Bitterli, R. Nesper, F. Kruumeick, H. U. Nissen, et al., Angew. Chem. Int. Ed. **37**, 1263–1265 (1998)
8. A. Rothschild, G. L. Frey, M. Homyonfer, M. Rappaport, R. Tenne, Mater. Res. Innov. **3**, 145–149 (1999)
9. A. Rothschild, J. Sloan, R. Tenne, J. Amer. Chem. Soc. **122**, 5169-5179 (2000)
10. J. Cumings, A. Zettl, Chem. Phys. Lett. **316**, 211 (2000)
11. A. Rubio, J. L. Corkill, M. L. Cohen, Phys. Rev. B **49**, 5081–5084 (1994)
12. N. G. Chopra, A. Zettl, Solid State Commun. **105**, 490-496 (1998)
13. A. Rothschild, S. R. Cohen, R. Tenne, Appl. Phys. Lett. **75**, 4025–4027 (1999)
14. L. Pauling, Proc. Nat. Acad. Sci. **16**, 578–582 (1930)
15. W. M. Gelbart, A. Ben-Shaul, D. Roux (Eds.), *Micelles, Membranes, Microemulsions and Monolayers* (Springer, New York 1994)
16. W. Shenton, T. Douglas, M. Young, G. Stubbs, S. Mann, Adv. Mater. **11**, 253–256 (1999)
17. X. Blase, A. Rubio, S. G. Louie, M. L. Cohen, Europhys. Lett. **28**, 335 (1994)
18. M. Remskar, Z. Skraba, F. Cléton, R. Sanjinés, F. Lévy, Appl. Phys. Lett. **69**, 351–353 (1996)
19. M. Remskar, Z. Skraba, M. Regula, C. Ballif, R. Sanjinés, F. Lévy, Adv. Mater. **10**, 246–249 (1998)
20. M. Remskar, Z. Skraba, F. Cléton, R. Sanjinés, F. Lévy, Surf. Rev. Lett. **5**, 423–426 (1998)
21. M. Remskar, Z. Skraba, C. Ballif, R. Sanjinés, F. Lévy, Surf. Sci. **435**, 637–641 (1999)
22. M. Remskar, Z. Skraba, R. Sanjinés, F. Lévy, Appl. Phys. Lett. **74**, 3633–3635 (1999)
23. C. M. Zelenski, P. K. Dorhout, J. Am. Chem. Soc. **120**, 734–742 (1998)
24. Z. Weng-Sieh, K. Cherrey, N. G. Chopra, X. Blase, Y. Miyamoto, A. Rubio, M. L. Cohen, S. G. Louie, A. Zettl, R. Gronsky, Phys. Rev. B **51**, 11229 (1995)

25. N. G. Chopra, J. Luyken, K. Cherry, V. H. Crespi, M. L. Cohen, S. G. Louie, A. Zettl, Science **269**, 966 (1995)
26. C. R. Martin, Acc. Chem. Res. **28**, 61–68 (1995)
27. J. Sloan, J. L. Hutchison, R. Tenne, Y. Feldman, M. Homyonfer, T. Tsirlina, J. Solid State Chem. **144**, 100–117 (1999)
28. E. Iguchi, J. Solid State Chem. **23**, 231–239 (1978)
29. L. Margulis, G. Salitra, R. Tenne, M. Talianker, Nature **365**, 113–114 (1993)
30. H. Nakamura, Y. Matsui, J. Am. Chem. Soc. **117**, 2651–2652 (1995)
31. M. E. Spahr, P. Stoschitzki-Bitterli, R. Nesper, O. Haas, P. Novak, J. Electrochem. Soc. **146**, 2780–2783 (1994)
32. F. Krumeich, H.-J. Muhr, M. Niederberger, F. Bieri, B. Schnyder, R. Nesper, J. Am. Chem. Soc. **121**, 8324–8331 (1999)
33. P. M. Ajayan, O. Stephan, P. Redlich, C. Colliex, Nature **375**, 564–567 (1995)
34. A. Leist, S. Stauf, S. Löken, E. W. Finckh, S. Lüdtke, K. K. Unger, W. Assenmacher, W. Mader, W. Tremel, J. Mater. Chem. **8**, 241–244 (1998)
35. M. Homyonfer, Y. Mastai, M. Hershfinkel, V. Volterra, J. L. Hutchison, R. Tenne, J. Am. Chem. Soc. **118**, 7804–7808 (1996)
36. Y. Mastai, M. Homyonfer, A. Gedanken, G. Hodes, Adv. Mater. **11**, 1010–1013 (1999)
37. K. S. Suslick, S.-B. Choe, A. A. Cichovlas, M. W. Grinstaff, Nature **353**, 414–416 (1991)
38. T. J. Mason, J. P. Lorimer, D. J. Walton, Ultrasonics **28**, 333–337 (1990)
39. A. Durant, J. L. Deplancke, R. Winand, and J. Reisse, Tetrahedron Lett. **36**, 4257–4260 (1995)
40. R. Tenne, M. Homyonfer, Y. Feldman, Chem. Mater. **10**, 3225–3238 (1998)
41. S. Avivi, Y. Mastai, G. Hodes, A. Gedanken, J. Am. Chem. Soc. **121**, 4196–4199 (1999)
42. K. S. Suslick, D. A. Hammerton, R. E. Cline, Jr, J. Am. Chem. Soc. **108**, 5641–5642 (1986)
43. S. Avivi, Y. Mastai, A. Gedanken, J. Am. Chem. Soc. **122**, 4331-4334 (2000)
44. Y. Rosenfeld-Hacohen, E. Grunbaum, R. Tenne, J. Sloan, J. L. Hutchison, Nature (London) **395**, 336 (1998)
45. W. Stöber, A. Fink, E. Bohn, J. Colloid Interface Sci. **26**, 62–69 (1968)
46. H. Nakamura, Y. Matsui, J. Amer. Chem. Soc. **117**, 2651–2652 (1995)
47. T. Kasuga, M. Hiramatsu, A. Hoson, T. Sekino, K. Niihara, Langmuir **14**, 3160–3163 (1998)
48. C. N. R. Rao, B. C. Satishkumar, A. Govindaraj, Chem. Commun. 1581–1582 (1997)
49. M. R. Ghadiri, J. R. Granja, R. A. Milligan, D. E. McRee, N. Khazanovich, Nature (London) **366**, 324–327 (1993)
50. A. Loiseau, F. Willaime, N. Demoncy, G. Hug, H. Pascard, Phys. Rev. Lett. **76**, 4737 (1996)
51. M. Terrones, W. K. Hsu, H. Terrones, J. P. Zhang, S. Ramos, J. P. Hare, K. Prassides, A. K. Cheetham, H. W. Kroto, D. R. M. Walton, Chem. Phys. Lett. **259**, 568–573 (1996)
52. F. Banhart et al., Chem. Phys. Lett. **269**, 349 (1996)
53. K. Suenaga, C. Colliex, N. Demoncy, A. Loiseau, H. Pascard, F. Willaime, Science **278**, 653 (1997)
54. Y. Zhang, K. Suenaga, S. Iijima, Science **281**, 973 (1998)

55. A. Thess, R. Lee, P. Nikolaev, H. Dai, P. Petit, J. Robert, C. Xu, Y. H. Lee, S. G. Kim, A. G. Rinzler, D. T. Colbert, G. E. Scuseria, D. Tománek, J. E. Fischer, R. E. Smalley, Science **273**, 483–487 (1996)
56. D. P. Yu et al., Appl. Phys. Lett. **72**, 1966 (1998)
57. J. C. Charlier, S. Iijima, chapter 4 in this volume
58. Y. K. Kwon, Y. H. Lee, S. G. Kim, P. Jund, D. Tománek, R. E. Smalley, Phys. Rev. Lett. **79**, 2065 (1997)
59. O. Stephan, P. M. Ajayan, C. Colliex, Ph. Redlich, J. M. Lambert, P. Bernier, P. Lefin, Science **266**, 1683 (1994)
60. J. Kouvetakis, T. Sasaki, C. Chen, R. Hagiwara, M. Lerner, K. M. Krishnan, N. Bartlett, Synthetic Mater. **34**, 1 (1989)
61. M. Homyonfer, B. Alperson, Yu. Rosenberg, L. Sapir, S. R. Cohen, G. Hodes, R. Tenne, J. Am. Chem. Soc. **119**, 2693–2698 (1997)
62. P. A. Lee (Ed.), *Physics and Chemistry of Materials with Low Dimensional Structures*, (Reidel, Dordrecht, New York 1976) p. 447
63. J. A. Wilson, A. D. Yoffe, Adv. Phys. **18**, 193–335 (1969)
64. J. Moser, F. Lévy, F. Busy, J. Vac. Sci. Technol. A **12**, 494–500 (1994)
65. L. Margulis, P. Dluzewski, Y. Feldman, R. Tenne, J. Microscopy **181**, 68–71 (1996)
66. R. Tenne, Adv. Mater. **7**, 965–995 (1995)
67. P. A. Parilla, A. C. Dillon, K. M. Jones, G. Riker, D. L. Schulz, D. S. Ginley, M. J. Heben, Nature (London) **397**, 114 (1999)
68. Y. Miyamoto, A. Rubio, S. G. Louie, M. L. Cohen, Phys. Rev. B **50**, 18360 (1994)
69. Y. Miyamoto, A. Rubio, M. L. Cohen, S. G. Louie, Phys. Rev. B **50**, 4976–4979 (1994)
70. Y. Miyamoto, A. Rubio, S. G. Louie, M. L. Cohen, Phys. Rev. B **50**, 18360–18366 (1994)
71. F. Jensen, H. Toftlund, Chem. Phys. Lett. **201**, 95–98 (1993)
72. O. Stéphan, Y. Bando, A. Loiseau, F. Willaime, N. Shramcherko, T. Tamiya, T. Sato, Appl. Phys. A **67**, 107–111 (1998)
73. S. Louie, chapter 6 in this volume
74. M. Cote, M. L. Cohen, D. J. Chadi, Phys. Rev. B **58**, R4277–R4280 (1998)
75. G. Seifert, H. Terrones, M. Terrones, G. Jungnickel, T. Frauenheim, Solid State Commun. **114**, 245–248 (2000)
76. E. Bucher, in *Physics and Chemistry of Materials with Layered Structures*,Vol. 14, A. Aruchamy (Ed.) (Kluwer Academic, New York 1992) pp. 1–8 1992
77. G. L. Frey, S. Elani, M. Homyonfer, Y. Feldman, R. Tenne, Phys. Rev. B **57**, 6660–6671 (1998)
78. G. L. Frey, R. Tenne, M. J. Matthews, M. S. Dresselhaus, G. Dresselhaus, J. Mater. Res. **13**, 2412–2417 (1998)
79. F. Consadori, R. F. Frindt, Phys. Rev. B **2**, 4893–4896 (1970)
80. M. W. Peterson, A. J. Nozik, in *Physics and Chemistry of Materials with Low Dimensional Structures*, Vol. 14, A. Aruchamy (Ed.) (Kluwer Academic, New York 1992) pp. 297–317
81. G. L. Frey, R. Tenne, M. J. Matthews, M. S. Dresselhaus, G. Dresselhaus, Phys. Rev. B **60**, 2883–2893 (1999)
82. D. J. Srolovitz, S. A. Safran, M. Homyonfer, R. Tenne, Phys. Rev. Lett. **74**, 1779–1782 (1995)

83. Y. Feldman, G. L. Frey, M. Homyonfer, V. Lyakhovitskaya, L. Margulis, H. Cohen, J. L. Hutchison, R. Tenne, J. Am, Chem. Soc. **118**, 5362–5367 (1996)
84. I. L. Singer, in *Fundamentals of Friction: Macroscopic and Microscopic Processes*, I. L. Singer, H. M. Pollock (Eds.) (Kluwer, Dordrecht 1992) p. 237
85. F. P. Bowden, D. Tabor, in *Friction: An Introduction to Tribology* (Anchor, New York 1973)
86. L. Rapoport, Yu. Bilik, Y. Feldman, M. Homyonfer, S. R. Cohen, R. Tenne, Nature (London) **387**, 791–793 (1997)
87. Y. Golan, C. Drummond, M. Homyonfer, Y. Feldman, R. Tenne, J. Israelachvily, Adv. Mater. **11**, 934–937 (1999)
88. L. Rapoport, Y. Feldman, M. Homyonfer, H. Cohen, J. Sloan, J. L. Hutchinson, R. Tenne, Wear **225-229**, 975–982 (1999)

# Electronic Properties, Junctions, and Defects of Carbon Nanotubes

Steven G. Louie[1,2]

[1] Department of Physics, University of California Berkeley
Berkeley, CA 94720, USA
[2] Materials Sciences Division, Lawrence Berkeley National Laboratory
Berkeley, CA 94720, USA
louie@jungle.berkeley.edu

**Abstract.** The nanometer dimensions of the carbon nanotubes together with the unique electronic structure of a graphene sheet make the electronic properties of these one-dimensional structures highly unusual. This chapter reviews some theoretical work on the relation between the atomic structure and the electronic and transport properties of single-walled carbon nanotubes. In addition to the ideal tubes, results on the quantum conductance of nanotube junctions and tubes with defects will be discussed. On-tube metal-semiconductor, semiconductor-semiconductor, and metal-metal junctions have been studied. Other defects such as substitutional impurities and pentagon-heptagon defect pairs on tube walls are shown to produce interesting effects on the conductance. The effects of static external perturbations on the transport properties of metallic nanotubes and doped semiconducting nanotubes are examined, with the metallic tubes being much less affected by long-range disorder. The structure and properties of crossed nanotube junctions and ropes of nanotubes have also been studied. The rich interplay between the structural and the electronic properties of carbon nanotubes gives rise to new phenomena and the possibility of nanoscale device applications.

Carbon nanotubes are tubular structures that are typically several nanometers in diameter and many microns in length. This fascinating new class of materials was first discovered by *Iijima* [1] in the soot produced in the arc-discharge synthesis of fullerenes. Because of their nanometer dimensions, there are many interesting and often unexpected properties associated with these structures, and hence there is the possibility of using them to study new phenomena and employing them in applications [2,3,4]. In addition to the multi-walled tubes, single-walled nanotubes [5,6,7], and ropes of close-packed single-walled tubes have been synthesized [8]. Also, carbon nanotubes may be filled with foreign materials [9,10] or collapsed into flat, flexible nanoribbons [11]. Carbon nanotubes are highly unusual electrical conductors, the strongest known fibers, and excellent thermal conductors. Many potentially important applications have been explored, including the use of nanotubes as nanoprobe tips [12], field emitters [13,14], storage or filtering media [15], and nanoscale electronic devices [16,17,18,19,20,21,22,23,24]. Further, it has been found that nanotubes may also be formed with other layered materials [25,26,27,28,29,30,31,32,33,34,35,36]. In particular, BN, $BC_3$, and other

M. S. Dresselhaus, G. Dresselhaus, Ph. Avouris (Eds.): Carbon Nanotubes,
Topics Appl. Phys. **80**, 113–145 (2001)

$B_xC_yN_z$ nanotubes have been theoretically predicted [25,26,27,28,29] and experimentally synthesized [30,31,32,33,34,35].

In this contribution, we focus on a review of some selected theoretical studies on the electronic and transport properties of carbon nanotube structures, in particular, those of junctions, impurities, and other defects. Structures such as ropes of nanotubes and crossed nanotubes are also discussed.

The organization of the Chapter is as follows. Section 1 contains an introduction to the geometric and electronic structure of ideal single-walled carbon nanotubes. Section 2 gives a discussion of the electronic and transport properties of various on-tube structures. Topics presented include on-tube junctions, impurities, and local defects. On-tube metal-semiconductor, semiconductor-semiconductor, and metal-metal junctions may be formed by introducing topological structural defects. These junctions have been shown to behave like nanoscale device elements. Other defects such as substitutional impurities and Stone-Wales defects on tube walls also are shown to produce interesting effects on the conductance. Crossed nanotubes provide another means to obtain junction behavior. The crossed-tube junctions, nanotube ropes, and effects of long-range disorder are the subjects of Section 3. Intertube interactions strongly modify the electronic properties of a rope. The effects of long-range disorder on metallic nanotubes are quite different from those on doped semiconducting tubes. Finally, a summary and some conclusions are given in Section 4.

# 1 Geometric and Electronic Structure of Carbon Nanotubes

In this section, we give an introduction to the structure and electronic properties of the single-walled carbon nanotubes (SWNTs). Shortly after the discovery of the carbon nanotubes in the soot of fullerene synthesis, single-walled carbon nanotubes were synthesized in abundance using arc discharge methods with transition metal catalysts [5,6,7]. These tubes have quite small and uniform diameter, on the order of one nanometer. Crystalline ropes of single-walled nanotubes with each rope containing tens to hundreds of tubes of similar diameter closely packed have also been synthesized using a laser vaporization method [8] and other techniques, such as arc-discharge and CVD techniques. These developments have provided ample amounts of sufficiently characterized samples for the study of the fundamental properties of the SWNTs. As illustrated in Fig. 1, a single-walled carbon nanotube is geometrically just a rolled up graphene strip. Its structure can be specified or indexed by its circumferential periodicity [37]. In this way, a SWNT's geometry is completely specified by a pair of integers $(n, m)$ denoting the relative position $\boldsymbol{c} = n\boldsymbol{a}_1 + m\boldsymbol{a}_2$ of the pair of atoms on a graphene strip which, when rolled onto each other, form a tube.

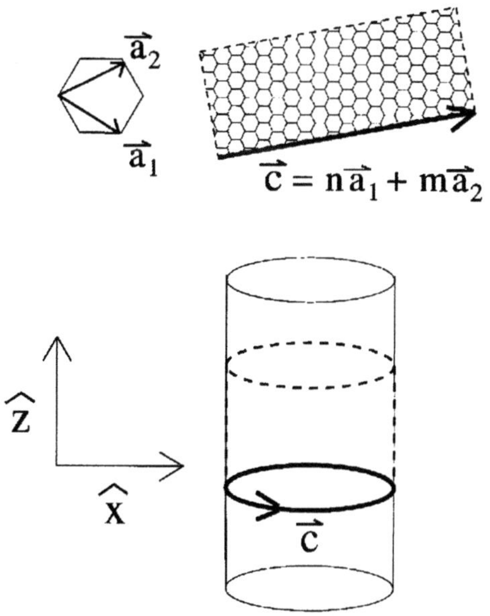


**Fig. 1.** Geometric structure of an $(n, m)$ single-walled carbon nanotube

Theoretical calculations [16,39,40,41] have shown early on that the electronic properties of the carbon nanotubes are very sensitive to their geometric structure. Although graphene is a zero-gap semiconductor, theory has predicted that the carbon nanotubes can be metals or semiconductors with different size energy gaps, depending very sensitively on the diameter and helicity of the tubes, i.e., on the indices $(n, m)$. As seen below, the intimate connection between the electronic and geometric structure of the carbon nanotubes gives rise to many of the fascinating properties of various nanotube structures, in particular nanotube junctions.

The physics behind this sensitivity of the electronic properties of carbon nanotubes to their structure can be understood within a band-folding picture. It is due to the unique band structure of a graphene sheet, which has states crossing the Fermi level at only 2 inequivalent points in $k$-space, and to the quantization of the electron wavevector along the circumferential direction. An isolated sheet of graphite is a zero-gap semiconductor whose electronic structure near the Fermi energy is given by an occupied $\pi$ band and an empty $\pi^*$ band. These two bands have linear dispersion and, as shown in Fig. 2, meet at the Fermi level at the $K$ point in the Brillouin zone. The Fermi surface of an ideal graphite sheet consists of the six corner $K$ points. When forming a tube, owing to the periodic boundary conditions imposed in the circumferential direction, only a certain set of $\boldsymbol{k}$ states of the planar graphite sheet is allowed. The allowed set of $k$'s, indicated by the lines in Fig. 2, depends on the diameter and helicity of the tube. Whenever the allowed $k$'s include the point $K$, the system is a metal with a nonzero density of states at the Fermi level, resulting in a one-dimensional metal with 2 linear dispersing

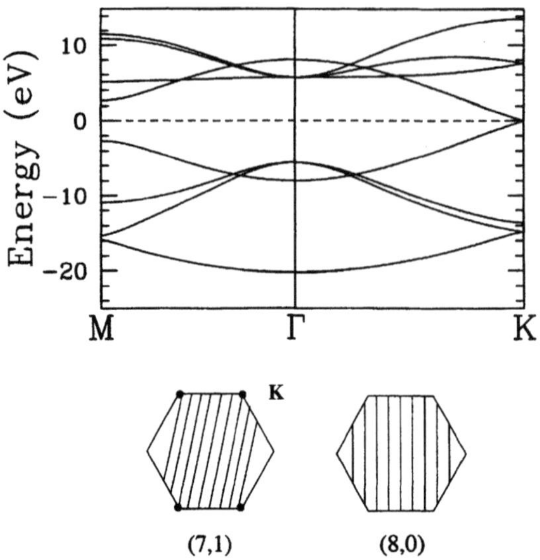


**Fig. 2.** (*Top*) Tight-binding band structure of graphene (a single basal plane of graphite). (*Bottom*) Allowed $\boldsymbol{k}$-vectors of the (7,1) and (8,0) tubes (*solid lines*) mapped onto the graphite Brillouin zone

bands. When the point $K$ is not included, the system is a semiconductor with different size energy gaps. It is important to note that the states near the Fermi energy in both the metallic and the semiconducting tubes are all from states near the $K$ point, and hence their transport and other properties are related to the properties of the states on the allowed lines. For example, the conduction band and valence bands of a semiconducting tube come from states along the line closest to the $K$ point.

The general rules for the metallicity of the single-walled carbon nanotubes are as follows: $(n, n)$ tubes are metals; $(n, m)$ tubes with $n-m = 3j$, where $j$ is a nonzero integer, are very tiny-gap semiconductors; and all others are large-gap semiconductors. Strictly within the band-folding scheme, the $n-m = 3j$ tubes would all be metals, but because of tube curvature effects, a tiny gap opens for the case where $j$ is nonzero. Hence, carbon nanotubes come in three varieties: large-gap, tiny-gap, and zero-gap. The $(n, n)$ tubes, also known as armchair tubes, are always metallic within the single-electron picture, being independent of curvature because of their symmetry. As the tube radius $R$ increases, the band gaps of the large-gap and tiny-gap varieties decreases with a $1/R$ and $1/R^2$ dependence, respectively. Thus, for most experimentally observed carbon nanotube sizes, the gap in the tiny-gap variety which arises from curvature effects would be so small that, for most practical purposes, all the $n-m = 3j$ tubes can be considered as metallic at room temperature. Thus, in Fig. 2, a (7,1) tube would be metallic, whereas a (8,0) tube would be semiconducting.

This band-folding picture, which was first verified by tight-binding calculations [38,39,40], is expected to be valid for larger diameter tubes. However, for a small radius tube, because of its curvature, strong rehybridization among the $\sigma$ and $\pi$ states can modify the electronic structure. Experimentally, nanotubes with a radius as small as 3.5 Å have been produced. *Ab initio* pseudopotential local density functional (LDA) calculations [41] indeed revealed that sufficiently strong hybridization effects can occur in small radius nanotubes which significantly alter their electronic structure. Strongly modified low-lying conduction band states are introduced into the band gap of insulating tubes because of hybridization of the $\sigma^*$ and $\pi^*$ states. As a result, the energy gaps of some small radius tubes are decreased by more than 50%. For example, the (6,0) tube which is predicted to be semiconducting in the band-folding scheme is shown to be metallic. For nanotubes with diameters greater than 1 nm, these rehybridization effects are unimportant. Strong $\sigma$–$\pi$ rehybridization can also be induced by bending a nanotube [42].

Energetically, *ab initio* total energy calculations have shown that carbon nanotubes are stable down to very small diameters. Figure 3 depicts the calculated strain energy per atom for different carbon nanotubes of various diameters [41]. The strain energy scales nearly perfectly as $d^{-2}$ where $d$ is the tube diameter (solid curve in Fig. 3), as would be the case for rolling a classical elastic sheet. Thus, for the structural energy of the carbon nanotubes, the elasticity picture holds down to a subnanometer scale. The elastic constant may be determined from the total energy calculations. This result has been used to analyze collapsed tubes [11] and other structural properties of nanotubes. Also shown in Fig. 3 is the energy/atom for a (6,0) carbon strip. It has an energy which is well above that of a (6,0) tube because of the dangling bonds on the strip edges. Because in general the energy per atom of

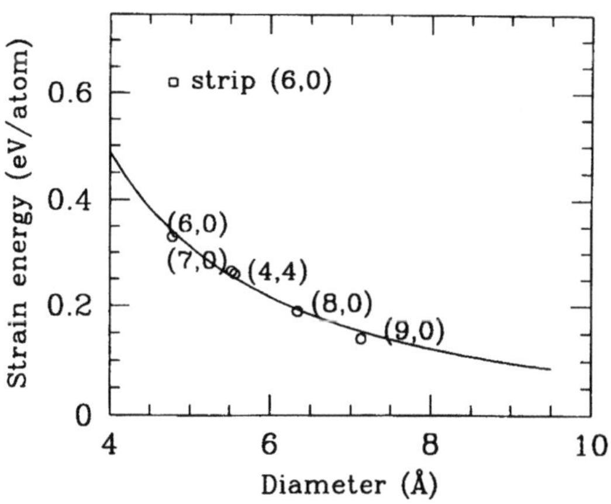


**Fig. 3.** Strain energy/atom for carbon nanotubes from *ab initio* total energy calculations [44]

a strip scales as $d^{-1}$, the calculation predicts that carbon nanotubes will be stable with respect to the formation of strips down to below 4 Å in diameter, in agreement with classical force-field calculations [43].

There have been many experimental studies on carbon nanotubes in an attempt to understand their electronic properties. The transport experiments [19,20,45,46,47] involved both two- and four-probe measurements on a number of different tubes, including multiwalled tubes, bundles of single-walled tubes, and individual single-walled tubes. Measurements showed that there are a variety of resistivity behaviors for the different tubes, consistent with the above theoretical picture of having both semiconducting and metallic tubes. In particular, at low temperature, individual metallic tubes or small ropes of metallic tubes act like quantum wires [19,20]. That is, the conduction appears to occur through well-separated discrete electron states that are quantum-mechanically coherent over distances exceeding many hundreds of nanometers. At sufficiently low temperature, the system behaves like an elongated quantum dot.

Figure 4 depicts the experimental set up for such a low temperature transport measurement on a single-walled nanotube rope from [20]. At a few degrees Kelvin, the low-bias conductance of the system is suppressed for voltages less than a few millivolts, and there are dramatic peaks in the conductance as a function of gate voltage that modulates the number of electrons in the rope (Fig. 5). These results have been interpreted in terms of single-electron charging and resonant tunneling through the quantized energy levels of the nanotubes. The data are explained quite well using the band structure of the conducting electrons of a metallic tube, but these electrons are confined to a small region defined either by the contacts or by the sample length, thus leading to the observed quantum confinement effects of Coulomb blockade and resonant tunneling.

There have also been high resolution low temperature Scanning Tunneling Microscopy (STM) studies, which directly probe the relationship between the structural and electronic properties of the carbon nanotubes [48,49]. Figure 6 is a STM image for a single carbon nanotube at 77 K on the surface of a rope. In these measurements, the resolution of the measurements allowed for the identification of the individual carbon rings. From the orientation of the carbon rings and the diameter of the tube, the geometric structure of the tube depicted in Fig. 6 was deduced to be that of a (11,2) tube. Measurement of the normalized conductance in the Scanning Tunneling Spectroscopy (STS) mode was then used to obtain the Local Density Of States (LDOS). Data on the (11,2) and the (12,3) nanotubes gave a constant density of states at the Fermi level, showing that they are metals as predicted by theory. On another sample, a $(14,-3)$ tube was studied. Since $14+3$ is not equal to 3 times an integer, it ought to be a semiconductor. Indeed, the STS measurement gives a band gap of 0.75 eV, in very good agreement with calculations.

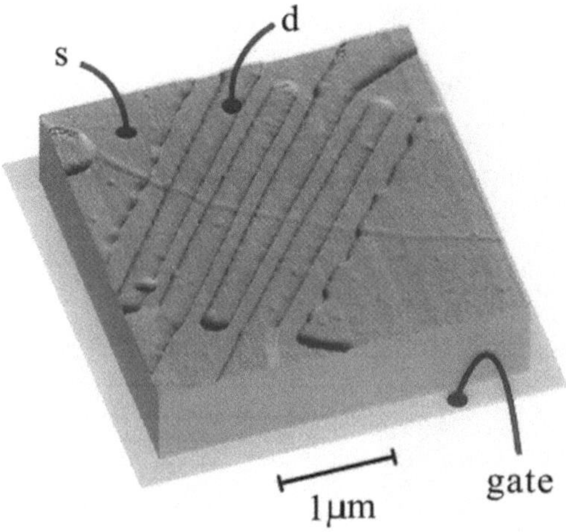


**Fig. 4.** Experimental set-up for the electrical measurement of a single-walled nanotube rope, visible as the diagonal curved line [20]

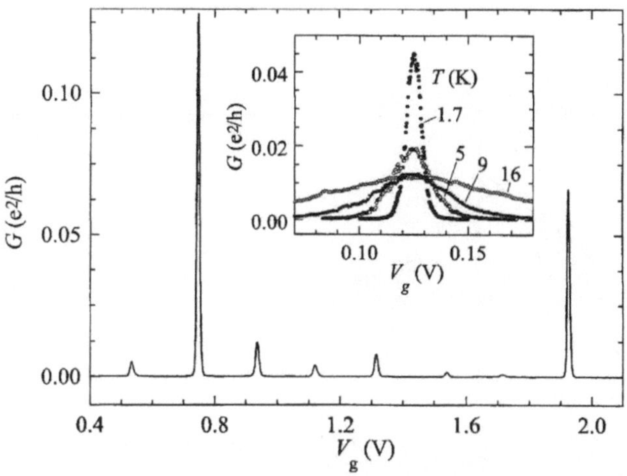


**Fig. 5.** Measured conductance of a single-walled carbon nanotube rope as a function of gate voltage [20]

The electronic states of the carbon nanotubes, being band-folded states of graphene, lead to other interesting consequences, including a striking geometry dependence of the electric polarizability. Figure 7 presents some results from a tight-binding calculation for the static polarizabilities of carbon nanotubes in a uniform applied electric field [50]. Results for 17 single-walled tubes of varying size and chirality, and hence varying band gaps, are given. The unscreened polarizability $\alpha_0$ is calculated within the random phase approximation. The cylindrical symmetry of the tubes allows the polarizability

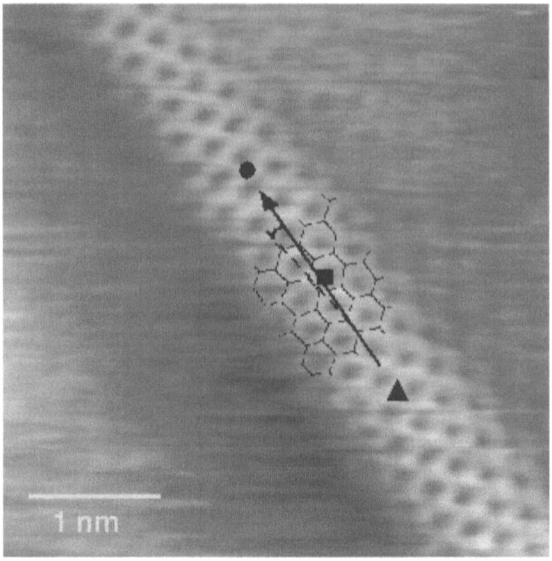


**Fig. 6.** STM images at 77 K of a single-walled carbon nanotube at the surface of a rope [49]

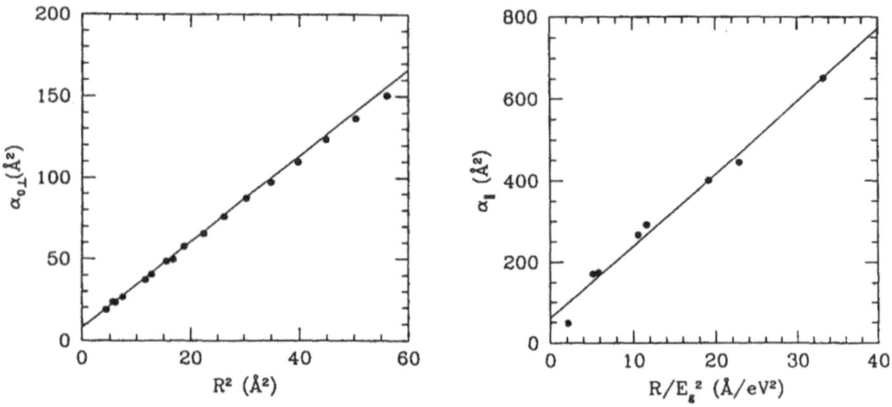


**Fig. 7.** Calculated static polarizability of single wall carbon nanotube, showing results both for $\alpha_{0\perp}$ vs $R^2$ on the left and $\alpha_{\parallel}$ vs $R/E_{\rm g}^2$ on the right [50]

tensor to be divided into components perpendicular to the tube axis, $\alpha_{0\perp}$, and a component parallel to the tube axis, $\alpha_{0\parallel}$. Values for $\alpha_{0\perp}$ predicted within this model are found to be totally independent of the band gap $E_{\rm g}$ and to scale linearly as $R^2$, where $R$ is the tube radius. The latter dependence may be understood from classical arguments, but the former is rather unexpected. The insensitivity of $\alpha_{0\perp}$ to $E_{\rm g}$ results from selection rules in the dipole matrix elements between the highest occupied and the lowest unoccupied states of these tubes. On the other hand, Fig. 7 shows that $\alpha_{0\parallel}$ is proportional to $R/E_g^2$, which is consistent with the static dielectric response of standard insulators. Also, using arguments analogous to those for $C_{60}$ [51,52], local field effects relevant to the screened polarizability tensor $\alpha$ may be included classically, resulting in a saturation of $\alpha_\perp$ for large $\alpha_{0\perp}$, but leaving $\alpha_\parallel$ unaffected. Thus, in general, the polarizability tensor of a carbon nanotube is expected to be highly anisotropic with $\alpha_\parallel \gg \alpha_\perp$. And the polarizability of small gap tubes is expected to be greatly enhanced among tubes of similar radii.

Just as in electronic states, the phonon states in carbon nanotubes are quantized into phonon subbands. This has led to a number of interesting phenomena [2,3] which are discussed elsewhere in this volume [53]. Here we mention several of them. It has been shown that twisting motions of a tube can lead to the opening up of a minuscule gap at the Fermi level, leading to the possibility of strong coupling between the electronic states and the twisting modes or twistons [54]. The heat capacity of the nanotubes is also expected to show a dimensionality dependence. Analysis [55] shows that the phonon contributions dominate the heat capacity, with single-walled carbon nanotubes having a $C_{\rm ph} \sim T$ dependence at low temperature. The temperature below which this should be observable decreases with increasing nanotube radius $R$, but the linear $T$ dependence should be accessible to experimental investigations with presently available samples. In particular, a tube with a 100 Å radius should have $C_{\rm ph} \sim T$ for $T < 7\,$K. Since bulk graphite has $C_{\rm ph} \sim T^{2-3}$, a sample of sufficiently small radius tubes should show a deviation from graphitic behavior. Multi-walled tubes, on the other hand, are expected to show a range of behavior intermediate between $C_{\rm ph} \sim T$ and $C_{\rm ph} \sim T^{2-3}$, depending in detail on the tube radii and the number of concentric walls.

In addition to their fascinating electronic properties, carbon nanotubes are found to have exceptional mechanical properties [56]. Both theoretical [57,58,59,60,61,62,63] and experimental [64,65] studies have demonstrated that they are the strongest known fibers. Carbon nanotubes are expected to be extremely strong along their axes because of the strength of the carbon-carbon bonds. Indeed, the Young's modulus of carbon nanotubes has been predicted and measured to be more than an order of magnitude higher than that of steel and several times that of common commercial carbon fibers. Similarly, BN nanotubes are shown [66] to be the world's strongest large-gap insulating fiber.

## 2 Electronic and Transport Properties of On-Tube Structures

In this section, we discuss the electronic properties and quantum conductance of nanotube structures that are more complex than infinitely long, perfect nanotubes. Many of these systems exhibit novel properties and some of them are potentially useful as nanoscale devices.

## 2.1 Nanotube Junctions

Since carbon nanotubes are metals or semiconductors depending sensitively on their structures, they can be used to form metal-semiconductor, semiconductor-semiconductor, or metal-metal junctions. These junctions have great potential for applications since they are of nanoscale dimensions and made entirely of a single element. In constructing this kind of on-tube junction, the key is to join two half-tubes of different helicity seamlessly with each other, without too much cost in energy or disruption in structure. It has been shown that the introduction of pentagon-heptagon pair defects into the hexagonal network of a single carbon nanotube can change the helicity of the carbon nanotube and fundamentally alter its electronic structure [16,17,18,67,68,69,70,71]. This led to the prediction that these defective nanotubes behave would as the desired nanoscale metal-semiconductor Schottky barriers, semiconductor heterojunctions, or metal-metal junctions with novel properties, and that they could be the building blocks of nanoscale electronic devices.

In the case of nanotubes, being one-dimensional structures, a local topological defect can change the properties of the tube at an infinitely long distance away from the defect. In particular, the chirality or helicity of a carbon nanotube can be changed by creating topological defects into the hexagonal network. The defects, however, must induce zero net curvature to prevent the tube from flaring or closing. The smallest topological defect with minimal local curvature (hence less energy cost) and zero net curvature is a pentagon-heptagon pair [16,17,18,67,68,69,70,71]. Such a pentagon-heptagon defect pair with its symmetry axis nonparallel to the tube axis changes the chirality of a $(n, m)$ tube by transferring one unit from $n$ to $m$ or vice versa. If the pentagon-heptagon defect pair is along the $(n, m)$ tube axis, then one unit is added or subtracted from $m$. Figure 8 depicts a (8,0) carbon tube joined to a (7,1) tube via a 5-7 defect pair. This system forms a quasi-1D semiconductor/metal junction, since within the band-folding picture the (7,1) half tube is metallic and the (8,0) half tube is semiconducting.

Figures 9 and 10 show the calculated local density of states (LDOS) near the (8,0)/(7,1) junction. These results are from a tight-binding calculation for the $\pi$ electrons [16]. In both figures, the bottom panel depicts the density of states of the perfect tube, with the sharp features corresponding to the van Hove singularities of a quasi-1D system. The other panels show the calculated LDOS at different distances away from the interface, with cell 1 being the closest to the interface in the semiconductor or side and ring 1 the closest to the interface in the metal side. Here, "cell" refers to a one unit cell of the tube and "ring" refers to a ring of atoms around the circumference. These results illustrate the spatial behavior of the density of states as it transforms from

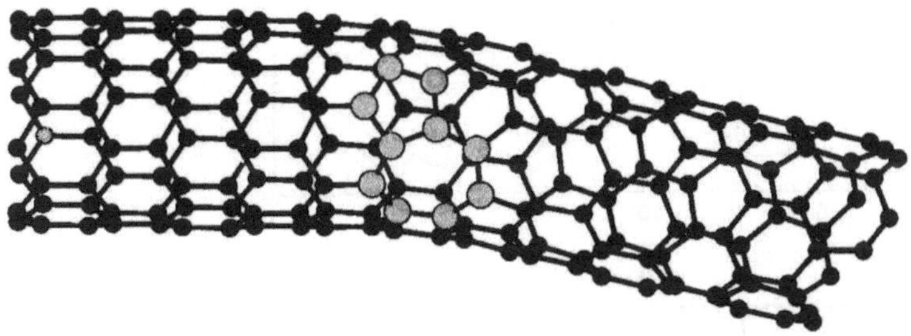

**Fig. 8.** Atomic structure of an (8,0)/(7,1) carbon nanotube junction. The large light-gray balls denote the atoms forming the heptagon-pentagon pair [16]

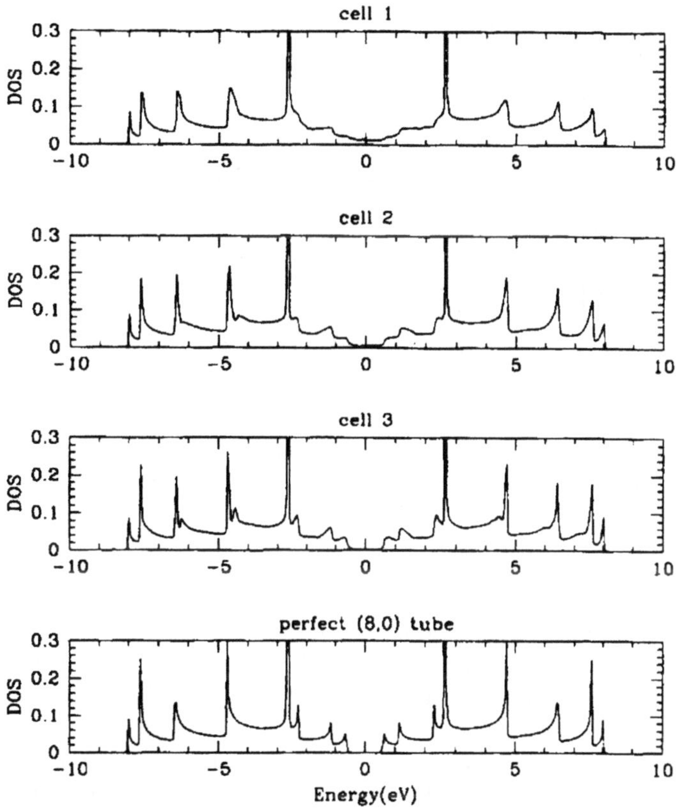


**Fig. 9.** Calculated LDOS of the (8,0)/(7,1) metal-semiconductor junction at the semiconductor side. From top to bottom, LDOS at cells 1, 2, and 3 of the (8,0) side. Cell 1 is at the interface [16]

that of a metal to that of a semiconductor across the junction. The LDOS very quickly changes from that of the metal to that of the semiconductor within a few rings of atoms as one goes from the metal side to the semiconductor side.

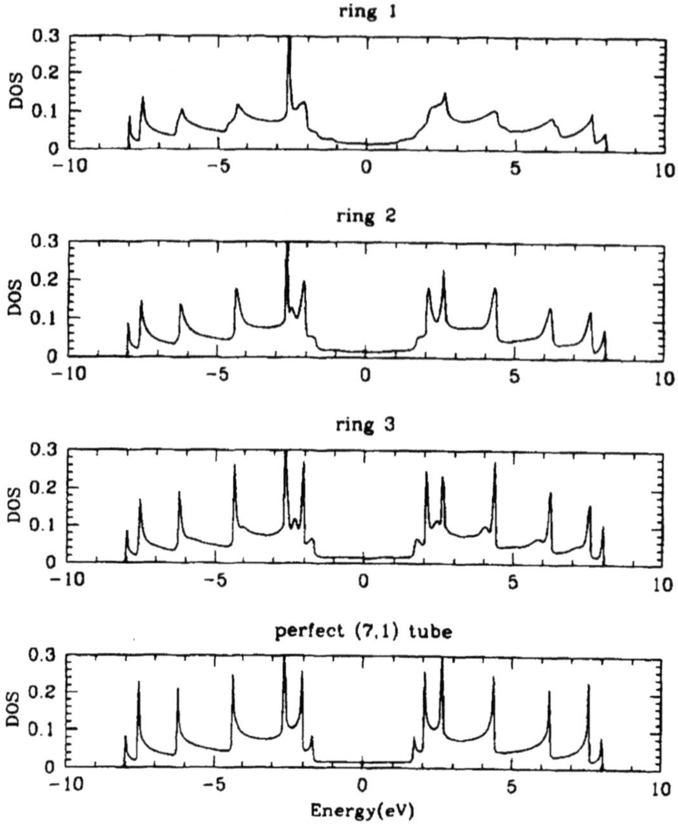


**Fig. 10.** Calculated LDOS of the (8,0)/(7,1) metal-semiconductor junction at the metal side. From top to bottom, the LDOS at rings 1, 2, and 3 of the (7,1) side. Ring 1 is at the interface [16]

As the interface is approached, the sharp van Hove singularities of the metal are diluted. Immediately on the semiconductor side of the interface, a different set of singular features, corresponding to those of the semiconductor tube, emerges. There is, however, still a finite density of states in the otherwise bandgap region on the semiconductor side. These are metal induced gap states [72] which decay to zero in about a few Å into the semiconductor. Thus, the electronic structure of this junction is very similar to that of a bulk metal-semiconductor junction, such as Al/Si, except it has a nanometer cross-section and is made out of entirely the element carbon.

Similarly, semiconductor-semiconductor and metal-metal junctions may be constructed with the proper choices of tube diameters and pentagon-heptagon defect pairs. For example, by inserting a 5-7 pair defect, a (10,0) carbon nanotube can be matched to a (9,1) carbon nanotube [16]. Both of these tubes are semiconductors, but they have different bandgaps. The (10,0)/(9,1) junction thus has the electronic structure of a semiconductor heterojunction. In this case, owing to the rather large structural distortion at the interface, there are interesting localized interface states at the junction. Theoretical studies have also been carried out for junctions of B–C–N nanotubes [73], showing very similar behaviors as the carbon case, and for other geometric arrangements, such as carbon nanotube T-junctions, where one tube joins to the side of another tube perpendicularly to form a "T" structure [74].

Calculations have been carried out to study the quantum conductance of the carbon nanotube junctions. Typically these calculations are done within the Landauer formalism [75,76]. In this approach, the conductance is given in terms of the transmission matrix of the propagating electron waves at a given energy. In particular, the conductance of metal-metal nanotube junctions is shown to exhibit a quite interesting new effect which does not have an analog in bulk metal junctions [67]. It is found that certain configurations of pentagon-heptagon pair defects in forming the junction completely stop the flow of electrons, while other arrangements permit the transmission of current through the junction. Such metal-metal junctions thus have the potential for use as nanoscale electrical switches. This phenomenon is seen in the calculated conductance of a (12,0)/(6,6) carbon nanotube junction in Fig. 11. Both the (12,0) and (6,6) tubes are metallic within the tight-binding model, and they can be matched perfectly to form a straight junction. However, the conductance is zero for electrons at the Fermi level, $E_F$. This peculiar effect is not due to a lack of density of states at $E_F$. As shown in Fig. 11, there is finite density of states at $E_F$ everywhere along the whole length of the total system for this junction. The absence of conductance arises from the fact that there is discrete rotational symmetry along the axis of the combined tube. But, for electrons near $E_F$, the states in one of the half tubes are of a different rotational symmetry from those in the other half tube. As an electron propagates from one side to the other, the electron encounters a symmetry gap and is completely reflected at the junction.

The same phenomenon occurs in the calculated conductance of a (9,0)/(6,3) metal–metal carbon nanotube junction. However, in forming this junction, there are two distinct ways to match the two halves, either symmetrically or asymmetrically. In the symmetric matched geometry, the conductance is zero at $E_F$ for the same symmetry reason as discussed above (Fig. 12). But, in the asymmetric matched geometry, the discrete rotational symmetry of the total system is broken and the electrons no longer have to preserve their rotational quantum number as they travel across the junction. The conductance for this case is now nonzero. Consequently, in some situations, bent junctions can conduct better than straight junctions for the nanotubes. This leads to the possibility of using these metal-metal or other similar junctions as nanoswitches or strain gauges, i.e., one can imagine using some symmetry breaking mechanisms such as electron-photon, electron-phonon or

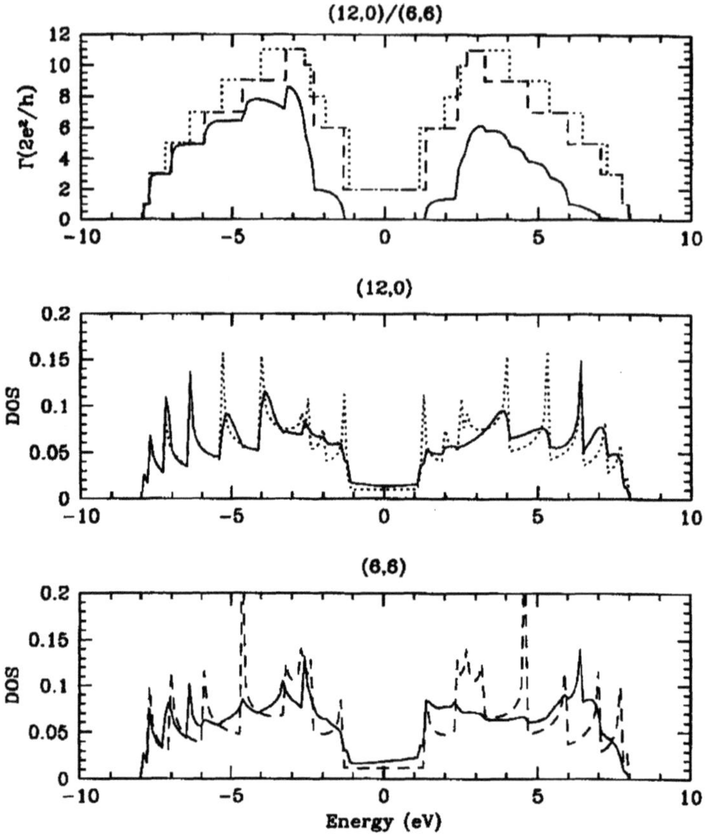


**Fig. 11.** Calculated results for the (12,0)/(6,6) metal–metal junction. *Top:* conductance of a matched tube (*solid line*), a perfect (12,0) tube (*dashed line*), and a perfect (6,6) tube (*dotted line*). *Center:* LDOS at the interface on the (12,0) side (*full line*) and of the perfect (12,0) tube (*dotted line*). *Bottom:* LDOS at the interface on the (6,6) side (*full line*) and of the perfect (6,6) tube (*dashed line*) [67]

mechanical deformation to switch a junction from a non-conducting state to a conducting state [67].

Junctions of the kind discussed above may be formed during growth, but they can also be generated by mechanical stress [77]. There is now considerable experimental evidence of this kind of on-tube junction and device behavior predicted by theory. An experimental signature of a single pentagon-heptagon pair defect would be an abrupt bend between two straight sections of a nanotube. Calculations indicate that a single pentagon-heptagon pair would induce bend angles of roughly 0–15 degrees, with the exact value depending on the particular tubes involved. Several experiments have reported sightings of localized bends of this magnitude for multiwalled carbon nanotubes [23,78,79]. Having several 5-7 defect pairs at a junction would allow

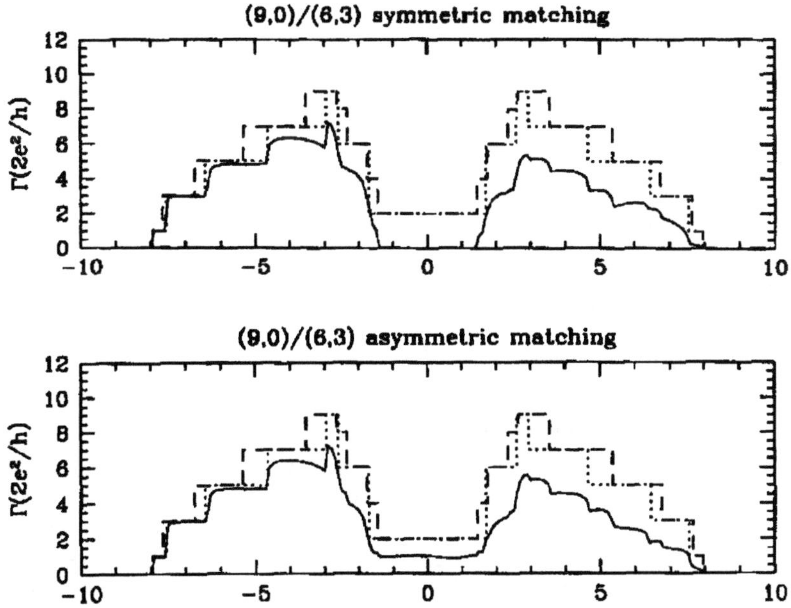


**Fig. 12.** Calculated conductance of the (9,0)/(6,3) junction – matched system (*solid line*), perfect (6,3) tube (*dotted line*), and perfect (9,0) tube (*dashed line*) [67]

the joining of tubes of different diameters and add complexity to the geometry. The first observation of nonlinear junction-like transport behavior was made on a rope of SWNTs [22], where the current-voltage properties were measured along a rope of single-walled carbon nanotubes using a scanning tunneling microscopy tip and the behavior shown in Fig. 13 was found in some samples. At one end of the tube, the system behaves like a semimetal showing a typical I–V curve of metallic tunneling, but after some distance at the other end it becomes a rectifier, presumably because a defect of the above type has been introduced at some point on the tube. A more direct measurement was carried out recently [23]. A kinked single-walled nanotube lying on several electrodes was identified and its electrical properties in the different segments were measured. The kink was indicative of two half tubes of different chiralities joined by a pentagon-heptagon defect pair. Figure 14 shows the measured I–V characteristics of a kinked nanotube. The inset is the I–V curve for the upper segment showing that this part of the tube is a metal; but the I–V curve across the kink shows a rectifying behavior indicative of a metal–semiconductor junction.

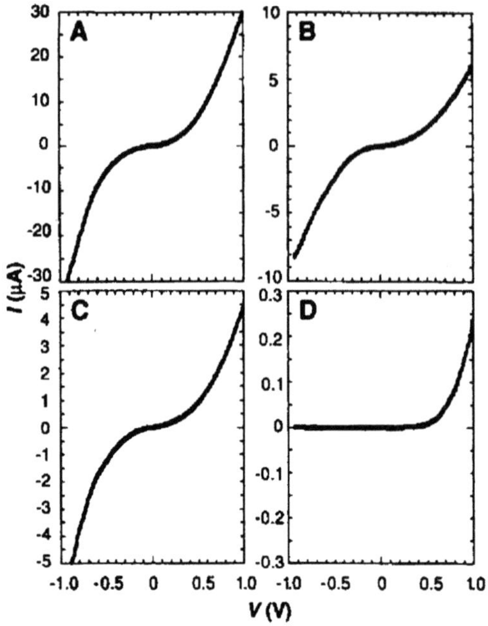


**Fig. 13.** Current–voltage characteristic measured along a rope of single-walled carbon nanotubes. Panels A, B, C, and D correspond to successive different locations on the rope [22]

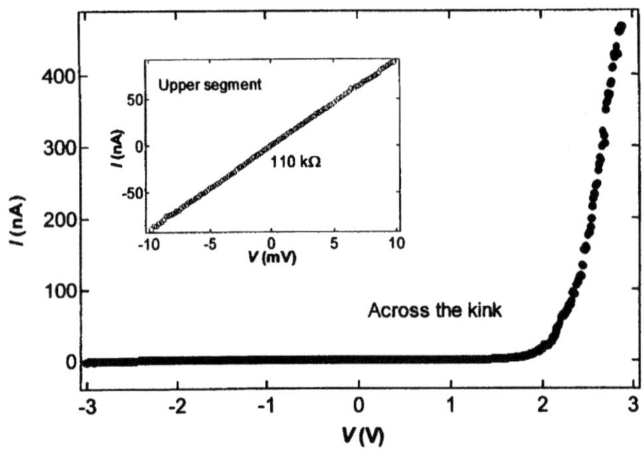


**Fig. 14.** Measured current-voltage characteristic of a kinked single-walled carbon nanotube [23]

## 2.2 Impurities, Stone-Wales Defects, and Structural Deformations in Metallic Nanotubes

An unanswered question in the field has been why do the metallic carbon nanotubes have such long mean free paths. This has led to consideration of the effects of impurities and defects on the conductance of the metallic nanotubes. We focus here on the (10,10) tubes; however, the basic physics is the same for all $(n, n)$ tubes. In addition to tight-binding studies, there are now first-principles calculations on the quantum conductance of nanotube structures based on an *ab initio* pseudopotential density functional method with a wavefunction matching technique [80,81]. The advantages of the *ab initio* approach are that one can obtain the self-consistent electronic and geometric structure in the presence of the defects and, in addition to the conductance, obtain detailed information on the electronic wavefunction and current density distribution near the defect.

Several rather surprising results have been found concerning the effects of local defects on the quantum conductance of the $(n, n)$ metallic carbon nanotubes [81]. For example, the maximum reduction in the conductance due to a local defect is itself often quantized, and this can be explained in terms of resonant backscattering by quasi-bound states of the defect. Here we discuss results for three simple defects: boron and nitrogen substitutional impurities and the bond rotation or Stone-Wales defect. A Stone-Wales defect corresponds to the rotation of one of the bonds in the hexagonal network by 90°, resulting in the creation of a quite low energy double 5-7 defect pair, without changing the overall helicity of the tube.

Figure 15 depicts several results for a (10,10) carbon nanotube with a single boron substitutional impurity. The top panel is the calculated conductance as a function of the energy of the electron. For a perfect tube, the conductance (indicated here by the dashed line) is 2 in units of the quantum of conductance, $2e^2/h$, since there are two conductance channels available for the electrons near the Fermi energy. For the result with the boron impurity, a striking feature is that the conductance is virtually unchanged at the Fermi level of the neutral nanotube. That is, the impurity potential does not scatter incoming electrons of this energy. On the other hand, there are two dips in the conductance below $E_F$. The amount of the reduction at the upper dip is one quantum unit of conductance and its shape is approximately Lorentzian. In fact, the overall structure of the conductance is well described by the superposition of two Lorentzian dips, each with a depth of 1 conductance quantum. These two dips can be understood in terms of a reduction in conductance due to resonant backscattering from quasi-bound impurity states derived from the boron impurity.

The calculated results thus show that boron behaves like an acceptor with respect to the first lower subband (i.e., the first subband with energy below the conduction states) and forms two impurity levels that are split off from the top of the first lower subband. These impurity states become resonance states or quasi-bound states due to interaction with the conduction states. The impurity states can be clearly seen in the calculated LDOS near the boron impurity (middle panel of Fig. 15). The two extra peaks correspond to the two quasi-bound states. The LDOS would be a constant for a perfect tube in the region between the van Hove singularity of the first lower subband and that of the first upper subband. Because a $(n, n)$ tube with a substitutional impurity still has a mirror plane perpendicular to the tube axis, the defect states have

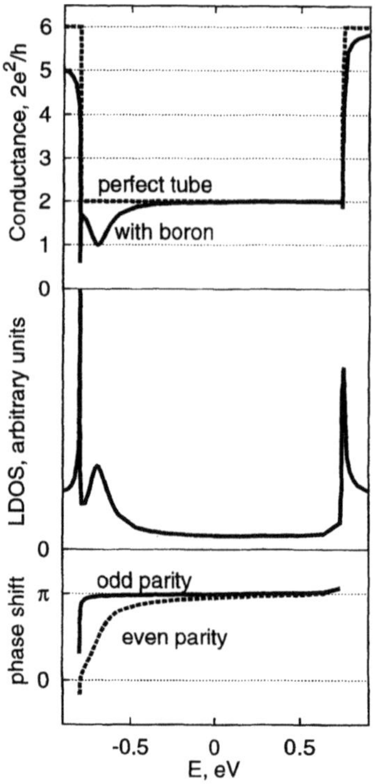


**Fig. 15.** Energy dependence of the calculated conductance, local density of states, and phase shifts of a (10,10) carbon nanotube with a substitutional boron impurity [81]

definite parity with respect to this plane. The upper energy state (broader peak) in Fig. 15 has even parity and the lower energy state (narrower peak) has odd parity, corresponding to $s$-like and $p$-like impurity states, respectively.

The conductance behavior in Fig. 15 may be understood by examining how electrons in the two eigen-channels interact with the impurity. At the upper dip, an electron in one of the two eigen-channels is reflected completely (99.9%) by the boron impurity, but an electron in the other channel passes by the impurity with negligible reflection (0.1%). The same happens at the lower dip but with the behavior of the two eigen-channels switched. The bottom panel shows the calculated scattering phase shifts. The phase shift of the odd parity state changes rapidly as the energy sweeps past the lower quasi-bound state level, with its value passing through $\pi/2$ at the peak position of the quasi-bound state. The same change occurs to the phase shift of the even parity state at the upper impurity-state energy. The total phase shift across a quasi-bound level is $\pi$ in each case, in agreement with the Friedel sum rule. The picture is that an incoming electron with energy exactly in resonance with the impurity state is being scattered back totally in one of the channels but not the other. This explains the exact reduction of one quantum of conductance at the dip. The upper-energy impurity state has a large binding energy (over 0.1 eV) with respect to the first lower subband and

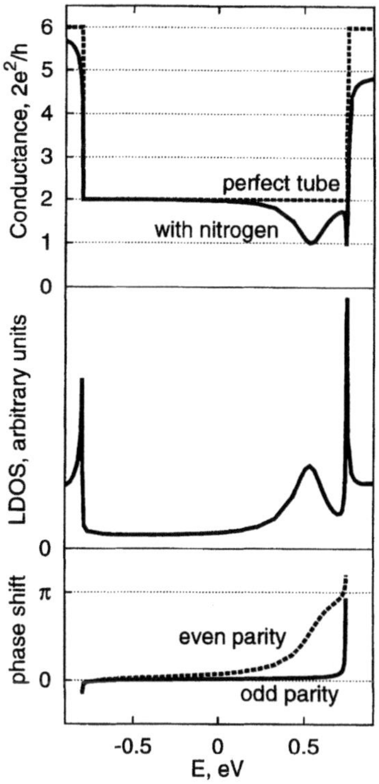


**Fig. 16.** Calculated conductance, local density of states and phase shifts of a (10,10) carbon nanotube with a substitutional nitrogen impurity [81]

hence is quite localized. It has an approximate extent of ~10 Å, whereas the lower impurity state has an extent of ~250 Å.

The results for a nitrogen substitutional impurity on the (10,10) tube are presented in Fig. 16. Nitrogen has similar effects on the conductance as boron, but with opposite energy structures. Again, the conductance at the Fermi level is virtually unaffected, but there are two conductance dips above the Fermi level just below the first upper subband. Thus, the nitrogen impurity behaves like a donor with respect to the first upper subband, forming an $s$-like quasi-bound state with stronger binding energy and a $p$-like state with weaker binding energy. As in the case of boron, the reduction of one quantum unit of conductance at the dips is caused by the fact that, at resonance, the electron in one of the eigen-channels is reflected almost completely by the nitrogen impurity but the electron in the other channel passes by the impurity with negligible reflection. The LDOS near the nitrogen impurity shows two peaks corresponding to the two quasi-bound states. The phase shifts of the two eigen channels show similar behavior as in the boron case.

For a (10,10) tube with a Stone-Wales or double 5-7 pair defect, the calculations also find that the conductance is virtually unchanged for the states at the Fermi energy. Thus these results show that the transport properties of the neutral $(n, n)$ metallic carbon nanotube are very robust with respect to

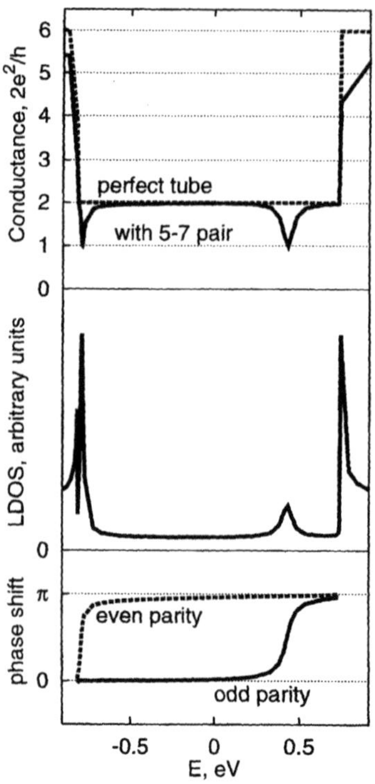


**Fig. 17.** Energy dependence of the calculated conductance, local density of states, and phase shifts of a (10,10) carbon nanotube with a Stone-Wales defect [81]

these kinds of intra-tube local defects. As in the impurity case, there are two dips in the quantum conductance in the conduction band energy range, one above and one below the Fermi level. These are again due to the existence of defect levels, and the reduction at the two dips is very close to one quantum of conductance for the same reason, as discussed above. The symmetry of the Stone-Wales defect in this case does not cause mixing between the $\pi$ and $\pi^*$ bands, and these two bands remain as eigen channels in the defective system. The lower dip is due to a complete reflection of the $\pi^*$ band and the upper dip is due to complete reflection of the $\pi$ band. This implies that the conductance of the nanotube, when there are more than one double 5-7 pair defect, would not sensitively depend on their relative positions, but only on their total numbers, as long as the distance between defects is far enough to be able to neglect inter-defect interactions. The analysis of the phase shifts show that the lower quasi-bound state is even with respect to a mirror plane perpendicular to the tube axis, while the upper quasi-bound state is odd with respect to the same plane.

The conductance of nanotubes can also be affected by structural deformations. Two types of deformations involving bending or twisting the nanotube structure have been considered in the literature. It was found that a smooth bending of the nanotube does not lead to scattering [54], but formation of a local kink induces strong $\sigma$–$\pi$ mixing and backscattering similar to that discussed earlier for boron impurity [82]. Twisting has a much stronger effect [54]. A metallic armchair $(n, n)$ nanotube upon twisting develops a bandgap which scales linearly with the twisting angle up to the critical angle at which the tube collapses into a ribbon [82].

## 3 Nanotube Ropes, Crossed-Tube Junctions, and Effects of Long-Range Perturbations

Another interesting carbon nanotube system is that of ropes of single-walled carbon nanotubes which have been synthesized in high yield [8]. These ropes, containing up to tens to hundreds of single-walled nanotubes in a close-packed triangular lattice, are made up of tubes of nearly uniform diameter, close to that of the (10,10) tubes (Fig. 18). Because of the rather weak interaction between these tubes, a naive picture would be that the packing of individual metallic nanotubes into ropes would not change their electronic properties significantly. Theoretical studies [83,84,85] however showed that this is not the case for a rope of (10,10) carbon nanotubes. A broken symmetry of the (10,10) nanotube caused by interactions between tubes in a rope induces formation of a pseudogap in the density of states of about 0.1 eV. The existence of this pseudogap alters many of the fundamental electronic properties of the rope.

**Fig. 18.** Perspective view of a model of a rope of (10,10) carbon nanotubes

### 3.1 Ropes of Nanotubes

As discussed above, an isolated $(n, n)$ carbon nanotube has two linearly dispersing conduction bands which cross at the Fermi level forming two "Dirac" points, as schematically presented in Fig. 19a. This linear band dispersion in a one-dimensional system gives rise to a finite and constant density of electronic states at the Fermi energy. Thus, an $(n, n)$ tube is a metal within

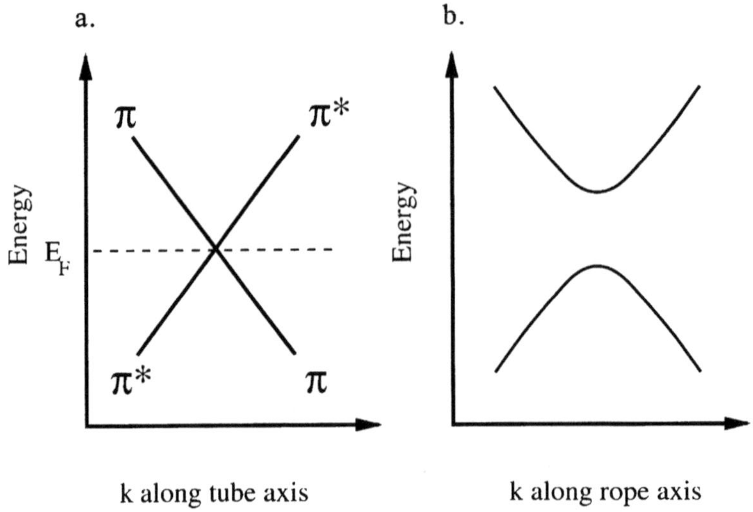


**Fig. 19.** Band crossing and band repulsion. (**a**) Schematic band structure of an isolated $(n, n)$ carbon nanotube near the Fermi energy. (**b**) Repulsion of bands due to the breaking of mirror symmetry

the one-electron picture. The question of interest is: How does the electronic structure change when the metallic tubes are bundled up to form a closely packed two-dimensional crystal, as in the case of the (10,10) ropes. In the calculation, a large (10,10) rope is modeled by a triangular lattice of (10,10) tubes infinitely extended in the lateral directions. For such a system, the electronic states, instead of being contained in the 1-D Brillouin zone of a single tube, are now extended to a three-dimensional irreducible Brillouin zone wedge. If tube-tube interactions are negligibly small, the electronic energy band structure along any line in the wedge parallel to the rope axis would be exactly the same as the band dispersion of an isolated tube. In particular, at the $k$-wavevector corresponding to the band crossing point, there will be a two-fold degenerate state at the Fermi energy. This allowed band crossing is due to the mirror symmetry of the (10,10) tube. For a tube in a rope, this symmetry is however broken because of intertube interactions. The broken symmetry causes a quantum level repulsion and opens up a gap almost everywhere in the Brillouin zone, as schematically shown in Fig. 19b.

The band repulsion resulting from the broken-symmetry strongly modifies the DOS of the rope near the Fermi energy compared to that of an isolated (10,10) tube. The calculated DOS is presented in Fig. 20a. Shown are the results for two cases: aligned and misaligned tubes in the rope. In both cases, there is a pseudogap of the order of 0.1 eV in the density of states. Examination of the electronic structure reveals that the system is a semimetal with both electron and hole carriers. The existence of the pseudogap in the rope makes the conductivity and other transport properties of the metallic rope

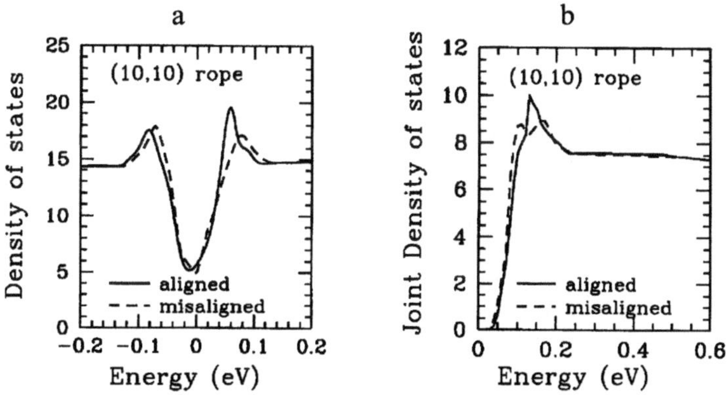


**Fig. 20.** (**a**) Calculated density of states for a rope of misaligned (10,10) carbon nanotubes (*broken line*) and aligned tubes (*solid line*). The Fermi energy is at zero. (**b**) Calculated joint density of states for a rope of misaligned (*broken line*) and aligned (*solid line*) (10,10) tubes. Results are in units of states per meV per atom [83,84]

significantly different from those of isolated tubes, even without considering the effect of local disorder in low dimensions. Since the DOS increases rapidly away from the Fermi level, the carrier density of the rope is sensitive to temperature and doping. The existence of both electron and hole carriers leads to qualitatively different thermopower and Hall-effect behaviors from those expected for a normal metal. The optical properties of the rope are also affected by the pseudogap. As illustrated by the calculated Joint Density of States (JDOS) in Fig. 20b, there would be a finite onset in the infrared absorption spectrum for a large perfectly ordered (10,10) rope, where one can assume $k$-conserving optical transitions. In the case of high disorder, an infrared experiment would more closely reflect the DOS rather than the JDOS. For most actual samples, the fraction of (10,10) carbon nanotubes (compared with other nanotubes of the same diameter) in the experimentally synthesized ropes appears to be small. However, the conclusion that broken symmetry induces a gap in the $(n, n)$ tubes is a general result which is of relevance for tubes under any significant asymmetric perturbations, such as those due to structural deformations or external fields.

### 3.2 Crossed-Tube Junctions

The discussion of nanotube junctions in Sect. 2 is focused on the on-tube junctions, i.e., forming a junction by joining two half tubes together. These systems are extremely interesting, but difficult to synthesize in a controlled manner at this time. Another way to form junctions is to have two tubes crossing each other in contact [86] (Fig. 21). This kind of crossed-tube junction is much easier to fabricate and control with present experimental techniques.

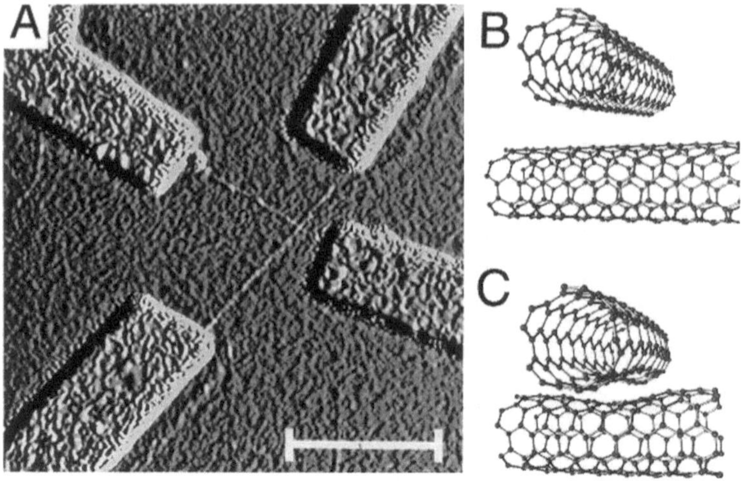


**Fig. 21.** AFM image of a crossed SWNT device (**A**). Calculated structure of a crossed (5,5) SWNT junction with a force of 0 nN (**B**) and 15 nN (**C**) [86]

When two nanotubes cross in free space, one expects that the tubes at their closest contact point will be at a van der Waals distance away from each other and that there will not be much intertube or junction conductance. However, as shown by *Avouris* and coworkers [87], for two crossed tubes lying on a substrate, there is a substantial force pressing one tube against the other due to the substrate attraction. For a crossed-tube junction composed of SWNTs with the experimental diameter of 1.4 nm, this contact force has been estimated to be about 5 nN [87]. This substrate force would then be sufficient to deform the crossed-tube junction and lead to better junction conductance.

In Fig. 21, panel A is an AFM image of a crossed-tube junction fabricated from two single-walled carbon nanotubes of 1.4 nm in diameter with electrical contacts at each end [86]. Panels B and C show the calculated structure corresponding to a (5,5) carbon nanotube pressed against another one with zero and 15 nN force, respectively. Because of the smaller diameter of the (5,5) tube, a larger contact force is required to produce a deformation similar to that of the experimental crossed-tube junction. The calculation was done using the *ab initio* pseudopotential density functional method with a localized basis [86]. As seen in panel C, there is considerable deformation, and the atoms on the different tubes are much closer to each other. At this distance, the closest atomic separation between the two tubes is 0.25 nm, significantly smaller than the van der Waals distance of 0.34 nm.

For the case of zero contact force (panel B in Fig. 21), the calculated intra-tube conductance is virtually unchanged from that of an ideal, isolated metallic tube, and the intertube conductance is negligibly small. However, when the tubes are under a force of 15 nN, there is a sizable intertube or junction conductance. As shown in Fig. 22, the junction conductance at the

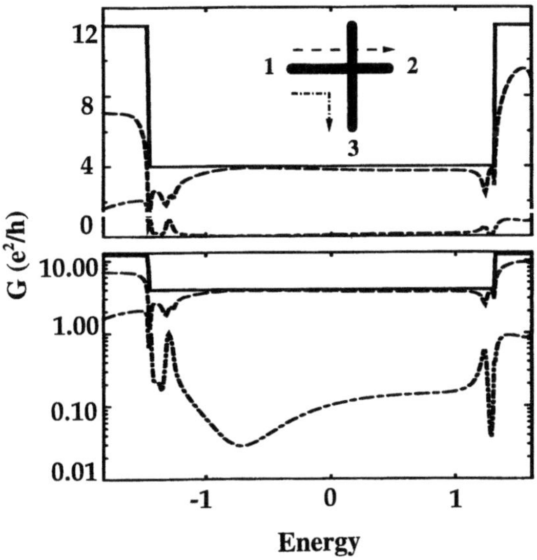


**Fig. 22.** Calculated conductance (expressed in units of $e^2/h$) of a crossed (5,5) carbon nanotube junction with a contact force of 15 nN on a linear (*top*) and log (*bottom*) scale. The dashed (*dotted-dashed*) curve corresponds to the intra-tube (intertube) conductance [86]

Fermi energy is about 5% of a quantum unit of conductance $G_0 = 2e^2/h$. The junction conductance is thus very sensitive to the force or distance between the tubes.

Experimentally, the conductance of various types of crossed carbon nanotube junctions has been measured, including metal-metal, semiconductor-semiconductor, and metal-semiconductor crossed-tube junctions. The experimental results are presented in Fig. 23. For the metal-metal crossed-tube junctions, a conductance of 2 to 6% of $G_0$ is found, in good agreement with the theoretical results. Of particular interest is the metal-semiconductor case in which experiments demonstrated Schottky diode behavior with a Schottky barrier in the range of 200–300 meV, which is very close to the value of 250 meV expected from theory for nanotubes with diameters of 1.4 nm [86].

### 3.3 Effects of Long-Range Disorder and External Perturbations

The effects of disorder on the conducting properties of metal and semiconducting carbon nanotubes are quite different. Experimentally, the mean free path is found to be much longer in metallic tubes than in doped semiconducting tubes [19,20,21,24,88]. This result can be understood theoretically if the disorder potential is long range. As discussed below, the internal structure of the wavefunction of the states connected to the sublattice structure of graphite lead to a suppression of scattering in metallic tubes, but not

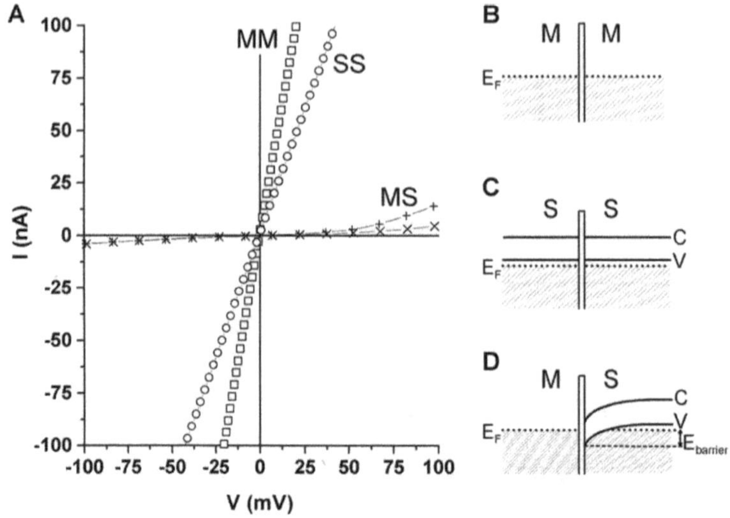


**Fig. 23.** Current–voltage characteristics of several crossed SWNT junctions [86] (see text)

in semiconducting tubes. Figure 24 shows the measured conductance for a semiconducting nanotube device as a function of gate voltage at different temperatures. [88]. The diameter of the tube as measured by AFM is 1.5 nm, consistent with a single-walled tube. The complex structure in the Coulomb blockade oscillations in Fig. 24 is consistent with transport through a number of quantum dots in series. The temperature dependence and typical charging energy indicates that the tube is broken up into segments of length of about 100 nm. Similar measurements on intrinsic metal tubes, on the other hand, yield lengths that are typically a couple of orders of magnitude longer [19,20,21,24,88].

Theoretical calculations have been carried out to examine the effects of long-range external perturbations [88]. In the calculation, to model the perturbation, a 3-dimensional Gaussian potential of a certain width is centered on one of the atoms on the carbon nanotube wall. The conductance with the perturbation is computed for different Gaussian widths, but keeping the integrated strength of the potential the same. Some typical tight-binding results are presented in Fig. 25. The solid lines show the results for the conductance of a disorder-free tube, while the dashed and the dot-dashed lines are, respectively, for a single long-range ($\sigma$=0.348 nm, $\Delta V$= 0.5 eV) and a short-range ($\sigma$=0.116 nm, $\Delta V$= 10 eV) scatterer. Here $\Delta V$ is the shift in the on-site energy at the potential center. The conduction bands (i.e., bands crossing the Fermi level) of the *metallic* tube are unaffected by the long-range scatterer, unlike the lower and upper subbands of both the *metallic* and *semiconducting* tubes, which are affected by both long- and short-range scatterers. All

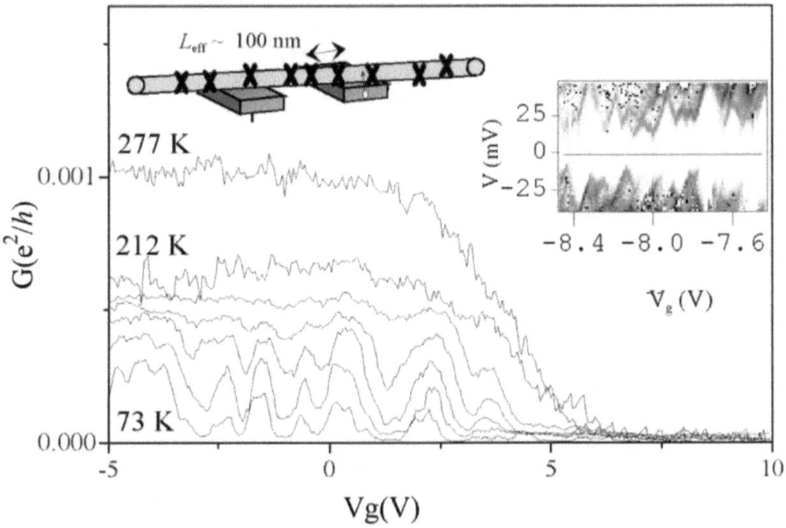


**Fig. 24.** Conductance vs. gate voltage $V_g$ for a semiconducting single-walled carbon nanotube at various temperatures. The *upper insert* schematically illustrates the sample geometry and the *lower insert* shows $dI/dV$ vs. $V$ and $V_g$ plotted as a gray scale [88]

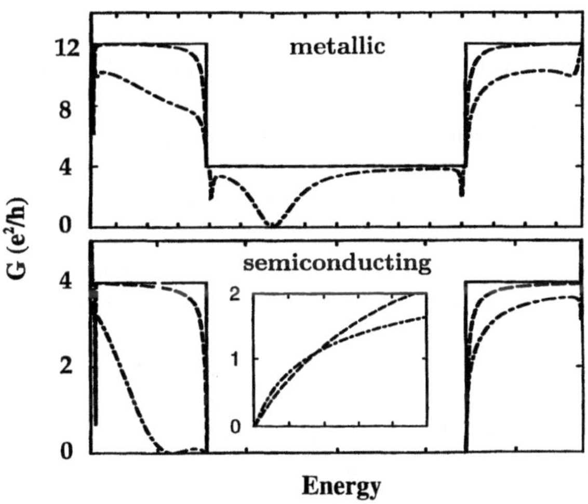


**Fig. 25.** Tight-binding calculation of the conductance of a (**a**) metallic (10,10) tube and (**b**) semiconducting (17,0) tube in the presence of a Gaussian scatterer. The energy scale on the abscissa is 0.2 eV per division in each graph [88]

subbands are influenced by the short-range scatterer. The inset shows an expanded view of the onset of conduction in the semiconducting tube at positive $E$, with each division corresponding to 1 meV. Also, the sharp step edges

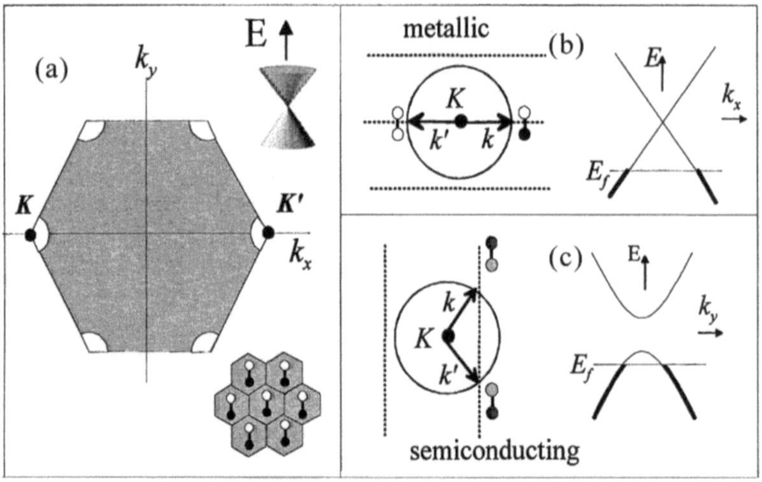


**Fig. 26.** (**a**) Filled states (*shaded*) in the first Brillouin zone of a p-type graphene sheet. There are two carbon atoms per unit cell (*lower right inset*). The dispersions of the states near $E_F$ are cones whose vertices are located at the corner points of the Brillouin zone. The Fermi circle, defining the allowed $\boldsymbol{k}$ vectors, and the band dispersions are shown in (**b**) and (**c**) for a metallic and a semiconducting tube, respectively [88]

in the calculated conductance of the perfect tubes are rounded off by both types of perturbations.

Both the experimental and theoretical findings strongly suggest that long-range scattering is suppressed in the metallic tubes. One can actually understand this qualitatively from the electronic structure of a graphene sheet [89,90]. The graphene structure has two atoms per unit cell. The properties of electrons near the Fermi energy are given by those states near the corner of the Brillouin zone (Fig. 26). If we look at the states near this point and consider them in terms of a $\boldsymbol{k}$-vector away from the corner $K$ point, then they can be described by a Dirac Hamiltonian. For these states, the wavefunctions can be written in terms of a product of a plane wave component (with a vector $\boldsymbol{k}$) and a pseudo-spin which describes the bonding character between the two atoms in the unit cell. The interesting result is that this pseudo-spin points along $\boldsymbol{k}$. For example, if the state at $\boldsymbol{k}$ is bonding, then the state at $-\boldsymbol{k}$ is antibonding in character. Within this framework, one can work out the scattering between the allowed states in a carbon nanotube due to long-range disorder, i.e., disorder with Fourier components $V(q)$ such that $q \ll K$. This, for example, will be the case for scattering by charged trap states in the substrate (oxide traps). In this case, the disorder does not couple to the pseudo-spin portion of the wavefunction, since the disorder potential is approximately constant on the scale of the interatomic distance. The resulting matrix element between states is then [89,90]: $|\langle k'|V(r)|k\rangle|^2 = |V(k-k')|^2 \cos^2[(1/2)\theta_{k,k'}]$, where $\theta_{k,k'}$

is the angle between the initial and final states. The first term in $V(k - k')$ is the Fourier component at the difference in $k$ values of the initial and final envelope wavefunctions. The cosine term is the overlap of the initial and final spinor states.

For a metallic tube Fig. 26b, backscattering in the conduction band corresponds to scattering between $\boldsymbol{k}$ and $-\boldsymbol{k}$. Such scattering is forbidden, because the molecular orbitals of these two states are orthogonal. In semiconducting tubes, however, the situation is quite different Fig. 26c. The angle between the initial and final states is less than $\pi$, and scattering is thus only partially suppressed by the spinor overlap. As a result, semiconducting tubes should be sensitive to long-range disorder, while metallic tubes should not. However, short-range disorder which has Fourier components $q \sim K$ will couple the molecular orbitals together and lead to scattering in all of the subbands. These theoretical considerations agree well with experiment and with the detailed calculations discussed above. Long-range disorder due to, e.g., localized charges near the tube, breaks the semiconducting tube into a series of quantum dots with large barriers, resulting in a dramatically reduced conductance and a short mean free path. On the other hand, metallic tubes are insensitive to this disorder and remain near-perfect 1D conductors.

## 4 Summary

This paper gives a short review of some of our theoretical understanding of the structural and electronic properties of single-walled carbon nanotubes and of various structures formed from these nanotubes. Because of their nanometer dimensions, the nanotube structures can have novel properties and yield unusual scientific phenomena. In addition to the multi-walled carbon nanotubes, single-walled nanotubes, nanotube ropes, nanotube junctions, and non-carbon nanotubes have been synthesized.

These quasi-one-dimensional objects have highly unusual electronic properties. For the perfect tubes, theoretical studies have shown that the electronic properties of the carbon nanotubes are intimately connected to their structure. They can be metallic or semiconducting, depending sensitively on tube diameter and chirality. Experimental studies using transport, scanning tunneling, and other techniques have basically confirmed the theoretical predictions. The dielectric responses of the carbon nanotubes are found to be highly anisotropic in general. The heat capacity of single-wall nanotubes is predicted to have a characteristic linear $T$ dependence at low temperature.

On-tube metal-semiconductor, semiconductor-semiconductor, and metal-metal junctions may be formed by introducing topological structural defects, and these junctions have been shown to behave like nanoscale device elements. For example, different half-tubes may be joined with 5-member ring/7-member ring pair defects to form a metal-semiconductor Schottky barrier. The calculated electronic structure of these junctions is very similar to that of standard metal-semiconductor interfaces, and in this sense, they are molecular level devices composed of the single element, carbon. Recent experimental measurements have confirmed the existence of such Schottky barrier behavior in nanotube ropes and across kinked nanotube junctions. Similarly, 5-7 defect pairs in different carbon and non-carbon nanotubes can produce semiconductor-semiconductor and metal-metal junctions. The existence of metal-metal nanotube junctions in which the conductance is suppressed for symmetry reasons has also been predicted. Thus, the carbon nanotube junctions may be used as nanoscale electronic elements.

The influence of impurities and local structural defects on the conductance of carbon nanotubes has also been examined. It is found that local defects in general form well defined quasi-bound states even in metallic nanotubes. These defect states give rise to peaks in the LDOS and reduce the conductance at the energy of the defect levels by a quantum unit of conductance via resonant backscattering. The theoretical studies show that, owing to the unique electronic structure of the graphene sheet, the transport properties of $(n, n)$ metallic tubes appear to be very robust against defects and long-range perturbations near $E_\mathrm{F}$. Doped semiconducting tubes are much more susceptible to long-range disorder. These results explain the experimental findings of the long coherence length in metallic tubes and the large difference in mean free path between the metallic and doped semiconducting tubes. For nanotube ropes, intertube interactions are shown to alter the electronic structure of $(n, n)$ metallic tubes because of broken symmetry effects, leading to a pseudogap in the density of states and to semimetallic behavior. Crossed-tube junctions have also been fabricated experimentally and studied theoretically. These systems show significant intertube conductance for metal-metal junctions and exhibit Schottky behavior for metal-semiconductor junctions when the tubes are subjected to contact force from the substrate.

The carbon nanotubes are hence a fascinating new class of materials with many unique and desirable properties. The rich interplay between the geometric and electronic structure of the nanotubes has given rise to many interesting, new physical phenomena. At the practical level, these systems have the potential for many possible applications.

### Acknowledgement

This work was supported by NSF Grant No. DMR95-20554 and by the U.S. DOE under Contract No. DE-AC03-76SF00098.

## References

1. S. Iijima, Nature (London) **354**, 56 (1991)
2. M. S. Dresselhaus, G. Dresselhaus, P. C. Eklund, *Science of Fullerenes and Carbon Nanotubes* (Academic, New York 1996)

3. P. M. Ajayan, T. W. Ebbesen, Rep. Prog. Phys. **60**, 1025 (1997)
4. C. Dekker, Phys. Today **52**, 22 (1999)
5. S. Iijima, T. Ichihashi, Nature (London) **363**, 603 (1993)
6. D. S. Bethune, C. H. Kiang, M. S. de Vries, G. Gorman, R. Savoy, J. Vazquez, R. Beyers, Nature (London) **363**, 605 (1993)
7. P. M. Ajayan, J. M. Lambert, P. Bernier, L. Barbedette, C. Colliex, J. M. Planeix, Chem. Phys. Lett. **215**, 509 (1993)
8. T. Guo, C.-M. Jin, R. E. Smalley, Chem. Phys. Lett. **243**, 49–54 (1995)
9. M. R. Pederson, J. Q. Broughton, Phys. Rev. Lett. **69**, 2689 (1992)
10. P. M. Ajayan, S. Iijima, Nature (London) **361**, 333 (1993)
11. N. G. Chopra, L. X. Benedict, V. H. Crespi, M. L. Cohen, S. G. Louie, A. Zettl, Nature (London) **377**, 135 (1995)
12. H. Dai, E. W. Wong, C. M. Lieber, Nature (London) **384**, 147 (1996)
13. W. A. de Heer, A. Châtelain, D. Ugarte, Science **270**, 1179 (1995) ibid p. 1119
14. A. G. Rinzler, J. H. Hafner, P. Nikolaev, L. Lou, S. G. Kim, D. Tománek, P. Nordlander, D. T. Colbert, R. E. Smalley, Science **269**, 1550 (1995)
15. A. C. Dillon, K. M. Jones, T. A. Bekkedahl, C. H. Kiang, D. S. Bethune, M. J. Heben, Nature (London) **386**, 377–379 (1997)
16. L. Chico, V. H. Crespi, L. X. Benedict, S. G. Louie, M. L. Cohen, Phys. Rev. Lett. **76**, 971–974 (1996)
17. Ph. Lambin, A. Fonseca, J. P. Vigneron, J. B. Nagy, A. A. Lucas, Chem. Phys. Lett. **245**, 85–89 (1995)
18. R. Saito, G. Dresselhaus, M. S. Dresselhaus, Phys. Rev. B **53**, 2044–2050 (1996)
19. S. J. Tans, M. H. Devoret, H. Dai, A. Thess, R. E. Smalley, L. J. Geerligs, C. Dekker, Nature (London) **386**, 474–477 (1997)
20. M. Bockrath, D. H. Cobden, P. L. McEuen, N. G. Chopra, A. Zettl, A. Thess, R. E. Smalley, Science **275**, 1922–1924 (1997)
21. R. Martel, T. Schmidt, H. R. Shea, T. Hertel, Ph. Avouris, Appl. Phys. Lett. **73**, 2447 (1998)
22. P. G. Collins, A. Zettl, H. Bando, A. Thess, R. E. Smalley, Science **278**, 5335 (1997)
23. Zhen Yao, H. W. C. Postma, L. Balents, C. Dekker, Nature (London) **402**, 273 (1999)
24. S. J. Tans, R. M. Verschueren, C. Dekker, Nature **393**, 49–52 (1998)
25. X. Blase, A. Rubio, S. G. Louie, M. L. Cohen, Europhys. Lett. **28**, 335 (1994)
26. Y. Miyamoto, A. Rubio, M. L. Cohen, S. G. Louie, Phys. Rev. B **50**, 18360 (1994)
27. Y. Miyamoto, A. Rubio, S. G. Louie, M. L. Cohen, Phys. Rev. B **50**, 4976 (1994)
28. X. Blase, A. Rubio, S. G. Louie, M. L. Cohen, Phys. Rev. B **51**, 6868 (1995)
29. Y. Miyamoto, M. L. Cohen, S. G. Louie, Solid State Comm. **102**, 605 (1997)
30. Z. Weng-Sieh, K. Cherrey, N. G. Chopra, X. Blase, Y. Miyamoto, A. Rubio, M. L. Cohen, S. G. Louie, A. Zettl, R. Gronsky, Phys. Rev. B **51**, 11229 (1995)
31. O. Stephan, P. M. Ajayan, C. Colliex, Ph. Redlich, J. M. Lambert, P. Bernier, P. Lefin, Science **266**, 1683 (1994)
32. N. G. Chopra, J. Luyken, K. Cherry, V. H. Crespi, M. L. Cohen, S. G. Louie, A. Zettl, Science **269**, 966 (1995)
33. A. Loiseau, F. Willaime, N. Demoncy, G. Hug, H. Pascard, Phys. Rev. Lett. **76**, 4737 (1996)

34. K. Suenaga, C. Colliex, N. Demoncy, A. Loiseau, H. Pascard, F. Willaime, Science **278**, 653 (1997)
35. P. Gleize, S. Herreyre, P. Gadelle, M. Mermoux, M. C. Cheynet, L. Abello, J. Materials Science Letters **13**, 1413 (1994)
36. See also the chapter by R. Tenne, A. Zettl in this volume
37. See also the chapter by R. Saito, H. Kataura in this volume
38. N. Hamada, S. Sawada, A. Oshiyama, Phys. Rev. Lett. **68**, 1579–1581 (1992)
39. R. Saito, M. Fujita, G. Dresselhaus, M. S. Dresselhaus, Appl. Phys. Lett. **60**, 2204–2206 (1992)
40. J. W. Mintmire, B. I. Dunlap, C. T. White, Phys. Rev. Lett. **68**, 631–634 (1992)
41. X. Blase, L. X. Benedict, E. L. Shirley, S. G. Louie, Phys. Rev. Lett. **72**, 1878 (1994)
42. A. Rochefort, D. S. Salahub, Ph. Avouris, Chem. Phys. Lett. **297**, 45 (1998)
43. S. I. Sawada, N. Hamada, Solid State Commun. **83**, 917–919 (1992)
44. X. Blase, S. G. Louie, unpublished
45. L. Langer, V. Bayot, E. Grivei, J. P. Issi, J. P. Heremans, C. H. Olk, L. Stockman, C. Van Haesendonck, Y. Bruynseraede, Phys. Rev. Lett. **76**, 479–482 (1996)
46. T. W. Ebbesen, H. J. Lezec, H. Hiura, J. W. Bennett, H. F. Ghaemi, T. Thio, Nature (London) **382**, 54–56 (1996)
47. H. Dai, E. W. Wong, C. M. Lieber, Science **272**, 523–526 (1994)
48. J. W. G. Wildöer, L. C. Venema, A. G. Rinzler, R. E. Smalley, C. Dekker, Nature (London) **391**, 59–62 (1998)
49. T. W. Odom, J. L. Huang, P. Kim, C. M. Lieber, Nature (London) **391**, 62–64 (1998)
50. L. X. Benedict, S. G. Louie, M. L. Cohen, Phys. Rev. B **52**, 8541 (1995)
51. G. F. Bertsch, A. Bulgac, D. Tománek, Y. Wang, Phys. Rev. Lett. **67**, 2690 (1991)
52. B. Koopmans, PhD Thesis, University of Groningen, (1993)
53. See also the chapter by J. Hone in this volume.
54. C. L. Kane, E. J. Mele, Phys. Rev. Lett. **78**, 1932 (1997)
55. L. X. Benedict, S. G. Louie, M. L. Cohen, Solid State Commun. **100**, 177–180 (1996)
56. See also the chapter by B. Yakobson, Ph. Avouris in this volume
57. D. H. Robertson, D. W. Brenner, J. W. Mintmire, Phys. Rev. B **45**, 12592 (1992)
58. R. S. Ruoff, D. C. Lorents, Carbon **33**, 925 (1995)
59. J. M. Molina, S. S. Savinsky, N. V. Khokhriakov, J. Chem. Phys. **104**, 4652 (1996)
60. B. I. Yakobson, C. J. Brabec, J. Bernholc, Phys. Rev. Lett. **76**, 2411 (1996)
61. C. F. Cornwell, L. T. Wille, Solid State Commun. **101**, 555 (1997)
62. S. Iijima, C. J. Brabec, A. Maiti, J. Bernholc, J. Chem. Phys. **104**, 2089 (1996)
63. J. P. Lu, Phys. Rev. Lett. **79**, 1297 (1997)
64. M. M. J. Treacy, T. W. Ebbesen, J. M. Gibson, Nature (London) **381**, 678 (1996)
65. E. W. Wong, P. E. Sheehan, C. M. Lieber, Science **277**, 1971 (1997)
66. N. G. Chopra, A. Zettl, Solid State Comm. **105**, 297 (1998)
67. L. Chico, L. X. Benedict, S. G. Louie, M. L. Cohen, Phys. Rev. B **54**, 2600 (1996)

68. B. I. Dunlap, Phys. Rev. B **49**, 5643 (1994)
69. J.-C. Charlier, T. W. Ebbesen, Ph. Lambin, Phys. Rev. B **53**, 11108 (1996)
70. T. W. Ebbesen, T. Takada, Carbon **33**, 973 (1995)
71. Ph. Lambin, L. Philippe, J.-C. Charlier, J. P. Michenaud, Synth. Met. **2**, 350–356 (1996)
72. S. G. Louie, M. L. Cohen, Phys. Rev. B **13**, 2461 (1976)
73. X. Blase, J. C. Charlier, A. de Vila, R. Car, Appl. Phys. Lett. **70**, 197 (1997)
74. M. Menon, D. Srivastava, Phys. Rev. Lett. **79**, 4453–4456 (1997)
75. R. Landauer, Philos. Mag. **21**, 863 (1970)
76. D. S. Fisher, P. A. Lee, Phys. Rev. B **23**, 6851 (1981)
77. M. Nardelli, B. I. Yakobson, J. Bernholc, Phys. Rev. Lett. **81**, 4656 (1998)
78. N. Koprinarov, M. Marinov, G. Pchelarov, M. Konstantinove, R. Stefanov, Phys. Rev. Lett. **99**, 2042 (1996)
79. A. Zettl, private Communications
80. H. J. Choi, J. Ihm, Phys. Rev. B **59**, 2267 (1999)
81. H. J. Choi, J. Ihm, S. G. Louie, M. L. Cohen, Phys. Rev. Lett. **84**, 2917 (2000)
82. A. Rochefort, Ph. Avouris, F. Lesage, R. R. Salahub, Phys. Rev. B **60**, 13824 (1999)
83. P. Delaney, H. J. Choi, J. Ihm, S. G. Louie, M. L. Cohen, Nature (London) **391**, 466 (1998)
84. P. Delaney, H. J. Choi, J. Ihm, S. G. Louie, M. L. Cohen, Phys. Rev. B **60**, 7899 (1999)
85. Y. K. Kwon, S. Saito, D. Tománek, Phys. Rev. B **58**, R13314 (1998)
86. M. S. Fuhrer, J. Nygard, L. Shih, M. Forero, Y. G. Yoon, M. S. C. Mazzone, H. J. Choi, J. Ihm, S. G. Louie, A. Zettl, P. L. McEuen, Science **288**, 494 (2000)
87. I. V. Hertel, R. E. Walkup, P. Avouris, Phys. Rev. B **58**, 13870 (1998)
88. P. L. McEuen, M. Bockrath, D. H. Cobden, Y. G. Yoon, S. G. Louie, Phys. Rev. Lett. **83**, 5098 (1999)
89. T. Ando, T. Nakkanishi, R. Saito, J. Phys. Soc. Jpn. **67**, 2857 (1998)
90. T. Ando, T. Nakkanishi, J. Phys. Soc. Jpn. **67**, 1704 (1998)

# Electrical Transport Through Single-Wall Carbon Nanotubes

Zhen Yao[1], Cees Dekker[1], and Phaedon Avouris[2]

[1] Department of Applied Physics, Delft University of Technology
Lorentzweg 1, 2628 CJ Delft, The Netherlands
yao@qt1.tn.tudelft.nl
dekker@qt.tn.tudelft.nl

[2] IBM Watson Research Laboratory
Yorktown Heights, NY 10598, USA
avouris@us.ibm.com

**Abstract.** We present a brief review of the phenomenal progress in electrical transport measurements in individual and ropes of single-wall carbon nanotubes in the past few years. Nanotubes have been made into single-electron transistors, field-effect transistors, and rectifying diodes. A number of interesting mesoscopic transport phenomena have been observed. More significantly, nanotubes exhibit strong electron–electron correlation effects, or so-called Luttinger liquid behavior, associated with their one-dimensional nature.

Electrical transport through Single-Wall Carbon Nanotubes (SWNTs) has generated considerable interest in the past few years (for earlier reviews, see [1,2]). This has been largely stimulated by many proposed applications of SWNTs in future nanoscale electronic devices based on their unique electronic properties and nanometer sizes. From the fundamental physics point of view, SWNTs provide a nearly perfect model system for one-dimensional (1D) conductors, in which electron–electron correlations have a profound influence on the properties of conduction electrons. They offer several clear advantages over other 1D systems. For example, comparing with semiconductor quantum wires, SWNTs are atomically uniform and well-defined; the strong confinement around the circumference leads to a large spacing between 1D subbands ($\sim$1 eV for a $\sim$1 nm tube in contrast to $\sim$10 meV for typical semiconductor quantum wires), which means that the 1D nature is retained up to room temperature and well above. Comparing with other molecular wires, nanotubes are structurally robust and chemically inert. Moreover, because of the tubular structure, the Peierls distortion which normally makes other molecular wires semiconducting, becomes energetically unfavorable in carbon nanotubes – the lattice energy cost of rearranging the carbon atoms around the whole circumference is large while the gain in electronic energy is low since there are only two subbands at the Fermi energy.

Electrical characterization of individual SWNT molecules has been made possible by advances in both bottom-up chemical synthesis of the these materials and modern top-down lithographic techniques for making electrical

M. S. Dresselhaus, G. Dresselhaus, Ph. Avouris (Eds.): Carbon Nanotubes,
Topics Appl. Phys. **80**, 147–171 (2001)

contacts. The real breakthrough came when Smalley's group at Rice University developed a laser ablation method to grow high quality SWNTs in bulk quantity [3]. Since then, transport measurements have been carried out by a number of research groups. We will limit our discussion to results obtained on individual tubes or small individual ropes of SWNTs. The effect of doping as well as transport in multiwall nanotubes are described in other chapters of this book.

This paper is organized as follows. In Sect. 1, two of us (Z.Y. and C.D.) will describe transport measurements performed on isolated individual SWNTs. Semiconducting nanotubes can be distinguished from metallic ones from gate voltage dependence measurements at room temperature. At low temperatures, a number of interesting mesoscopic phenomena have been observed such as single-electron charging, resonant tunneling through discrete energy levels and proximity-induced superconductivity. At relatively high temperatures, tunneling conductance into the nanotubes displays a power-law suppression as a function of temperature and bias voltage, which is consistent with the physics of the 1D Luttinger liquid. Metallic nanotubes are able to carry a remarkably high current density and the main scattering mechanism for high energy electrons is due to optical or zone-boundary phonons. In addition, we will discuss devices for potential electronic applications such as junctions, rectifying diodes and electromechanical devices. In Sect. 2, one of us (Ph.A.) will discuss the importance of scattering and coherent-backscattering processes in the low-temperature transport in ropes and rings of SWNTs. Magneto-resistance measurements of coherence lengths, evidence for dephasing involving electron–electron interactions, zero-bias anomalies due to strong electron correlation, and carrier localization will be discussed. Finally, intertube transport within a rope will be addressed.

# 1 Transport in Individual Nanotubes

In this section, we discuss electrical measurements of individual SWNTs or thin ropes in which the charge transport is dominated by individual tubes within these ropes.

## 1.1 Device Geometry and Room-Temperature Characterization

We describe the procedure to electrically contact the nanotubes and typical device geometries. By employing a nearby gate, semiconducting nanotubes can be made into field-effect transistors that operate at room temperature.

### 1.1.1 Electrical Contacts

Making electrical contacts to carbon nanotubes is in principle straightforward. A small amount of raw SWNT material, which usually consists of highly entangled ropes, is first ultrasonically dispersed in some organic solvent, typically dichloroethane, and then spun onto an oxidized silicon wafer on which a large array of electrodes have been fabricated with conventional electron-beam lithography, metal evaporation and lift-off [4]. An Atomic Force Microscope (AFM) is then used to determine whether there is an individual SWNT or a small rope (by apparent height) bridging two or more electrodes. Alternatively, the electrodes are aligned and fabricated on top of a nanotube that has been identified by AFM using predefined alignment markers on the substrate [5]. However, this latter procedure often results in nanotubes being cut between the electrodes due to electron beam damage during the lithography process. Figure 1 is a schematic of a typical three-terminal device geometry for transport measurement. The AFM image shows an individual SWNT molecule lying across two metal electrodes, which are used as source and drain, respectively. A variety of metals have been used for electrical leads including gold and platinum. A typical electrode width is on the order of 100 nm and the source-drain spacing varies between ~100 nm and ~1 μm. In most experiments, another nearby electrode or a doped silicon substrate underneath the $SiO_2$ is used as a gate to electrostatically modulate the carrier density of the nanotube under study.

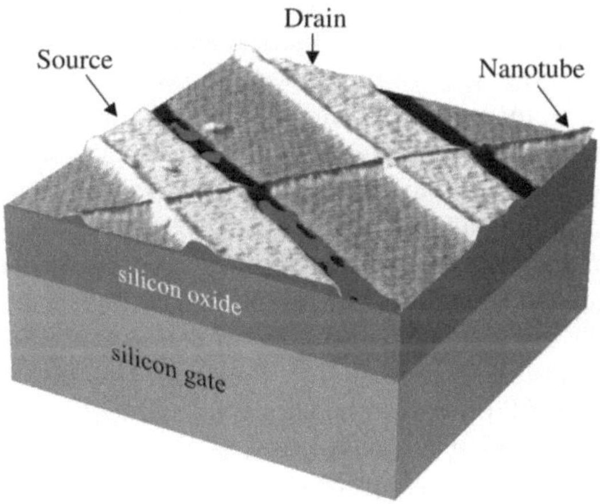


**Fig. 1.** Typical device geometry for electrical transport measurement

Room temperature transport characteristics fall into two distinct types. The first type of nanotubes shows no or weak gate voltage dependence of the linear-response conductance. These nanotubes are identified as the metallic type. A strong gate dependence is observed for the second type of devices (as we will show below), which indicates that these tubes are semiconducting.

For metallic nanotubes, the measured two-terminal resistance is often dominated by the contact resistance between the nanotubes and the metal electrodes. In early transport measurements, the nanotubes typically formed a tunnel barrier of high resistance of ~ 1 MΩ with the electrodes. However, metallic nanotubes have an ideal intrinsic two-terminal resistance of only $h/4e^2$ or $6.5\,\mathrm{k}\Omega$ since there are only two propagating subbands crossing at the Fermi energy. Thus in these measurements, the bias voltage dropped almost entirely across the contacts, and tunneling phenomena dominated the transport. The high contact resistance is likely due to a combination of extrinsic factors such as granularity of the contacts and contamination at the interface, and intrinsic ones as have been considered theoretically by several authors [6,7,8]. In order to observe the intrinsic transport properties of metallic nanotubes and certain transport phenomena, low-resistance electrical contacts to the nanotubes must be used. *Soh* et al. [9] have made such contacts by evaporating metal on top of the nanotubes grown directly on a silicon chip with a chemical vapor deposition method. The measured two-terminal resistance was as low as $\sim 10\,\mathrm{k}\Omega$. Similar reduction in contact resistance was achieved by using planarized and briefly annealed gold electrodes [10]. In a third approach, a laser pulse was used to weld the two ends of suspended nanotubes into gold electrodes [11].

### 1.1.2 Field-Effect Transistors with Semiconducting Nanotubes

Figure 2 shows typical room-temperature current-voltage ($I$–$V$) characteristics for semiconducting nanotubes [12,13,14,15]. By sweeping the gate voltage from a positive value to negative, the $I$–$V$ curve is changed from highly nonlinear insulating behavior with a large gap to linear metallic behavior, and the linear-response conductance is increased by many orders of magnitude (see inset). The $I$–$V$ characteristics indicate that the nanotubes are hole-doped semiconductors and the devices behave as p-type Field-Effect Transistors (FETs). However, the exact doping and transport mechanism is unclear. Scanning tunneling spectroscopy measurements indicated that nanotubes are typically hole-doped by the underlying metal surface due to the high work function for the metal, and that the valence band edge of semiconducting tubes is pinned to the Fermi energy of the metal [16]. Based on this observation, *Tans* et al. explained the FET operation using a semiclassical band-bending model [12]. In addition, *Martel* et al. measured a large hole density that led them to suggest that the nanotubes are doped with acceptors, as a result of their processing [13]. They also suggested that transport is diffusive. Interestingly, it has been proposed that semiconducting nanotubes are more sensitive to disorder than their metallic counterparts [17,18]. By using a conducting tip in an AFM as a local gate, *Tans* et al. [19] and Bachtold et al. [20] found significant potential fluctuations along semiconducting nanotubes. However, the microscopic origin of the disorder remains to be sorted out.

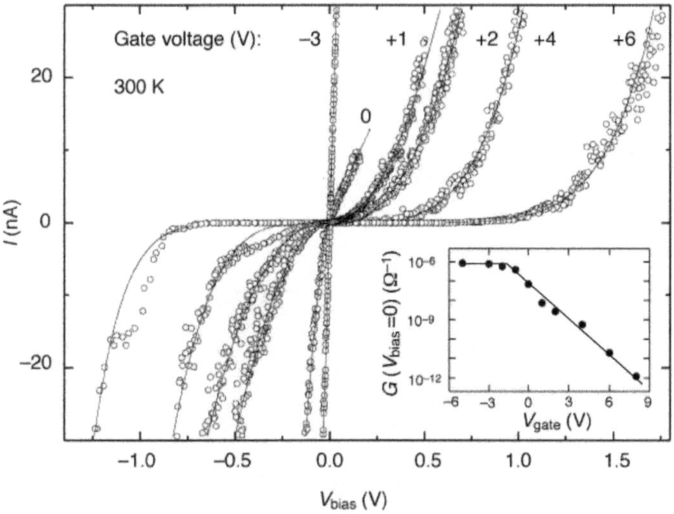


**Fig. 2.** Transport characteristics for a field-effect transistor employing semiconducting nanotubes (from [12])

## 1.2 Low-Temperature Mesoscopic Electron Transport

Carbon nanotubes have displayed a variety of transport phenomena at low temperatures which used to be manifest in mesoscopic metallic conductors or semiconductor quantum wires and dots. Detailed energy level spectra and spin states in metallic nanotubes were studied in the Coulomb blockade regime. Phase-coherent transport was employed to show proximity-induced superconductivity.

### 1.2.1 Coulomb Blockade and Transport Spectroscopy

For metallic nanotubes exhibiting high contact resistance with the electrical leads, the low temperature transport is dominated by the Coulomb blockade effect [4,5,21,22,23,24]. Figure 3a shows an example of the linear-response conductance versus gate voltage for a metallic nanotube rope measured at 100 mK in a dilution refrigerator. It exhibits a quasi-periodic sequence of sharp peaks separated by zero-conductance regions, which signifies Coulomb blockade single-electron charging behavior. Such phenomena have been well-studied in semiconductor quantum dots and small metallic grains [25,26]. Coulomb blockade occurs at low temperature when the total capacitance $C$ of a conducting island (which is weakly coupled to source and drain leads through tunnel barriers with a resistance larger than the quantum resistance $h/e^2 \approx 26\,\mathrm{k}\Omega$), is so small that adding even a single electron requires an electrostatic energy $E_\mathrm{c} = e^2/2C$ that is larger than the thermal energy $k_\mathrm{B}T$. For an estimate, the capacitance of a nanotube at a distance $z$ away from a con-

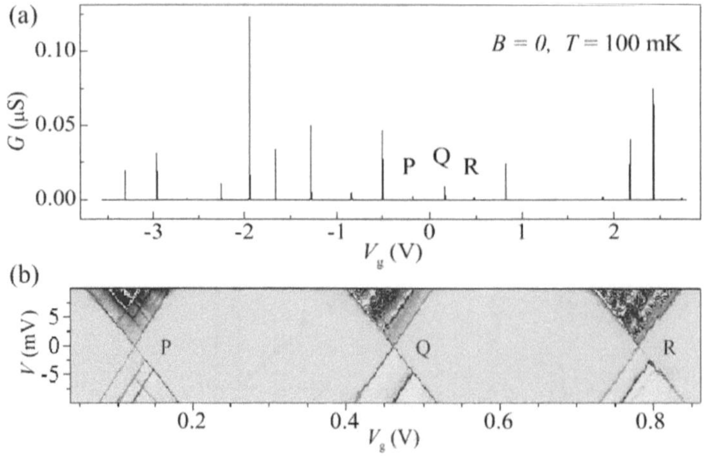


**Fig. 3.** Typical low-temperature Coulomb blockade measurements (from [2]). (**a**) Linear-response conductance as a function of gate voltage. (**b**) Greyscale plot of differential conductance versus bias and gate voltages

ducting substrate is $C = 2\pi\epsilon_r\epsilon_0 L/\ln(2z/r)$, with $\epsilon_r$ being the average dielectric constant of the environment, $r$ and $L$ the nanotube's radius and length. Using $\epsilon_r \approx 2$ (for comparison, $\epsilon_r = 3.9$ for $SiO_2$), $z = 300\,\text{nm}$ and $r = 0.7\,\text{nm}$, the charging energy is $E_c \approx 5\ \text{meV}/L\,(\mu\text{m})$. Thus for a typical $\sim$1 µm long tube, Coulomb blockade would set in below $\sim$50 K ($k_B T = 5\,\text{meV}$). For small electrode spacing, the total capacitance is often dominated by the capacitance to the leads. Another relevant energy scale in the Coulomb blockade regime is the level splitting due to the finite size of the nanotube. A simple 'particle-in-a-box' estimate gives $\Delta E = hv_F/4L \approx 1\ \text{meV}/L\,(\mu\text{m})$, where $h$ is the Planck's constant, $v_F = 8.1 \times 10^5\,\text{m/s}$ the Fermi velocity in the nanotube and a factor of 2 has been introduced to account for the two subbands near the Fermi energy. Note that both $E_c$ and $\Delta E$ scale inversely with length (up to a logarithmic factor), and the ratio $E_c/\Delta E$ is thus roughly independent of length, i.e., the level spacing is always a small but appreciable fraction of the charging energy.

In Fig. 3a, each conductance peak represents the addition of an extra electron. The peak spacing is given by $\Delta V_g = (2E_c + \Delta E)/e\alpha$, where $\alpha = C_g/C$, with $C_g$ and $C$ being the capacitance, to the gate and the total capacitance respectively, converts the gate voltage into the corresponding electrostatic potential change in the nanotube. At low temperatures, the height of the conductance peak varies inversely with temperature whereas the peak width is proportional to the temperature [4,5]. This shows that transport takes place via resonant *tunneling* through discrete energy levels and the electronic wavefunctions are extended between the contacts. It is remarkable that metallic nanotubes are coherent quantum wires at mK temperature.

For a fixed gate voltage, the current shows a stepwise increase with increasing bias voltage, yielding the excited-state spectrum. Each step in the current (or peak in the differential conductance) is associated with a new higher-lying energy level that enters the bias window. A typical plot of the differential conductance $dI/dV$ as a function of both bias voltage and gate voltage is shown in Fig. 3b. Within each of the diamonds (the full diamonds are not displayed due to the finite bias window), the number of electrons on the nanotube is fixed and the current is blockaded. The boundary of each diamond represents the transition between $N$ and $N+1$ electrons and the parallel lines outside the diamonds correspond to excited states. Such a plot is well-understood within the constant-interaction model, in which the capacitance is assumed to be independent of the electronic states. However, Tans et al. reported significant deviations from this simple picture [22]. In their transport spectroscopy, changes in slopes or kinks are observed in transition lines bordering the Coulomb blockade region. These kinks can be explained by a gate-voltage-induced transition between different electronic states with varying capacitances, which could result from state-dependent screening properties or different charge density profiles for different single-particle states.

The ground-state spin configuration in a nanotube was revealed by studying the transport spectrum in a magnetic field $B$. Experiments by *Cobden* et al. [21] showed that the level spectrum is split by the Zeeman energy $g\mu_B B$, where $\mu_B$ is the Bohr magneton, and the $g$ factor is found to be 2 indicating the absence of orbital effects as expected for nanotubes; the total spin of the ground state alternates between 0 and 1/2 as successive electrons are added, demonstrating simple shell-filling, or even-odd, effect, i.e., successive electrons occupy the levels in spin-up and spin-down pairs. In contrast, *Tans* et al. [22] found that the spin degeneracy has already been lifted at zero magnetic field and all the electrons enter the nanotubes with the same spin direction. This behavior can be explained by a model in which spin-polarized states can result from spin flips induced by the gate voltage. Indeed a microscopic model was proposed by *Oreg* et al. [27] considering nonuniform gate coupling to the nanotube due to the screening of the source and drain electrodes. Further experimental and theoretical work is needed to address this intriguing issue.

### 1.2.2 Superconducting Proximity Effect

A normal metal in close contact with a superconductor will acquire superconductivity within a characteristic lengthscale given by the phase coherence length in the normal conductor. This proximity effect has been extensively studied in macroscopic samples and patterned two-dimensional electron gas structures. Carbon nanotubes are particularly interesting for investigating this phenomenon due to their long phase coherence lengths. Moreover, novel phenomena were predicted for 1D quantum wires in contact with superconductors [28].

*Kazumov* et al. [11] observed Josephson supercurrents in individual nanotubes and ropes of nanotubes suspended between two superconducting bilayer electrodes (gold/rhenium and gold/tantalum) when the devices were cooled below the superconducting transition temperature of the contacts. The nanotubes were embedded in the molten gold top layer (which becomes superconducting due to the proximity effect) to ensure low nanotube–electrode junction resistance – a key to observing the proximity effect. For ropes of nanotubes, the maximum supercurrent agreed with theoretical predictions of $\pi\Delta/eR_{\mathrm{N}}$ where $\Delta$ is the superconducting energy gap and $R_{\mathrm{N}}$ the normal-state resistance of the junctions. However, the measurement on an individual nanotube gave a supercurrent that is 40 times higher than expected – a surprising result that has yet to be explained. In contrast, *Morpurgo* et al. [29] did not observe supercurrents through nanotubes sandwiched between two niobium contacts ($T_{\mathrm{c}} \sim 9.2$ K) despite the fact that their samples had lower normal-state resistances. At 4.2 K, a backgate was used to tune the niobium-nanotube interface transparency. When the transparency is high, the transport across the interface is dominated by Andreev reflection processes (in which an incident electron is converted into a Cooper pair leaving a reflected hole in the normal region) which results in a dip around zero bias in the differential resistance. When the transparency is tuned low, the normal tunneling process dominates, resulting in a peak in differential resistance. The absence of a supercurrent and subharmonic gap structure due to multiple Andreev reflections, and the emergence of a sharp zero-bias differential-resistance peak superimposed on the Andreev dip at below ~4 K seem to point to possible electron–electron correlation effects in carbon nanotubes.

## 1.3 Electron–Electron and Electron–Phonon Interactions

At high temperature where the Coulomb blockade is unimportant, tunneling into a metallic nanotube is consistent with predictions for tunneling into a Luttinger liquid which is formed due to strong electron interactions. This is described in 1.3.1. Section 1.3.2 treats the effects of electron–phonon interactions at high electric fields. Large-bias current-voltage characteristics using low-resistance contacts suggest that the main scattering mechanism for high-energy electrons is due to optical or zone-boundary phonons.

### 1.3.1 Electron Correlations and Luttinger-Liquid Behavior

It has been known for decades that electron–electron interactions are of paramount importance in 1D metals [30,31]. As a result, electrons form a correlated ground state called the Luttinger Liquid (LL), which is characterized by some exotic properties such as low-energy charge and spin excitations that propagate with different velocities, and a tunneling Density of States (DOS) that is suppressed as a power-law function of energy, i.e., $\rho(E) \propto |E - E_{\mathrm{F}}|^{\alpha}$. SWNTs are truly 1D conductors and thus are expected to behave as LLs. This has been theoretically considered by a number of authors [32,33,34]. The strength of electron interactions in an LL is described by the so-called Luttinger parameter $g$. For non-interacting electrons $g = 1$, whereas for repulsive Coulomb interactions $g < 1$. An estimate of the $g$ parameter in nanotubes is given by $g = (1 + 4E_c/\Delta E)^{-1/2}$, where $E_c$ and $\Delta E$ are the charging energy and level spacing as defined earlier. Using previous estimates, we obtain $g \approx 0.22$. Thus a carbon nanotube is a strongly correlated system. Interestingly, tunneling into the end of an LL is more strongly suppressed than into the bulk, due to the decreased ability of a charge to spread away from the end compared to the bulk. In carbon nanotubes, the exponent $\alpha$ in the tunneling DOS is related to $g$ as $\alpha_{\text{bulk}} = (g^{-1} + g - 2)/8$ and $\alpha_{\text{end}} = (g^{-1} - 1)/4$ respectively. Since $g$ is much less than unity, $\alpha_{\text{end}}$ is nearly twice as large as $\alpha_{\text{bulk}}$.

Power-law transport characteristics associated with the LL behavior have been indeed observed in metallic ropes of nanotubes as well as individual tubes [35,36]. Interestingly, for observing this behavior it is helpful to have poor contact between the electrodes and the tubes. The tunneling conductance in the linear-response regime $G$ should vary as $G(T) \propto T^{\alpha}$ (for $eV \ll k_BT$), i.e., opposite to the behavior expected for conventional metals. In addition, the differential conductance $dI/dV$ at large bias ($eV \gg k_BT$) should vary as $dI/dV \propto V^{\alpha}$. Figures 4a and b show typical measurements of *Bockrath* et al. [35] of bulk tunneling into an individual rope lying on top of the electrodes. The power-law behavior is complicated by the Coulomb blockade effect which sets in at low temperature causing additional suppression in the conductance. However, this can be corrected from the known temperature dependence for the Coulomb blockade, as shown by the dashed lines in (a). Fitting of the power-law temperature and voltage dependences yields the bulk-tunneling exponent which falls within the range expected for a Luttinger parameter $g \approx 0.2$–$0.3$. The data also exhibit the expected universal scaling behavior [35,39], i.e., the measured $dI/dV$ at different temperatures can be collapsed onto a single curve by plotting $(dI/dV)/T^{\alpha}$ against $eV/k_BT$ (as shown in Fig. 4b). Signatures for end tunneling were observed in samples with top contacts [35], where the lithography process cuts the tubes between the electrodes resulting in making contacts to the tube ends.

The LL behavior was recently investigated in transport measurements across a naturally occurring kink junction between two metallic nanotube segments (shown in Fig. 4c) [37]. The kink is associated with a pentagon-heptagon defect pair in the nanotube which acts as a tunnel barrier [38] separating two LLs. To a first-order approximation, the tunneling conductance across the kink is proportional to the product of the end-tunneling DOS on both sides and therefore varies again as a power law of energy, but with an exponent twice as large as the end tunneling [31,39], which makes the power-law behavior particularly pronounced in this case. Both $G$–$T$ (Fig. 4(c)) and $dI/dV$–$V$ measurements across the kink junction give the same power-law ex-

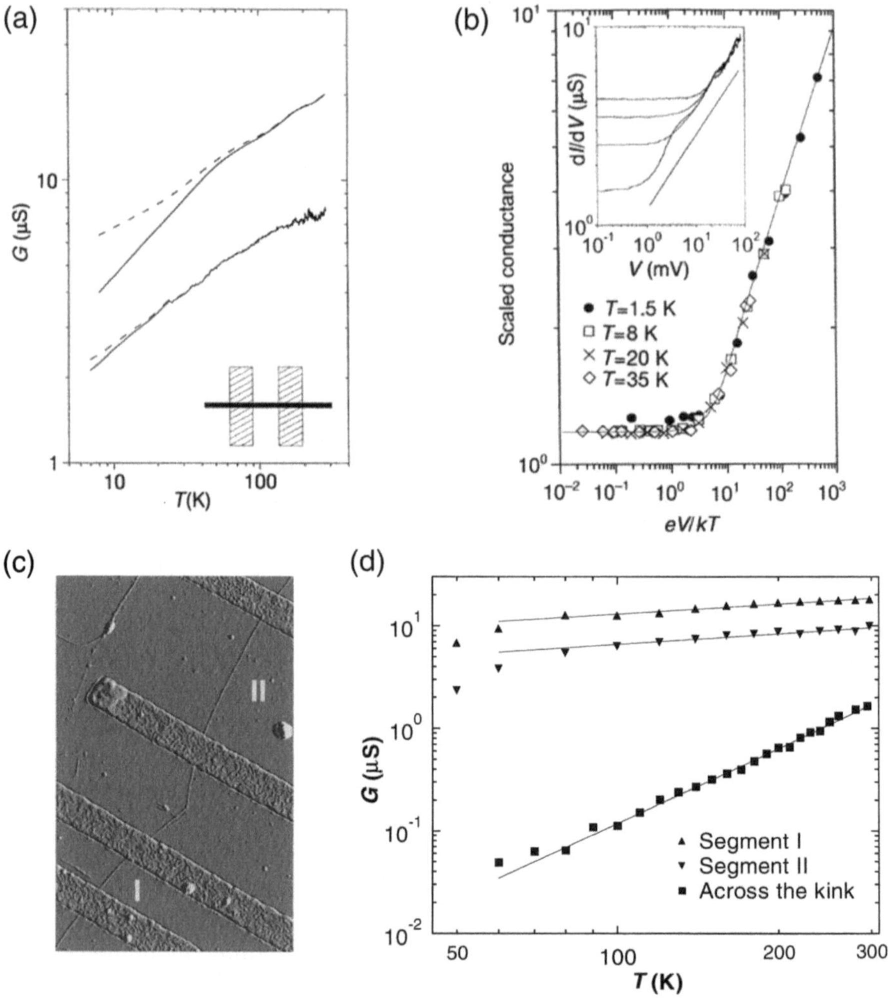


**Fig. 4.** Luttinger liquid behavior in carbon nanotubes. (**a**) and (**b**) plot $G$ vs $T$ and scaled $dI/dV$ vs $eV/k_{\mathrm{B}}T$ corresponding to bulk tunneling on double logarithmic scale (from [35]). (**c**) AFM image of a nanotube kink. (**b**) $G$ vs $T$ across the kink showing tunneling between the ends of two Luttinger liquids (from [37])

ponent of ~2, consistent with end-to-end tunneling between two LLs. Moreover, both end-to-end tunneling and bulk tunneling, which were observed in the same sample, gave the same $g$ parameter, thus providing strong evidence for the LL model.

It appears that the LL model can also explain the transport phenomena in a variety of other structures such as buckled and crossed nanotubes [36] (see Sect. 1.4). The fact that the LL effects persist even up to room temperature may have important implications for future devices based on nanotubes, i.e., the electron-correlation effects will have to be included in designing these molecular devices.

### 1.3.2 Large-Bias Transport and Phonon Scattering

Electron–phonon interactions at high electric fields were investigated by *Yao* et al. [10] using low-resistance contacts. In contrast to high-contact-resistance samples where tunneling across the contacts dominates the transport, a bias voltage applied between two low-resistance contacts establishes an electric field inside the nanotubes which accelerates the electrons, enabling studies of intrinsic transport properties of high-energy electrons. Shown in Fig. 1.3.1 are $I$–$V$ curves measured up to 5 V at different temperatures. The curves essentially overlap with each other. Remarkably, the current at 5 V exceeds 20 μA, which corresponds to a current density of more than $10^9\,\mathrm{A/cm^2}$. The current seems to saturate at large bias. It turns out that the resistance, $R = V/I$, can be fit very well with a simple linear function of $V$ (right inset): $R = R_0 + V/I_0$, where $R_0$ and $I_0$ are constants with $I_0$ being the extrapolated saturation current.

The behavior can be explained by considering optical or zone-boundary phonons as the main scattering mechanism for high-energy electrons. As shown in the schematic in the left inset to Fig. 1.3.1, assuming that the electron–phonon coupling is strong enough, once an electron gains enough energy to emit an optical or zone-boundary phonon of energy $\hbar\Omega$, it is immediately backscattered. A steady state is then established in which the right moving electrons are populated to an energy $\hbar\Omega$ higher than the left moving ones, leading to a saturation current $I_0 = (4e/h)\hbar\Omega$. Taking a typical phonon energy of $\hbar\Omega = 0.16\,\mathrm{eV}$ [40], this leads to a saturation current of 25 μA which agrees favorably with experiments. In this picture, the mean free path $\ell_\Omega$ for backscattering phonons is just the distance an electron must travel to reach the phonon threshold: $\ell_\Omega = \hbar\Omega L/eV$ with $L$ being the electrode spacing. This may be combined with a constant elastic scattering term with mean

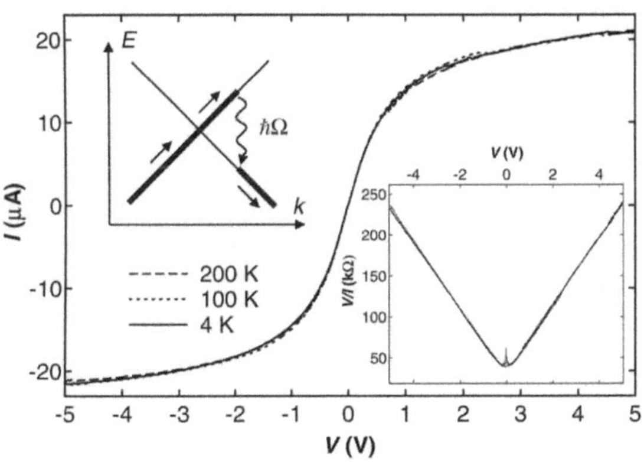


**Fig. 5.** High-field $I$–$V$ characteristics in metallic nanotubes at different temperatures. The *right inset* plots $V/I$ vs $V$. The *left inset* shows a schematic of the phonon emission model

free path $\ell_e$ to obtain a resistance $R = (h/4e^2)L(\ell_e^{-1} + \ell_\Omega^{-1})$, which then has exactly the same form as the simple phenomenological function. Further work is expected to address the heat dissipation issue and the role of electron-correlation effects.

## 1.4 Nanotube Junctions and Electromechanical Effects

Nanotube junctions constitute building blocks for integrated nanotube-based electronic devices. Local junctions can be formed by introducing topological defects into individual nanotube molecules, between two crossed nanotubes, or by mechanically deforming nanotubes.

### 1.4.1 Intramolecular Junctions

Intramolecular nanotube junctions have been theoretically proposed for a long time [41,42,43]. The basic idea is that by introducing a pentagon and a heptagon into the hexagonal carbon lattice, two pieces of carbon nanotubes with different atomic and electronic structures can be seamlessly fused together to create metal-metal, metal-semiconductor and semiconductor-semiconductor junctions within individual nanotube molecules. Such junctions generally have the shape of a sharp kink. *Yao* et al. [37] have measured isolated kink junctions connected to metal electrodes. A metal-metal junction has been discussed in the previous subsection. The inset to Fig. 6 shows the AFM image of another kinked nanotube lying across three electrodes. The left segment is identified as metallic. The transport characteristic across the kink, in strong contrast, shows highly rectifying behavior with nonlinear and asymmetric $I$–$V$ curve, which can be strongly modulated by a voltage applied to the backgate. The kink is thus unambiguously associated with a metal-semiconductor nanotube heterojunction.

The details of the observed $I$–$V$ characteristics are not well-understood. For example, neither the bias polarity nor the onset of conduction can be simply explained in terms of the Schottky barrier at the junction interface that one would expect from p-type doping behavior and band bending. To model the data, one in principle has to properly take into account the effect of the metal electrodes and in particular the unique screening properties in 1D conductors, which have been theoretically investigated by *Leonard* et al. [44] and *Odintsov* [45]. In nanotubes, the screening occurs by rings of charges or dipole rings if image charges are taken into account. The potential induced by a dipole ring falls off as $1/z^2$ as a function of distance $z$ which makes screening much less effective. This introduces novel length scales in the charge transfer phenomena in nanotube junctions. A better understanding is important for the designing and functioning of nanotube devices, or more generally, 1D molecular devices.

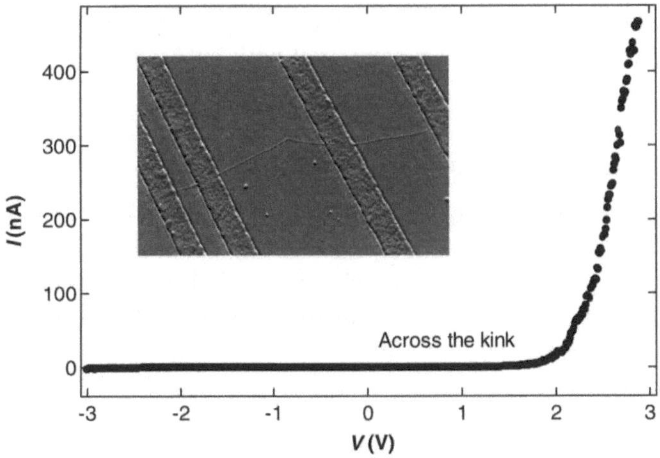


**Fig. 6.** $I$–$V$ for a rectifying diode from a metal-semiconductor heterojunction. The inset shows an AFM image of the device (from [37])

### 1.4.2 Crossed Junctions

While it remains a challenge to controllably grow nanotube intramolecular junctions for possible electronic applications, a potentially viable alternative would be to construct junctions between two different nanotubes. *Fuhrer* et al. [46] have measured devices consisting of two naturally occurring crossed nanotubes with electrical contacts at each end of each nanotube. Both metal-metal and semiconductor-semiconductor junctions exhibit high tunneling conductances on the order of $0.1\,e^2/h$. Theoretical study indicates that the contact force between the tubes is responsible for the high transmission probability of the junctions. Metal-semiconductor junctions showed asymmetry in the $I$–$V$ curves and these results appeared to be understood well from the formation of a Schottky barrier at the junction. A novel three-terminal rectifying behavior was demonstrated in this crossed geometry.

*Lefebvre* et al. [47] have measured a circuit consisting of two crossed SWNT ropes. The junction $I$–$V$ was found to be highly nonlinear with a 0.2 V gap, which enabled the top rope to be used as a local gate at small voltages. Gate sweeps at low temperature showed that the bottom rope behaves as double quantum dots in series, separated by a tunnel barrier at the junction which could be introduced from electrostatic screening or from mechanical deformation by the top rope.

Crossed nanotube junctions can also be created by mechanical manipulation using the tip of an atomic force microscope [36,47]. For example, *Postma* et al. [36] used an AFM tip to first cut a nanotube which was lying across two electrodes, and then push one segment on top of the other. Transport across such a junction exhibited power-law characteristics consistent with bulk-to-bulk tunneling between two Luttinger liquids. It is interesting to note that the results suggested that the contact force or the mechanical deformation was quite small in this case compared with naturally occurring crossed junctions.

### 1.4.3 Electromechanical Effects

The coupling between structural distortions and electrical properties of carbon nanotubes represents another possibility for creating nanotube junctions and promises potential applications in nano-electromechanical devices. Earlier on it was realized that nanotubes can be distorted as a result of the van der Waals interaction between the nanotubes and the substrates, and that controlled deformation could be easily induced by the use of the tip of an AFM [48].

The effects of bending on the electronic properties of the nanotubes have been theoretically investigated by a number of authors [8,49,50,51]. In the $\pi$-electron approximation, the electronic properties of metallic armchair nanotubes remain essentially unchanged upon small bending deformation [49]. However, for strongly bent carbon nanotubes, $\sigma$–$\pi$ hybridization effects have to be taken into account. Electronic structure calculations involving both $s$ and $p$ electrons showed that strong bending introduces localized density of states [50] that may lead to localization. Further calculations showed that drastic decrease in the conductance of metallic tubes occurs upon buckling [8,51].

Experimentally, *Bezryadin* et al. [23] found that local bending of a nanotube near the electrodes causes local barriers and breaks the nanotube into multiple quantum dots at low temperature. AFM proves especially useful in controllably introducing mechanical deformation. *Postma* et al. [36] measured the transport across a buckled nanotube manipulated from an initially straight nanotube by an AFM. The buckle was found to act as a tunnel barrier and the transport showed end-to-end tunneling behavior between two Luttinger liquids. Recently *Tombler* et al. [52] conducted in situ transport measurement of suspended nanotubes which were reversibly bent by an AFM tip. They found that conductance of the nanotubes can be decreased by two orders of magnitude upon a bending angle up to 14°. Simulations showed that the large suppression in conductance originates from the highly deformed region near the AFM tip where local bonding is changed from $sp^2$ to nearly $sp^3$ configuration.

## 1.5 Summary

In this section, we have reviewed recent experimental results on electrical transport through individual single-wall carbon nanotubes. Nanotubes prove to be a remarkably versatile system for exploring both fundamental physics and promising electronic device applications. Future experiments are expected to reveal other interesting aspects of 1D physics, such as spin-charge separation associated with a Luttinger liquid. It would be of interest to explore transport using ferromagnetic and superconducting contacts and to study the charge dynamics by shot noise measurements across a single impurity.

# 2 Transport in Single Wall Nanotube Ropes

The primary products of the synthesis of SWNTs are aggregates of individual tubes forming structures akin to ropes of fibers [3,53]. The driving forces for the formation of these ropes are the strong van der Waals interactions between the individual tubes which lead to the formation of a regular triangular lattice of nanotubes within the rope [3,53]. It is from these ropes that individual SWNTs are prepared usually by sonication. Many studies and applications of SWNTs have been based on ropes, and their findings are presented in the different chapters of this book. Some transport studies involving SWNT ropes that have been discussed in terms of the behavior of individual SWNTs [2,5,11,21,35] have already been discussed in Sect. 1. Here we focus on aspects of electrical transport in SWNT ropes that were not discussed in the studies of transport in individual nanotubes, and on inter-nanotube transport within ropes. Specifically, we will concentrate on issues such as the influence of long coherence lengths and the effects of backscattering on transport in these systems, and the effects of magnetic field on transport.

## 2.1 Scattering and Localization Phenomena

Knowledge of the transport mechanism is an essential requirement if nanotubes are to be used in nanoelectronics. The nature of transport determines fundamental properties such as the amount and location of energy dissipation and the speed of signal propagation. The transport mechanism, however, is not determined only by the intrinsic properties of the nanotube itself, but also by the presence of scatterers such as defects, impurities, structural distortions, tube-tube interactions in ropes, substrate charges, and the nature of the contacts to the macroscopic leads. Thus, while a single metallic nanotube should ideally have a conductance of $4e^2/h$ [38], a much lower conductance is usually observed. Experimentally, it is not always possible to disentangle the multiple contributions to the measured resistance of a nanotube.

### 2.1.1 Weak Localization

To illustrate the possible complex origin of a non-ideal conductance, we present in Fig. 7 the results of a calculation of the conductance of a (6,6) metallic tube connected to two model metal pads. The clean nanotube is doped with first one and then two oxygen atoms. The two O atoms are seen to act like a potential well or a Fabry–Perot interferometer for Fermi level electrons. Constructive interference between incident electrons and electrons reflected from the second O atom is responsible for the partial "transparencies" seen when the separation between the O atoms is an integral multiple of $\pi/k_F$. Figure 7 also gives the computed reduction in the conductance due to the imperfect coupling at the metal–nanotube contacts [54].

One powerful experimental technique with which the nature and importance of collisions within conductors can be evaluated involves Magneto-Resistance (MR) measurements. So far this technique has been applied on multi-wall nanotubes (see chapter by Forro and Schoenenberger in this book) and on SWNT mats [56] and rings [57].

In the following we will discuss MR, and illustrate a number of different electrical transport phenomena using the behavior of SWNT rings as examples. SWNT rings are formed as a result of the van der Waals interactions between the two end segments of the same nanotube. This adhesive interaction can overcome the increased strain generated by bending the nanotube into a circular loop (see chapter by Yakobson and Avouris in this book). Stable rings composed of SWNT ropes have been prepared at high yields [58,59]. Figure 8a shows low temperature MR measurements on a 0.82μm diameter SWNT ring. The magnetic field is normal to the plane of the ring. It is seen that the resistance of the ring decreases with increasing magnetic field strength. The observation of such *negative* MR is considered as an indication that the conductor is in a state of *weak localization* [61,62,63]. Negative MR was also observed in the case of multi-wall nanotubes [60,55] and SWNT

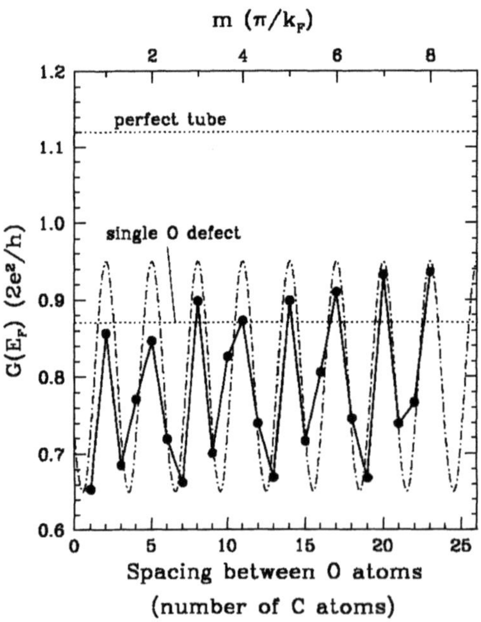


**Fig. 7.** Calculated conductance at $E_F$ of a (6,6) nanotube doped with one and then two oxygen atoms as a function of the distance between these atoms along the axis of the nanotube. The reduction in the conductance due to the partially reflective contacts to gold electrodes, to which the nanotube is end-bonded, is indicated by the top dotted line

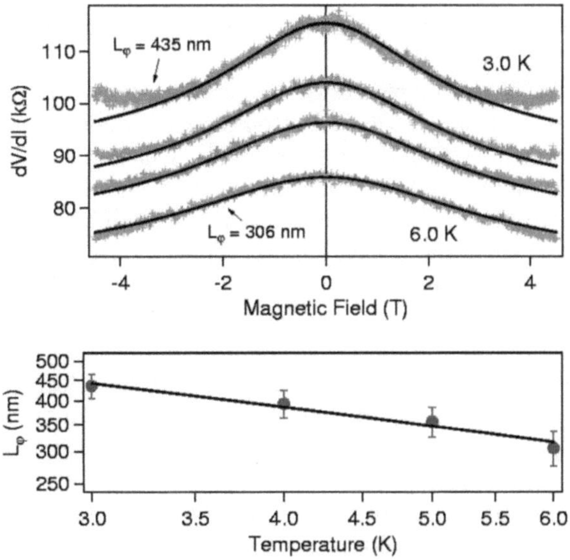


**Fig. 8.** *Top:* Differential resistance $\mathrm{d}V/\mathrm{d}I$ of a single-wall nanotube ring as a function of the strength of a magnetic field perpendicular to the plane of the ring. The probe current was 10 pA. The four curves (in gray) were obtained, from top to bottom, at 3.00, 4.00, 5.00 and 6.00 K. The *solid black lines* are fits to one-dimensional weak localization theory. *Bottom:* The dependence of the coherence length on temperature. The line is a fit to $L_\varphi \propto T^{-1/3}$ [57]

mats [56], and was also interpreted in terms of Weak Localization (WL). WL implies that $L_\mathrm{e} \ll L_\phi < L$, where $L_\mathrm{e}$, $L_\phi$, and L are the elastic mean free path, coherence length, and nanotube length, respectively. WL involves *coherent back-scattering* processes. Especially important in this respect are self-intersecting electron (hole) trajectories. As a result of time-reversal invariance, an electron entering such a trajectory can be thought of as splitting into counter-propagating waves (conjugate waves) which can interfere when they meet again at the origin. In the absence of dephasing processes the interference will be constructive, leading to an enhanced back-scattering and an increase in the resistance of the nanotube. However, if the electron were to undergo dephasing collisions during its travel around the path, then the resistance increase will be lower, or absent in the case of complete dephasing. By measuring the contribution to the total resistance of coherent back-scattering processes one can then determine the importance of inelastic collisions in the nanotube. By applying the external magnetic field, time-reversal invariance is destroyed, the conjugate waves acquire opposite phase shifts and the resulting destructive interference eliminates the resistance due to coherent back-scattering.

The $\mathrm{d}I/\mathrm{d}V$ vs. $H$ curves shown in Fig. 8 can be fit (dark curves) to the equations describing weak localization to extract the coherence length $L_\phi$ as a function of temperature. The coherence lengths increase as the temperature is reduced, reaching about 450 nm at 3 K [57].

Information on the nature of the interactions that lead to dephasing in the nanotube rings was obtained from the observed temperature dependence of $L_\phi$. It is important to remember that dephasing involves inelastic collisions, while elastic collisions involve only momentum relaxation. As Fig. 8 shows, $L_\varphi \propto T^{-1/3}$ [57]. This behavior is characteristic of dephasing by weakly inelastic electron–electron interactions [64,65]. The fact that these interactions dominate the dephasing mechanism at low temperatures is not surprising, given that carbon nanotubes are strongly correlated electron systems (see discussion in Sect. 1.3). Most importantly, defects and disorder not only affect back-scattering but also increase the strength of electron–electron interactions [64]. MR studies have also shown that electron–electron interactions dominate the low temperature dephasing in multi-wall nanotubes [60] and other low-dimensional systems [66].

Inelastic electron–electron interactions in SWNT ropes were also studied by photo-exciting the SWNT ropes and following the relaxation in real time using femtosecond time-resolved photoemission [67]. It was found that the electron system thermalized very fast, within about 200 fs, as a result of electron–electron interactions (electron–phonon relaxation was found to be much slower) to yield a distribution that was well described by a Fermi–Dirac distribution.

Another manifestation of strong electron–electron interactions involves the appearance of a Zero-Bias Anomaly (ZBA) or Fermi Level Singularity (FLS) due to the blocking of tunneling by Coulomb interactions. Figure 9 shows such an anomaly in the case of a SWNT ring appearing as a resistance peak near zero bias.

Under appropriate conditions, the tunneling of an electron from a metal lead into the nanotube can be thought of as a two step process involving a fast transit throught the tunnel barrier followed by a slower spreading of the electron charge. Before spreading, the electron cloud interacts repulsively

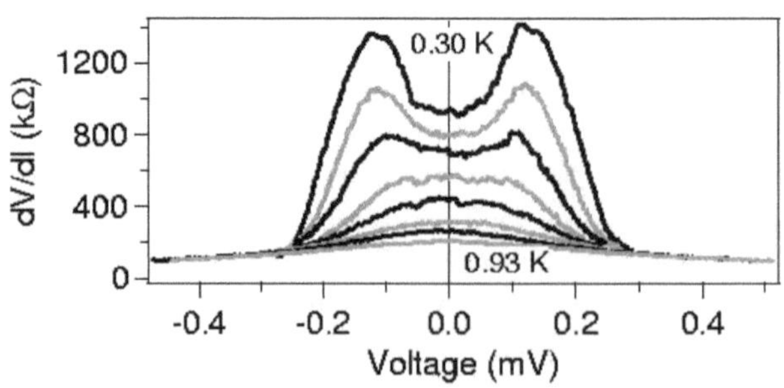


**Fig. 9.** Differential resistance of a SWNT ring as a function of applied bias at 0.30, 0.40, 0.50, 0.63, 0.70, 0.81 and 0.93 K. The curves are not offset and were measured at zero magnetic field [57]

with itself and the other electrons producing a Coulomb barrier (not Coulomb blockade) that dissipates as the charge spreads. It is this Coulomb barrier that is responsible for the ZBA, for the power law dependence of the conductance on the voltage bias which was discussed in Sect. 1.3 in the context of Luttinger liquids, and the difference between end and bulk coupling of nanotubes and leads. It should be noted, however, that the ZBA and the associated power laws are not uniquely associated with the Luttinger liquid state. In fact, they have first been observed and are well documented in the case of disordered Fermi liquids [68,69].

While a Luttinger liquid description is appropriate for an isolated non-interacting nanotube, it may not be appropriate for a rope. As *Komnik* and *Egger* [70] have pointed out, for a rope to be in a Luttinger liquid state three conditions must be met: (a) only one tube is contacted by the leads, (b) most tubes in the rope are not metallic, and (c) electron tunneling between different metallic tubes in the rope can be neglected. Given the fragility of the Luttinger liquid state, and the fact that the theory of Luttinger liquids has not been developed so far to account for the effects of magnetic fields as reported here, the observations on nanotube ropes and rings have been discussed within the Fermi liquid model [57]. In any case, in both models the origin of the phenomena is strong electron correlation. A semiclassical model that describes Coulomb inhibition of tunneling and the resulting power law, is widely applicable, and involves experimentally observed quantities such as the conductance of the sample has been developed by *Levitov* and *Shytov* [69].

The temperature dependence of the resistance of isolated nanotubes in the absence of a magnetic field depends on the transport mechanism. A ballistic conductor will have a temperature independent conductance, while electron–phonon scattering will lead to a conductance linearly dependent on temperature. If, on the other hand, the conductance is dominated by tunneling at the contacts, then a power law dependence is expected. Given the weak electron–phonon coupling observed in SWNT ropes [67], the temperature dependence of the conductance of nanotubes at low temperature is expected to be weak and to be determined by electronic interactions. The resistance of a metallic system in the weak localization state is known to increase with decreasing temperature. This behavior reflects the increasing coherence length as the temperature is reduced which in turn increases the importance of coherent back-scattering processes. The conductance can be written as [63]: $1/R_{H-0} = \sigma_0 - C_0 T^{-p/2}$, where $\sigma_0$ and $C_0$ are sample specific constants and the value of the exponent $p$ depends on the dephasing mechanism. For the dominant dephasing mechanism that involves electron–electron collisions $p = 2/3$. The dependence of the resistance of the SWNT ring on temperature is shown in Fig. 10a along with the fit to the above equation. It should be noted that the equation describing the resistance in WL is completely analogous to that predicted for a Luttinger liquid with scatterers [71].

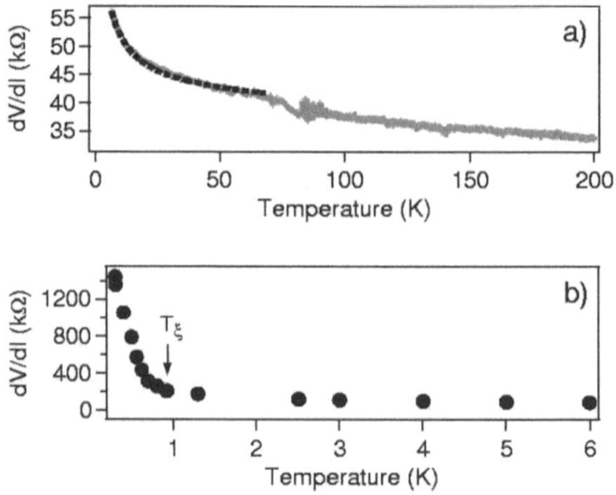


**Fig. 10.** (**a**) Temperature dependence of the zero-field SWNT ring resistance from 6-200 K (*gray curve*), and fit to one-dimensional weak localization theory (*dashed line*). (**b**) Ring resistance from 0.3 to 6 K [57]

### 2.1.2 Spin Effects

Spin effects on the transport properties of nanotubes are of particular interest given the proposed nanotubes' Luttinger liquid nature which implies a spin–charge separation. Anomalous thermopower and resistance behavior of SWNTs ropes at low temperatures has been interpreted in terms of a Kondo-type interaction between the magnetic moments of magnetic impurities from the catalyst and the spin of the $\pi$ electrons of the nanotube [72]. Returning to Fig. 9, we note that at very low temperatures (below about 0.7 K) the ZBA in Fig. 9 develops a cusp form. The central dip in resistance is evidence of *anti-localization*, i.e. enhanced forward scattering [62] produced, most likely, by spin–orbit scattering of $\pi$ electrons by the high-Z gold atoms of the leads [57]. A Kondo origin of this effect has also been proposed [73,74].

### 2.1.3 Strong Localization

The increasing coherence length of nanotubes upon cooling can lead to the formation of a strongly localized state. This will occur when the coherence lifetime $\tau_\phi$ becomes so long that the width of the coherent state $\hbar/\tau_\varphi$ becomes smaller than the energy separation between coherent states [75]. Localization in the context of nanotubes, but ignoring interference effects, was studied theoretically by *Kostyrko* et al. and *Roche* and *Saito* [76,77]. These theories are expected to be valid in cases where disorder is dominant over electron interactions. Experimentally strong localization has been observed in SWNT mats [78] and SWNT rings [57]. As Fig. 10b shows, continued cooling of the ring below 3 K leads at about 1 K to a sharp transition to a state of high resistance [57]. Below this point transport becomes thermally activated with the resistance following a $R \propto \exp(T_0/T)$ dependence with $T_0 \approx 0.8\ K$. This behavior is characteristic of a conductor in a strongly localized state.

This conclusion is further supported by corresponding changes in magneto-transport [57]. The effect of the magnetic field is now stronger, and the MR curves cannot be fit by weak localization theory. Side bands to the central MR peak appear, and based on the fact that their detailed structure is altered by mild thermal annealing and re-cooling [79], these bands are ascribed to *conductance fluctuations*, a low temperature phenomenon involving defect sites.

From knowing the localization temperature, the strong localization length $L_\xi$ can be estimated as the coherence length $L_\phi$ at the localization temperature (about 1 K). For this ring a localization length of $L_\xi$=750 nm is obtained [57]. This localization length implies a mean free path in the range of 250-350 nm [76,75] in good agreement with mean free path of about 300 nm deduced in studies of individual SWNTs [10].

## 2.2 Inter-Tube Transport in Nanotube Ropes

By analogy with graphite where coupling between the graphene layers produces a 3D conductor, one would expect that coupling between the nanotubes in a rope may lead to some degree of electronic delocalization within the rope. A number of studies of the electrical properties of ropes [5,35,57] have assumed a weak delocalization within the rope. This weak interaction is primarily due to the fact that current synthetic routes produce a mixture of tubes with different diameters and chiralities which make the ropes inhomogeneous. *Maarouf* et al. [80] explored theoretically the inter-tube transport process using a tight-binding model and concluded that the need to conserve crystal momentum along the tube axis limits inter-tube transport in ropes containing tubes with diffferent chiralities.

An experiment that allows the measurement of the weak inter-tube transport was performed by *Stahl* et al. [81] In this study, surface defects were deliberately introduced along the surface of a rope by ion bombardment, and the damaged areas were then covered by gold electrodes. At high temperatures the current flows through the damaged tubes that are in direct contact with the gold leads. As the temperature is decreased the two-terminal resistance rises as a result of localization in the damaged tubes. When their resistance becomes the same as the inter-tube (transfer) resistance, the current switches its path to an undamaged bulk nanotube. This is reflected by a sharp decreas in the four-terminal resistance. (Fig. 11) Analysis of the experimental results suggests that inter-tube transport involves electron tunneling with a penetration depth of about 1.25 nm.

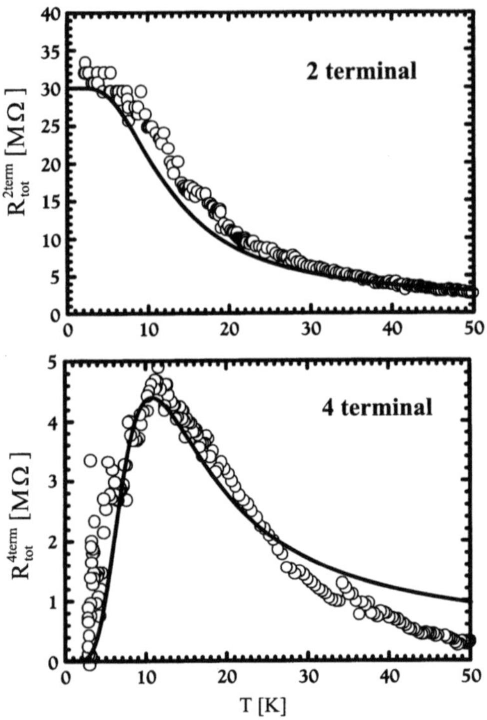


**Fig. 11.** Temperature dependence of the two-terminal and four-terminal resistance of a single-wall nanotube rope intentionally damaged directly under the gold contacts used to measure the resistance. The maximum in the resistance measured by the four-terminal measurement marks the point at which the resistance for inter-tube transport becomes equal to the resistance for transport along the tube at the same temperature

## 3 Conclusion

We may conclude that defects in the nanotube structure can have opposite effects on the electronic delocalization along the nanotube axis and perpendicular to it. Back-scattering from defects and disorder-enhanced electron–electron interactions can lead to localization within an individual tube, while relaxation of momentum conservation in a defective tube enhances inter-tube delocalization in a rope.

## References


1. C. Dekker: Phys. Today **52**, No. 5, 22 (1999)
2. J. Nygård, D. H. Cobden, M. Bockrath, P. L. McEuen, P. E. Lindelof: Appl. Phys. A **69**, 297 (1999)
3. A. Thess, R. Lee, P. Nikolaev, H. Dai, P. Petit, J. Robert, C. H. Xu, Y. H. Lee, S. G. Kim, A. G. Rinzler, D. T. Colbert, G. E. Scuseria, D. Tomanek, J. E. Fischer, R. E. Smalley: Science **273**, 483 (1996)
4. S. J. Tans, M. H. Devoret, H. Dai, A. Thess, R. E. Smalley, L. J. Geeligs, C. Dekker: Nature **386**, 474 (1997)
5. M. Bockrath, D. H. Cobden, P. L. McEuen, N. G. Chopra, A. Zettl, A. Thess, R. E. Smalley: Science **275**, 1922 (1997)

6. J. Tersoff: Appl. Phys. Lett. **74**, 2122 (1999)
7. M. P. Anantram, S. Datta, Y. Xue: Phys. Rev. **61**, 14219 (2000)
8. A. Rochefort, Ph. Avouris, F. Lesage, D. R. Salahub: Phys. Rev. B **60**, 13824 (1999)
9. H. T. Soh, C. F. Quate, A. F. Morpurgo, C. M. Marcus, J. Kong, H. Dai: Appl. Phys. Lett. **75**, 627 (1999)
10. Z. Yao, C. L. Kane, C. Dekker: Phys. Rev. Lett. **84**, 2941 (2000)
11. A.Yu. Kazumov, R. Deblock, M. Kociak, B. Reulet, H. Bouchiat, I. I. Khodos, Yu.B. Gorbatov, V. T. Volkov, C. Journet, M. Burghard: Science **284**, 1508 (1999)
12. S. J. Tans, A. R. M. Verschueren, C. Dekker: Nature **393**, 49 (1998)
13. R. Martel, T. Schmidt, H. R. Shea, T. Hertel, Ph. Avouris: Appl. Phys. Lett. **73**, 2447 (1998)
14. R. D. Antonov, A. T. Johnson: Phys. Rev. Lett. **83**, 3274 (1999)
15. C. Zhou, J. Kong, H. Dai: Appl. Phys. Lett. **76**, 1597 (2000)
16. J. W. G. Wildöer, L. C. Venema, A. G. Rinzler, R. E. Smalley, C. Dekker: Nature **391**, 59 (1998)
17. T. Ando, T. Nakanishi, R. Saito: J. Phys. Soc. Jpn. **67**, 1704 (1997)
18. P. L. McEuen, M. Bockrath, D. H. Cobden, Y.-G. Yoon, S. G. Louie, Phys. Rev. Lett. **83**, 5098 (1999)
19. S. J. Tans, C. Dekker: Nature, **404**, 834 (2000)
20. A. Bachtold, M. S. Fuhrer, S. Plyasunov, M. Forero, E. H. Anderson, A. Zettl, P. L. McEuen: Phys. Rev. Lett. **84**:(26), 6082-6085 (2000)
21. D. H. Cobden, M. Bockrath, P. L. McEuen, A. G. Rinzler, R. E. Smalley: Phys. Rev. Lett. **81**, 681 (1998)
22. S. J. Tans, M. H. Devoret, R. J. A. Groeneveld, C. Dekker: Nature **394**, 761 (1998)
23. A. Bezryadin, A. R. M. Verschueren, S. J. Tans, C. Dekker: Phys. Rev. Lett. **80**, 4036 (1998)
24. H. W. Ch. Postma, Z. Yao, C. Dekker: J. Low Temp. Phys. **118**, 495 (2000)
25. H. Grabert, M. H. Devoret (Eds.): *Single Electron Tunneling: Coulomb Blockade Phenomena in Nanostructure* (Plenum, New York 1992)
26. L. P. Kouwenhoven, L. L. Sohn, G. Schön (Eds.): *Mesoscopic Electron Transport* (Kluwer Academic, Dordrecht 1997)
27. Y. Oreg, K. Byczuk, B. L. Halperin: Phys. Rev. Lett. **85**:(2), 365-368 (2000)
28. R. Fazio, F. W. J. Hekking, A. A. Odintsov: Phys. Rev. Lett. **74**, 1843 (1995)
29. A. F. Morpurgo, J. Kong, C. M. Marcus, H. Dai: Science **286**, 263 (1999)
30. J. Voit: Rep. Prog. Phys. **57**, 977 (1995)
31. M. P. A. Fisher, L. I. Glazman: In *Mesoscopic Electron Transport*, L. P. Kouwenhoven, L. L. Sohn, G. Schön (Eds.) (Kluwer Academic, Dordrecht 1997)
32. R. Egger, A. O. Gogolin: Phys. Rev. Lett. **79**, 5082 (1997)
33. C. L. Kane, L. Balents, M. P. A. Fisher: Phys. Rev. Lett. **79**, 5086 (1997)
34. H. Yoshioka, A. A. Odintsov: Phys. Rev. Lett. **82**, 374 (1999)
35. M. Bockrath, D. H. Cobden, J. Lu, A. G. Rinzler, R. E. Smalley, L. Balents, P. L. McEuen: Nature **397**, 598 (1999)
36. H. W. Ch. Postma, M. de Jonge, Z. Yao, C. Dekker: Phys. Rev. Lett **62**:(16), R10653-R10656 (2000)
37. Z. Yao, H. W. Ch. Postma, L. Balents, C. Dekker: Nature **402**, 273 (1999)

38. L. Chico, L. X. Benedict, S. G. Louie, M. L. Cohen: Phys. Rev. B **54**, 2600 (1996)
39. L. Balents: Phys. Rev. Lett. **61**:(7), 4429-4432 (2000)
40. M. S. Dresselhaus, G. Dresselhaus, P. C. Eklund: *Science of Fullerenes and Carbon Nanotubes* (Academic, San Diego 1996)
41. L. Chico, V. H. Crespi, L. X. Benedict, S. G. Louie, M. L. Cohen: Phys. Rev. Lett. **76**, 971 (1996)
42. Ph. Lambin, A. Fonseca, J. P. Vigneron, J. B. Nagy, A. A. Lucas: Chem. Phys. Lett. **245**, 85 (1995)
43. R. Saito, G. Dresselhaus, M. S. Dresselhaus: Phys. Rev. B **53**, 2044 (1996)
44. F. Léonard, J. Tersoff: Phys. Rev. Lett. **83**, 5174 (1999)
45. A. A. Odintsov: Phys. Rev. Lett. **85**:(1), 150-153 (2000)
46. M. S. Fuhrer, J. Nygård, L. Shih, M. Forero, Y.-G. Yoon, M. S. C. Mazzoni, H. J. Choi, J. Ihm, S. G. Louie, A. Zettl, P. L. McEuen: Science **288**, 494 (2000)
47. J. Lefebvre, J. F. Lynch, M. Llaguno, M. Radosavljevic, A. T. Johnson: Appl. Phys. Lett. **75**, 3014 (1999)
48. T. Hertel, R. Martel, Ph. Avouris: J. Phys. Chem. B **103**, 910 (1998); Phys. Rev. B **58**, 13780 (1998)
49. C. L. Kane, E. J. Mele: Phys. Rev. Lett. **78**, 1932 (1997)
50. A. Rochefort, D. R. Salahub, Ph. Avouris: Chem. Phys. Lett. **297**, 45 (1998)
51. M. Buongiorno Nardelli, J. Bernholc: Phys. Rev. B **60**, R16338 (1999)
52. T. W. Tombler, C. Zhou, L. Alexseyev, J. Kong, H. Dai, L. Liu, C. S. Jayanthi, M. Tang, S.-Y. Wu: Nature **405**:(6788), 769-772 (2000)
53. C. Journet, W. K. Maser, P. Bernier, A. Loisseau, M. Lamydela, S. Lefrant, P. Deniard, R. Lee, J. E. Fisher, Nature **388**, 756 (1997)
54. A. Rochefort, P. Avouris, F. Lesage, D. R. Salahub: Phys. Rev. B-condensed matter **60**:(19), 13824-13830 (1999)
55. L. Langer, V. Bayot, E. Grivei, J. P. Issi, J. P. Heremans, C. H. Olk, L. Stockman, C. Van Haesendonck, Y. Bruynseraede, Phys. Rev. Lett. **76**, 497 (1996)
56. G. T. Kim, E. S. Choi, D. C. Kim, D. S. Suh, Y. W. Park, K. Liu, G. Duesberg, S. Roth, Phys. Rev. B **58**, 16064 (1998)
57. H. R. Shea, R. Martel, Ph. Avouris, Phys. Rev. Lett. **84**, 4441 (2000)
58. R. Martel, H. R. Shea, Ph. Avouris, Nature **398**, 299 (1999)
59. R. Martel, H. R. Shea, Ph. Avouris, J. Phys. Chem. B **103**, 7551 (1999)
60. A. Bachtold, C. Strunk, J. P. Salvetat, J. M. Bonard, L. Forro, T. Nussbaumer, C. Schonenberger, Nature **397**, 673 (1999)
61. S. Datta: *Electronic Transport in Masoscopic Systems* (Cambridge Univ. Press, Cambridge 1995)
62. G. Bergmann, Phys. Rep. **107**,1 (1984)
63. P. A. Lee, T. V. Ramakrishnan, Rev. Mod. Phys. **57**, 287 (1985)
64. B. L. Altshuler, A. G. Aronov, D. E. Khmelnitsky, J. Phys. C **15**, 7369 (1982)
65. B. L. Altshuler, A. G. Aronov, M. E. Gerhernson, Yu. V. Sharvin, Sov. Sci. Rev. A, Phys. Rev. **9**, 233 (1987)
66. Y. Imry, *Introduction to Mesoscopic Physics*(Oxford Univ. Press, Oxford 1997)
67. T. Hertel, G. Moos, Phys. Rev. Lett., **84**, 5002 (2000)
68. B. L. Altshuler, A. G. Aronov, P. A. Lee, Phys. Rev. Lett. **44**, 1288 (1980)
69. L. S. Levitov, A. V. Shytov, JETP Lett. **66**, 214 (1997)
70. A. Komnik, R. Egger, in *Electronic Properties of Novel Materials - Science and Technology of Nanostructures*, H. Kuzmany, J. Fink, S. Roth (Eds.) (Am. Inst. Phys., New York 1999)

71. R. Egger, A. O. Gogolin, Eur. Phys. J. B **3**, 281 (1998)
72. L. Grigorian, G. U. Sumanasekera, A. L. Loper, S. L. Fang, J. K. Allen, P. C. Eklund, Phys. Rev. B **60**, R11309 (1999)
73. Ph. Avouris, H. R. Shea, R. Martel, In *Electronic Properties of Novel Materials - Science and Technology of Nanostructures*, H. Kuzmany, J. Fink, S. Roth (Eds.) (Am. Inst. Phys., New York, 1999)
74. K. Harigawa, J. Phys. Soc. Jpn., **69**, 316 (2000)
75. D. J. Thouless, Phys. Rev. Lett. **39**, 1167 (1977)
76. T. Kostyrko, M. Bartkowiak, and G. D. Mahan, Phys. Rev. B **60**, 10735 (1999); Phys. Rev. B **59** 3241 (1999)
77. S. Roch, G. Dresselhaus, M. S. Dresselhaus, R. Saito, Phys. Rev. Lett. **62**, 16092-16099 (2000)
78. M. S. Furher, M. L. Cohen, A. Zettl, V. Crespi, Solid. State Commun. **109**, 105 (1999)
79. P. A. Lee, A. D. Stone, H. Fukuyama, Phys. Rev. B **35**, 1039 (1987)
80. A. A. Maarouf, C. L. Kane, E. J. Mele, Phys. Rev. B **61**, 11156 (2000)
81. H. Stahl, J. Appenzeller, R. Martel, Ph. Avouris, B. Lengeler, Phys. Rev. Lett., accepted for publication (2000)

# Scanning Probe Microscopy Studies of Carbon Nanotubes

Teri Wang Odom[1], Jason H. Hafner[1], and Charles M. Lieber[1,2]

[1] Department of Chemistry, Harvard University
Cambridge, MA 02138, USA
{teri,jason,cml}@cmliris.harvard.edu
[2] Division of Engineering and Applied Sciences, Harvard University
Cambridge, MA 02138, USA

**Abstract.** This paper summarizes scanning probe microscopy investigations of the properties and manipulation of carbon nanotubes, and moreover, the fabrication and utilization of nanotubes as novel tips for probe microscopy experiments. First, scanning tunneling microscopy and spectroscopy measurements that elucidate (1) the basic relationship between Single-Walled Carbon Nanotube (SWNT) atomic structure and electronic properties, (2) the one-dimensional band structure of nanotubes, (3) localized structures in SWNTs, and (4) the electronic behavior of finite-size SWNTs are discussed. Second, atomic force microscopy investigations of the manipulation of nanotubes on surfaces to obtain information about nanotube-surface interactions and nanotube mechanical properties, and to create nanotube device structures are reviewed. Lastly, the fabrication, properties and application of carbon nanotube probe microscopy tips to ultrahigh resolution and chemically sensitive imaging are summarized. Prospects for future research are discussed.

Carbon nanotubes are currently the focus of intense interest worldwide. This attention to carbon nanotubes is not surprising in light of their promise to exhibit unique physical properties that could impact broad areas of science and technology, ranging from super strong composites to nanoelectronics [1,2,3]. Recent experimental studies have shown that carbon nanotubes are the stiffest known material [4,5] and buckle elastically (vs. fracture) under large bending or compressive strains [5,6]. These mechanical characteristics suggest that nanotubes have significant potential for advanced composites, and could be unique force transducers to the molecular world. Moreover, the remarkable electronic properties of carbon nanotubes offer great intellectual challenges and the potential for novel applications. For example, theoretical calculations first predicted that Single-Walled Carbon Nanotubes (SWNTs) could exhibit either metallic or semiconducting behavior depending only on diameter and helicity [7,8,9]. This ability to display fundamentally distinct electronic properties without changing the local bonding, which was recently experimentally demonstrated through atomically resolved Scanning Tunneling Microscopy (STM) measurements [10,11] sets nanotubes apart from all other nanowire materials [12,13].

M. S. Dresselhaus, G. Dresselhaus, Ph. Avouris (Eds.): Carbon Nanotubes,
Topics Appl. Phys. **80**, 173–211 (2001)

Scanning probe microscopies, such as STM and Atomic Force Microscopy (AFM), have been exploited to interrogate the electrical and mechanical properties of individual 1D nanostructures such as carbon nanotubes, and moreover, nanotubes have been incorporated as tips in scanning probe microscopies to enable these techniques to image with unprecedented sensitivity. In this paper, we will review scanning probe microscopy investigations of the fundamental properties and manipulation of carbon nanotubes, and the fabrication and use of nanotubes as novel probe tips. The basic structure of the review is as follows. First, we discuss scanning tunneling microscopy and spectroscopy measurements addressing (1) the basic relationship between SWNT atomic structure and electronic properties, (2) the one-dimensional band structure of nanotubes, (3) localized structures in SWNTs, and (4) the electronic behavior of finite size SWNTs. Second, we review AFM investigations of the manipulation of nanotubes on surfaces to obtain information about nanotube-surface interactions and nanotube mechanical properties, and to create nanotube device structures. Lastly, we discuss the fabrication and properties of carbon nanotube probe microscopy tips and the application of this new generation of probes to ultrahigh resolution and chemically sensitive imaging. Prospects for future research are discussed.

## 1 Expectations from Theory

SWNTs can be viewed as an extension in one-dimension (1D) of different fullerene molecular clusters or as a strip cut from an infinite graphene sheet and rolled up to form a tube (Fig. 1a). Major characteristics of their electronic properties can be built up from relatively simply Hückel-type models using p($\pi$) atomic orbitals. The diameter and helicity of a SWNT are uniquely characterized by the roll-up vector $\boldsymbol{C}_h = n\boldsymbol{a}_1 + m\boldsymbol{a}_2 = (n, m)$ that connects crystallographically equivalent sites on a two-dimensional (2D) graphene sheet, where $\boldsymbol{a}_1$ and $\boldsymbol{a}_2$ are the graphene lattice vectors and n and m are integers. The limiting, achiral cases, $(n, 0)$ zigzag and $(n, n)$ armchair are indicated with dashed lines in Fig. 1b.

Electronic band structure calculations predict that the (n,m) indices determine whether a SWNT will be a metal or a semiconductor [7,8,9]. To understand this unique ability to exhibit distinct electronic properties within an all-carbon $sp^2$ hybridized network, it is instructive to consider the 2D energy dispersion of graphite. Graphite is a semi-metal or zero-gap semiconductor whose valence and conduction bands touch and are degenerate at six $K(k_F)$ points; these six positions define the corners of the first Brillouin zone. As a finite piece of the 2D graphene sheet is rolled up to form a 1D tube, the periodic boundary conditions imposed by $\boldsymbol{C}_h$ can be used to enumerate the allowed 1D subbands-the quantized states resulting from radial confinement-as follows:

$$\boldsymbol{C}_h \boldsymbol{k} = 2\pi q \,, \tag{1}$$

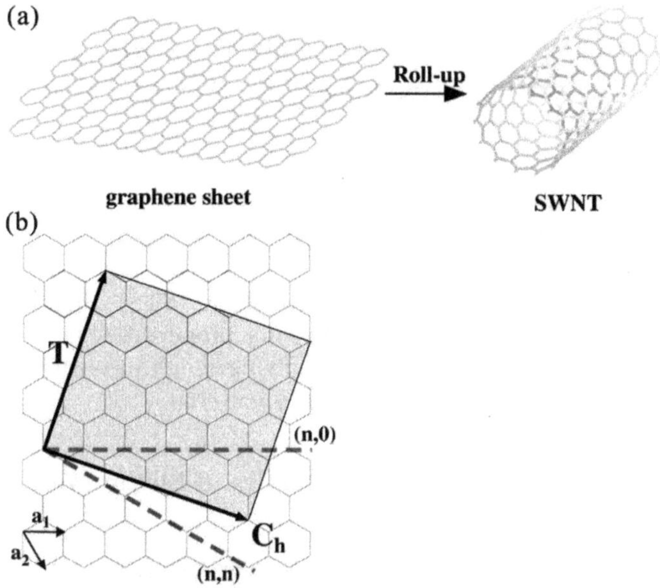


**Fig. 1.** Schematic diagrams of a SWNT and roll-up vector. (**a**) Graphene strip rolled into a seamless tube. (**b**) 2D graphene sheet illustrating lattice vectors $\boldsymbol{a}_1$ and $\boldsymbol{a}_2$, and the roll-up vector $\boldsymbol{C}_h = n\boldsymbol{a}_1 + m\boldsymbol{a}_2$. The limiting, achiral cases of $(n,0)$ zigzag and $(n,n)$ armchair are indicated with dashed lines. The translation vector $\boldsymbol{T}$ is along the nanotube axis and defines the 1D unit cell. The *shaded, boxed* area represents the unrolled unit cell formed by $\boldsymbol{T}$ and $\boldsymbol{C}_h$. The diagram is constructed for $(n,m) = (4,2)$

where $q$ is an integer. If one of these allowed subbands passes through one of the $K$ points, the nanotube will be metallic, and otherwise semiconducting. Thus to first order, zigzag $(n,0)$ or chiral $(n,m)$ SWNTs are metallic when $(n-m)/3$ is an integer and otherwise semiconducting.

Independent of helicity, the energy gaps of the semiconducting $(n,0)$ and $(n,m)$ tubes should depend inversely on diameter. In addition, the finite curvature of the tubes also leads to mixing of the $\pi/\sigma$ bonding and $\pi^*/\sigma^*$ antibonding orbitals in carbon. This mixing should cause the graphene band crossing ($k_F$) to shift away from the $K$ point and should produce small gaps in $(n,0)$ and $(n,m)$ metallic tubes with the magnitude of the gap depending inversely with the square of the diameter [7,14]. However, $(n,n)$ armchair tubes are expected to be truly metallic since $k_F$ remains on the subband of the nanotube [15].

Away from the $K$ point, signature features in the Density of States (DOS) of a material appear at the band edges, and are commonly referred to as van Hove Singularities (VHS). These singularities are characteristic of the dimension of a system. In three dimensions, VHS are kinks due to the increased degeneracy of the available phase space, while in two dimensions the VHS appear as stepwise discontinuities with increasing energy. Unique to one-dimensional systems, the VHS are manifested as peaks. Hence, SWNTs and other 1D materials are expected to exhibit spikes in the DOS due to the 1D nature of their band structure.

## 2 Scanning Tunneling Microscopy Studies of Electronic Properties of Nanotubes

Scanning Tunneling Microscopy (STM) and Spectroscopy (STS) offer the potential to probe these predictions about the electronic properties of carbon nanotubes since these techniques are capable of resolving simultaneously the atomic structure and electronic density of states of a material.

### 2.1 Atomic Structure and Electronic Properties of Single-Walled Carbon Nanotubes

Atomically resolved images of in situ vapor-deposited carbon structures believed to be Multi-Walled Carbon Nanotubes (MWNTs) were first reported by Sattler and Ge [16]. Bias-dependent imaging [17] and STS investigations [18] of independently characterized arc generated MWNTs suggested that some fraction of MWNTs produced by the arc method were semiconducting, and in these semiconducting nanotubes, the STS data suggested that the energy gap depended inversely on diameter. Subsequent STM and STS studies of MWNTs and SWNTs have provided indications of different structures and structure-dependent electronic properties, but have not revealed an explicit relationship between structure and electronic properties. The failure of these previous studies to elucidate clearly the expected diameter and helicity dependent electronic properties of nanotubes can be attributed in part to the lack of pure SWNT samples, since (1) the electronic band structure of MWNTs is considerably more complex than SWNTs, and (2) relatively pure samples are required to carry out unambiguous STM and STS measurements.

The development of techniques to produce and purify relatively large quantities of SWNTs has made possible definitive testing of the remarkable predicted electronic properties of nanotubes [19,20,21]. Indeed, seminal STM and STS measurements of the atomic structure and electronic properties of purified SWNTs by *Wildöer* et al. [10] and *Odom* et al. [11] have shown that the electronic properties do indeed depend sensitively on diameter and helicity. In both of these studies, the SWNTs were grown by laser vaporization, ultrasonically suspended in organic solvents, and then deposited by spin coating onto Au (111) substrates. Subsequent STM imaging studies were carried out at low-temperature, in ultra-high vacuum STMs.

### 2.1.1 Carbon Nanotube Atomic Structure

A large scale STM image of individual tubes and small ropes containing a number of individual SWNTs is shown in Fig. 2a. A high-resolution image of a SWNT (Fig. 2b) exhibits a graphite-like honeycomb lattice, thus enabling the determination of the $(n, m)$ indices from the image. The $(n, m)$ indices were obtained from the experimentally measured values of the chiral angle and diameter. The chiral angle was measured between the zigzag $(n, 0)$ direction (the dashed line connecting sites separated by 0.426 nm) and the tube axis.

The angle measurements were confined to the tops of the atomically resolved nanotubes, which minimizes contributions from the sides of the highly curved tubes, and over distances at least 20 nm to eliminate possible twist-deformation contributions. The SWNT diameters were determined from the projected widths of the nanotube images after deconvoluting the tip contribution to the image. This approach yields a more robust diameter than that determined from the cross-sectional height, since the apparent height is highly dependent upon imaging conditions.

Atomically resolved images of isolated SWNTs on a Au (111) substrate are shown in Fig. 3a,b. The measured chiral angle and diameter of the tube in Fig. 3a constrain the $(n, m)$ indices to either (12,3) or (13,3). Note that a (12,3) tube is expected to be metallic, while a (13,3) tube should be semiconducting. On the other hand, the chiral angle and diameter of the SWNT in Fig. 3b constrain the indices to $(14, -3)$. This tube has helicity opposite to the SWNT in Fig. 3a.

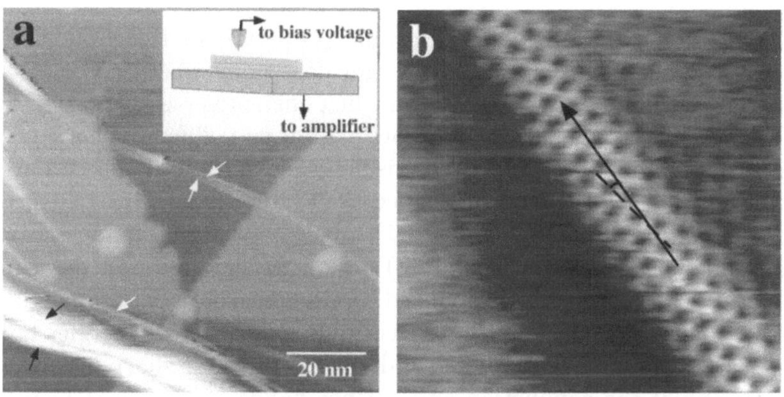


**Fig. 2.** STM images of nanotubes. (**a**) Large area showing several small bundles and isolated SWNTs on a stepped Au (111) surface. The *white arrows* indicate individual SWNTs and the *black arrows* point to small ropes of SWNTs. (*inset*) Schematic diagram of the STM experiment [13]. (**b**) SWNT on the surface of a rope. The *solid, black arrow* highlights the tube axis and the *dashed line* indicates the zigzag direction [11]

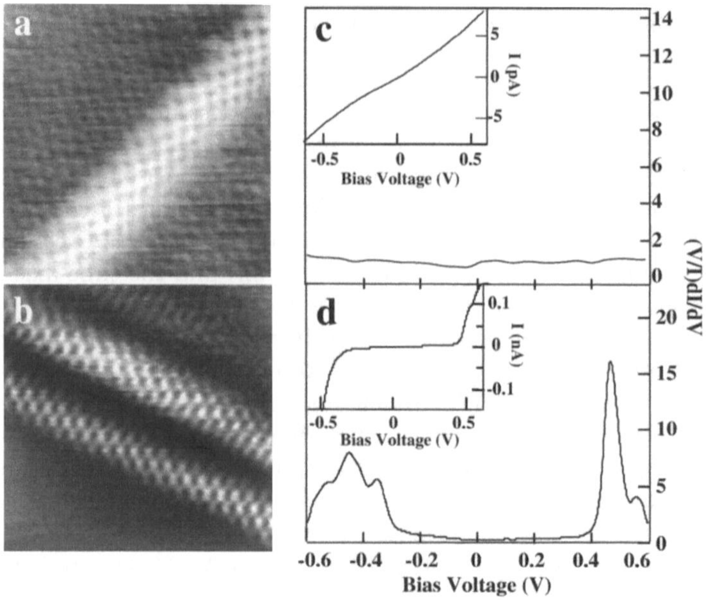


**Fig. 3.** STM imaging and spectroscopy of individual nanotubes. (**a, b**) Constant current images of isolated nanotubes. The Au (111) lattice is clearly seen in (**a**). (**c, d**) Calculated normalized conductance, $(V/I)\mathrm{d}I/\mathrm{d}V$ and measured $I$–$V$ (*inset*) recorded on the nanotubes in (**a, b**) [11,29]

### 2.1.2 Spectroscopy: Metals and Semiconductors

Subsequent characterization of the electronic properties of the atomically resolved tubes by tunneling spectroscopy can determine whether the electronic properties depend on structure. Tunneling current versus voltage data recorded along the two tubes discussed above exhibit very different characteristics (Fig. 3c and 3d), and the LDOS that is determined from these $I$–$V$ data sets are quite distinct. For the tube assigned as (12,3) or (13,3), the LDOS is finite and constant between −0.6 and +0.6 V. This behavior is characteristic of a metal, and thus shows that the (12,3) indices provide the best description for the tube. Moreover, the normalized conductance data determined for the (14, −3) tube exhibit an absence of electronic states at low energies but sharp increases in the LDOS at −0.325 and +0.425 V. These sharp increases are characteristic of the conduction and valence bands of a semiconductor, and thus confirm our expectation that (14, −3) indices correspond to a semiconducting SWNT. These key measurements first verified the unique ability of SWNTs to exhibit fundamentally different electronic properties with only subtle variations in structure [7,8,9].

In addition, the semiconducting energy gaps ($E_g$) of SWNTs are predicted to depend inversely on the tube diameter, d, and to be independent of helicity. A summary of the energy gaps obtained by *Odom* et al. [11] for tubes with diameters between 0.7 and 1.1 nm is shown in Fig. 4. Significantly, these results and those obtained by *Wildöer* et al. [10] for tubes with larger diameters between 1 and 2 nm show the expected $1/d$ dependence. Moreover, these results can be used to obtain a value for the nearest neighbor overlap integral ($\gamma_o$) used in tight-binding calculations of the electronic properties by fitting to $E_g = 2\gamma_o a_{C-C}/d$, where $a_{C-C}$ is 0.142 nm. The values obtained from the one parameter fit by [11] and [10] respectively, 2.5 eV and 2.7 eV, are in good agreement with the reported values in the literature that range from 2.4–2.9 eV [1,22,23].

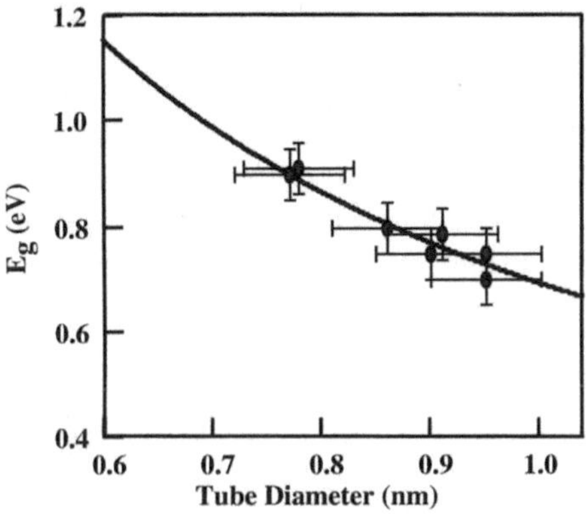


**Fig. 4.** Summary of energy gap ($E_g$) versus tube diameter obtained in [11]. The *solid line* corresponds to a fit described in the text

## 2.2 One-Dimensional Band Structure of Nanotubes

The characterization of semiconducting and metallic SWNTs with subtle changes in structure confirms the remarkable electronic behavior of the nanotubes, and represents a significant step forward in understanding these 1D materials. In addition, the ability to probe simultaneously atomic structure and electronic properties provides a unique opportunity to investigate further several interesting properties of these 1D materials. These properties include, for example, the detailed DOS of the nanotubes, the role of symmetry breaking distortions on the electronic character of nanotubes, and the electronic properties of defect and end electronic states. Several of these interesting issues are addressed in the following sections and are compared quantitatively with theory.

By making tunneling spectroscopy measurements over an extended energy range, the one-dimensional nature of the energy bands are observed as sharp peaks in the DOS [10,24,25]. When spectroscopic measurements are made on atomically-resolved nanotubes, it is also possible to compare the experimental DOS quantitatively with that resulting from a simple $\pi$-only tight-binding calculation. *Kim* et al. [25] reported the first detailed experimental comparison with theory on a metallic tube with indices (13,7). This nanotube is the upper isolated tube that rests on the Au surface shown in Fig. 5a. Current vs. voltage measurements exhibited a linear response at $V = 0$ as expected for a metal and showed steps at larger voltages that correspond to a series of sharp peaks in the $dI/dV$. These peaks correspond to the VHS resulting from the extremal points in the 1D energy bands.

A direct comparison of these experimental data to the theoretical electronic band structure calculated by a $\pi$-only tight-binding model was made [25]. Significantly, the spectroscopy data show good agreement with the calculated DOS for the (13,7) tube (Fig. 5b). The agreement between the VHS positions determined from the calculations and $dI/dV$ data are especially good below $E_F$, where the first seven peaks correspond well. The peak splitting due to the anisotropy around $K$ is also reproduced in the $dI/dV$. Notably, the experimental gap between the first VHS in this metallic tube, $E_g^m \sim 1.6\,\text{eV}$, is in agreement with predictions for metallic tubes [26]; that is, $E_g^m = 6\gamma_o a_{C-C}/d = 1.6\,\text{eV}$, where $\gamma_o = 2.5\,\text{eV}$, the value determined from the semiconducting energy gap data [11]. Above the Fermi energy some deviation between the experimental data and calculations exist, but the observed differences may be due to band repulsion, which arises from curvature-induced hybridization [27].

*Kim* et al. [25] also compared their results to a published $\sigma + \pi$ calculation for a (13,7) SWNT [28] and a $\pi$-only calculation for a closely related

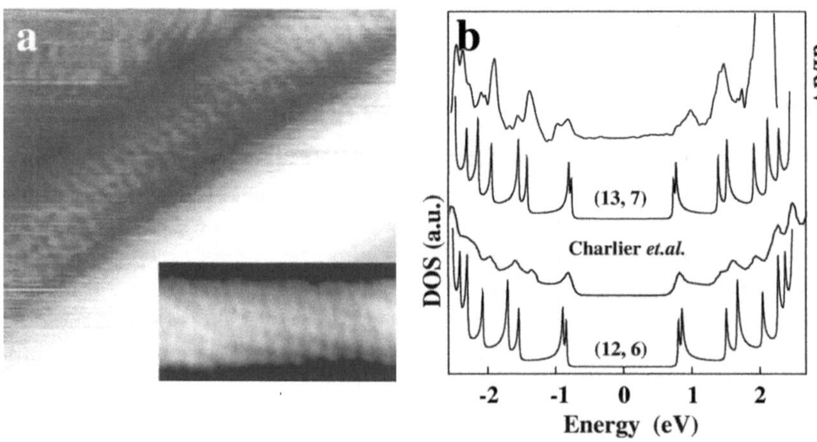


**Fig. 5.** STM imaging and spectroscopy on a metallic SWNT. (**a**) Tunneling spectra were recorded on the isolated upper tube. The *inset* shows an atomic resolution image of this tube. A portion of a hexagonal lattice is overlaid to guide the eye. (**b**) Comparison of the DOS obtained from experiment (*upper curve*) and a $\pi$-only tight-binding calculation for the (13,7) SWNT (*second curve from top*). The broken vertical lines indicate the positions of VHS in the tunneling spectra after consideration of thermal broadening convolution. The calculated DOS for a (12,6) tube is included for comparison [25]

set of indices. Although detailed comparison is difficult due to the large DOS broadening, all peaks within ± 2 eV match well with the π-only calculation. This comparison suggests that curvature-induced hybridization is only a small perturbation within the experimental energy scale ($| V | < 2\,\mathrm{V}$) for the (13,7) tube. The sensitivity of the VHS to variations in the $(n, m)$ indices was investigated by calculating the DOS of the next closest metallic SWNT to the experimental diameter and angle; that is, a (12,6) tube. Significantly, the calculated VHS for this (12,6) tube deviate much more from the experimental DOS peaks than in the case of the (13,7) tube. It is worth noting that the poor agreement in this case demonstrates that subtle variations in diameter and helicity do produce experimentally distinguishable changes in the DOS.

Van Hove singularities in the electronic DOS of semiconducting nanotubes have also been observed [10,29]. *Odom* et al. [29] have characterized spectroscopically a small-diameter (10,0) semiconducting nanotube (Fig. 6a), and directly compared the DOS with a tight-binding calculation.

The normalized conductance exhibits relatively good agreement with the calculated (10,0) DOS below $E_{\mathrm{F}}$ but poorer agreement above (Fig. 6b). However, the π-only DOS calculation does not include $\pi/\sigma$ and $\pi^*/\sigma^*$ mixing due to curvature. This hybridization of $\pi/\sigma$ orbitals is believed to produce more pronounced effects on the conduction band [27], and this might explain the observed deviations. Additional work is needed to resolve this point. These results show clearly that the VHS peaks in the electronic band structure, which are characteristic of 1D systems, can be measured experimentally and agree well with the DOS calculated using π-only tight-binding models.

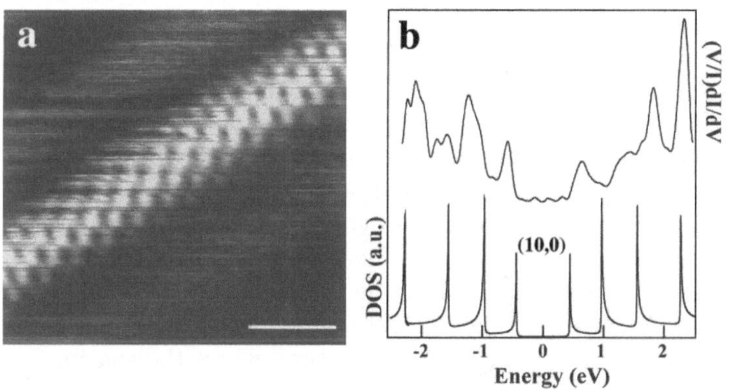


**Fig. 6.** STM image and spectroscopy of a semiconducting nanotube. (**a**) Image of a SWNT on the surface of a rope. (**b**) Comparison of the DOS obtained from experiment (*upper curve*) and calculation for the (10,0) SWNT (*lower curve*) [29]

### 2.3 Localized Structures in Nanotubes

The above results focus on the properties of defect-free, nearly infinite SWNTs. Nanotubes can, however, exhibit a variety of structural defects. For example, the insertion of a pentagon-heptagon (5-7) pair in the hexagonal network [30,31] can produce a kink in the tubular nanotube structure. Also, more gradual bends can result from mechanical distortions and bending. In addition, structural deformations such as bends, twisting, and collapse have been observed occasionally in these seemingly infinite carbon cylinders [32,33,34]. These defects may develop during growth, processing, deposition, or following an interaction with surface features [35]. The electronic properties of localized SWNT structures, such as bends and ends [36,37,38], are essential to proposed device applications. Below selected examples of structures that have been characterized with atomically-resolved imaging and spectroscopy are described.

#### 2.3.1 Bent and Twisted Nanotubes

*Odom* et al. [29] recently reported a kink in an atomically-resolved SWNT rope arising from a mechanical distortion (Fig. 7). The bend angle defined by this kink is approximately 60°. Tunneling spectroscopy was used to characterize the electronic properties of the uppermost nanotube in the bent rope. $I$–$V$ measurements were performed at the positions indicated by the symbols in Fig. 7a, and their corresponding $\mathrm{d}I/\mathrm{d}V$ are displayed in Fig. 7b.

The positions of the van Hove peaks indicate that the tube is metallic. Significantly, the data also show new features at low bias voltages on either side of the bend. These peaks are likely due to the presence of the bend, since five nanometers away (+) from the kink, the sharpness and prominence of these features have greatly diminished. In addition, *Avouris* et al. [40] reported that STS spectra taken near a kink in a semiconducting nanotube also showed an increased DOS at low energies. Notably, recent calculations on bends in armchair tubes show similar low energy features in the DOS for similar bend angles [39] observed by [29] and [40].

The bend region observed by Odom and co-workers was further investigated using bias-dependent STM imaging. On the right side of the bend, a superstructure on the tube is observed at the biases of the localized peaks (Fig. 7c,d ). Figure 7c shows stripes parallel to the zigzag direction of the tube and Fig. 7d displays a triangular ring structure, where the spacing between nearest-neighbor rings is ca. 0.42 nm (the zigzag spacing). These new electronic features could be due to electron scattering and interference at the defect site [41]. Although the bias voltage, 0.45 V, at which Fig. 7e was imaged is not at a prominent peak in the $\mathrm{d}I/\mathrm{d}V$, some electronic structure can be seen extending $\sim$ 1.5 nm to the right of the bend. However, this additional structure diminishes and an unperturbed atomic lattice is observed, consistent with the spectroscopic measurements. Further experimental and

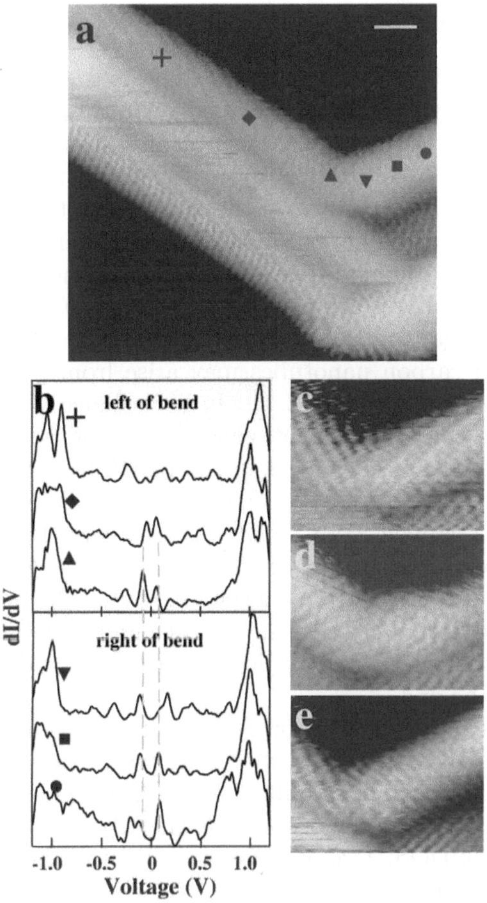


**Fig. 7.** STM image and spectroscopy of a bend in a rope of SWNTs. (**a**) Image of $\sim 60°$ bend. The symbols correspond to locations where $I$–$V$ were measured. The scale bar is 1 nm. (**b**) Differential conductance calculated from the locations indicated in (**a**). The *upper* portion of the graph is spectroscopy performed on the left side of the bend over 5 nm. The *lower* portion of the graph is spectroscopy performed on the right side of the bend over 2 nm. The *dashed lines* highlight the low-energy features. (**c**–**e**) STM images recorded at bias voltages of $-0.15$, 0.15, and 0.45 V, respectively [29]

computational work is needed to elucidate clearly such interesting observations.

Besides bends in nanotubes, twisting of individual tubes within ropes has been reported by *Clauss* et al. [33]. Large-scale nanotube twists may result from mechanical interactions during deposition upon surfaces, be introduced during the growth process and frozen by shear forces, or result from different helicity tubes attempting to align their hexagonal lattices within ropes. *Clauss* et al. [33] observed anomalous lattice orientations upon careful inspection of many tubes, namely, that the armchair direction is on average perpendicular to the tube axis, and that the average angle between the zigzag and armchair direction is greater than 90°. This apparent distortion from an equilibrium conformation can be explained if the imaged nanotubes are of the armchair-type with a twist distortion of several degrees, or from distortions contributed by the finite size and asymmetry of the STM tip.

### 2.3.2 SWNT Ends: Structure and Electronic Properties

Another example of localized geometric structures in nanotubes is the ends. Analogous to the surface states of a 3D crystal and the edge state of a 2D electron gas, end states are expected at the end of the 1D electron system. The ends of a 1D electronic system can be considered as the "surface" of the 1D bulk. Both resonant and localized states are possible at the ends of nanotubes. Resonant end states are expected for metallic nanotubes because there are no gaps in the 1D band structure of metallic SWNTs to localize the end states. In the same way, localized end states are possible for semiconducting nanotubes since they exhibit energy gaps in their DOS.

The end states associated with carbon nanotubes may arise from pentagons in a capped end or an open nanotube [42,43]. In accordance with Euler's rule, a capped end should contain six pentagons. The presence of these topological defects can cause dramatic changes in the LDOS near the end of the nanotube. *Kim* et al. [25] reported the first detailed investigation of the electronic character of a capped SWNT end (Fig. 8a), with bulk indices $(13, -2)$. The expected metallic behavior of the $(13, -2)$ tube was confirmed in $(V/I)\mathrm{d}I/\mathrm{d}V$ data recorded away from the end (Fig. 8c). Significantly, spectroscopic data recorded at and close to the SWNT end show two distinct peaks at 250 and 500 mV that decay and eventually disappear in the bulk DOS recorded far from the tube end.

To investigate the origin of these new spectroscopic features, tight-binding calculations were carried out for a $(13, -2)$ model tube terminated with different end caps (Fig. 8b). Both models exhibit a bulk DOS far from the end (lower curve in Fig. 8d); however, near the nanotube ends the LDOS show pronounced differences from the bulk DOS: Two or more peaks appear above $E_{\mathrm{F}}$, and these peaks decay upon moving away from the end to the bulk. These models were chosen to illustrate the relatively large peak differences for caps closed with isolated versus adjacent pentagons. The LDOS obtained for cap I shows excellent agreement with the measured LDOS at the tube end, while cap II does not (Fig. 8d). The positions of the two end LDOS peaks as well as the first band edge of cap I match well with those from the experimental spectra. These results suggest that the arrangement of pentagons is responsible for the observed DOS peaks at the SWNT ends, and are thus similar to conclusions drawn from measurements on MWNTs that were not atomically resolved [43].

Besides characterization of capped ends in metallic tubes, *Avouris* and coworkers [40] reported spectroscopic data on an atomically resolved semiconducting SWNT and its end. Interestingly, as tunneling spectra were recorded along the tube axis to the end, the Fermi level position shifted to the center of the energy gap. This is the first reported evidence of band-bending behavior observed by STM spectroscopy in individual nanotubes. Future studies could provide important and much needed information addressing the nature of nanotube–metal contacts.

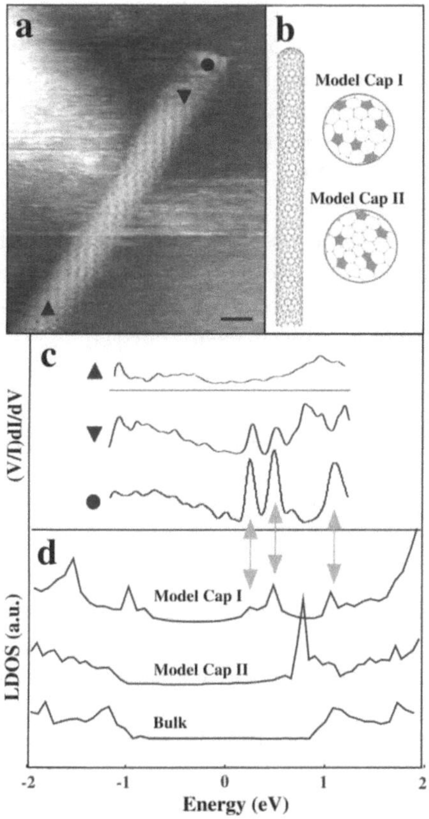


**Fig. 8.** STM image and spectroscopy of a SWNT end. (**a**) Image of a nanotube end. The symbols correspond to the locations where the tunneling spectra in (**c**) were recorded. The scale bar is 1 nm. (**b**) A model (13, −2) SWNT recorded two different cap configurations; the pentagons are shaded gray. (**c**) Experimental tunneling spectra from the end •, near the end ▼, and far from the end ▲. (**d**) LDOS obtained from tight-binding calculations on capped (13, −2) tubes for caps I and II. Similar features in • and cap I are highlighted by gray arrows. The bulk DOS for both cap models are identical and is shown in the lowest curve [25]

## 2.4 Finite-Size SWNTs

The studies reviewed above have focused on SWNTs that have always retained characteristic features of a periodic 1D system. What happens when this 1D system is made increasingly smaller? Conceptually, as the length of a SWNT is reduced, one ultimately will reach the limit of a fullerene molecular cluster–a 0D object. In this regard, studies of finite-size SWNTs offer a unique opportunity to probe the connection between and evolution of electronic structure in periodic molecular systems. Investigations of finite-sized effects in SWNTs are also important to the future utilization of nanotubes in device applications. Low-temperature transport experiments on metallic SWNTs have shown that μm long tubes behave as islands in single electron transistors, with an island energy level spacing characteristic of the 1D particle-in-a-box states [44,45]. Since the coulomb charging energy $E_c \propto 1/L$ ($L$ is the nanotube length), shorter nanotubes allow the working temperature of such devices to increase. In addition, finite-size effects should be visible at room temperature if $\Delta E > k_B T$; thus a resonant tunneling device may be conceived with nanotubes whose lengths are less than 50 nm.

To first order, the 1D energy levels and spacing may be described by either quantization of the metallic band structure or by recollection of the textbook particle in a 1D well. To first order, the bulk metallic nanotube band structure is characterized by two linear bands ($\pi$ and $\pi^*$) that cross the Fermi energy, and these bands contribute a finite, constant DOS at low energies. Confinement of the electrons due to reduced axial lengths produces a discretization $\Delta k = \pi/L$ on the crossing bands. The intersection of $\Delta k$ and the linear bands in the zone-folding scheme results in an energy level spectrum. An alternative, simpler analysis of this problem is to consider the finite-length nanotube as a 1D particle-in-a-box, whose well-known eigenvalues ($E$) are $E = \hbar^2 k^2/2\,\mathrm{m}$. The energy level spacing is easily derived:

$$\Delta \mathrm{E} = \hbar^2 k_{\mathrm{F}} \Delta k/m = h v_{\mathrm{F}}/2L \approx 1.67\,\mathrm{eV/L(nm)} \tag{2}$$

where $h$ is Planck's constant and $v_{\mathrm{F}} = 8.1 \times 10^5\,\mathrm{m/s}$ is the Fermi velocity for graphene.

### 2.4.1 Quantum Effects in Carbon Nanotubes

STM can probe the transition from 1D delocalized states to molecular levels since voltage pulses can first be used to systematically cut nanotubes into short lengths [29,46] (Fig. 9). Subsequently, these finite-size nanotubes can be characterized spectroscopically.

*Venema* et al. [47] first reported investigations of quantum size effects in a $\sim 30\,\mathrm{nm}$ metallic, armchair nanotube shortened by STM voltage pulses. Current vs. voltage measurements carried out near the middle of the tube showed an irregular step-like behavior. The steps in the spectra observed over a small voltage range ($\pm 0.2\,\mathrm{V}$) correspond to quantized energy levels entering the bias window as the voltage is increased, and the irregularity in the step spacing is due to coulomb charging effects competing with the 1D level spacing. Remarkably, they discovered that compilation of 100 consecutive $I$–$V$ measurements spaced 23 pm apart into a spectroscopic map exhibited $\mathrm{d}I/\mathrm{d}V$ peaks which varied periodically with position along the tube axis (Fig. 10a). This periodic variation in $\mathrm{d}I/\mathrm{d}V$ as a function of position along the tube, 0.4 nm,

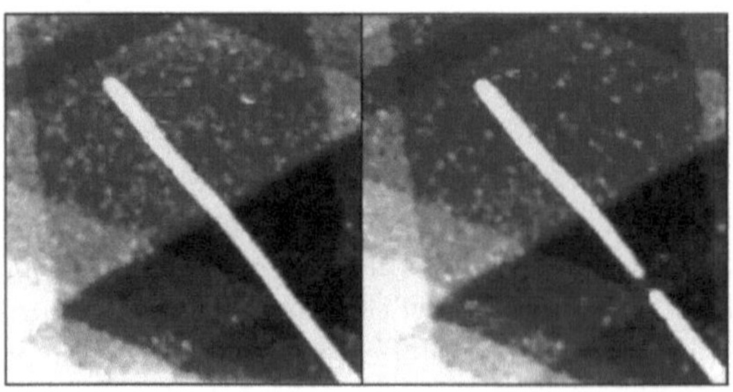

**Fig. 9.** SWNTs before and after a voltage pulse was applied to cut the nanotube [46]

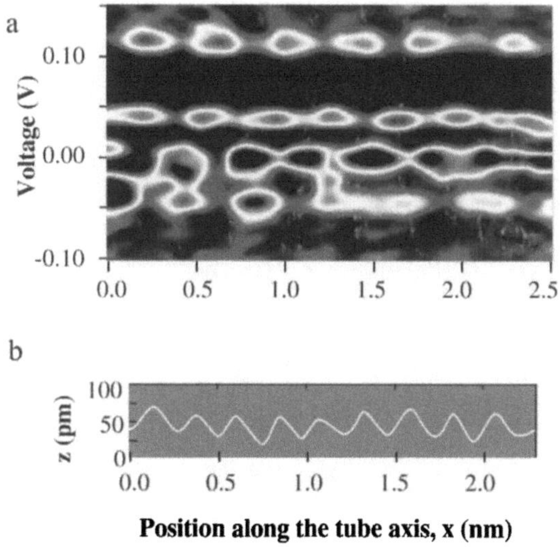


**Fig. 10.** STM line scans. (**a**) Spectroscopic image compiled from 100 d$I$/d$V$ measurements. The periodicity is determined from the square of the amplitude of the electron wavefunction at discrete energies. (**b**) Topographic line profile of atomic corrugation in a shortened armchair nanotube [47]

is different from the lattice constant $a_0 = 0.25\,\mathrm{nm}$ (Fig. 10b), and can be described by the electronic wavefunctions in the nanotube. Since d$I$/d$V$ is a measure of the squared amplitude of the wavefunction, they were able to fit the experimental d$I$/d$V$ with the trial function $A\sin^2(2\pi x/\lambda + \phi) + B$. This enabled the separation between the d$I$/d$V$ peaks to be correlated with half the Fermi wavelength $\lambda_F$. The calculated value for $\lambda_F = 3a_0 = 0.74\,\mathrm{nm}$, determined from the two linear bands crossing at $k_F$, is in good agreement with experimental observations. Hence discrete electron standing waves were observed in short armchair nanotubes. It is also worth noting that the observed widths of the nanotubes probed in these investigations, $\sim 10\,\mathrm{nm}$, are significantly larger than expected for a single SWNT. This suggests that it is likely the measurements were on ropes of SWNTs. In this regard, it will be important in the future to assess how tube-tube interactions perturb the quantum states in a single SWNT.

It is possible that additional features in the electronic structure of finite-sized nanotubes may appear in lengths nearly an order of magnitude shorter. To this end, *Odom* et al. [29] have studied quantum size effects in both chiral metallic and semiconducting tubes. STM images of nanotubes shortened to six and five nanometers, respectively, are shown in Fig. 11a,b. The $I$–$V$ measurements show a step-wise increase of current over a two-volt bias range for both tubes, and the observed peaks in the $(V/I)$d$I$/d$V$ were attributed to resonant tunneling through discrete energy levels resulting from the finite length of the SWNT. To first order, analysis of the peak spacing for

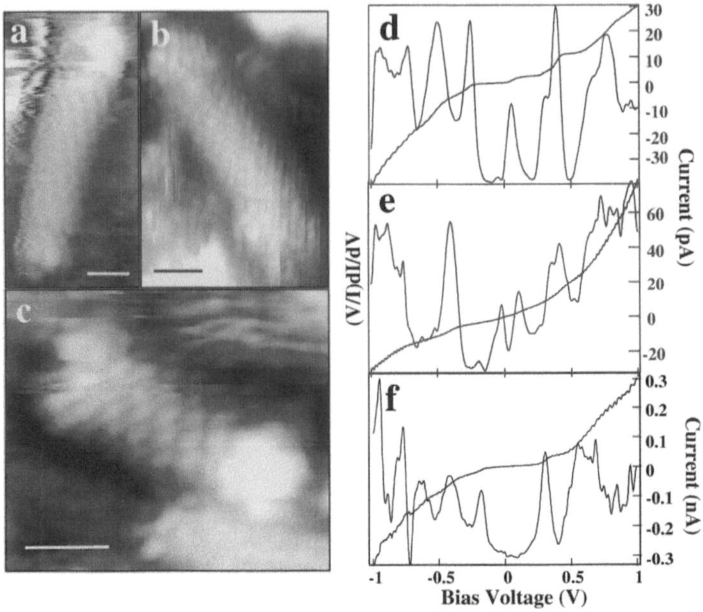


**Fig. 11.** STM imaging and spectroscopy of finite-size SWNTs. (**a–c**) SWNTs cut by voltage pulses and shortened to lengths of 6 nm, 5 nm, and 3 nm, respectively. (**d–f**) Averaged normalized conductance and I–V measurements performed on the nanotubes in (**a–c**), respectively. Six I–V curves were taken along the tube length and averaged together since the spectra were essentially indistinguishable [29]

the finite-sized nanotubes (Fig. 11d,e) agrees with the simple 1D particle-in-a-box model. The former tube that is six nanometers long exhibits a mean peak spacing of approximately 0.27 eV. A six nanometer tube within this 1D box model would have an average level spacing $\Delta E \sim 1.67\,\text{eV}/6 = 0.28\,\text{eV}$. For the latter tube with its shorter length, the observed peak spacing is also wider, as expected from this model.

In addition, an atomically resolved SWNT, only three nanometers long was investigated (Fig. 11c). The normalized conductance of this short piece shown in Fig. 11f appears, however, quite different from the expected $1.67\,\text{eV}/3 = 0.55\,\text{eV}$ energy level spacing for a nanotube three nanometers long. This is not surprising due to the limitations of this simple model, and the need for a more detailed molecular model to explain adequately the electronic structure is evident. Ab initio calculations of SWNT band structure have recently shown that the energy level spacing of finite-size tubes may be considerably different from that predicted from a Hückel model due to the asymmetry and shifting of the linear bands crossing at $E_F$ [48]. In addition, several molecular computational studies have predicted that nanotubes less than four nanometers long should open a HOMO–LUMO gap around $E_F$, although its magnitude varies greatly among different calculation methods [49,50]. These studies have been performed on finite-sized, open-ended, achiral $(n, 0)$ zigzag and $(n, n)$ armchair tubes. In quantum chemistry calculations, symmetry considerations are important, and in this regard, chiral nanotubes may exhibit drastically different electronic characteristics compared with achiral ones. Clearly, more sophisticated molecular and first principle calculations are required to fully understand nanotubes at such ultra-short length scales.

In contrast to the metallic nanotubes and other nanoscale systems [51], no significant length dependence is observed in finite-sized semiconducting nanotubes down to five nanometers long [52]. Namely, tunneling spectroscopy data obtained from the center of the shortened tube showed a striking resemblance to the spectrum observed before cutting. That is, the positions of the valence and conduction bands are nearly identical before and after cutting. Spectra taken at the ends also exhibited the same VHS positions, and a localized state near $\sim$0.2 eV, which is attributed to dangling bonds, was also observed. It is possible that long-length scale disorder and very short electron mean free paths ($\sim$2 nm) in semiconducting tubes [53] may account for the similar electronic behavior observed in short and long nanotubes. This suggests that detailed studies should be carried out with even shorter tube lengths. However, recent ab initio calculations on the electronic structure of semiconducting nanotubes with lengths of two-three nanometers also seem robust to reduced-length effects and merely reproduce the major features of the bulk DOS (i.e. the energy gap) [54].

### 2.4.2 Coulomb Charging in SWNTs

In the experiments discussed above, the finite-sized nanotubes remained in good contact with the underlying substrate after cutting, and the voltage drop was primarily over the vacuum tunnel junction. If the nanotubes are weakly coupled to the surface, a second barrier for electron tunneling is created and these nanotubes may behave as coulomb islands and exhibit coulomb blockade and staircase features in their I–V [55]. The investigation of finite-sized nanotubes in the presence of charging effects is interesting since both effects scale inversely with length and thus can be probed experimentally, in contrast to 3D metal quantum dots [56,57].

*Odom* et al. [29] recently reported the first detailed investigation of the interplay between these two effects in nanotubes at ultra-short length scales and compared the tunneling spectra with a modified semi-classical theory for single-electron tunneling. The tunneling current vs. voltage of the nanotube in Fig. 12a exhibits suppression of current at zero bias as well as relatively sharp, irregular step-like increases at larger $|V|$ (Fig. 12b), reminiscent of the coulomb blockade and staircase [55]. Similar to [47], the irregularities in the conductance peak spacing and amplitude are attributed to contributions from the discrete level spacing of the finite-sized nanotube [29].

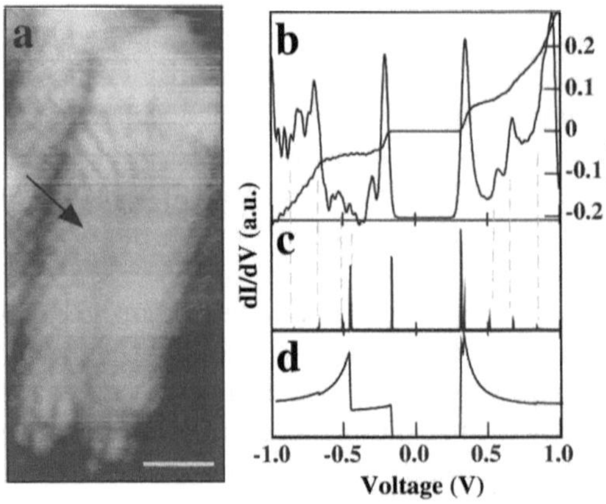


**Fig. 12.** Charging effects in finite-sized SWNTs. (**a**) STM image of SWNTs shortened by voltage pulses. The scale bar is 1 nm. (**b**) Measured $I$–$V$ and $\mathrm{d}I/\mathrm{d}V$ performed on the tube indicated by an arrow in (**a**). (**c**) Numerical derivative of the $I$–$V$ curve calculated by a semi-classical calculation including the level spacing of a 1D box. (**d**) Same calculation as in (**c**) except the nanotube DOS is treated as continuous [29]

To compare directly the complex tunneling spectra with calculations, *Odom* and co-workers [29] modified a semi-classical double junction model [56] to include the level spacing of the nanotube quantum dot, $\Delta E \sim 1.67\,\mathrm{eV}/7 = 0.24\,\mathrm{eV}$. The capacitance of a SWNT resting on a metal surface may be approximated by [58]

$$C = 2\pi\varepsilon L / \ln[d + (d^2 - R^2)^{1/2}]/R\,, \tag{3}$$

where $d$ is the distance from the center of the nanotube to the surface, and $\varepsilon$ is $8.85 \times 10^{-3}\,\mathrm{aF/nm}$. Estimating $d \sim 1.9\,\mathrm{nm}$, the geometric capacitance for the nanotube in Fig. 12a is 0.21 aF. The calculated $\mathrm{d}I/\mathrm{d}V$ that best fits the tunneling conductance is shown in Fig. 12c, and yields a Au-tube capacitance $C_1 = 0.21 \pm 0.01\,\mathrm{aF}$, in good agreement with the capacitance estimated by the geometry of the nanotube. In contrast, if the calculation neglected the level spacing of the nanotube island, only the blockade region is reproduced well (Fig. 12d). These studies demonstrate that it is possible also to study the interplay of finite size effects and charging effects in SWNT quantum dots of ultra-short lengths.

## 2.5 Future Directions

Scanning tunneling microscopy and spectroscopy have been used to characterize the atomic structure and tunneling density of states of individual SWNTs and SWNT ropes. Defect-free SWNTs exhibit semiconducting and metallic behavior that depends predictably on helicity and diameter. In addition, the 1D VHS in the DOS for both metallic and semiconducting tubes have been characterized and compare reasonably well with tight-binding calculations. Lastly, studies show that it is possible to access readily a regime of "0D" behavior, where finite length produces quantization along the tube axis, which opens up future opportunities to probe, for example, connections between extended and molecular systems. In short, much of the fascinating overall structural and electronic properties of SWNTs are now in hand – but this really only has scratched the surface. Future work addressing the role of defects and other structural perturbations, coupling to metal and magnetic systems, the connection between extended and finite size/molecular clusters, as well as other directions will help to define further the fundamental physics of these systems and define emerging concepts in nanotechnology.

# 3 Manipulation of Nanotubes with Scanning Probe Microscopies

Scanning probe microscopies not only can be utilized to probe the fundamental electronic properties of carbon nanotubes but also to manipulate them. The ability to manipulate carbon nanotubes on surfaces with scanning probe techniques enables investigations of nanotube/surface interactions, mechanical properties and deformations, and controllable assembly of nanotube electrical devices. Examples of experiments that have exploited the ability of AFM to connect from the macroscopic world to the nanometer scale and thereby elucidate new behavior and physical properties of nanotubes are discussed below.

## 3.1 Manipulation of Nanotubes on Surfaces

The interaction between nanotubes and their underlying substrate is sensitive to the details of the particular surface. In order for an AFM tip to controllably manipulate nanotubes across a surface, it is crucial that the tip/nanotube interaction be greater than the adhesion/friction force between the nanotube and the substrate. Depending on the experimental goals, one might, for example, require surfaces with high friction, in order to pin and manipulate the nanotubes into desired configurations. However, if one wants to measure certain physical properties such as stress and strain, a low frictional surface is necessary to decouple the surface contributions from the intrinsic mechanical properties of the nanotube.

*Hertel* et al. [59] investigated the interaction between a H-passivated Si (100) surface and deposited MWNTs using non-contact mode AFM. For example, the AFM image of nanotubes in Fig. 13 illustrates one type of elastic deformation that may occur when nanotubes are deposited onto a surface. At the crossing point of the two overlapping MWNTs, the measured height is six nanometers less than the sum of the individual diameters. In addition, the width of the upper nanotube shortly before and after crossing the lower

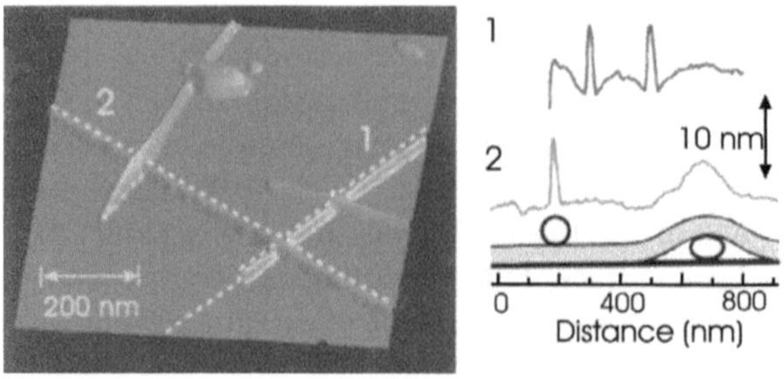


**Fig. 13.** AFM non-contact mode image of several overlapping MWNTs. The upper tubes are seen to wrap around the lower ones which are slightly compressed [59]

tube appears broadened. This observed axial and radial deformation of the tubes may be explained by a strong surface/nanotube interaction. As the upper nanotube is forced to bend over the lower nanotube, the strain energy increases; however, this is compensated for by a gain in binding energy between the surface and the lower nanotube, which attempts to maximize its contact area with the substrate. The strength of the attractive force between the nanotubes and the surface may be estimated using a 1D model where the profile along the tube axis is determined by the balance of strain and adhesion energy. In this framework, *Hertel* and co-workers [59] found that the binding energy is determined primarily by van der Waals interactions and approaches up to $0.8 \pm 0.3\,\mathrm{eV/Å}$for multi-wall nanotubes 10 nm in diameter on hydrogen-terminated silicon.

An important consequence of this large binding energy is that nanotubes will tend to distort and conform to the substrate topography. In addition, nanotubes can be pinned in highly strained configurations on different substrates after manipulation with an AFM tip [5,59,60]. *Wong* et al. [5] and *Falvo* et al. [60] exploited high substrate/nanotube friction in order to apply lateral stresses at specified locations along a MWNT to produce translations and bends (Fig. 14).

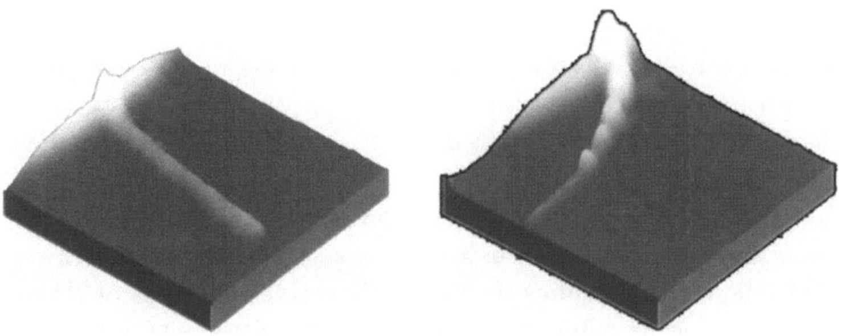

**Fig. 14.** AFM tapping-mode images ($0.9\mu m \times 0.9\mu m$) of a 4.4 nm-diameter MWNT before and after bending on an oxidized Si substrate. After bending, buckling occurs along the nanotube axis [5]

When bent to large angles, the nanotubes exhibited raised features, which correspond to locations along the nanotube where the tube has buckled. This buckling behavior was shown to be reversible in both experimental studies. The motion of nanotubes in contact with a surface can occur via rolling, sliding or a combination of these processes. The details of this behavior are expected to depend sensitively on the substrate as well as where the AFM tip contacts the nanotube to induce motion. *Falvo* et al. [61] found that if the AFM tip pushes the MWNTs from the end on a mica or graphite surface, a single stick-slip peak in the lateral force trace is observed. These peaks are attributed to the pinning force between the nanotube and the substrate that must be overcome before motion can proceed. However, when the nanotube is manipulated from its side on mica, the resulting motion is an in-plane rotation of the nanotube about a pivot point dependent upon the position of the AFM tip. This sliding behavior is illustrated in Fig. 15. However, if the

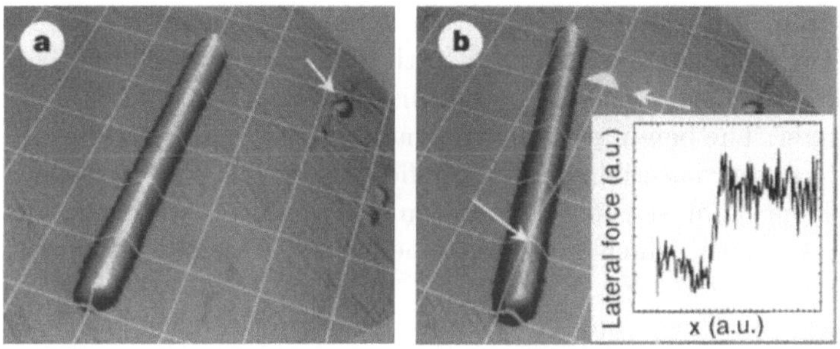


**Fig. 15.** Sliding a multi-walled carbon nanotube. (**a**) Tube in its original position. Grid lines are overlaid so that one of the grid axes corresponds to the original orientation of the tube axis. (**b**) Tube's orientation after AFM manipulation. The pivot point and push point are indicated by the bottom and top arrows, respectively. Inset shows the lateral force trace during a sliding manipulation [61]

nanotube was pushed from the side on a graphite surface, new behavior was observed: a lateral stick-slip motion with the absence of in-plane rotation. Hence the nanotube appeared to undergo a stick-slip rolling motion, which was topographically verified due to the asymmetrically shaped nanotube cap. Upon comparison of the lateral force measurements for rolling and sliding, *Falvo* et al. [61] discovered that the slip-stick peaks in rolling are higher than the lateral force needed to sustain sliding, although quantitative force values were not reported. Thus unlike macroscopic systems where rolling is preferred over sliding, the energy cost for rolling in nanoscale systems is larger than that of the sliding cases.

## 3.2 Nanotube Mechanical Properties

Taking advantage of low substrate/nanotube friction, *Wong* et al. [5] measured directly the bending and buckling forces for MWNTs on single-crystal $MoS_2$. They devised a flexible method that uses a combination of conventional lithography to pin one end of the nanotube, and the AFM tip to locate and probe the nanotube region protruding from the static contact. As the AFM scanned normal to the nanotube, the lateral force vs. displacement ($\boldsymbol{F}$–$\boldsymbol{d}$) curves at varying distances from the pinning point was recorded, and information on the Young's modulus, toughness, and strength of the nanotube were obtained. The $\boldsymbol{F}$–$\boldsymbol{d}$ curves recorded on the MWNTs showed several important points (Fig. 16). The initial location at which the lateral $\boldsymbol{F}$ increased in each scan was approximately the same, and thus it was possible to conclude that the nanotube deflection is elastic. Second, the lateral force recorded in each of the individual scans increased linearly once the tip contacted the nanotube ($\boldsymbol{F} = k\boldsymbol{d}$), and $k$ decreased for scan lines recorded at increasingly large distances from the pinning point. From simple nanobeam mechanics formulae, the Young's modulus of MWNTs with diameters from 26 to 76 nm was determined to be $1.28 \pm 0.59$ TPa, and this value was found to be independent of diameter. The bending strength of nanotubes was also determined by recognizing that the material softens significantly at the buckling point; that is, the buckling point is taken as a measure of the bend strength. The average bending strength, which was determined directly from $\boldsymbol{F}$–$\boldsymbol{d}$ curves, was found to be $14.2 \pm 8.0$ GPa.

The approach of *Wong* and co-workers [5] was extended by *Walters* et al. [62] to assess the tensile strength of SWNT ropes. In these latter measurements, the rope was freely suspended across a trench, and then deflected with the AFM. Lateral force vs. displacement curves recorded with the AFM tip at several different vertical heights across the center of the tube

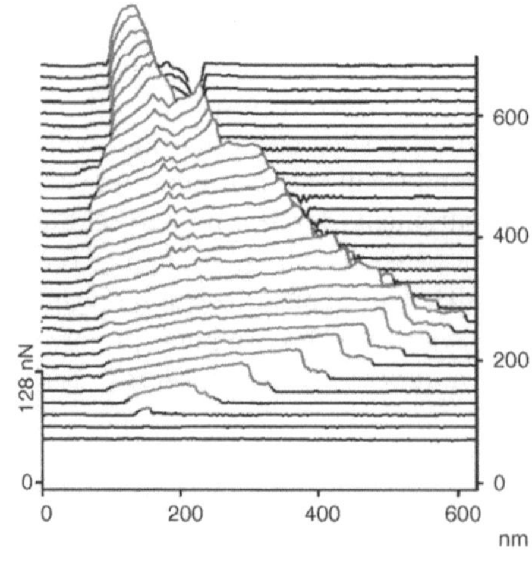


**Fig. 16.** Surface plot showing the $\boldsymbol{F}$–$\boldsymbol{d}$ response of a 32.9 nm diameter MWNT recorded with a normal load of 16.4 nN. The nanotube is pinned by a SiO pad beyond the top of the image. The data were recorded in water to minimize the nanotube-$MoS_2$ friction force [5]

were modeled reasonably well as an elastic string. The resulting fit to the experimental data yielded a maximum elastic strain of $5.8 \pm 0.9\%$. To compare this value with conventional, bulk materials, *Walters* et al. [62] calculated the yield strength for SWNT ropes, using a Young's modulus of 1.25 TPa for SWNTs [63], and found it to be $45 \pm 7$ GPa, over 20 times the yield strength of high-strength steels.

## 3.3 Making Nanotube Devices

The ability to manipulate carbon nanotubes on surfaces is the first step to controllably assemble them into electrical circuits and nanoscale devices. This toolkit has currently been extended to encompass tube translation, rotation, cutting, and even putting nanotubes on top of each other. Recently, *Roschier* et al. [64] and *Avouris* et al. [65] exploited nanotube translation and rotation by an AFM probe to create nanoscale circuits where a MWNT served as the active element. *Roschier* et al. [64] reported the fabrication of a MWNT single electron device (Fig. 17) and characterized its electrical properties. In addition, *Avouris* and co-workers [65] have fabricated a variety of structures to investigate the maximum current that can be passed through individual MWNTs, and the possibility of room-temperature electronics. In this latter regard they were able to fabricate a working field-effect transistor from a semiconducting nanotube.

*Lefebvre* et al. [66] reported the first tapping mode AFM manipulation of SWNT ropes to create crossed nanotube junctions. Their device fabrication strategy was similar to the manipulation method of [64,65], where translation

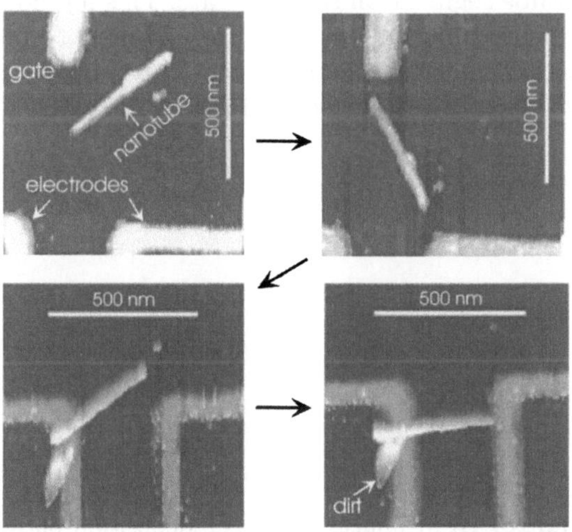


**Fig. 17.** AFM manipulation of a MWNT over pre-defined electrodes. The 410 nm long MWNT, the side gate, and the electrode structure are marked in the first frame. The last frame represents the measured configuration [64]

and rotation of nanotubes on surfaces occurred in small increments, typically 10 nm and 5 degrees at a time; however, instead of moving the nanotube over already-deposited electrodes, electrical contact was made to the nanotubes after the manipulation. The top nanotube and degenerate Si substrate can both serve as gate electrodes for the lower nanotube, and the current measured at 5 K through the lower nanotube exhibited current peaks reminiscent of two quantum dots in series. The authors speculate that the origin of the junction responsible for the double-dot behavior may either be mechanical (a combination of tube-tube and lower tube-surface interactions) or electrostatic (potential applied between lower tube and upper tube). In any case, the resulting local perturbation changes the nanotube band structure to preclude electron propagation along the lower tube. Interestingly, the tunnel barrier formed by the top bundle on the lower one can be tuned by the substrate voltage, and in the limit of strong coupling, the lower nanotube exhibits charging behavior of a single quantum dot.

As discussed above, manipulation of nanotubes on surfaces allows control in device fabrication of nanotube circuits, although this approach can also damage the nanotubes. Recently, *Cheung* et al. [67] reported a novel method of AFM manipulation and controlled deposition to create nanotube nanostructures. Their technique eliminates the laborious steps of incremental pushes and rotations as well as the unknown tube damaged caused by AFM manipulation across surfaces. Individual SWNTs and ropes were grown from Si-AFM tapping mode tips by chemical vapor deposition [68,69]. The SWNTs are deposited from the AFM tip to a pre-defined position on the substrate by three simple steps: (i) biasing the tip against the surface, (ii) scanning the nanotube tip along a set path, and (iii) then applying a voltage pulse to disconnect the tip from the nanotube segment on the substrate. This method can produce straight structures since tube-surface forces do not need to be overcome, and in addition, complex junctions between nanotubes may be created since the nanotube may be deposited at specified angles (Fig. 18).

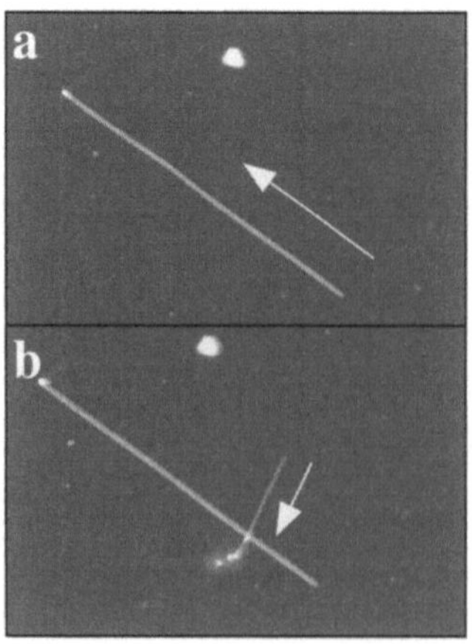


**Fig. 18.** AFM images of SWNT deposition onto a substrate. (**a**) A SWNT deposited along the direction of the arrow. (**b**) A cross SWNT structure made by a second nanotube lithography step. The images in (**a**) and (**b**) are 2μm × 1.3μm [67]

# 4 Nanotube Probe Microscopy Tips

In Atomic Force Microscopy (AFM), the probe is typically a sharp pyramidal tip on a micron-scale cantilever that allows measurement of the tip-sample force interactions and surface topography. Since the resulting image is a convolution of the structure of the sample and tip, sample features are either broadened by the tip, or narrowed in the case of trenches (Fig. 19). A well-characterized tip is therefore essential for accurately interpreting an image, and moreover, the size of the tip will define the resolution of the image. Well-characterized nanostructures, such as a carbon nanotube, may prove to be the ideal AFM tip.

Integrated AFM cantilever-tip assemblies are fabricated from silicon (Si) or silicon nitride ($Si_3N_4$) [70]. The tips on these assemblies are pyramidal in shape, have cone angles of 20–30 degrees and radii of curvature of 5–10 nm (Si) or 20–60 nm ($Si_3N_4$). Several techniques have been developed to improve these geometrical factors such as oxide sharpening, focused ion beam (FIB) milling, electron beam deposition of carbon, as well as improvements in the original tip formation processes. Despite these technological advances, there remain important limitations. Variation in tip-to-tip properties can be quite large, and will always be difficult to control at the scale relevant to high-resolution structural imaging. In addition, tips wear during scanning [71,72], making it quite difficult to account for tip contributions to image broadening. The problems of tip wear increase for sharper tips due to the higher pressure at the tip-sample interface.

Consider the ideal AFM tip. It should have a high aspect ratio with 0° cone angle, have a radius as small as possible with well-defined and reproducible molecular structure, and be mechanically and chemically robust such that it retains its structure while imaging in air or fluid environments. Carbon nanotubes are the only known material that can satisfy all of these criteria, and thus have the potential to create ideal probes for AFM imaging. For example, nanotubes have exceptional mechanical properties. Recent calculations [73] and experimental measurements [4,5,63] of SWNT and MWNT Young's moduli yield values ranging 1–2 TPa, demonstrating that nanotubes are stiffer than any other known material. The extremely high Young's modu-

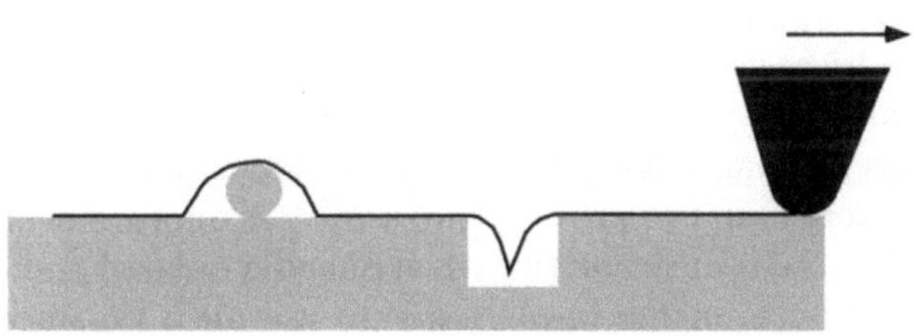

**Fig. 19.** Tip-sample convolution effect. The black line represents the path of the tip as it scans over the sample. The finite size of the imaging tip will broaden raised features and restrict access to recessed features

lus of nanotubes is critical to the creation of high aspect ratio, sub-nanometer radius tips with high resolution– if the modulus were significantly smaller, the amplitude of thermal vibrations would degrade the resolution of tips. In addition, carbon nanotubes buckle elastically under large loads unlike conventional materials which either fracture or plastically deform. Experimental studies in which nanotubes were used as AFM tips [74] and others which measured the deflection of nanotubes pinned at one end to a surface [5] revealed that the buckling is elastic. Both types of experiments demonstrated that nanotubes can be bent close to 90° many times without observable damage, and thus should be highly robust probes for AFM imaging.

### 4.1 Mechanical Assembly of Nanotube Tips

In the first demonstration of carbon nanotube AFM tips, *Dai* et al. [74] manually attached MWNTs to the pyramids of conventional tips. In this process, micromanipulators are used to control the positions of a commercial cantilever-tip assembly and nanotubes while viewing in an optical microscope. Micromanipulators allow the fabrication of nanotube tips that are well aligned for AFM imaging; that is, they are parallel to the tip axis and therefore perpendicular to the sample surface. However, the attached nanotube tips are typically too long to permit high-resolution imaging. The vibration amplitude at the end of the tip, $X_{\mathrm{tip}}$, can be readily estimated by equating $\frac{1}{2}k_{\mathrm{B}}T$ with the potential energy of the fundamental bending mode. This yields

$$X_{\mathrm{tip}} = \sqrt{\frac{4k_{\mathrm{B}}TL^3}{3\pi Er^4}} \tag{4}$$

where $k_{\mathrm{B}}$ is Boltzmann's constant, $T$ is the temperature in Kelvin, $L$ is the length, $E$ is the Young's modulus and $r$ is the radius of the nanotube tip. The length of a nanotube tip can be decreased to reduce the amplitude of vibration to a level where it does not affect resolution by electrical etching on a conductive surface [74,75].

Mechanically assembled MWNT tips have demonstrated several important features. First, the high aspect ratio of the tips enabled more accurate images of structures with steep sidewalls such as silicon trenches [74]. Second, these studies revealed that tip-sample adhesion could be greatly reduced due to the small size and cylindrical geometry of the nanotube [74,75], which allows imaging at lower cantilever energies. Third, they clearly demonstrated the elastic buckling property of nanotubes, and thus their robustness.

However, MWNT tips were found to provide only a modest improvement in resolution compared with standard silicon tips when imaging isolated amyloid fibrils [76]. The clear route to higher resolution is to use SWNTs, since they typically have 0.5–2 nm radii. Unfortunately, bulk SWNT material consists of bundles approximately 10 nm wide containing up to hundreds of nanotubes each, and thus cannot provide enhanced resolution unless single tubes or small numbers of tubes are exposed at the bundle ends. *Wong* et al. [77,78] attached these bundles to silicon AFM tips, and adjusted their lengths by the electrical etching procedure described above for optimal imaging. This etching process was found to occasionally produce very high-resolution tips that likely resulted from the exposure of only a small number of SWNTs at the apex. It was not possible in these studies to prepare individual SWNT tips for imaging.

The mechanical assembly production method is conceptually straightforward but also has several limitations. First, it inherently leads to the selection of thick bundles of nanotubes since these are easiest to observe in the optical microscope. Recently, *Nishijima* et al. [79] mechanically assembled nanotube tips inside a Scanning Electron Microscope (SEM). The use of a SEM still limits assembly to nanotube bundles or individual tubes with diameters greater than 5–10 nm, and moreover, increases greatly the time required to make one tip. Second, well-defined and reproducible tip etching procedures designed to expose individual SWNTs at the tip apex do not exist. Third, a relatively long time is required to attach each nanotube to an existing cantilever. This not only inhibits carrying out the research needed to develop these tips, but also precludes mass production required for general usage.

## 4.2 CVD Growth of Nanotube Tips

All of the problems associated with mechanical assembly potentially can be solved by direct growth of nanotubes onto AFM tips by catalytic Chemical Vapor Deposition (CVD). In the CVD synthesis of carbon nanotubes, metal catalyst particles are heated in a gas mixture containing hydrocarbon or CO. The gas molecules dissociate on the metal surface and carbon is adsorbed into the catalyst particle. When this carbon precipitates, it nucleates a nanotube of similar diameter to the catalyst particle. Hence, CVD allows control over nanotube size and structure including the production of SWNTs [80], with radii as low as 4 Å [81].

The central issues in the growth of nanotube AFM tips by CVD are (1) how to align the nanotubes at the tip such that they are well positioned for imaging and (2) how to ensure there is only one nanotube or nanotube bundle at the tip apex. Two approaches [68,69] for CVD nanotube tip growth have been developed (Fig. 20). First, *Hafner* et al. [68] grew nanotube tips by CVD from pores created in silicon tips. Electron microscopy revealed that MWNTs grew from pores in the optimal orientation for imaging in these initial studies. These "pore-growth" CVD nanotube tips were shown to exhibit the favorable mechanical and adhesion properties found earlier with manually assembled nanotube tips. In addition, the ability to produce thin, individual nanotube tips has enabled improved resolution imaging [68] of isolated proteins. More recent studies of the pore-growth nanotube tips by *Cheung* et al. [82] have focused on well-defined iron oxide nanocrystals as catalysts. This effort has enabled the controlled growth of thin SWNT bundles

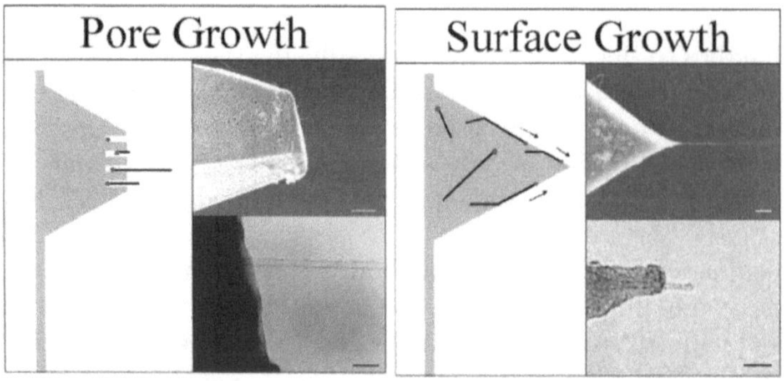


**Fig. 20.** CVD nanotube tip growth methods. The left panel illustrates pore growth. (*top*) The SEM image shows a flattened, porous AFM tip with a MWNT protruding from the flattened area. Scale bar is 1 µm. (*bottom*) The TEM image demonstrates that the tip consists of a thin, single MWNT. Scale bar is 20 nm. The right panel illustrates the surface growth technique. (*top*) The SEM image demonstrates that nanotubes are steered towards the tip. Scale bar is 100 nm. (*bottom*) The TEM image reveals that there is an individual SWNT at the tip. Scale bar is 10 nm [68,69]

1–3 nm in diameter from pores made at the silicon tip ends. The pore-growth method has demonstrated the great potential of CVD to produce controlled diameter nanotube tips, although it still has some limitations. In particular, the preparation of a porous layer can be time consuming and may not place individual SWNTs at the optimal location on the flattened apex.

In a second approach, *Hafner* et al. [69] grew CVD SWNTs directly from the pyramids of silicon cantilever-tip assemblies. The "surface growth" approach has exploited the trade-off between the energy gain of the nanotube-surface interaction and energy cost to bend nanotubes to grow SWNTs from the silicon pyramid apex in the ideal orientation for high resolution imaging. Specifically, when a growing nanotube reaches an edge of a pyramid, it can either bend to align with the edge or protrude from the surface. The pathway followed by the nanotube is determined by a trade-off in the energetic terms introduced above: if the energy required to bend the tube and follow the edge is less than the attractive nanotube-surface energy, then the nanotube will follow the pyramid edge to the apex; that is, nanotubes are steered towards the tip apex by the pyramid edges. At the apex, the nanotube must protrude along the tip axis since the energetic cost of bending is too high. This steering of nanotubes to the pyramid apex has been demonstrated experimentally [69]. For example, SEM investigations of nanotube tips produced by the surface growth method show that a very high yield of tips contain nanotubes only at the apex, with very few protruding elsewhere from the pyramid. TEM analysis has demonstrated that tips consist of individual SWNTs and small SWNT bundles. In the case of the small SWNT bundles, the TEM images show that the bundles are formed by nanotubes coming together from different edges of the pyramid to join at the apex, thus confirming the surface growth model described above [69]. The surface-growth approach is also important since it provides a conceptual and practical framework for preparing individual SWNT tips by lowering the catalyst density on the surface such that only 1 nanotube reaches the apex [67].

The synthesis of carbon nanotube AFM tips by CVD clearly resolves the major limitations of nanotube tips that arise from the manual assembly method. Rather than requiring tedious mircomanipulation for each tip, it is possible to envision production of an entire wafer of SWNT tips. In addition to ease of production, CVD yields thin, individual SWNT tips that cannot be made by other techniques and represent perhaps the ultimate AFM probe for high-resolution, high aspect ratio imaging.

### 4.3 Resolution of Nanotube Probes

The high-resolution capabilities of nanotube tips have been evaluated using 5.7 nm diameter gold nanoparticle imaging standards. These are good standards due to their well-defined shape and size, and their incompressibility [83]. The effective tip radius can be calculated from particle images using a two-sphere model [84]. Mechanically assembled MWNT tips typically have shown radii as small as 6 nm, which is expected for the size of arc-produced MWNT material [76]. Manually assembled SWNT tips are composed of thick SWNT bundles made from 1.4 nm nanotubes. The etching procedure can produce high resolution on gold nanoparticles, down to 3.4 nm radius [76,77], although more often the resolution is lower. Pore-growth CVD MWNT have radii ranging 3-6 nm when measured from gold nanoparticle images [68], which is better resolution than manually assembled tips because they are relatively thin and always consist of individual nanotubes. Both the pore-growth [82] and surface growth [69,67] SWNT bundles have shown tip radii of less than 4 nm, reflecting the very thin SWNT bundle or individual SWNT tip structures possible with these methods. These results are summarized in Table 1.

**Table 1.** Radius of curvature for differently prepared nanotube tips

| Nanotube Type | Tip Radius |
|---|---|
| Mech. Assembled MWNT | $> 6$ nm |
| Mech. Assembled SWNT | $> 3.5$ nm |
| Pore Growth MWNT | 3.5-6 nm |
| Pore Growth SWNT | 2-4 nm |
| Surface Growth SWNT | 2-4 nm |
| Low Density Surface Growth | $< 2$ nm |

Another significant point regarding nanotube tips is that the range of tip radii measured by AFM has agreed with the range measured by TEM. These results demonstrate that the imaging has been carried out with well-defined probes, which should be amenable to tip deconvolution techniques. Lastly, the levels of resolution achieved with the nanotubes obtained on gold nanoparticles nearly 6 nm in height, rather than on 1–2 nm features which at times indicate high resolution with tenuous asperities on microfabricated Si and $Si_3N_4$ tips.

## 4.4 Applications in Structural Biology

To determine the potential of nanotube tips in structural biology, DNA and several well-characterized proteins have been imaged. DNA was imaged by manually assembled MWNT tips in air [79] and in fluid [85]. The fluid imaging experiments produced a measured height of 2 nm, the expected value based on the intrinsic DNA diameter, and the resolution in these studies was on the order of 3.5 to 5 nm. These values for the resolution are consistent with that expected for multiwalled nanotube material, but are also similar to the best values observed with microfabricated tips [86,72].

Studies of isolated proteins provide a more stringent test of the capabilities of probe tips, and demonstrate clearly the advantages of nanotube tips. Pore-growth MWNT CVD tips have been used to image isolated immunoglobulin-M (IgM) antibody proteins [68]. IgM is a ca. 1 MDa antibody protein with a pentameric structure. It has not been crystallized for X-ray diffraction, but electron microscopy has elucidated the basic features of the pentameric structure [87]. Room temperature studies with pore-growth CVD MWNT nanotube tips [68] have clearly shown the pentameric structure, including five external pairs of antigen binding fragments (Fab domains) and five internal Fc fragments (Fig. 21).

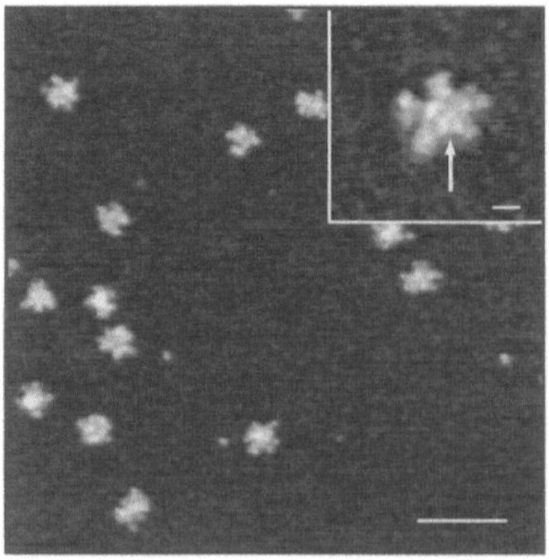

**Fig. 21.** Images of immunoglobulin-M (IgM) taken by CVD MWNT tips. Many IgM molecules are seen in various conformations due to their inherent flexibility. Scale bar is 100 nm. The inset shows an example of a well-oriented IgM molecule with the potential J-loop structure highlighted by the white arrow. Scale bar is 10 nm [68]

In addition, these images occasionally exhibited a structure connecting two of the five Fc domains that could correspond to the joining (J) loop [68]. Because the exact structure of this region is still unclear [87], these investigations have shown the potential of nanotube probes to reveal new structural features on large proteins that cannot be crystallized for diffraction studies.

The resolution of the smaller diameter CVD SWNT tips has been further tested in studies of a smaller protein, GroES, which is a component of the GroEL/GroES chaperoning system. GroES is a hollow dome shaped heptamer that is approximately 8 nm in outer diameter [88]. The seven 10 kD subunits each consist of a core b-barrel with a b-hairpin loop at the top and bottom. The top b-hairpins point inward to form the top of the dome, while the bottom hairpins are disordered when not in contact with GroEL [89]. CVD SWNT tip images of individual, well-separated GroES molecules on mica [82] reveal that it is possible to resolve the seven-fold symmetry [89] as shown in Fig. 22. These results have demonstrated clearly the ability of the present CVD nanotube tips to achieve sub-molecular resolution on isolated protein systems.

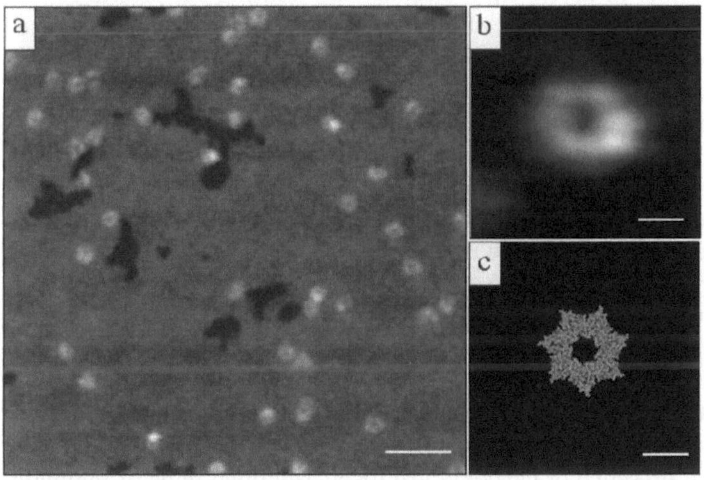


**Fig. 22.** GroES images taken by CVD SWNT tips. (**a**) A field of many GroES are shown, some displaying the pore side and some displaying the dome. Scale bar is 50 nm. (**b**) A high-resolution image showing the heptameric structure. (**c**) The crystal structure for (**b**) is shown for comparison. [82] Scale bar in (**b**) and (**c**) is 5 nm

### 4.5 Applications in Chemical Force Microscopy and Force Spectroscopy

In Chemical Force Microscopy (CFM), the AFM tip is modified with specific chemical functional groups [90]. This enables the tip to generate contrast dependent on the chemical properties of the sample from the friction signal in contact mode or the phase lag signal in tapping mode simultaneously with topography [91]. Functionalized tips have also been employed in force spectroscopy. In this mode of operation the tip is brought into contact with a surface, then retracted. The forces applied to the tip during retraction are due to the interactions of tip and sample molecules. Force spectroscopy has been used to measure a variety of interactions including the intermolecular adhesion between fundamental chemical groups [90,92,93,94,96], the unfolding of protein molecules [97], antigen–antibody interactions [98], and DNA stretching and unbinding [95].

Despite the progress made in chemically sensitive imaging and force spectroscopy using silicon and silicon nitride tips, these probes have important limitations. First, the tips have a large radius of curvature making it difficult to control the number of active tip molecules and limiting the lateral resolution. Second, the orientation and often the spatial location of the attached molecules cannot be controlled, leading to uncertainty in the reaction coordinate for force spectroscopy, and increased non-specific interactions. Carbon nanotube tips can overcome these limitations. They have small radii of curvature for higher resolution and can be specifically modified only at their very ends, creating fewer active molecular sites localized in a relatively controlled orientation. Modified SWNT tips could lead to subnanometer resolution in chemical contrast and binding site recognition.

Nanotube tips etched in air are expected to have carboxyl groups at their ends based on bulk studies of oxidized nanotubes [99], although conventional analytical techniques have insufficient sensitivity to observe this for isolated tubes. Chemical species present at the ends of nanotube tips can be studied with great sensitivity by measuring the adhesion of a nanotube tip on chemically well-defined self assembled monolayers (SAMs). *Wong* et al. [75,78] demonstrated the presence of carboxyl groups at the open ends of manually assembled MWNT and SWNT tips by measuring force titrations as shown in Fig. 23. In the force titration, the adhesion force between a nanotube tip and a SAM surface terminating in hydroxyl groups is recorded as a function of solution pH, thus effectively titrating ionizable groups on the tip [95,96]. Significantly, force titrations recorded between pH 2 and 9 with MWNT and SWNT tips were shown to exhibit well-defined drops in the adhesion force at ca. pH 4.5 that are characteristic of the deprotonation of a carboxylic acid.

*Wong* et al. [75,78] also modified assembled SWNT and MWNT bundle tips' organic and biological functionality by coupling organic amines to form amide bonds as outlined in Fig. 23a. Nanotube tips modified with benzylamine, which exposes nonionizable, hydrophobic functional groups at the

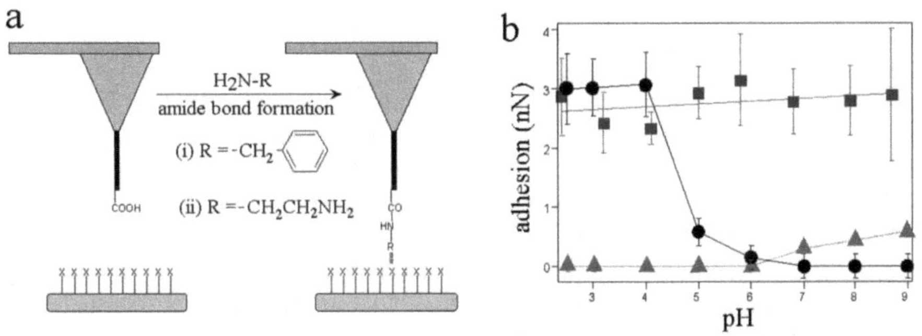


**Fig. 23.** Chemically modified nanotube tips. (**a**) Schematic diagram of the bond configuration of -COOH and amine functionalization. (**b**) Force titration data shows the expected adhesion dependence on pH for basic (▲), acidic (•) , and neutral (■) tip functionality [75]

tip end, yielded the expected pH-independent interaction force on hydroxyl-terminated monolayers (Fig. 23). Moreover, force titrations with ethylene diamine-modified tips exhibit no adhesion at low pH and finite adhesion above pH 7 (Fig. 23b), consistent with our expectations for exposed basic amine functionality that is protonated and charged at low pH and neutral at high pH.

Covalent reactions localized at nanotube tip ends represent a powerful strategy for modifying the functionality of the probe. However, the linking atoms that connect the tip and active group introduce conformational flexibility that may reduce the ultimate resolution. In an effort to develop a chemically sensitive probe without conformation flexibility, *Wong* et al. [100] explored the modification of the tips during the electrical etching process in the presence of $O_2$, $H_2$, or $N_2$. Significantly, force titrations carried out on tips modified in $O_2$, $N_2$ and $H_2$ exhibited behavior consistent with the incorporation of acidic, basic and hydrophobic functionality, respectively, at the tip ends.

*Wong* et al. [75,78] used functionalized nanotube probes obtain chemically sensitive images of patterned monolayer and bilayer samples [75,78]. Tapping mode images recorded with -COOH and benzyl terminated tips exhibit greater phase lag on the -COOH and -$CH_3$ sample regions, respectively. The "chemical resolution" of functionalized manually assembled MWNT and SWNT tips was tested on partial lipid bilayers [78]. Significantly, these studies showed that an assembled SWNT tip could detect variations in chemical functionality with resolution down to 3 nm, which is the same as the best structural resolution obtained with this type of tip. This resolution should improve with CVD SWNT tips, and recent studies bear this idea out [101].

Lastly, modified nanotube probes have been used to study ligand-receptor binding/unbinding with control of orientation, and to map the position of ligand-receptor binding sites in proteins and on cell surfaces with nanometer or better resolution. To illustrate this point, *Wong* et al. [75] examined the biotin-streptavidin interaction, which is a model ligand-receptor system that has been widely studied [104]. Biotin-modified nanotube tips were used to probe the receptor binding site on immobilized streptavidin as shown in Fig. 24a [75]. Force spectroscopy measurements show well-defined binding force quanta of ca. 200 pN per biotin-steptavidin pair (Fig. 24b). A key feature of these results compared to previous work [102,103], which relied on nonspecific attachment of biotin to lower resolution tips, is the demonstration that a single active ligand can be localized at the end of a nanotube tip using well-defined covalent chemistry. With the current availability of individual SWNT tip via surface CVD growth, it is now possible to consider the direct mapping of ligand binding sites for a wide range of proteins.

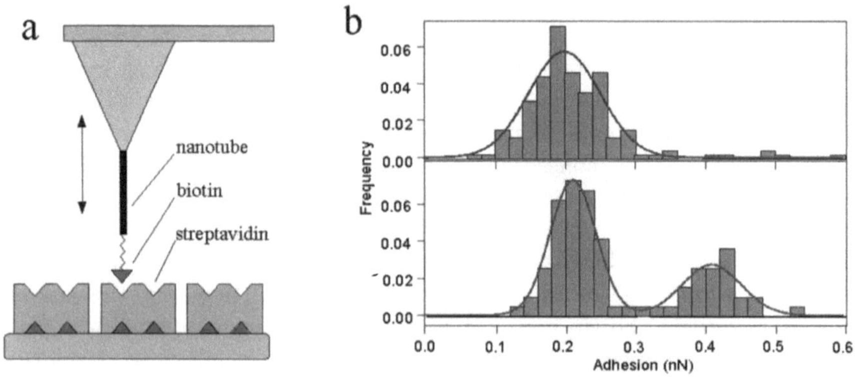


**Fig. 24.** Nanotube tips in force spectroscopy. (**a**) Nanotube tips are functionalized with biotin and their adhesion is studied against a surface coated with streptavidin. (**b**) The quantized adhesion measurements indicate that single and double biotin-streptavidin binding events are being observed [75]

## 5 Conclusions

Scanning probe microscopies have contributed significantly to understanding a wide range of properties of carbon nanotubes. STM and STS have been used to characterize the atomic structure and tunneling density of states of individual SWNTs and SWNT ropes. Studies of defect-free SWNTs have demonstrated semiconducting and metallic behavior that depends predictably on helicity and diameter. STM and spectroscopy measurements have defined the 1D VHS in the DOS for both metallic and semiconducting tubes, and comparisons to tight-binding calculations have shown good agreement with $\pi$-only calculations. Deviations from "simple" $\pi$-only models also suggest that further work will be necessary to understand fully how tube-tube interactions, which can produce broken symmetry, and curvature effects perturb the electronic structure of SWNTs. STM has also been used to characterize local structure and electronic properties of SWNT bends and ends. These studies have shown the presence of sharp spectroscopic features that in many cases can be understood well using $\pi$-only models, although more subtle features, which may reflect electron scattering, will require more detailed experimental and theoretical focus to unravel. The characterization of electronic features at SWNT ends also has implications to understanding and developing the chemical reactivity of this material and to efficiently couple nanotubes for electron transport. In addition, STM has been used to probe the electronic properties of finite length nanotubes. These studies show that it is possible to access readily a regime of "0D" behavior–where finite length produces quantization along the tube axis. These results suggest a number of future opportunities to probe, for example, connections between extended and molecular systems.

AFM has also proven to be a valuable tool for assessing the mechanical properties of nanotubes and for manipulating nanotubes into new structures. AFM has been used to assess fundamental energetics of nanotube-surface interactions and the frictional properties as nanotubes are slid and/or rolled on surfaces. The high force sensitivity of AFM has also been exploited to assess the Young's modulus of nanotubes and to determine the bending and tensile strengths of MWNTs and SWNTs, respectively. In addition, AFM has been exploited as a tool for positioning nanotube precisely to form nanoscale electronic devices. Lastly, carbon nanotubes have been used as novel probe microscopy tips. CVD methods have been demonstrated to produce well-defined MWNT and SWNT tips in an orientation optimal for imaging. AFM studies using these nanotube tips have demonstrated their robustness and high resolution. This new generation of probes has demonstrated ultrahigh resolution and chemically sensitive imaging capabilities, and is expected to have a significant impact on nanoscale research in Biology, Chemistry and Physics. As one looks to the future, we believe that probe microscopy studies of and with nanotubes will be rewarded with answers to many fundamental scientific problems, and moreover, will push many emerging concepts in nanotechnologies.

## References


1. M. S. Dresselhaus, G. Dresselhaus, *Science of Fullerenes and Carbon Nanotubes* (Academic, San Diego 1996)
2. B. I. Yakobson, R. E. Smalley, Amer. Sci. **85**, 324 (1997)
3. C. Dekker, Phys. Today **52**, 22 (1999)
4. M. M. J. Treacy, T. W. Ebbesen, J. M. Gibson, Nature **381**, 678 (1996)
5. E. W. Wong, P. E. Sheehan, C. M. Lieber, Science **277**, 1971 (1997)
6. B. I. Yakobson, C. J. Brabec, J. Bernholc, Phys. Rev. Lett.**76**, 2511 (1996)
7. N. Hamada, S. Sawada, A. Oshiyama, Phys. Rev. Lett. **68**, 1579 (1992)

8. J. W. Mintmire, B. I. Dunlap, C. T. White, Phys. Rev. Lett.**68**, 631 (1992)
9. R. Saito, M. Fujita, G. Dresselhaus, M. S. Dresselhaus, Appl. Phys. Lett. **60**, 2204 (1992)
10. J. W. G. Wildöer, L. C. Venema, A. G. Rinzler, R. E. Smalley, C. Dekker, Nature **391**, 59 (1998)
11. T. W. Odom, J.-L. Huang, P. Kim, C. M. Lieber, Nature **391**, 62 (1998)
12. C. M. Lieber, Solid State Comm. **107**, 607 (1998)
13. J. T. Hu, T. W. Odom, C. M. Lieber, Accts. Chem. Res. **32**, 435 (1999)
14. C. L. Kane, E. J. Mele, Phys. Rev. Lett. **78**, 1932 (1997)
15. V. H. Crespi, M. L. Cohen, A. Rubio, Phys. Rev. Lett. **79**, 2093 (1997)
16. M. Ge, K. Sattler, Science **260**, 515 (1993)
17. Z. Zhang, C. M. Lieber, Appl. Phys. Lett. **62**, 2972 (1993)
18. C. H. Olk, J. P. Heremans, J. Mater. Res. **9**, 259 (1994)
19. A. Thess, R. Lee, P. Nikolaev, H. Dai, P. Petit, J. Rober, C. Zu, Y. H. Lee, S. G. Kim, A. G. Rinzler, D. T. Colbert, G. E. Scuseria, D. Tomanek, J. E. Fischer, R. E. Smalley, Science **273**, 483 (1996)
20. C. Journet, W. K. Maser, P. Bernier, A. Loiseau, M. Lamy de la Chapelle, S. Lefrant, P. Deniard, R. Lee and J. E. Fischer, Nature **388**, 756 (1997)
21. T. Guo, P. Nikolaev, A. Thess, D. T. Colbert, R. E. Smalley, Chem. Phys. Lett. **243**, 49 (1995)
22. C. T. White, J. W. Mintmire, Nature **394**, 29 (1998)
23. C. T. White, D. H. Robertson, J. W. Mintmire, Phys. Rev. B **47**, 5485 (1993)
24. T. W. Odom, J.-L. Huang, P. Kim, M. Ouyang, C. M. Lieber, J. Mater. Res. **13**, 2380 (1998)
25. P. Kim, T. W. Odom, J.-L. Huang, C. M. Lieber, Phys. Rev. Lett. **82**, 1225 (1999)
26. A. Rubio, Appl. Phys. A. **68**, 275 (1999)
27. X. Blasé, L. X. Benedict, E. L. Shirley, S. G. Louie, Phys. Rev. Lett. **72**, 1878 (1994)
28. J. C. Charlier, Ph. Lambin, Phys. Rev. B **57**, R15037 (1998)
29. T. W. Odom, J.-L. Huang, P. Kim, C. M. Lieber, J. Phys. Chem. B **104**, 2794 (2000)
30. V. Meunier, L. Henrard, Ph. Lambin, Phys. Rev. B **57**, 2596 (1998)
31. J.-C. Charlier, T. W. Ebbeson, Ph. Lambin, Phys. Rev. B **53**, 11108 (1996)
32. T. Hertel, R. E. Walkup, Ph. Avouris, Phys. Rev. B **58**, 13870 (1998)
33. W. Clauss, D. J. Bergeron, A. T. Johnson, Phys. Rev. B **58**, 4266 (1998)
34. N. G. Chopra, L. X. Benedict, V. H. Crespi, M. L. Cohen, S. G. Louie, A. Zettl, Nature **377**, 135 (1995)
35. P. Lambin, V. Meunier, L. P. Biro, Carbon **36**, 701 (1998)
36. P. H. Lambin, A. A. Lucas, J. C. Charlier, J. Phys. Chem. Solids **58**, 1833 (1997)
37. L. Chico, V. H. Crespi, L. X. Benedict, S. G. Louie, M. L. Cohen, Phys. Rev. Lett. **76**, 971 (1996)
38. J. Han, M. P. Anantram, R. L. Jaffe, J. Kong, H. Dai, Phys. Rev. B **57**, 14983 (1998)
39. A. Rochefort, D. R. Salahub, Ph. Avouris, Chem. Phys. Lett. **297**, 45 (1998)
40. Ph. Avouris, R. Martel, H. Ikeda, M. Hersam, H. R. Shea, A. Rochefort, in *Science and Application of Nanotubes*, D. Tomanek, R. J. Enbody (Eds.) (Kluwer Academic / Plemun Publishers, New York, 2000).

41. C. L. Kane, E. J. Mele, Phys. Rev. B **59**, R12759 (1999)
42. R. Tamura, M. Tsukada, Phys. Rev. B **52**, 6015 (1995)
43. D. L. Carroll, P. Redlich, P. M. Ajayan, J. C. Charlier, X. Blasé, A. DeVita, R. Car, Phys. Rev. Lett. **78**, 2811 (1997)
44. M. Bockrath, D. H. Cobden, P. L. McEuen, N. G. Chopra, A. Zettl, A. Thess, R. E. Smalley, Science **275**, 1922 (1997)
45. S. J. Tans, M. H. Devoret, H. Dai, A. Thess, R. E. Smalley, L. J. Geerligs, C. Dekker, Nature **386**, 474 (1997)
46. L. C. Venema, J. W. G. Wildoer, H. L. J. Temminck Tunistra, C. Dekker, A. Rinzler, R. E. Smalley, Appl. Phys. Lett. **71**, 2629 (1997)
47. L. C. Venema, J. W. G. Wildoer, J. W. Janssen, S. J. Tans, H. L. J. Temminck Tuinstra, L. P. Kouwenhoven, C. Dekker, Science **283**, 52 (1999)
48. A. Rubio, D. Sanchez-Portal, E. Attach, P. Ordoejon, J. M. Soler, Phys. Rev. Lett. **82**, 3520 (1999)
49. L. G. Bulusheva, A. V. Okotrub, D. A. Romanov, D. Tomanek, J. Phys. Chem. A **102**, 975 (1998)
50. A. Rochefort, D. R. Salahub, Ph. Avouris, J. Phys. Chem. B **103**, 641 (1999)
51. S. A. Empedocles, D. J. Norris, M. G. Bawendi, Phys. Rev. Lett. **77**, 3873 (1996)
52. T. W. Odom, J.-L. Huang, C. M. Lieber, unpublished
53. P. L. McEuen, M. Bockrath, D. H. Cobden, Y.-G. Yoon, S. G. Louie, Phys. Rev. Lett. **83**, 5098 (1999)
54. R. A. Jishi, J. Bragin, L. Lou, Phys. Rev. B **59**, 9862 (1999)
55. H. Grabert, M. H. Devort, *Single Charge Tunneling* (Plenum, New York, 1992)
56. A. E. Hanna, M. Tinkham, Phys. Rev. B **44**, 5919 (1991)
57. D. C. Ralph, C. T. Black, M. Tinkham, Phys. Rev. Lett. **74**, 3241 (1995)
58. M. H. Nayfeh, M. K. Brussel, *Electricity and Magnetism* (John Wiley and Sons, New York, 1985)
59. T. Hertel, R. Martel, Ph. Avouris, J. Phys. Chem. **102**, 910 (1998)
60. M. R. Falvo, G. J. Clary, R. M. Taylor II, V. Chi, F. P. Brooks Jr, S. Washburn, R. Superfine, Nature **389**, 582 (1997)
61. M. R. Falvo, R. M. Taylor, A. Helser, V. Chi, F. P. Brooks, S. Washburn, R. Superfine, Nature **397**, 236 (1999)
62. D. A. Walters, L. M. Ericson, M. J. Casavant, J. Liu, D. T. Colbert, K. A. Smith, R. E. Smalley, Appl. Phys. Lett. **74**, 3803 (1999)
63. A. Krishnan, E. Dujardin, T. W. Ebbesen, P. N. Yianilos, M. M. J. Treacy, Phys. Rev. B **58**, 14013 (1998)
64. L. Roschier, J. Penttilä, M. Martin, P. Hakonene, M. Paalanen, U. Tapper, E. I. Kauppinen, C.Journet, P. Bernier, Appl. Phys. Lett. **75**, 728 (1999)
65. Ph. Avouris, T. Hertel, R. Martel, T. Schmidt, H. R. Shea, R. E. Walkup, Appl. Surf. Sci. **141**, 201 (1999)
66. J. Lefebvre, J. F. Lynch, M. Llanguno, M. Radosavljevic, A. T. Johnson, Appl. Phys. Lett. **75**, 3014 (1999)
67. C. L. Cheung, J. H. Hafner, T. W. Odom, K. Kim, C. M. Lieber, Appl. Phys. Lett. **76**, 3136 (2000)
68. J. H. Hafner, C.-L. Cheung, C. M. Lieber, Nature **398**, 761 (1999)
69. J. H. Hafner, C.-L. Cheung, C. M. Lieber, J. Am. Chem. Soc. **121**, 9750 (1999)
70. T. R. Albrecht, S. Akamine, T. E. Carber, C. F. Quate, J. Vac. Sci. Tech. A **8**, 3386 (1989)

71. H. G. Hansma, D. E. Laney, M. Bezanilla, R. L. Sinsheimer, P. K. Hansma, Biophys. J. **68**, 1672 (1995)
72. Y. L. Lyubchenko, L. S. Schlyakhtenko, Proc. Natl. Acad. Sci. Am. **94**, 496 (1997)
73. J. P. Lu, Phys. Rev. Let. **79**, 1297 (1997)
74. H. Dai, J. H. Hafner, A. G. Rinzler, D. T. Colbert, R. E. Smalley, Nature **384**, 147 (1996)
75. S. S. Wong, E. Joselevich, A. T. Woolley, C.-L. Cheung, C. M. Lieber, Nature **394**, 52 (1998)
76. S. S. Wong, J. D. Harper, P. T. Lansbury, C. M. Lieber, J. Am. Chem. Soc. **120**, 603 (1998)
77. S. S. Wong, A. T. Woolley, T. W. Odom, J.-L. Huang, P. Kim, D. V. Vezenov, C. M. Lieber, Appl. Phys. Lett. **73**, 3465 (1998)
78. S. S. Wong, A. T. Woolley, E. Joselevich, C. M. Lieber, J. Am. Chem. Soc **120**, 8557 (1998)
79. H. Nishijima, S. Kamo, S. Akita, Y. Nakayama, Appl. Phys. Lett. **74**, 4061 (1999)
80. J. H. Hafner, M. J. Bronikowski, B. R. Azamian, P. Nikolaev, A. G. Rinzler, D. T. Colbert, K. Smith, R. E. Smalley, Chem. Phys. Lett. **296**, 195 (1998)
81. P. Nikolaev, M. J. Bronikowski, R. K. Bradley, F. Rohmund, D. T. Colbert, K. A. Smith, R. E. Smalley, Chem. Phys. Lett. **313**, 91 (1999)
82. C.-L. Cheung, J. H. Hafner, C. M. Lieber, Proc. Natl. Acad. Sci., **97**, 3809 (2000)
83. J. Vesenka, S. Manne, R. Giberson, T. Marsh, E. Henderson, Biophys. J. **65**, 992 (1993)
84. C. Bustamante, D. Keller, G. Yang, Curr. Opin. Stuct. Biol. **3**, 363 (1993)
85. J. Li, A. M. Cassell, H. Dai, Surf. InterFac. Anal. **28**, 8 (1999)
86. W. Han, S. M. Lindsay, M. Klakic, R. E. Harrington, Nature **386**, 563 (1997)
87. S. J. Perkins, A. S. Nealis, B. J. Sutton, A. Feinstein, J. Mol. Biol. **221**, 1345 (1991)
88. P. B. Sigler, Z. Xu, H. S. Rye, S. G. Burston, W. A. Fenton, A. L. Horwich, Annu. Rev. Biochem. **67**, 581 (1998)
89. J. F. Hunt, A. J. Weaver, S. J. Landry, L. Gierasch, J. Deisenhofer, Nature **379**, 37 (1996)
90. C. D. Frisbie, L. F. Rozsnyai, A. Noy, M. S. Wrighton, C. M. Lieber, Science **265**, 2071 (1994)
91. A. Noy, C. H. Sanders, D. V. Vezenov, S. S. Wong, C. M. Lieber, Langmuir **14**, 1508 (1998)
92. J. B. D. Green, M. T. McDermott, M. D. Porter, L. M. Siperko, J. Phys. Chem. **99**, 10960 (1995)
93. A. Noy, C. D. Frisbie, L. F. Rozsnyai, M. S. Wrighton, C. M. Lieber, J. Am. Chem. Soc. **117**, 7943 (1995)
94. A. Noy, D. V. Vezenoz, C. M. Lieber, Annu. Rev. Mater. Sci. **27**, 381 (1997)
95. A. Noy, D. V. Vezenov, J. F. Kayyem, T. J. Meade, C. M. Lieber, Chem. Biol. **4**, 519 (1997)
96. D. V. Vezenov, A. Noy, L. F. Rozsnyai, C. M. Lieber, J. Am. Chem. Soc. **119**, 2006 (1997)
97. M. Rief, M. Gautel, F. Oesterhelt, J. M. Fernandez, H. E. Gaub, Science **276**, 1109 (1997)

98. P. Hinterdorfer, W. Baumgertner, H. J. Gruber, K. Schilcher, H. Schindler, Proc. Natl. Acad. Sci. Am. **93**, 3477 (1996)
99. H. Hiura, T. W. Ebbesen, K. Tanigaki, Adv. Mater. **7**, 275 (1995)
100. S. S. Wong, A. T. Woolley, E. Joselevich, C. M. Lieber, Chem. Phys. Lett. **306**, 219 (1999)
101. C.-L. Cheung, L. Chen, C. M. Lieber, unpublished results (2000)
102. E.-L. Florin, V. T. Moy, H. E. Gaub, Science **264**, 415 (1994)
103. G. U. Lee, D. A. Kidwell, R. J. Colton, Langmuir **10**, 354 (1994)
104. O. Livnah, E. A. Bayer, M. Wilchek, J. L. Sussman, Proc. Natl. Acad. Sci. Am. **90**, 5076 (1993)

# Optical Properties and Raman Spectroscopy of Carbon Nanotubes

Riichiro Saito[1] and Hiromichi Kataura[2]

[1] Department of Electronic-Engineering, The University of Electro-Communications
1-5-1, Chofu-gaoka, Chofu, Tokyo 182-8585, Japan
rsaito@tube.ee.uec.ac.jp
[2] Department of Physics, Tokyo Metropolitan University
1-1 Minami-Ohsawa, Hachioji, Tokyo 192-0397, Japan
kataura@phys.metro-u.ac.jp

**Abstract.** The optical properties and the resonance Raman spectroscopy of single wall carbon nanotubes are reviewed. Because of the unique van Hove singularities in the electronic density of states, resonant Raman spectroscopy has provided diameter-selective observation of carbon nanotubes from a sample containing nanotubes with different diameters. The electronic and phonon structure of single wall carbon nanotubes are reviewed, based on both theoretical considerations and spectroscopic measurements.

The quantum properties of Single-Wall Carbon Nanotubes (SWNTs) depend on the diameter and chirality, which is defined by the indices $(n, m)$ [1,2]. Chirality is a term used to specify a nanotube structure, which does not have mirror symmetry. The synthesis of a SWNT sample with a single chirality is an ultimate objective for carbon nanotube physics and material science research, but this is still difficult to achieve with present synthesis techniques. On the other hand, the diameter of SWNTs can now be controlled significantly by changing the furnace growth temperature and catalysts [3,4,5,6]. Thus, a mixture of SWNTs with different chiralities, but with a small range of nanotube diameters is the best sample that can be presently obtained. Resonance Raman spectroscopy provides a powerful tool to investigate the geometry of SWNTs for such samples and we show here that metallic and semiconducting carbon nanotubes can be separately observed in the resonant Raman signal.

In this paper, we first review theoretical issues concerning the electron and phonon properties of a single-walled carbon nanotube. We then describe the electronic and phonon density of states of SWNTs. In order to discuss resonant Raman experiments, we make a plot of the possible energies of optical transitions as a function of the diameter of SWNTs.

Then we review experimental issues concerning the diameter-controlled synthesis of SWNTs and Raman spectroscopy by many laser frequencies. The optical absorption measurements of SWNTs are in good agreement with the theoretical results.

M. S. Dresselhaus, G. Dresselhaus, Ph. Avouris (Eds.): Carbon Nanotubes,
Topics Appl. Phys. **80**, 213–246 (2001)

# 1 Theoretical Issues

The electronic structure of carbon nanotubes is unique in solid-state physics in the sense that carbon nanotubes can be either semiconducting or metallic, depending on their diameter and chirality [1,2]. The phonon properties are also remarkable, showing unique one-dimensional (1D) behavior and special characteristics related to the cylindrical lattice geometry, such as the Radial Breathing Mode (RBM) properties and the special twist acoustic mode which is unique among 1D phonon subbands.

Using the simple tight-binding method and pair-wise atomic force constant models, we can derive the electronic and phonon structure, respectively. These models provide good approximations for understanding the experimental results for SWNTs.

## 1.1 Electronic Structure and Density of States of SWNTs

We now introduce the basic definitions of the carbon nanotube structure and of the calculated electronic and phonon energy bands with their special Density of States (DOS). The structure of a SWNT is specified by the chiral vector $\mathbf{C}_h$

$$\mathbf{C}_h = n\mathbf{a}_1 + m\mathbf{a}_2 \equiv (n, m)\,, \tag{1}$$

where $\mathbf{a}_1$ and $\mathbf{a}_2$ are unit vectors of the hexagonal lattice shown in Fig. 1. The vector $\mathbf{C}_h$ connects two crystallographically equivalent sites $O$ and $A$ on a two-dimensional (2D) graphene sheet, where a carbon atom is located at each vertex of the honeycomb structure [7]. When we join the line $AB'$ to the parallel line $OB$ in Fig. 1, we get a seamlessly joined SWNT classified by the integers $(n, m)$, since the parallel lines $AB'$ and $OB$ cross the honeycomb lattice at equivalent points. There are only two kinds of SWNTs which have mirror symmetry: zigzag nanotubes $(n, 0)$, and armchair nanotubes $(n, n)$. The other nanotubes are called chiral nanotubes, and they have axial chiral symmetry. The general chiral nanotube has chemical bonds that are not

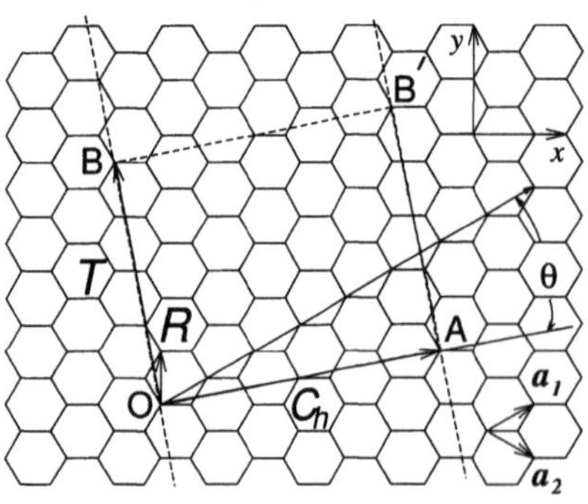


**Fig. 1.** The unrolled honeycomb lattice of a nanotube. When we connect sites $O$ and $A$, and sites $B$ and $B'$, a nanotube can be constructed. $\overrightarrow{OA}$ and $\overrightarrow{OB}$ define the chiral vector $\mathbf{C}_h$ and the translational vector $\mathbf{T}$ of the nanotube, respectively. The rectangle $OAB'B$ defines the unit cell for the nanotube. The figure is constructed for an $(n, m) = (4, 2)$ nanotube [2]

parallel to the nanotube axis, denoted by the chiral angle $\theta$ in Fig. 1. Here the direction of the nanotube axis corresponds to $OB$ in Fig. 1. The zigzag, armchair and chiral nanotubes correspond, respectively, to $\theta = 0°$, $\theta = 30°$, and $0 \leq |\theta| \leq 30°$. In a zigzag or an armchair nanotube, respectively, one of three chemical bonds from a carbon atom is parallel or perpendicular to the nanotube axis.

The diameter of a $(n, m)$ nanotube $d_t$ is given by

$$d_t = C_h/\pi = \sqrt{3}a_{C-C}(m^2 + mn + n^2)^{1/2}/\pi \tag{2}$$

where $a_{C-C}$ is the nearest-neighbor C–C distance (1.42 Å in graphite), and $C_h$ is the length of the chiral vector $\mathbf{C}_h$. The chiral angle $\theta$ is given by

$$\theta = \tan^{-1}[\sqrt{3}m/(m + 2n)]. \tag{3}$$

The 1D electronic DOS is given by the energy dispersion of carbon nanotubes which is obtained by zone folding of the 2D energy dispersion relations of graphite. Hereafter we only consider the valence $\pi$ and the conduction $\pi^*$ energy bands of graphite and nanotubes. The 2D energy dispersion relations of graphite are calculated [2] by solving the eigenvalue problem for a $(2 \times 2)$ Hamiltonian $\mathcal{H}$ and a $(2 \times 2)$ overlap integral matrix $\mathcal{S}$, associated with the two inequivalent carbon atoms in 2D graphite,

$$\mathcal{H} = \begin{pmatrix} \epsilon_{2p} & -\gamma_0 f(k) \\ -\gamma_0 f(k)^* & \epsilon_{2p} \end{pmatrix} \text{ and } \mathcal{S} = \begin{pmatrix} 1 & sf(k) \\ sf(k)^* & 1 \end{pmatrix} \tag{4}$$

where $\epsilon_{2p}$ is the site energy of the $2p$ atomic orbital and

$$f(k) = \mathrm{e}^{\mathrm{i}k_x a/\sqrt{3}} + 2\mathrm{e}^{-\mathrm{i}k_x a/2\sqrt{3}} \cos\frac{k_y a}{2} \tag{5}$$

where $a = |\mathbf{a}_1| = |\mathbf{a}_2| = \sqrt{3}a_{C-C}$. Solution of the secular equation $\det(\mathcal{H} - E\mathcal{S}) = 0$ implied by (4) leads to the eigenvalues

$$E_{g2D}^{\pm}(\boldsymbol{k}) = \frac{\epsilon_{2p} \pm \gamma_0 w(\boldsymbol{k})}{1 \mp s w(\boldsymbol{k})} \tag{6}$$

for the C-C transfer energy $\gamma_0 > 0$, where $s$ denotes the overlap of the electronic wave function on adjacent sites, and $E^+$ and $E^-$ correspond to the $\pi^*$ and the $\pi$ energy bands, respectively. Here we conventionally use $\gamma_0$ as a positive value. The function $w(\boldsymbol{k})$ in (6) is given by

$$w(\boldsymbol{k}) = \sqrt{|f(\boldsymbol{k})|^2} = \sqrt{1 + 4\cos\frac{\sqrt{3}k_x a}{2}\cos\frac{k_y a}{2} + 4\cos^2\frac{k_y a}{2}}. \tag{7}$$

In Fig. 2 we plot the electronic energy dispersion relations for 2D graphite as a function of the two-dimensional wave vector $\boldsymbol{k}$ in the hexagonal Brillouin zone in which we adopt the parameters $\gamma_0 = 3.013\,\mathrm{eV}$, $s = 0.129$ and $\epsilon_{2p} = 0$

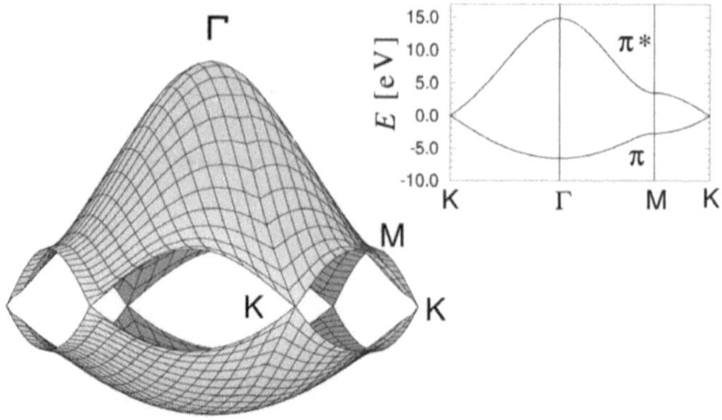


**Fig. 2.** The energy dispersion relations for 2D graphite with $\gamma_0 = 3.013\,\text{eV}$, $s = 0.129$ and $\epsilon_{2p} = 0$ in (6) are shown throughout the whole region of the Brillouin zone. The inset shows the energy dispersion along the high symmetry lines between the $\Gamma$, $M$, and $K$ points. The valence $\pi$ band (*lower part*) and the conduction $\pi^*$ band (*upper part*) are degenerate at the $K$ points in the hexagonal Brillouin zone which corresponds to the Fermi energy [2]

so as to fit both the first principles calculation of the energy bands of 2D turbostratic graphite [8,9] and experimental data [2,10]. The corresponding energy contour plot of the 2D energy bands of graphite with $s = 0$ and $\epsilon_{2p} = 0$ is shown in Fig. 3. The Fermi energy corresponds to $E = 0$ at the $K$ points.

Near the $K$-point at the corner of the hexagonal Brillouin zone of graphite, $w(\boldsymbol{k})$ has a linear dependence on $k \equiv |\boldsymbol{k}|$ measured from the $K$ point as

$$w(\boldsymbol{k}) = \frac{\sqrt{3}}{2}ka + \ldots, \qquad \text{for } ka \ll 1. \tag{8}$$

Thus, the expansion of (6) for small $k$ yields

$$E^{\pm}_{g2\mathrm{D}}(\boldsymbol{k}) = \epsilon_{2p} \pm (\gamma_0 - s\epsilon_{2p})w(\boldsymbol{k}) + \ldots\,, \tag{9}$$

so that in this approximation, the valence and conduction bands are symmetric near the $K$ point, independent of the value of $s$. When we adopt $\epsilon_{2p} = 0$ and take $s = 0$ for (6), and assume a linear $k$ approximation for $w(k)$, we get the linear dispersion relations for graphite given by [12,13]

$$E(k) = \pm\frac{\sqrt{3}}{2}\gamma_0 ka = \pm\frac{3}{2}\gamma_0 ka_{\mathrm{C-C}}\,. \tag{10}$$

If the physical phenomena under consideration only involve small $k$ vectors, it is convenient to use (10) for interpreting experimental results relevant to such phenomena.

The 1D energy dispersion relations of a SWNT are given by

$$E^{\pm}_{\mu}(k) = E^{\pm}_{g2\mathrm{D}}\left(k\frac{\mathbf{K}_2}{|\mathbf{K}_2|} + \mu\mathbf{K}_1\right), \quad \left(-\frac{\pi}{T} < k < \frac{\pi}{T}, \text{ and } \mu = 1, \cdots, N\right), \tag{11}$$

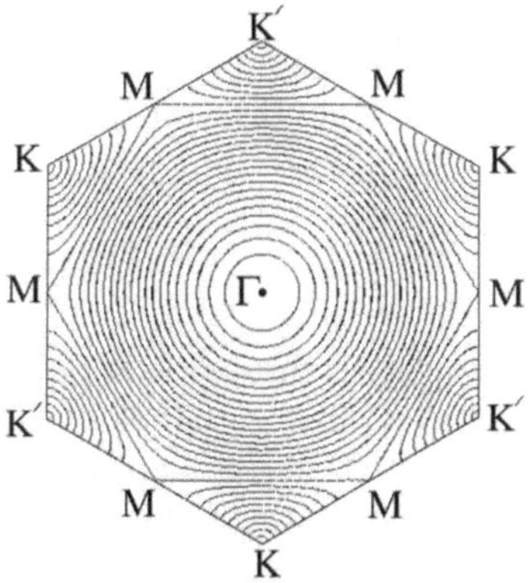


**Fig. 3.** Contour plot of the 2D electronic energy of graphite with $s = 0$ and $\epsilon_{2p} = 0$ in (6). The equi-energy lines are circles near the $K$ point and near the center of the hexagonal Brillouin zone, but are straight lines which connect nearest $M$ points. Adjacent lines correspond to changes in height (energy) of $0.1\gamma_0$ and the energy value for the $K$, $M$ and $\Gamma$ points are 0, $\gamma_0$ and $3\gamma_0$, respectively. It is useful to note the coordinates of high symmetry points: $K = (0, 4\pi/3a)$, $M = (2\pi/\sqrt{3}a, 0)$ and $\Gamma = (0,0)$, where $a$ is the lattice constant of the 2D sheet of graphite [11]

where $T$ is the magnitude of the translational vector $\mathbf{T}$, $k$ is a 1D wave vector along the nanotube axis, and $N$ denotes the number of hexagons of the graphite honeycomb lattice that lie within the nanotube unit cell (see Fig. 1). $T$ and $N$ are given, respectively, by

$$T = \frac{\sqrt{3}C_h}{d_\mathrm{R}} = \frac{\sqrt{3}\pi d_\mathrm{t}}{d_\mathrm{R}}, \quad \text{and} \quad N = \frac{2(n^2 + m^2 + nm)}{d_\mathrm{R}} . \tag{12}$$

Here $d_\mathrm{R}$ is the greatest common divisor of $(2n + m)$ and $(2m + n)$ for a $(n, m)$ nanotube [2,14]. Further $\mathbf{K}_1$ and $\mathbf{K}_2$ denote, respectively, a discrete unit wave vector along the circumferential direction, and a reciprocal lattice vector along the nanotube axis direction, which for a $(n, m)$ nanotube are given by

$$\begin{aligned} \mathbf{K}_1 &= \{(2n + m)\mathbf{b}_1 + (2m + n)\mathbf{b}_2\}/N d_\mathrm{R} \qquad \text{and} \\ \mathbf{K}_2 &= (m\mathbf{b}_1 - n\mathbf{b}_2)/N , \end{aligned} \tag{13}$$

where $\mathbf{b}_1$ and $\mathbf{b}_2$ are the reciprocal lattice vectors of 2D graphite and are given in $x, y$ coordinates by

$$\mathbf{b}_1 = \left(\frac{1}{\sqrt{3}}, 1\right)\frac{2\pi}{a}, \quad \mathbf{b}_2 = \left(\frac{1}{\sqrt{3}}, -1\right)\frac{2\pi}{a}. \tag{14}$$

The periodic boundary condition for a carbon nanotube $(n, m)$ gives $N$ discrete $k$ values in the circumferential direction. The $N$ pairs of energy dispersion curves given by (11) correspond to the cross sections of the two-dimensional energy dispersion surface shown in Fig. 2, where cuts are made on the lines of $k\mathbf{K}_2/|\mathbf{K}_2| + \mu\mathbf{K}_1$. In Fig. 4 several cutting lines near one of the $K$ points are shown. The separation between two adjacent lines and the length of the cutting lines are given by the $\mathbf{K}_1$ and $\mathbf{K}_2$ vectors of (13), respectively, whose lengths are given by

$$|\mathbf{K}_1| = \frac{2}{d_t}, \quad \text{and} \quad |\mathbf{K}_2| = \frac{2\pi}{T} = \frac{2d_R}{\sqrt{3}d_t}. \tag{15}$$

If, for a particular $(n, m)$ nanotube, the cutting line passes through a $K$ point of the 2D Brillouin zone (Fig. 4a), where the $\pi$ and $\pi^*$ energy bands of 2D graphite are degenerate (Fig. 2) by symmetry, then the 1D energy bands have a zero energy gap. Since the degenerate point corresponds to the Fermi energy, and the density of states are finite as shown below, SWNTs with a zero band gap are metallic. When the $K$ point is located between two cutting lines, the $K$ point is always located in a position one-third of the distance between two adjacent $\mathbf{K}_1$ lines (Fig. 4b) [14] and thus a semiconducting nanotube with a finite energy gap appears. The rule for being either a metallic or a semiconducting carbon nanotube is, respectively, that $n - m = 3q$ or $n - m \neq 3q$, where $q$ is an integer [2,8,15,16,17].

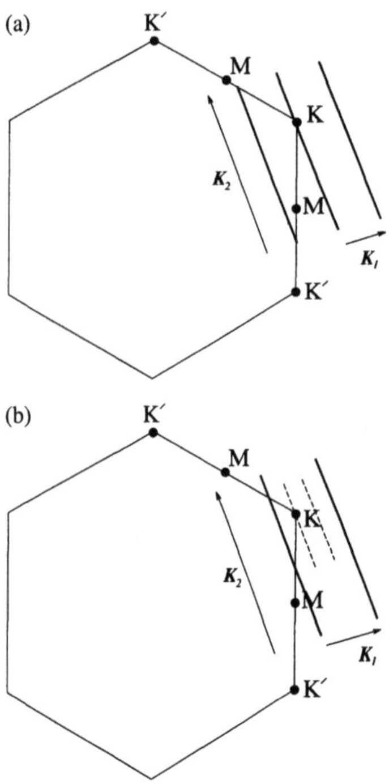


**Fig. 4.** The wave vector $k$ for one-dimensional carbon nanotubes is shown in the two-dimensional Brillouin zone of graphite (hexagon) as bold lines for (**a**) metallic and (**b**) semiconducting carbon nanotubes. In the direction of $\mathbf{K}_1$, discrete $k$ values are obtained by periodic boundary conditions for the circumferential direction of the carbon nanotubes, while in the direction of the $\mathbf{K}_2$ vector, continuous $k$ vectors are shown in the one-dimensional Brillouin zone. (**a**) For metallic nanotubes, the bold line intersects a $K$ point (corner of the hexagon) at the Fermi energy of graphite. (**b**) For the semiconductor nanotubes, the $K$ point always appears one-third of the distance between two bold lines. It is noted that only a few of the $N$ bold lines are shown near the indicated $K$ point. For each bold line, there is an energy minimum (or maximum) in the valence and conduction energy subbands, giving rise to the energy differences $E_{pp}(d_t)$

The 1D density of states (DOS) in units of states/C-atom/eV is calculated by

$$D(E) = \frac{T}{2\pi N} \sum_{\pm} \sum_{\mu=1}^{N} \int \frac{1}{\left|\frac{dE_{\mu}^{\pm}(k)}{dk}\right|} \delta(E_{\mu}^{\pm}(k) - E) dE, \tag{16}$$

where the summation is taken for the $N$ conduction (+) and valence (−) 1D bands. Since the energy dispersion near the Fermi energy (10) is linear, the density of states of metallic nanotubes is constant at the Fermi energy: $D(E_F) = a/(2\pi^2 \gamma_0 d_t)$, and is inversely proportional to the diameter of the nanotube. It is noted that we always have two cutting lines (1D energy bands) at the two equivalent symmetry points $K$ and $K'$ in the 2D Brillouin zone in Fig. 3. The integrated value of $D(E)$ for the energy region of $E_\mu(k)$ is 2 for any $(n, m)$ nanotube, which includes the plus and minus signs of $E_{g2D}$ and the spin degeneracy.

It is clear from (16) that the density of states becomes large when the energy dispersion relation becomes flat as a function of $k$. One-dimensional van Hove singularities (vHs) in the DOS, which are known to be proportional to $(E^2 - E_0^2)^{-1/2}$ at both the energy minima and maxima $(\pm E_0)$ of the dispersion relations for carbon nanotubes, are important for determining many solid state properties of carbon nanotubes, such as the spectra observed by scanning tunneling spectroscopy (STS), [18,19,20,21,22], optical absorption [4,23,24], and resonant Raman spectroscopy [25,26,27,28,29].

The one-dimensional vHs of SWNTs near the Fermi energy come from the energy dispersion along the bold lines in Fig. 4 near the $K$ point of the Brillouin zone of 2D graphite. Within the linear $k$ approximation for the energy dispersion relations of graphite given by (10), the energy contour as shown in Fig. 3 around the $K$ point is circular and thus the energy minima of the 1D energy dispersion relations are located at the closest positions to the $K$ point. Using the small $k$ approximation of (10), the energy differences $E_{11}^{M}(d_t)$ and $E_{11}^{S}(d_t)$ for metallic and semiconducting nanotubes between the highest-lying valence band singularity and the lowest-lying conduction band singularity in the 1D electronic density of states curves are expressed by substituting for $k$ the values of $|\mathbf{K}_1|$ of (15) for metallic nanotubes and of $|\mathbf{K}_1/3|$ and $|2\mathbf{K}_1/3|$ for semiconducting nanotubes, respectively [30,31], as follows:

$$E_{11}^{M}(d_t) = 6a_{C-C}\gamma_0/d_t \quad \text{and} \quad E_{11}^{S}(d_t) = 2a_{C-C}\gamma_0/d_t. \tag{17}$$

When we use the number $p$ ($p = 1, 2, \ldots$) to denote the order of the valence $\pi$ and conduction $\pi^*$ energy bands symmetrically located with respect to the Fermi energy, optical transitions $E_{pp'}$ from the $p$-th valence band to the $p'$-th conduction band occur in accordance with the selection rules of $\delta p = 0$ and $\delta p = \pm 1$ for parallel and perpendicular polarizations of the electric field with respect to the nanotube axis, respectively [23]. However, in the case of perpendicular polarization, the optical transition is suppressed by the depolarization effect [23], and thus hereafter we only consider the optical absorption of $\delta p = 0$. For mixed samples containing both metallic and semiconducting carbon nanotubes with similar diameters, optical transitions may appear with the following energies, starting from the lowest energy, $E_{11}^{\mathrm{S}}(d_{\mathrm{t}})$, $2E_{11}^{\mathrm{S}}(d_{\mathrm{t}})$, $E_{11}^{\mathrm{M}}(d_{\mathrm{t}})$, $4E_{11}^{\mathrm{S}}(d_{\mathrm{t}})$, ....

In Fig. 5, both $E_{pp}^{\mathrm{S}}(d_{\mathrm{t}})$ and $E_{pp}^{\mathrm{M}}(d_{\mathrm{t}})$ are plotted as a function of nanotube diameter $d_{\mathrm{t}}$ for all chiral angles at a given $d_{\mathrm{t}}$ value. [3,4,11]. This plot is very useful for determining the resonant energy in the resonant Raman spectra corresponding to a particular nanotube diameter. In this figure, we use the values of $\gamma_0 = 2.9eV$ and $s = 0$, which explain the experimental observations discussed in the experimental section.

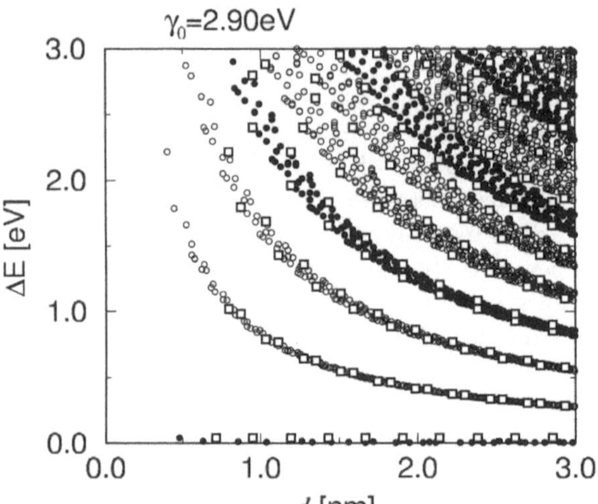


**Fig. 5.** Calculation of the energy separations $E_{pp}(d_{\mathrm{t}})$ for all $(n, m)$ values as a function of the nanotube diameter between $0.7 < d_{\mathrm{t}} < 3.0\,\mathrm{nm}$ (based on the work of *Kataura* et al. [3]). The results are based on the tight binding model of Eqs. (6) and (7), with $\gamma_0 = 2.9\,\mathrm{eV}$ and $s = 0$. The open and solid circles denote the peaks of semiconducting and metallic nanotubes, respectively. *Squares* denote the $E_{pp}(d_{\mathrm{t}})$ values for zigzag nanotubes which determine the width of each $E_{pp}(d_{\mathrm{t}})$ curve. Note the points for zero gap metallic nanotubes along the abscissa [11]

## 1.2 Trigonal Warping Effects in the DOS Windows

Within the linear $k$ approximation for the energy dispersion relations of graphite, $E_{pp}$ of (17) depends only on the nanotube diameter, $d_{\mathrm{t}}$. However, the width of the $E_{pp}$ band in Fig. 5 becomes large with increasing $E_{pp}$ [11].

When the value of $|\mathbf{K}_1| = 2/d_{\mathrm{t}}$ is large, which corresponds to smaller values of $d_{\mathrm{t}}$, the linear dispersion approximation is no longer correct. When we then plot equi-energy lines near the $K$ point (see Fig. 3), we get circular contours for small $k$ values near the $K$ and $K'$ points in the Brillouin zone, but for large $k$ values, the equi-energy contour becomes a triangle which connects the three $M$ points nearest to the $K$-point (Fig. 6). The distortion in Fig. 3 of the equi-energy lines away from the circular contour in materials with a 3-fold symmetry axis is known as the trigonal warping effect.

In metallic nanotubes, the trigonal warping effects generally split the DOS peaks for metallic nanotubes, which come from the two neighboring lines near the $K$ point (Fig. 6). For armchair nanotubes as shown in Fig. 6a, the two lines are equivalent to each other and the DOS peak energies are equal, while for zigzag nanotubes, as shown in Fig. 6b, the two lines are not equivalent, which gives rise to a splitting of the DOS peak. In a chiral nanotube the two lines are not equivalent in the reciprocal lattice space, and thus the splitting values of the DOS peaks are a function of the chiral angle.

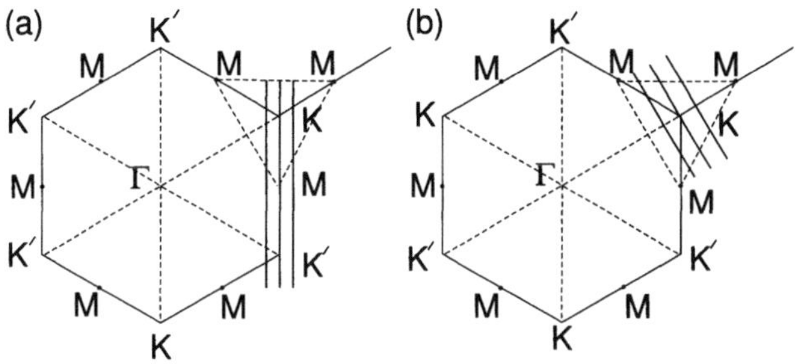


**Fig. 6.** The trigonal warping effect of the van Hove singularities. The three bold lines near the $K$ point are possible $k$ vectors in the hexagonal Brillouin zone of graphite for metallic (**a**) armchair and (**b**) zigzag carbon nanotubes. The minimum energy along the neighboring two lines gives the energy positions of the van Hove singularities

On the other hand, for semiconducting nanotubes, since the value of the $k$ vectors on the two lines near the K point contribute to different spectra, namely to that of $E_{11}^{\mathrm{S}}(d_{\mathrm{t}})$ and $E_{22}^{\mathrm{S}}(d_{\mathrm{t}})$, there is no splitting of the DOS peaks for semiconducting nanotubes. However, the two lines are not equivalent Fig. 4b, and the $E_{22}^{\mathrm{S}}(d_{\mathrm{t}})$ value is not twice that of $E_{11}^{\mathrm{S}}(d_{\mathrm{t}})$. It is pointed out here that there are two equivalent $K$ points in the hexagonal Brillouin zone denoted by $K$ and $K'$ as shown in Fig. 4, and the values of $E_{ii}^{\mathrm{S}}(d_{\mathrm{t}})$ are the same for the $K$ and $K'$ points. This is because the $K$ and $K'$ points are related to one another by time reversal symmetry (they are at opposite corners from each other in the hexagonal Brillouin zone), and because the chirality of a nanotube is invariant under the time-reversal operation. Thus, the DOS for semiconducting nanotubes will be split if very strong magnetic fields are applied in the direction of the nanotube axis.

The peaks in the 1D electronic density of states of the conduction band measured from the Fermi energy are shown in Fig. 7 for several *metallic* $(n, m)$ nanotubes, all having about the same diameter $d_t$ (from 1.31 nm to 1.43 nm), but having different chiral angles: $\theta = 0°$, 8.9°, 14.7°, 20.2°, 24.8°, and 30.0° for nanotubes (18,0), (15,3), (14,5), (13,7), (11.8), and (10,10), respectively. When we look at the peaks in the 1D DOS as the chiral angle is varied from the armchair nanotube (10,10) ($\theta = 30°$) to the zigzag nanotube (18,0) ($\theta = 0°$) of Fig. 7, the first DOS peaks around $E = 0.9$ eV are split into two peaks whose separation in energy (width) increases with decreasing chiral angle.

This theoretical result [11] is important in the sense that STS (scanning tunneling spectroscopy) [22] and resonant Raman spectroscopy experiments [25,27,28,29] depend on the chirality of an individual SWNT, and therefore trigonal warping effects should provide experimental information about the chiral angle of carbon nanotubes. *Kim* et al. have shown that the DOS of a (13, 7) metallic nanotube has a splitting of the lowest energy peak in their STS spectra [22], and this result provides direct evidence for the trigonal warping effect. Further experimental data will be desirable for a systematic study of this effect. Although the chiral angle is directly observed by scanning tunneling microscopy (STM) [32], corrections to the experimental observations are necessary to account for the effect of the tip size and shape and for the deformation of the nanotube by the tip and by the substrate [33]. We expect that the chirality-dependent DOS spectra are insensitive to such effects.

In Fig. 8 the energy dispersion relations of (6) along the $K$–$\Gamma$ and $K$–$M$ directions are plotted. The energies of the van Hove singularities corresponding

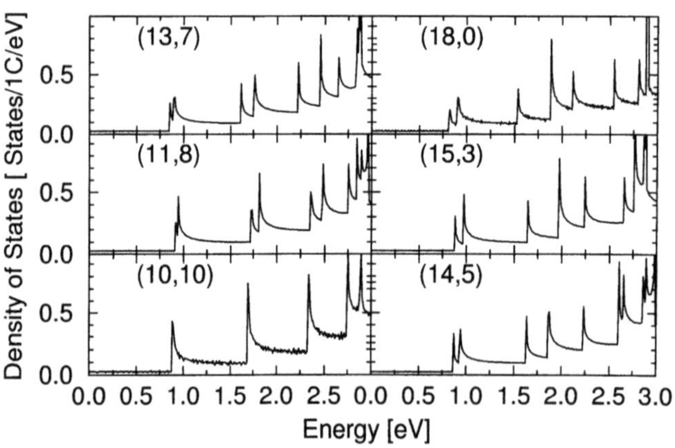


**Fig. 7.** The 1D electronic density of states vs energy for several metallic nanotubes of approximately the same diameter, showing the effect of chirality on the van Hove singularities: (10,10) (armchair), (11,8), (13,7), (14,5), (15,3) and (18,0) (zigzag). We only show the density of states for the conduction $\pi^*$ bands

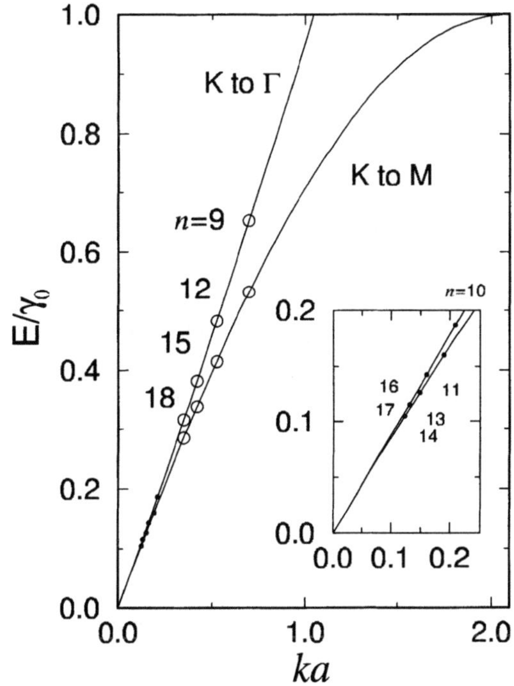


**Fig. 8.** Splitting of the DOS in zigzag nanotubes. Two minimum energy positions are found in the conduction band for zigzag nanotubes, $(n, 0)$ measured from the energy at the $K$ point. *Open circles* denote metallic carbon nanotubes for $k = |\mathbf{K}_1|$ vectors away from the $K$ point along the $K \to M$ and $K \to \Gamma$ lines, which are the directions of the energy minima (see Fig. 6). (The *inset* shows an expanded view of the figure at small $E/\gamma_0$ and small $ka$ for semiconducting nanotubes. The closed circles denote semiconducting carbon nanotubes for $k = |\mathbf{K}_1|/3$ vectors. ) Note that the maximum of the horizontal axis corresponds to the $M$ point, $ka = 2\pi/3$, which is measured from the $K$ point. A nanotube diameter of 1 nm corresponds to a (13,0) carbon nanotube

to the lowest 1D energy level are plotted for metallic (open circles) and semiconducting (closed circles) zigzag nanotubes $(n, 0)$ by putting $ka = |\mathbf{K}_1|a$ and $ka = |\mathbf{K}_1|a/3$, respectively. The corresponding energy separation is plotted in Fig. 5 as solid squares. In the case of $(3n + 1, 0)$ and $(3n - 1, 0)$ semiconducting zigzag nanotubes, $E_{11}^{\mathrm{S}}$ comes respectively, from the $K$–$\Gamma$ and $K$–$M$ lines, while $E_{22}^{\mathrm{S}}$ comes from $K$–$M$ and $K$–$\Gamma$ and so on. In the case of $(3n, 0)$ metallic zigzag nanotubes, the DOS peaks come from both $K$–$M$ and $K$–$\Gamma$. This systematic rule will be helpful for investigating the STS spectra in detail. Using Eqs. (7) and (15), the widths of $E_{11}^{\mathrm{M}}$ and $E_{11}^{\mathrm{S}}$, denoted by $\Delta E_{11}^{\mathrm{M}}$

and $\Delta E_{11}^{\mathrm{S}}$, respectively, are determined by the zigzag nanotubes, and are analytically given by

$$\Delta E_{11}^{\mathrm{M}}(d_{\mathrm{t}}) = 8\gamma_0 \sin^2\left(\frac{a}{2d_{\mathrm{t}}}\right), \quad \Delta E_{11}^{\mathrm{S}}(d_{\mathrm{t}}) = 8\gamma_0 \sin^2\left(\frac{a}{6d_{\mathrm{t}}}\right). \tag{18}$$

Although this trigonal warping effect is proportional to $(a/d_{\mathrm{t}})^2$, the terms in (18) are not negligible, since this correction is the leading term in the expressions for the width $\Delta E_{pp}(d_{\mathrm{t}})$, and the factor 8 before $\gamma_0$ makes this correction significant in magnitude for $d_{\mathrm{t}}$=1.4 nm. For example, $E_{11}(d_{\mathrm{t}})$ is split by about 0.18 eV for the metallic (18, 0) zigzag nanotube, and this splitting is large enough to be observable by STS experiments. Although the trigonal warping effect is larger for metallic nanotubes than for semiconducting nanotubes of comparable diameters, the energy difference of the third peaks $E_{33}^{\mathrm{S}}(d_{\mathrm{t}}) = 8\gamma_0 \sin^2(2a/3d_{\mathrm{t}})$ between the (17, 0) and (19, 0) zigzag nanotubes is about 0.63 eV, using an average $d_{\mathrm{t}}$ value of 1.43 nm, which becomes easily observable in the experiments. These calculations show that the trigonal warping effect is important for metallic single wall zigzag nanotubes with diameters $d_{\mathrm{t}} < 2$ nm. More direct measurements [22] of the chirality by the STM technique and of the splitting of the DOS by STS measurements on the *same* nanotube would provide very important confirmation of this prediction.

### 1.3 Phonon Properties

A general approach for obtaining the phonon dispersion relations of carbon nanotubes is given by tight binding molecular dynamics (TBMD) calculations adopted for the nanotube geometry, in which the atomic force potential for general carbon materials is used [25,34]. Here we use the scaled force constants from those of 2D graphite [2,14], and we construct a force constant tensor for a constituent atom of the SWNT so as to satisfy the rotational sum rule for the force constants [35,36]. Since we have $2N$ carbon atoms in the unit cell, the dynamical matrix to be solved becomes a $6N \times 6N$ matrix [35,37].

In Fig. 9 we show the results thus obtained for (a) the phonon dispersion relations $\omega(k)$ and (b) the corresponding phonon density of states for 2D graphite (left) and for a (10,10) armchair nanotube (right). For the $2N = 40$ carbon atoms per circumferential strip for the (10,10) nanotube, we have 120 vibrational degrees of freedom, but because of mode degeneracies, there are only 66 distinct phonon branches, for which 12 modes are non-degenerate and 54 are doubly degenerate. The phonon density of states for the (10,10) nanotube is close to that for 2D graphite, reflecting the zone-folded nanotube phonon dispersion. The same discussion as is used for the electronic structure can be applied to the van Hove singularity peaks in the phonon density of states of carbon nanotubes below a frequency of 400 cm$^{-1}$ which can be observed in neutron scattering experiments for rope samples.

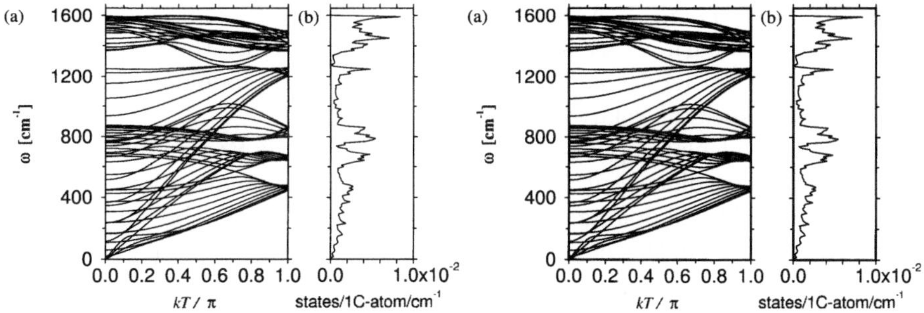


**Fig. 9.** (**a**) Phonon dispersion relations and (**b**) phonon DOS for 2D graphite (*left*) and for a (10,10) nanotube (*right*) [35]

There are four acoustic modes in a nanotube. The lowest energy acoustic modes are the Transverse Acoustic (TA) modes, which are doubly degenerate, and have $x$ and $y$ displacements perpendicular to the nanotube $z$ axis. The next acoustic mode is the "twisting" acoustic mode (TW), which has $\theta$-dependent displacements along the nanotube surface. The highest energy mode is the Longitudinal Acoustic (LA) mode whose displacements occur in the $z$ direction. The sound velocities of the TA, TW, and LA phonons for a (10,10) carbon nanotube, $v_{\mathrm{TA}}^{(10,10)}$, $v_{\mathrm{TW}}^{(10,10)}$ and $v_{\mathrm{LA}}^{(10,10)}$, are estimated as $v_{\mathrm{TA}}^{(10,10)}$=9.42 km/s, $v_{\mathrm{TW}}^{(10,10)}$ = 15.00 km/s, and $v_{\mathrm{LA}}^{(10,10)}$ =20.35 km/s, respectively. The calculated phase velocity of the in-plane TA and LA modes of 2D graphite are $v_{\mathrm{TA}}^{\mathrm{G}}$=15.00 km/s and $v_{\mathrm{LA}}^{\mathrm{G}}$=21.11 km/s, respectively. Since the TA mode of the nanotube has both an 'in-plane' and an 'out-of-plane' component, the nanotube TA modes are softer than the in-plane TA modes of 2D-graphite. The calculated phase velocity of the out-of-plane TA mode for 2D graphite is almost 0 km/s because of its $k^2$ dependence. The sound velocities that have been calculated for 2D graphite are similar to those observed in 3D graphite [10], for which $v_{\mathrm{TA}}^{\mathrm{G3D}} = 12.3$ km/s and $v_{\mathrm{LA}}^{\mathrm{G3D}} = 21.0$ km/s. The discrepancy between the $v_{\mathrm{TA}}$ velocity of sound for 2D and 3D graphite comes from the interlayer interaction between the adjacent graphene sheets.

The strongest low frequency Raman mode for carbon nanotubes is the Radial Breathing $A_{1g}$ mode (RBM) whose frequency is calculated to be 165 cm$^{-1}$ for the (10,10) nanotube. Since this frequency is in the silent region for graphite and other carbon materials, this $A_{1g}$ mode provides a good marker for specifying the carbon nanotube geometry. When we plot the $A_{1g}$ frequency as a function of nanotube diameter for $(n,m)$ in the range $8 \le n \le 10$, $0 \le m \le n$, the frequencies are inversely proportional to $d_{\mathrm{t}}$ [5,35], within only a small deviation due to nanotube-nanotube interaction in a nanotube bundle. Here $\omega_{(10,10)}$ and $d_{(10,10)}$ are, respectively, the frequency and diameter $d_{\mathrm{t}}$ of the (10,10) armchair nanotube, with values of $\omega_{(10,10)}$=165 cm$^{-1}$ and $d_{(10,10)}$=1.357 nm, respectively. However, when we adopt $\gamma_0$ = 2.90 eV, the resonant spectra becomes consistent when we take $\omega_{(10,10)} = 177\,\mathrm{cm}^{-1}$. As for the higher frequency Raman modes around $1590\,\mathrm{cm}^{-1}$ (G-band), we see some dependence on $d_t$, since the frequencies of the higher optical modes can be obtained from the zone-folded $k$ values in the phonon dispersion relation of 2D graphite [26].

Using the calculated phonon modes of a SWNT, the Raman intensities of the modes are calculated within the non-resonant bond polarization theory, in which empirical bond polarization parameters are used. [38] The bond parameters that we used in this chapter are $\alpha_{\parallel} - \alpha_{\perp} = 0.04\,\text{Å}^3$, $\alpha'_{\parallel} + 2\alpha'_{\perp} = 4.7\,\text{Å}^2$, and $\alpha'_{\parallel} - \alpha'_{\perp} = 4.0\,\text{Å}^2$, where $\alpha$ and $\alpha'$ denote the polarizability parameters and their derivatives with respect to bond length, respectively. [35] The eigenfunctions for the various vibrational modes are calculated numerically at the $\Gamma$ point ($k = 0$). When some symmetry-lowering effects, such as defects and finite size effects occur, phonon modes away from the $\Gamma$ point are observed in the Raman spectra. For example, the DOS peaks at $1620\,\mathrm{cm}^{-1}$ related to the highest energy of the DOS, and some DOS peaks related to $M$ point phonons can be strong. In general, the lower dimensionality causes a broadening in the DOS, but the peak positions do not change much. The $1350\,\mathrm{cm}^{-1}$ peaks (D-band) are known to be defect-related Raman peaks which originate from $K$ point phonons, and exhibit a resonant behavior [39].

# 2 Experiment Issues

For the experiments described below, SWNTs were prepared by both laser vaporization and electric arc methods. In the laser vaporization method, the second harmonic of the Nd:YAG laser pulse is focused on a metal catalyzed carbon rod located in a quartz tube filled with 500 Torr Ar gas, which is heated to 1200°C in an electric furnace. The laser-vaporized carbon and catalyst are transformed in the furnace to a soot containing SWNTs and nanoparticles containing catalyst species.

## 2.1 Diameter-Selective Formation of SWNTs

The diameter distribution of the SWNTs can be controlled by changing the temperature of the furnace. In the electric arc method, the dc arc between the catalyzed carbon anode and the pure carbon cathode produces SWNTs in He gas at 500 Torr. In the arc method, the diameters of the SWNTs are controlled by changing the pressure of the He gas. Increasing the temperature makes larger diameter SWNTs, while the higher ambient gas pressure, up to 760 Torr, makes a larger yield and diameter of SWNTs by the carbon arc method.

The diameter of SWNTs can be controlled, too, by adopting different catalysts and different relative concentrations of the catalyst species, such as NiY (4.2 and 1.0 at. %), NiCo (0.6 and 0.6 at. %), Ni (0.6 at. %) and RhPd (1.2 and 1.2 at. %), which have provided the following diameter distributions by the laser ablation method with a furnace temperature of 1150 to 1200°C, respectively, 1.24–1.58 nm, 1.06–1.45 nm, 1.06–1.45 nm and 0.68–1.00 nm [3,40]. The diameter distribution in each case was determined from TEM experiments and from measurement of the RBM frequencies using Raman spectroscopy and several different laser excitation energies. It is important to note that the determination of the frequency of the RBM does not provide a measurement of the nanotube chirality, though the diameter dependence is well observed by measurement of the RBM frequency. The diameter distribution is then obtained if the RBM of a (10,10) armchair nanotube is taken to be 165 $cm^{-1}$ RBM and $\gamma_0 = 2.75\,eV$ [3]. However, if we adopt the value of $\gamma_0 = 2.90\,eV$, the Raman signal is consistent with 177 $cm^{-1}$ for a (10,10) armchair nanotube. For these larger values of $\gamma_0$ and $\omega_{RBM}(10,10)$ the diameter distribution for each catalyst given above is shifted upward by 7%. Most of the catalysts, except for the RhPd, show very similar diameter distributions for both the laser vaporization and the electric arc methods at growth conditions giving the highest yield. In the case of the RhPd catalyst, however, no SWNTs are synthesized by the arc discharge method, in contrast to a high yield provided by the laser vaporization method.

## 2.2 Sample Preparation and Purification

SWNTs are not soluble in any solvent and they cannot be vaporized by heating at least up to 1450°C in vacuum. In order to measure the optical absorption of SWNTs, the sample can be prepared in two possible forms: one is a solution sample and the other is a thin film. *Chen* et al. made SWNT solutions by cutting and grinding the nanotube sample [41], and they successfully measured the optical absorption spectra of undoped and of doped SWNTs using a solution sample. Kataura and co-workers have developed a so called "spray method" for thin film preparation [42], whereby the soot containing SWNTs is dispersed in ethanol and then sprayed onto a quartz plate using a conventional air-brush which is normally obtained in a paint store. In this way, the thickness and the homogeneity of the thin film are controlled by the number of spraying and drying processes, but the thickness of the film ($\sim$ 300nm with 20% filling) is not precisely controlled.

In the case of the NiY catalyst, a web form of SWNTs which is predominantly in the bundle form is obtained by the electric arc method, and the resulting material can be easily purified by heating in air at 350°C for 30min and by rinsing out metal particles using hydrochloric acid. The purification is effective in removing the nanospheres (soot) and catalyst, and this is confirmed by TEM images and X-ray diffraction. The nanotube diameter distribution of the sample can be estimated by TEM observations [43,44], and the diameter distribution, thus obtained, is consistent with the distribution obtained using resonance Raman spectroscopy of the RBMs.

### 2.3 Diameter-Dependent Optical Absorption

In Fig. 10, the optical absorption spectra of an as-prepared and a purified SWNT thin film sample are shown, respectively, by the solid and dashed curves. Both samples are synthesized using the electric arc method and the NiY catalyst [3]. The three peaks appearing at 0.68, 1.2 and 1.7 eV correspond to the two semiconductor DOS peaks and the metallic DOS peaks discussed in the previous section. When we consider the distribution of nanotube diameters, only the first three peaks of the DOS spectra can be distinguished in relation to the calculation [45], which is consistent with the optical spectra shown in Fig. 10. Since there is no substantial difference in the spectra between the as-prepared and purified samples, we can conclude that the peaks come from the SWNTs. The dotted line denotes the photo-thermal deflection spectrum (PDS) for the same purified sample. The signal of the PDS data is proportional to the heat generated by multi-phonon processes involved in the recombination of the optically pumped electron-hole pairs, and thus the PDS spectra are considered to be free from light scattering by nano-particles [46]. Furthermore, since carbon black is used as a black body reference, the PDS reflects the difference in electronic states between SWNTs and amorphous carbon. These peak structures are more clearly seen in the PDS than in the absorption spectra, while the peak positions are almost the same as in the absorption spectra, which indicates that these peaks are not due to light scattering losses. Thus we understand that the residual nanospheres and metal

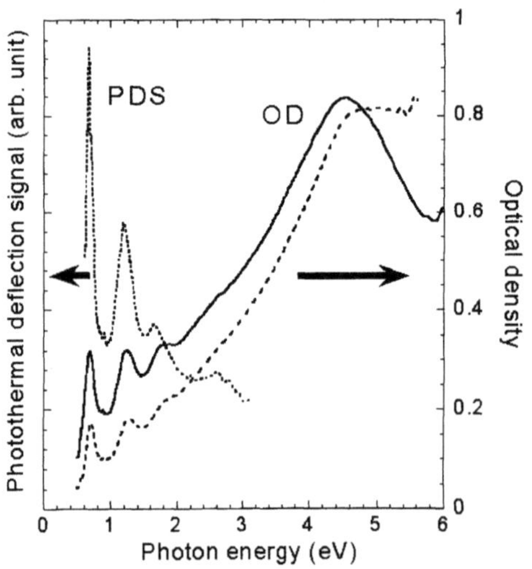


**Fig. 10.** Optical density in the absorption spectra of as-prepared (*solid line*) and purified (*dashed line*) SWNT thin film samples synthesized by the electric arc method using a NiY catalyst [3]. The photo-thermal deflection spectrum (PDS, *dotted line*) is also plotted for the same sample, and the spectral features of the PDS data are consistent with the absorption spectra

particles in the sample do not seriously affect the optical absorption spectrum in the energy region below 2 eV. This fact is confirmed by the observation of no change in the absorption spectra between purified and pristine samples in which the density of nanoparticles and catalysts are much different from each other.

The purified sample shows a large optical absorption band at 4.5 eV, which corresponds to the $\pi$-plasmon of SWNTs observed in the energy loss spectrum [47], which is not so clearly seen in the as-prepared sample. Figure 11 shows the optical absorption spectra of SWNTs with different diameter distributions associated with the use of four different catalysts [3]. For convenience, the large background due to the $\pi$ plasmon is subtracted. The inset shows the corresponding Raman spectra of the RBMs taken with 488 nm laser excitation. The diameter distributions can be estimated from the peak frequencies using the rule, $\omega_{\mathrm{RBM}} \propto (1/d_{\mathrm{t}})$, where $d_{\mathrm{t}}$ is the diameter of a SWNT that is in resonance with the laser photons [5,35]. Thus, higher lying Raman

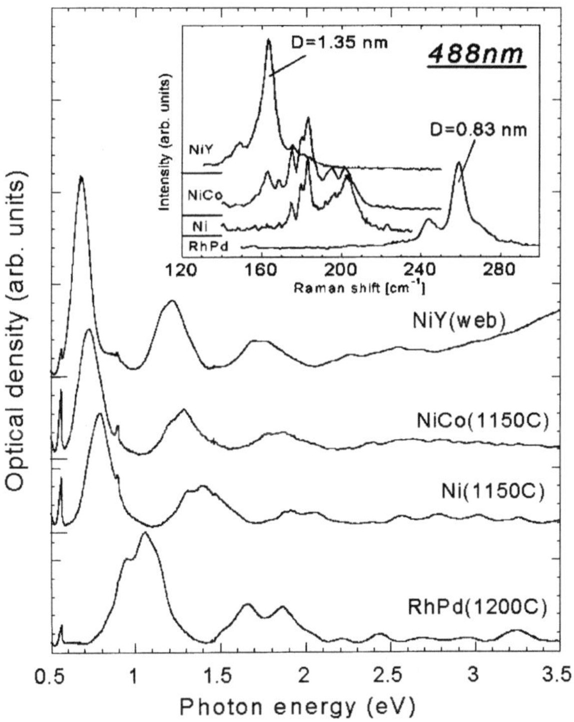


**Fig. 11.** Optical absorption spectra are taken for single wall nanotubes synthesized using four different catalysts, [3,4] namely NiY (1.24–1.58 nm), NiCo (1.06–1.45 nm), Ni (1.06–1.45 nm), and RhPd (0.68–1.00 nm). Peaks at 0.55 eV and 0.9 eV are due to absorption by the quartz substrate [3]. The *inset* shows the corresponding RBM modes of Raman spectroscopy obtained at 488 nm laser excitation with the same 4 catalysts

peaks indicate the presence of smaller diameter SWNTs in the sample. The nanotube diameter distribution can be estimated from the diameter $d_t$ dependence of $\omega_{RBM} \propto 1/d_t$, once the proportionality between $\omega_{RBM}$ and *one* $(n, m)$ nanotube is established, such as for the (10,10) nanotube. Information on the nanotube diameter distribution is available either by TEM or from measurement of the $\omega_{RBM}$ band for many laser excitation energies $E_{laser}$.

A method for determining $E_{11}(d_t)$ comes from optical spectra, where the measurements are made on ropes of SWNTs, so that appropriate corrections should be made for inter-tube interactions in interpreting the experimental data [29,48,49,50,51,52,53,54]. In interpreting the optical transmission data, corrections for the nonlinear $k$ dependence of $E(k)$ away from the $K$-point also needs to be considered. In addition, the asymmetry of the 1D electronic density of states singularities should be taken into account in extracting the energy $E_{pp}(d_t)$ from the absorption line shape. Furthermore, the diameter distributions of the nanotubes, as well as the difference in gap energies for nanotubes of different chiralities, but for a given $d_t$, should be considered in the detailed interpretation of the optical transmission data to yield a value for $\gamma_0$. The calculations given in Fig. 5 provide a firm basis for a more detailed analysis.

Another important issue to address here is the so-called antenna effect of nanotubes. Since the diameter of SWNTs is much smaller than the wave length of light, an effective medium theory or other model must be used for describing the dielectric function of the nanotubes within an aligned nanotube bundle, for nanotubes that have an arbitrary polarization with respect to the randomly oriented nanotube bundles, which collectively have an anisotropic $\varepsilon_1(\omega)$ and $\varepsilon_2(\omega)$. The optical measurements should determine such fundamental properties for SWNTs.

*Kazaoui* et al. [24] have reported optical absorption spectra for doped SWNTs as shown in Fig. 12, including both donor (Cs) and acceptor (Br), and they found that the intensity of the absorption peaks decreased, especially for the lower energy absorption peaks with increasing dopant concentration. In the undoped SWNTs, three peaks at 0.68 eV, 1.2 eV and 1.8 eV are found in the absorption spectra in Fig. 12. When the doping concentration $x$ in $M_xC$, (M = Cs, Br) is less than 0.005, the first peak at 0.68 eV decreases continuously in intensity with increasing $x$ without changing the intensity of the second and the third peaks. In subsequent doping in the range $0.005 < x < 0.04$, the two peaks of 0.68 eV and 1.2 eV decrease in intensity. At the high doping level shown in Fig. 12b, the peak at 1.8 eV smoothes out and new bands appears at 1.07 eV and 1.30 eV for $CBr_{0.15}$ and $CCs_{0.10}$, respectively. These doping-induced absorption peaks may come from the transition between conduction to conduction inter-subband transitions and from valence to valence inter-subband transitions, respectively, for donor and acceptor type SWNTs. The difference between the peak positions 1.07 eV and 1.30 eV for acceptor and donor type SWNTs, respectively, is consistent with the expected magni-

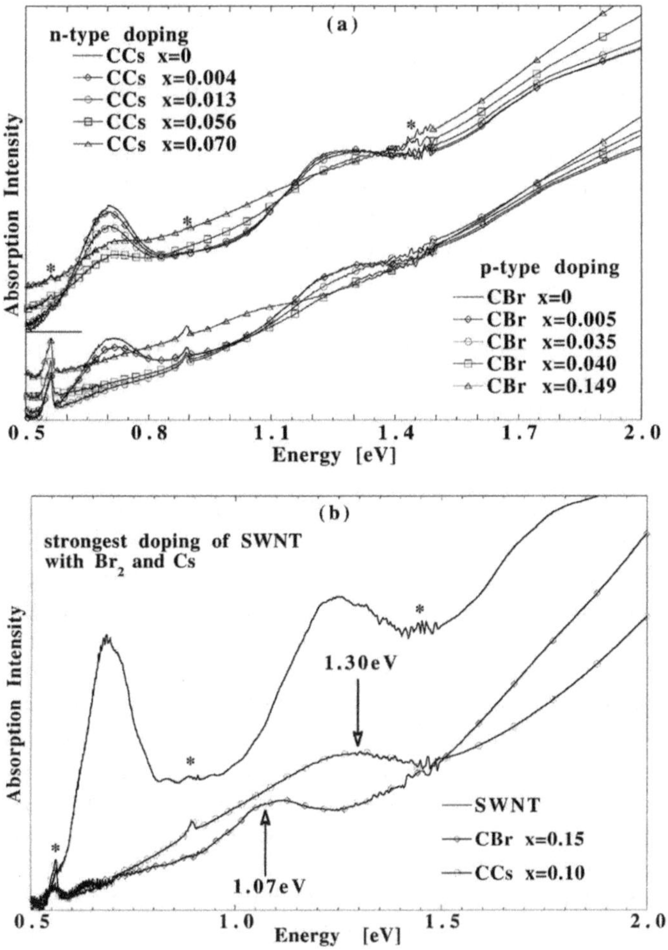


**Fig. 12.** (**a**) Optical absorption spectra for Cs and Br doped SWNT samples for various stoichiometries $x$ for $CCs_x$ and $CBr_x$. The entire set of spectra for the $CCs_x$ samples is offset for clarity with a short line indicating the 0 level. * in the figure indicates features coming from the quartz substrate and from spectrometer noise. (**b**) The absorption spectra for $CCs_x$ and $CBr_x$ for the almost saturated doping regime [24]

tude of the asymmetry between the $\pi$ and $\pi^*$ bands. However, the detailed assignments for the inter-subband transitions which are responsible for the doping-induced peaks are not clear within the rigid band model.

## 2.4 Diameter-Dependent Resonance Raman Scattering

In the resonance Raman effect, a large scattering intensity is observed when *either the incident or the* scattered light is in resonance with electronic transitions between vHs in the valence and conduction bands $E_{pp}(d_t)$ for a given nanotube $(n,m)$ [4,25,27,28,29,55,56,57,58]. In general, the size of the optical excitation beam is at least 1 μm in diameter, so that many nanotubes with a large variety of $(n,m)$ values are excited by the optical beam simultaneously, as is also the case for the optical absorption measurements discussed above. Since it is unlikely that any information on the nanotube chirality distribution is available experimentally, the assumption of equal a priori probability can be assumed, so that at a given diameter $d_t$ the resonance Raman effect is sensitive to the width of the $E_{pp}(d_t)$ inter-subband transitions plotted in Fig. 5.

In Fig. 13 are plotted the resonance Raman spectra for SWNT samples using (a) NiY and (b) RhPd catalysts. The left and right figures for each sample (see Fig. 13) show the Raman spectra the phonon energy region of the radial breathing mode and the tangential G-bands, respectively [4]. As a first approximation, the resonant laser energy for the RBM spectra, and the G-band Raman spectra are used to estimate the $E_{pp}(d_t)$ transition energies, as shown in Fig. 5, with the diameter distribution for each catalyst. When the nanotube diameter values of $d_t = 1.24$–$1.58$ nm and $d_t = 0.68$–$1.00$ nm are used for the NiY and RhPd catalyst samples, respectively, the resonance for the metallic nanotubes $E^{M}_{pp}(d_t)$ is seen in the laser energy region around 1.6–2.0 eV and 2.4–2.8 eV, respectively. Hereafter we call this region of laser energy, which is resonant with metallic nanotubes, the "metallic window". This metallic window for the Raman RBM intensity is consistent with the optical density of the third peaks as a function of laser excitation energy, as shown in Fig. 14, where for laser excitation energies greater than 1.5 eV, the optical density (absorption) and the Raman intensity of the RBMs are consistent both for the NiY and RhPd catalyzed samples.

The metallic window for a given diameter distribution of SWNTs is obtained by the third peak of the optical absorption, as discussed in the previous subsection, and more precisely by the appearance of Raman intensity at 1540 cm$^{-1}$ which can be seen only in the case of a rope sample containing metallic nanotubes, where the spectra are fit to a Breit–Wigner–Fano plot [4,27,28,29] as shown in Fig. 2.4.

It is pointed out here that the phonon energies of the G-band are large (0.2 eV) compared with the RBM phonon (0.02 eV), so that the resonant condition for the metallic energy window is generally different according to the difference between the RBM and G-band phonon energy. Furthermore, the resonant laser energies for phonon-emitted Stokes and phonon-absorbed anti-Stokes Raman spectra (see Sect. 2.5) are different from each other by twice the energy of the corresponding phonon. Thus when a laser energy is selected, carbon nanotubes with different diameters $d_t$ are resonant between the RBM and $G$ bands and between the Stokes and anti-Stokes Raman spectra, which will be described in more detail in the following subsection.

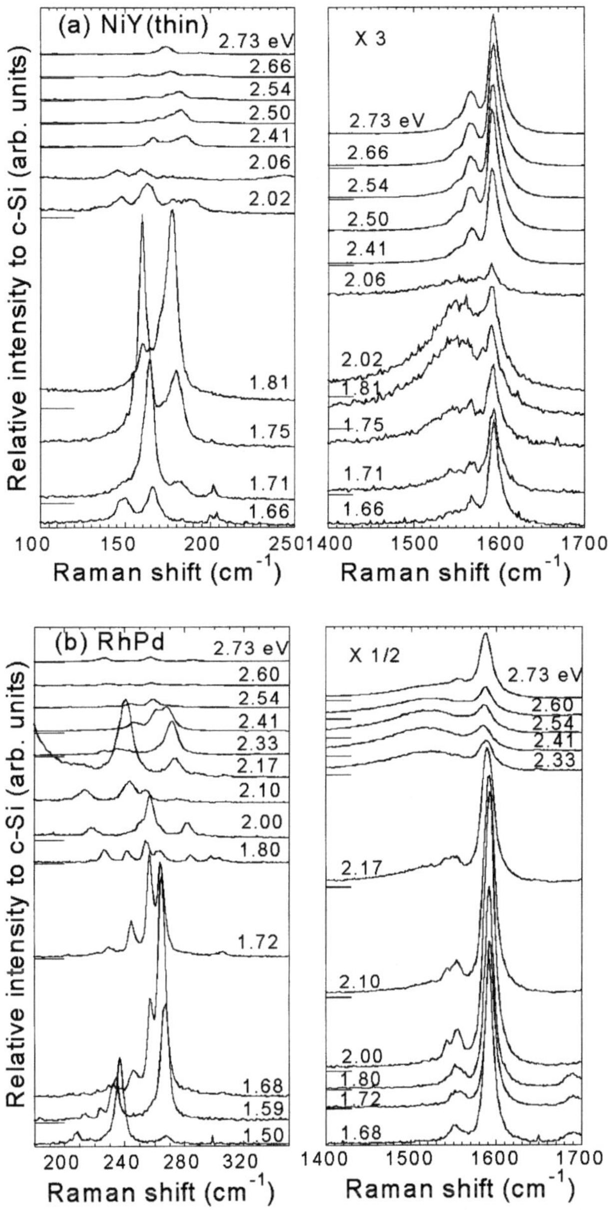


**Fig. 13.** Resonance Raman spectra for (**a**) NiY (*top*) and (**b**) RhPd (*bottom*) catalyzed samples. The *left* and *right* figures for each sample show Raman spectra in the phonon energy region of the RBM and the tangential G-bands, respectively [4]

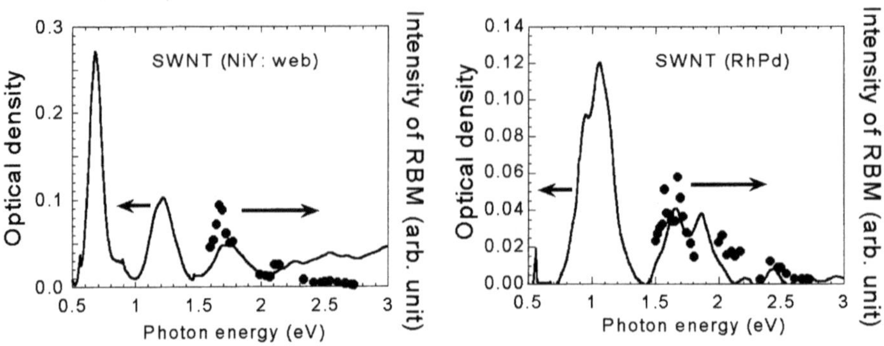


**Fig. 14.** Optical density of the absorption spectra (*left scale*) and the intensity of the RBM feature in the Raman spectra are plotted as a function of the laser excitation energy greater than 1.5 eV for NiY and RhPd catalyzed SWNT samples. The third peaks correspond to the metallic window [3]

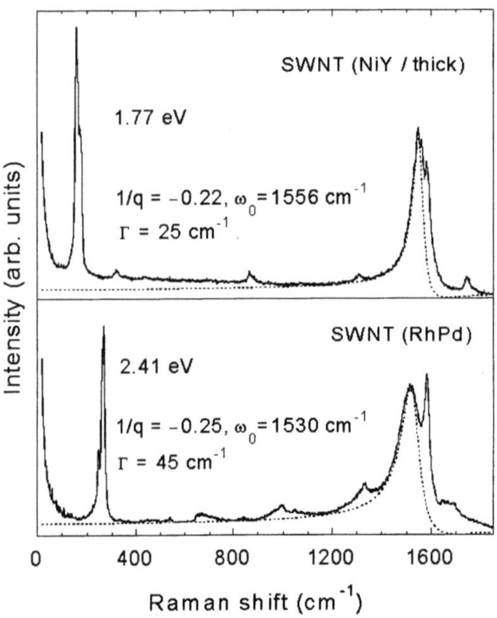


**Fig. 15.** Breit–Wigner–Fano plot for the Raman signals associated with the indicated G-band feature for the NiY and RhPd catalyzed samples [3]. The difference in the fitting parameters in the figures might reflect the different density of states at the Fermi level $D(E_F)$ which have been reported [59]

## 2.5 Stokes and Anti-Stokes Spectra in Resonant Raman Scattering

So far, almost all of the resonance Raman scattering experiments have been carried out on the Stokes spectra. The metallic window is determined experimentally as the range of $E_{\text{laser}}$ over which the characteristic Raman spectrum for metallic nanotubes is seen, for which the most intense Raman component is at $1540\,\text{cm}^{-1}$ [28]. Since there is essentially no Raman scattering intensity for semiconducting nanotubes at this phonon frequency, the intensity $I_{1540}$ provides a convenient measure for the metallic window. The normalized intensity of the dominant Lorentzian component for metallic nanotubes $\tilde{I}_{1540}$ (normalized to a reference line) has a dependence on $E_{\text{laser}}$ given by

$$\begin{aligned}\tilde{I}_{1540}(d_0) = \sum_{d_t} A \exp\left[\frac{-(d_t - d_0)^2}{\Delta d_t^2/4}\right] \\ \times [(E_{11}^{\mathrm{M}}(d_t) - E_{\text{laser}})^2 + \Gamma_e^2/4]^{-1} \\ \times [(E_{11}^{\mathrm{M}}(d_t) - E_{\text{laser}} \pm E_{\text{phonon}})^2 + \Gamma_e^2/4]^{-1},\end{aligned} \tag{19}$$

where $d_0$ and $\Delta d_t$ are, respectively, the mean diameter and the width of the Gaussian distribution of nanotube diameters within the SWNT sample, $E_{\text{phonon}}$ is the average energy (0.197 eV) of the tangential phonons and the + (−) sign in (19) refers to the Stokes (anti-Stokes) process, $\Gamma_e$ is a damping factor that is introduced to avoid a divergence of the resonant denominator, and the sum in Eq. (19) is carried out over the nanotube diameter distribution. Equation (19) indicates that the normalized intensity for the Stokes process $\tilde{I}_{1540}^{S}(d_0)$ is large when either the incident laser energy is equal to $E_{11}^{\mathrm{M}}(d_t)$ or when the scattered laser energy is equal to $E_{11}^{\mathrm{M}}(d_t)$ and likewise for the anti-Stokes process. Since the phonon energy is on the same order of magnitude as the width of the metallic window for nanotubes with diameters $d_t$, the Stokes and the anti-Stokes processes can be observed at different resonant laser energies in the resonant Raman experiment. The dependence of the normalized intensity $\tilde{I}_{1540}(d_0)$ for the actual SWNT sample on $E_{\text{laser}}$ is primarily sensitive [27,28,29] to the energy difference $E_{11}^{\mathrm{M}}(d_t)$ for the various $d_t$ values in the sample, and the resulting normalized intensity $\tilde{I}_{1540}(d_0)$ is obtained by summing over $d_t$.

In Fig. 16 we present a plot of the expected integrated intensities $\tilde{I}_{1540}(d_0)$ for the resonant Raman process for metallic nanotubes for both the Stokes (solid curve) and anti-Stokes (square points) processes. This figure is used to distinguishes 4 regimes for observation of the Raman spectra for Stokes and anti-Stokes processes shown in Fig. 17: (1) the semiconducting regime (2.19 eV), for which both the Stokes and anti-Stokes spectra receive contributions from semiconducting nanotubes, (2) the metallic regime (1.58 eV), where metallic nanotubes contribute to both the Stokes and anti-Stokes spectra, (3) the regime (1.92 eV), where metallic nanotubes contribute to the Stokes spectra and not to the anti-Stokes spectra, and (4) the regime (1.49 eV), where the metallic nanotubes contribute only to the anti-Stokes spectra and not to the Stokes spectra. The plot in Fig. 16 is for a nanotube diameter distribution $d_t = 1.49 \pm 0.20$ nm assuming $\gamma_e = 0.04$ eV. Equation (17) can be used to determine $\gamma_0$ from the intersection of the Stokes and anti-Stokes curves at 1.69 eV in Fig. 16, yielding a value of $\gamma_0 = 2.94 \pm 0.05$ eV [55,60].

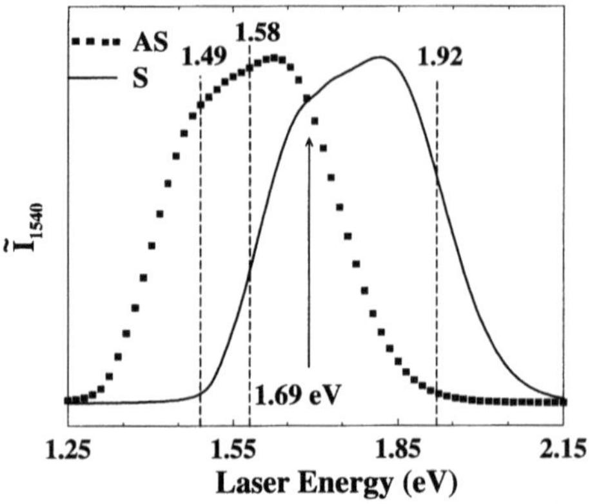


**Fig. 16.** Metallic window for carbon nanotubes with diameter of $d_t = 1.49 \pm 0.20$ nm for the Stokes (*solid line*) and anti-Stokes (*square points*) processes plotted in terms of the normalized intensity of the phonon component at $1540\,cm^{-1}$ for metallic nanotubes vs the laser excitation energy for the Stokes and the anti-Stokes scattering processes [60]. The crossing between the Stokes and anti-Stokes curves is denoted by the *vertical arrow*, and provides a sensitive determination of $\gamma_0$ [55,60]

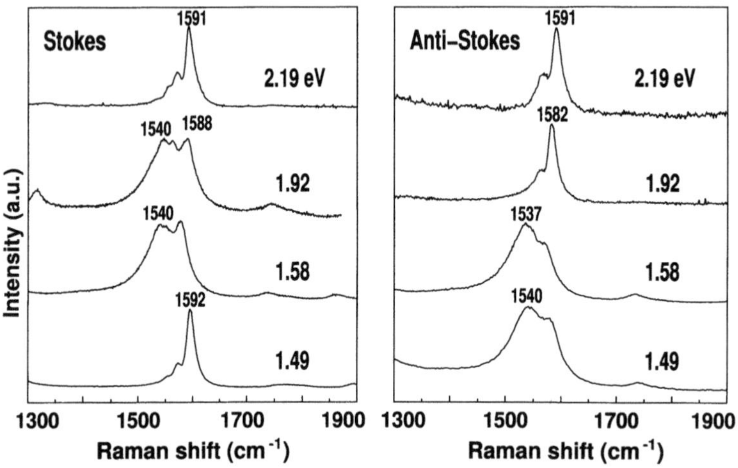


**Fig. 17.** Resonant Raman spectra for the Stokes and anti-Stokes process for SWNTs with a diameter distribution $d_t = 1.49 \pm 0.20$ nm [60]

## 2.6 Bundle Effects on the Optical Properties of SWNTs (Fano Effect)

Although the origin of the $1540\,cm^{-1}$ Breit–Wigner–Fano peak is not well explained, the Fano peaks are relevant to the bundle effect which is discussed in this subsection. This idea can be explained by the Raman spectra observed for the $Br_2$ doped SWNT sample. The frequency of the RBMs are shifted upon doping, and from this frequency shift the charge transfer of the electrons from the SWNTs to the $Br_2$ molecules can be measured [61]. This charge transfer enhances the electrical conductivity whose temperature dependence shows metallic behavior [62]. When SWNTs made by the arc method with the NiY catalyst are used, the undoped SWNT sample exhibits the RBM features around $170\,cm^{-1}$. When the SWNTs are doped by $Br_2$ molecules, new RBM peaks appear at around $240\,cm^{-1}$ when the laser excitation energy is greater than 1.8 eV, as shown in Fig. 18a. When the Raman spectra for the fully $Br_2$ doped sample are measured, new features at $260\,cm^{-1}$ are observed, but the peak at $260\,cm^{-1}$ disappears and a new peak at $240\,cm^{-1}$ can be observed for laser excitation energy greater than 1.96 eV (see Fig. 18) when the sample chamber is evacuated at room temperature, and the spectra for the undoped SWNTs are observed showing RBM peaks around $170\,cm^{-1}$. Since heating in vacuum up to 250°C is needed to remove the bromine completely, the evacuated sample at room temperature consists of a partially doped bundle and an easily undoped portion, which is identified with isolated SWNTs, not in bundles. Since the Fermi energy shifts downward in the acceptor-doped portion of the sample, no resonance Raman effect is expected in the excitation

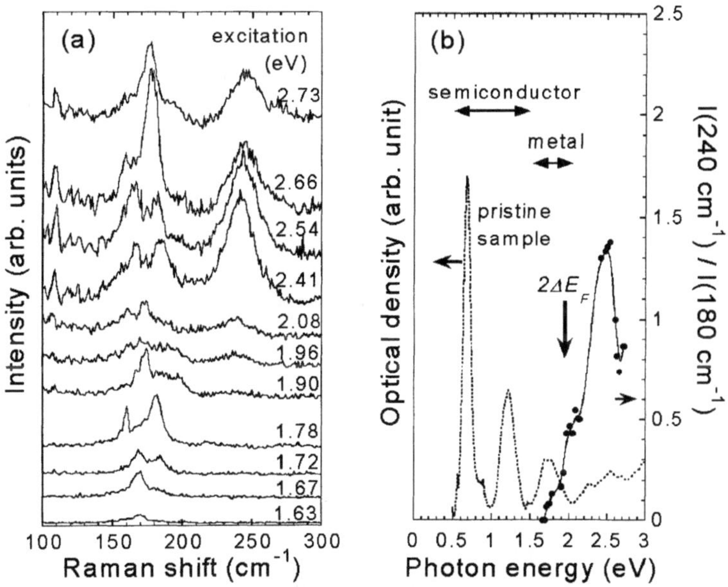


**Fig. 18.** (**a**) Resonance Raman spectra for bromine doped SWNTs prepared using a NiY catalyst. The sample is evacuated after full doping at room temperature. An additional peak around $240\,cm^{-1}$ can be seen for laser excitation energies greater than 1.96 eV. (**b**) (*left scale*) The optical density of the absorption spectra for pristine (undoped) SWNT samples and (*right scale*) the intensity ratio of the RBMs at $\sim 240\,cm^{-1}$ appearing only in the doped samples to the RBM at $\sim 180\,cm^{-1}$ for the undoped sample. The additional RBM peaks appear when the metallic window is satisfied [63]

energy range corresponding to the semiconductor first and second peaks and the metallic third peak in the optical absorption spectra. In fact, in Fig. 18b, the intensity ratio of the Raman peaks around 240 cm$^{-1}$ to that at 180 cm$^{-1}$ is plotted by solid circles and the curve connecting these points is shown in the figure as a function of laser excitation energy. Also shown in the figure is the corresponding optical absorption spectrum for the pristine (undoped) sample plotted by the dotted curve. The onset energy of the Raman peaks at 240 cm$^{-1}$ is consistent with the energy $2\Delta E_F$ which corresponds to the energy of the third metallic peak of the optical absorption. In fact, the optical absorption of the three peaks disappear upon $Br_2$ doping (Fig.12) [24,41]. The peaks of Raman intensity at 240 cm$^{-1}$ are relevant to resonant Raman scattering associated with the fourth or the fifth broad peaks of doped semiconductor SWNTs.

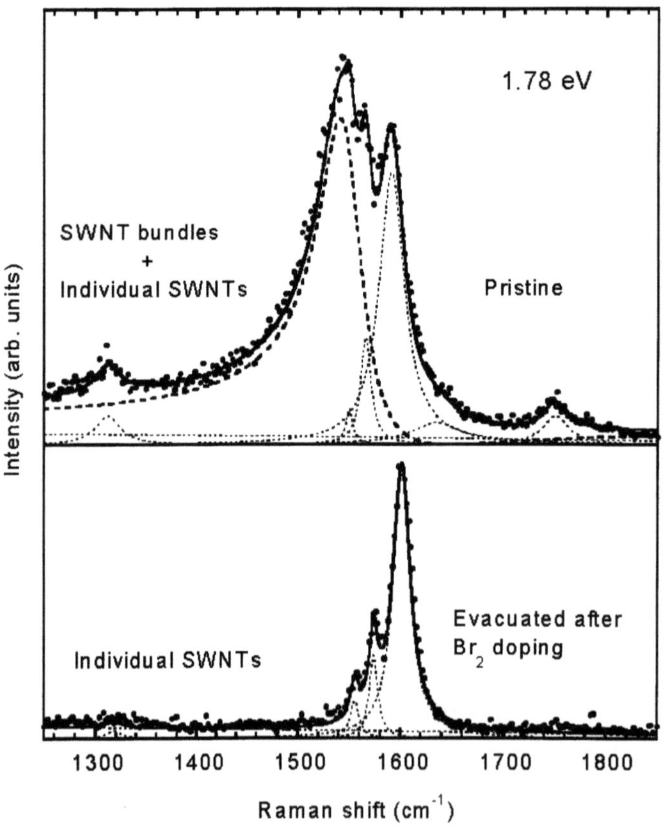


**Fig. 19.** The Raman Spectra for the undoped sample (*top*) and for the evacuated sample (*bottom*) after full $Br_2$ doping at room temperature. The Fano spectral feature at 1540 cm$^{-1}$ is missing in the spectrum for the evacuated sample [63]

For this evacuated sample, the G-band spectra with the laser energy 1.78 eV is shown in Fig. 19. This laser energy corresponds to an energy in the metallic window, but no resonance Raman effect is expected from the doped bundle portion, as discussed above. Thus the resonant Raman spectra should be observed only in metallic nanotubes in the undoped portion of the sample which is considered to contain only isolated SWNTs. Surprisingly there are no 1540 $cm^{-1}$ Fano-peaks for such an evacuated sample, although the undoped sample has a mixture 1590 $cm^{-1}$ and 1540 $cm^{-1}$ peaks, as shown in Fig. 19 for comparison. Thus it is concluded that the origin of the 1540 $cm^{-1}$ peaks is relevant to the nanotubes located within bundles. The interlayer interaction between layers of SWNTs is considered to be on the order of 5–50 $cm^{-1}$ [29,48,49,50,52,53,54] and thus the difference between 1590 $cm^{-1}$ and 1540 $cm^{-1}$ is of about the same order of magnitude as the interlayer interaction. One open issue awaiting solution is why the 1540 $cm^{-1}$ peaks are observed only when the metallic nanotube is within a bundle, and when the laser excitation is within the metallic window and corresponds to an interband transition contributing to the optical absorption. Thus the mechanism responsible for the 1540 $cm^{-1}$ peak is not understood from a fundamental standpoint.

## 2.7 Resonance Raman Scattering of MWNTs

Multi-walled carbon nanotubes (MWNTs) prepared by the carbon arc method are thought to be composed of a coaxial arrangement of concentric nanotubes. For example, $^{13}$C-NMR [64] and magnetoresistance measurements [65,66] show Aharonov–Bohm effects that are associated with the concentric tube structures. On the other hand, the thermal expansion measurements [67] and the doping effects [68] suggest that some kinds of MWNTs have scroll structures. If the RBMs, which are characteristic of SWNTs [35], are observed in MWNTs, the RBM Raman spectra might provide experimental evidence for the coaxial structure. In many cases, however, MWNTs have very large diameters compared with SWNTs even for the innermost layer of the nanotube, and no one has yet succeeded in observing the RBMs in large diameter MWNTs. Zhao and Ando have succeeded in synthesizing MWNTs with an innermost layer having a diameter less than 1.0 nm, by using an electric arc operating in hydrogen gas [69]. The spectroscopic observations on this sample revealed many Raman peaks in the low frequency region, which these RBM frequencies can be used to assign $(n, m)$ values for some constituent layers of MWNTs [70]. Since the resonance Raman effect can be observed in MWNTs (see Fig. 20), we can be confident that these low frequency features are associated with RBMs.

Several MWNT samples have been prepared by the carbon arc method using a range of hydrogen pressures from 30 to 120 Torr, and yielding good MWNT samples under all of these operating conditions. Relative yields depend on the hydrogen pressure and on the arc current [69], with the highest

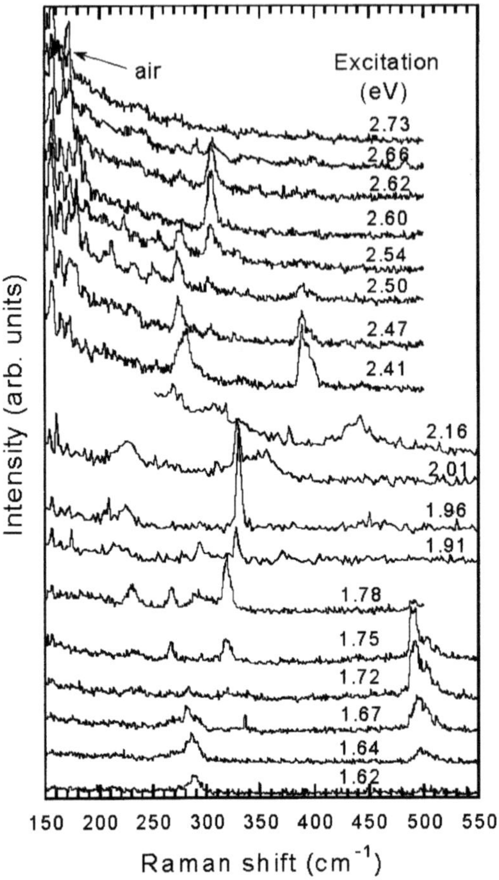


**Fig. 20.** The resonant Raman spectra of multi-wall carbon nanotubes with very small innermost diameters that grow preferentially using an electric arc in hydrogen gas [71]

yield of MWNTs being obtained at 60 Torr of hydrogen gas pressure. The sample purity, after purification of the sample, which was characterized using an infrared lamp, is over 90% MWNTs and the diameter distribution of the innermost shell was measured by TEM. Most of the MWNTs have diameters of the innermost shell of about 1.0 nm, and sometimes innermost diameters less than 0.7 nm were observed.

In Fig. 20 resonance Raman scattering of samples synthesized under different conditions have been measured, and RBM peaks have been observed from 200 to 500 cm$^{-1}$[63]. Peaks between 150 and 200 cm$^{-1}$ are due to the air. In fact these peaks of $O_2$ and $N_2$ are commonly observed not only for the MWNT sample but also for the quartz substrate and they are not observed in Ar gas. Peaks above 200 cm$^{-1}$ show very sharp resonances, which strongly suggest that these structures originate from the RBM vibrations of nanotubes. Resonance effects for each peak are similar to those of single-walled nanotubes. However, the peak frequencies are about 5% higher than those of single-walled nanotubes with the same diameter, which might be due to the inter-layer interaction. For example, the RBM peak at $280\,\mathrm{cm}^{-1}$ shows a maximum intensity at 2.41 eV. This is the same behavior as the peak at $268\,\mathrm{cm}^{-1}$ in SWNTs. This fact is consistent with the recent calculation of bundle effects on the RBM frequency of SWNTs which predict a 10% upshift in the mode frequency due to tube–tube interactions [53]. From the simple relationship between nanotube diameter and RBM frequency [35], the candidate nanotube shells for the peak at $490\,\mathrm{cm}^{-1}$ are (5,1), (6,0), (4,3) and (5,2) having RBM frequencies at 509.6, 472.8, 466.4 and $454.3\,\mathrm{cm}^{-1}$, respectively. If we take into account the 5% up-shift due to the interlayer interactions, the candidates are narrowed down to the nanotube shells (6,0), (4,3) and (5,2), which have diameters of 0.470, 0.477 and 0.489 nm, respectively, and these diameters are consistent with the TEM observations. It is very interesting that the (6,0) nanotube has the same structure as $D_{6\mathrm{h}}$ $\mathrm{C}_{36}$ which has $D_{6\mathrm{h}}$ symmetry [72]. However, we also have to consider the electronic states of the nanotube to clearly identify the resonance effect. By use of the zone-folding band calculation [2,8,16], assuming a transfer integral $\gamma_0 = 2.75\,\mathrm{eV}$, it is found that (6,0) and (5,2) are metallic nanotubes and have their lowest energy gap $E_{11}^{\mathrm{M}}$ at 4.0 eV. The resonance laser energy, where the RBM peak has a maximum intensity, occurs at 1.7 eV, and the peak at $490\,\mathrm{cm}^{-1}$ was assigned to the (4,3) nanotube which is a semiconductor, and has its lowest energy gap $E_{11}^{\mathrm{S}}$ at 1.6 eV. In the same way, the candidates (7,1) and (5,4) were considered for the Raman band at $388\,\mathrm{cm}^{-1}$. The nanotube (7,1) is metallic and the lowest energy gap $E_{11}^{\mathrm{M}}$ is at 3.4 eV, while the (5,4) nanotube should be semiconducting and is expected to have $E_{11}^{\mathrm{S}}$ and $E_{22}^{\mathrm{S}}$ at 1.28 and 2.52 eV, respectively. Thus, the peak at $388\,\mathrm{cm}^{-1}$ should be assigned to the nanotube shell (5,4) because of the resonance observed at 2.4 eV[71].

Finally we consider the interlayer interactions in MWNTs. The RBM band in Fig. 20 at $490\,\mathrm{cm}^{-1}$ is split into three peaks indicating the same resonance feature. These peaks cannot be explained by different nanotubes, since there are no other candidates available. The nanotube (5,1) is the only candidate having the nearest diameter and the nearest energy gap in the optical spectra. However, the calculated energy gap of a (5,1) nanotube is 1.7 eV, which is 0.1 eV wider than that for a (4,3) nanotube. If one of the peaks originates from a (5,1) nanotube, the resonance feature should be different from that for the other peaks. Further, the RBM frequency of a (5,1) nanotube becomes $534\,\mathrm{cm}^{-1}$, taking into account the 5% up-shift due to the inter-tube interaction in a nanotube bundle. Thus, it is proper to think that these three peaks are originating from the same nanotube. The possible reason for the splitting of this peak is the interlayer interaction. When the first layer is (4,3), then (10,7) is the best selection as the second layer, since the interlayer distance is 0.342 nm, which is a typical value for MWNTs [73]. The other nearest candidates for the second layers are (13,3), (9,8) and (11,6) having inter-layer distances 0.339, 0.339 and 0.347 nm, respectively. The interlayer distance for the (13,3) and (9,8) candidates are about the same (about 1% smaller) as the typical inter-layer distance, and but the interlayer distance for the (11,6) nanotube is 1.5% larger. The magnitude of the interlayer interaction should depend on the interlayer distance, and, consequently, the RBM frequency of the first layer may depend on the chiral index of the second layer. Indeed, the observed frequency separation between the split peaks is about 2%, which may be consistent with the difference in interlayer distances. The splitting into three RBM probably indicates that there are at least three kinds of second layers. Furthermore, this splitting cannot be explained by a scrolled structure for MWNTs. This strongly suggests that the MWNTs fabricated by the electric arc operating in hydrogen gas has a concentric structure. For the thinnest nanotube (4,3), the RBM frequency of the second layer is 191 $cm^{-1}$. This should be the highest RBM frequency of the second layer nanotube. Since the low frequency region is affected by signals from the air, Raman spectra were taken while keeping the sample in argon gas. However, no peak was observed below 200 $cm^{-1}$, suggesting that only the innermost nanotube has a significant Raman intensity. The innermost layer has only an outer nanotube as a neighbor, while the other nanotubes, except for the outermost layer, have both inner and outer nanotube neighbors.

The interlayer interaction probably broadens the one-dimensional band structure, in a like manner to the bundle effect in SWNTs [48,49,50,51,52,29] [53,54]. The band broadening decreases the magnitude of the joint density of states at the energy gap, leading to a decrease in the resonant Raman intensity of the second layer. On the other hand, the RBM frequency of the outermost layer is too low to measure because of its large diameter. Thus, RBMs are observed in MWNTs only for the innermost nanotubes. Theoretical calculations show that SWNTs with diameters smaller than $C_{60}$ show metallic behavior because of the hybridization effect of the $2p_z$ orbital with that of the $\sigma$ electron [74,75]. The hybridization effect lowers the energy of the conduction band and raises the energy of the valence band, which results in the semi-metallic nature of the electronic states. However, the electrostatic-conductance of two-probe measurement of MWNTs shows that semiconducting nanotubes seems to be dominant in this diameter region[76]. Thus it is necessary to investigate the electronic properties of SWNTs with diameters smaller than that of $C_{60}$.

## 3 Summary

In summary, the spectra of the DOS for SWNTs have a strong chirality dependence. Especially for metallic nanotubes, the DOS peaks are found to be split into two peaks because of the trigonal warping effect, while semiconducting nanotubes do not show a splitting. The width of the splitting becomes a maximum for the metallic zigzag nanotubes $(3n, 0)$, and is zero for armchair nanotubes $(n, n)$, which are always metallic. In the case of semiconducting nanotubes, the upper and lower bounds of the peak positions of $E_{11}^{S}(d_t)$ on the Kataura chart shown in Fig. 5 are determined by the values of $E_{11}^{S}(d_t)$ for the $(3n+1,0)$ or $(3n-1,0)$ zigzag nanotubes. The upper and lower bounds of the widths of the $E_{ii}^{\mathrm{S}}(d_{\mathrm{t}})$ curves alternate with increasing $i$ between the $(3n+1,0)$ and $(3n-1,0)$ zigzag nanotubes.

The existence of a splitting of the DOS spectra for metallic nanotubes should depend on the chirality which should be observable by STS/STM experiments, consistent with the experiments of *Kim* et al. [22]. The width of the metallic window can be observed in resonant Raman experiments, especially through the differences between the analysis for the Stokes and the anti-Stokes spectra. Some magnetic effects should be observable in the resonant Raman spectra because an applied magnetic field should perturb the 1D DOS for the nanotubes, since the magnetic field will break the symmetry between the $K$ and $K'$ points. The magnetic susceptibility, which has been important for the determination of $\gamma_0$ for 3D graphite [77,78], could also provide interesting results regarding a determination of $E_{pp}(d_{\mathrm{t}})$ for SWNTs, including the dependence of $E_{pp}(d_{\mathrm{t}})$ on $d_{\mathrm{t}}$.

Purification of SWNTs to provide SWNTs with a known diameter and chirality should be given high priority for future research on carbon nanotube physics. Furthermore, we can anticipate future experiments on SWNTs which could illuminate phenomena showing differences in the $E(k)$ relations for the conduction and valence bands of SWNTs. Such information would be of particular interest for the experimental determination of the overlap integral $s$ as a function of nanotube diameter. The discussion presented in this article for the experimental determination of $E_{pp}(d_{\mathrm{t}})$ depends on assuming $s=0$, in order to make direct contact with the tight-binding calculations. However, if $s \neq 0$, then the determination of $E_{pp}(d_{\mathrm{t}})$ would depend on the physical experiment that is used for this determination, because different experiments emphasize different $k$ points in the Brillouin zone. The results of this article suggest that theoretical tight binding calculations for nanotubes should also be refined to include the effect of $s \neq 0$. Higher order (more distant neighbor) interactions should yield corrections to the lowest order theory discussed here.

The $1540\,\mathrm{cm}^{-1}$ feature appears only in the Raman spectra for a metallic bundle, but not for semiconducting SWNTs nor for individual metallic SWNTs. The inter-tube interaction in MWNTs gives 5% higher RBM mode frequencies than in SWNT bundles, and the intertube-interaction effect between the MWNT innermost shell and its adjacent outer shell is important for splitting the RBM peaks of a MWNT sample.

## Acknowledgments

The authors gratefully acknowledge stimulating and valuable discussions with Profs. M.S. Dresselhaus and G. Dresselhaus for the writing of this chapter. R.S. and H.K. acknowledge a grant from the Japanese Ministry of Education (No. 11165216 and No. 11165231), respectively. R.S. acknowledges support from the Japan Society for the Promotion of Science for his visit to MIT.

H.K. acknowledges the Japan Society for Promotion of Science Research for support for the Future Program.

## References

1. M. S. Dresselhaus, G. Dresselhaus, P. C. Eklund, *Science of Fullerenes and Carbon Nanotubes* (Academic, New York 1996)
2. R. Saito, G. Dresselhaus, M. S. Dresselhaus, *Physical Properties of Carbon Nanotubes* (Imperial College Press, London, 1998)
3. H. Kataura, Y. Kumazawa, Y. Maniwa, I. Umezu, S. Suzuki, Y. Ohtsuka, Y. Achiba, Synth. Met. **103**, 2555 (1999)
4. H. Kataura, Y. Kumazawa, N. Kojima, Y. Maniwa, I. Umezu, S. Masubuchi, S. Kazama, X. Zhao, Y. Ando, Y. Ohtsuka, S. Suzuki, Y. Achiba In *Proc. of the Int. Winter School on Electronic Properties of Novel Materials (IWEPNM'99)*, H. Kuzmany, M. Mehring, J. Fink (Eds.) (American Institute of Physics, Woodbury 1999) AIP Conf. Proc. (in press)
5. S. Bandow, S. Asaka, Y. Saito, A. M. Rao, L. Grigorian, E. Richter, P. C. Eklund, Phys. Rev. Lett. **80**, 3779 (1998)
6. J. C. Charlier and S. Iijima, Chapter 4 in this volume
7. M. S. Dresselhaus, G. Dresselhaus, R. Saito, Phys. Rev. B **45**, 6234 (1992)
8. R. Saito, M. Fujita, G. Dresselhaus, M. S. Dresselhaus, Phys. Rev. B **46**, 1804 (1992)
9. G. S. Painter, D. E. Ellis, Phys. Rev. B **1**, 4747 (1970)
10. M. S. Dresselhaus, G. Dresselhaus, K. Sugihara, I. L. Spain, H. A. Goldberg, *Graphite Fibers and Filaments* , Vol. 5, *Springer Ser. Mater. Sci.* (Springer, Berlin, Heidelberg 1988)
11. R. Saito, G. Dresselhaus, M. S. Dresselhaus, Phys. Rev. B **61**, 2981 (2000)
12. P. R. Wallace, Phys. Rev. **71**, 622 (1947)
13. J. W. McClure, Phys. Rev. **104**, 666 (1956)
14. R. A. Jishi, D. Inomata, K. Nakao, M. S. Dresselhaus, G. Dresselhaus, J. Phys. Soc. Jpn. **63**, 2252 (1994)
15. N. Hamada, S. Sawada, A. Oshiyama, Phys. Rev. Lett. **68**, 1579 (1992)
16. R. Saito, M. Fujita, G. Dresselhaus, M. S. Dresselhaus, Appl. Phys. Lett. **60**, 2204 (1992)
17. K. Tanaka, K. Okahara, M. Okada, T. Yamabe, Chem. Phys. Lett. **191**, 469 (1992)
18. J. W. G. Wildöer, L. C. Venema, A. G. Rinzler, R. E. Smalley, C. Dekker, Nature (London) **391**, 59 (1998)
19. T. W. Odom, J. L. Huang, P. Kim, C. M. Lieber, Nature (London) **391**, 62 (1998)
20. T. W. Odom, J. L. Huang, P. Kim, M. Ouyang, C. M. Lieber, J. Mater. Res. **13**, 2380 (1998)
21. T. W. Odom, Private communication
22. P. Kim, T. Odom, J.-L. Huang, C. M. Lieber, Phys. Rev. Lett. **82**, 1225 (1999)
23. H. Ajiki, T. Ando, Physica B, Condensed Matter **201**, 349 (1994)
24. S. Kazaoui, N. Minami, R. Jacquemin, H. Kataura, Y. Achiba, Phys. Rev. B **60**, 13339 (1999)

25. A. M. Rao, E. Richter, S. Bandow, B. Chase, P. C. Eklund, K. W. Williams, M. Menon, K. R. Subbaswamy, A. Thess, R. E. Smalley, G. Dresselhaus, M. S. Dresselhaus, Science **275**, 187 (1997)
26. A. Kasuya, Y. Sasaki, Y. Saito, K. Tohji, Y. Nishina, Phys. Rev. Lett. **78**, 4434 (1997)
27. M. A. Pimenta, A. Marucci, S. D. M. Brown, M. J. Matthews, A. M. Rao, P. C. Eklund, R. E. Smalley, G. Dresselhaus, M. S. Dresselhaus, J. Mater. Res. **13**, 2396 (1998)
28. M. A. Pimenta, A. Marucci, S. Empedocles, M. Bawendi, E. B. Hanlon, A. M. Rao, P. C. Eklund, R. E. Smalley, G. Dresselhaus, M. S. Dresselhaus, Phys. Rev. B **58**, R16016 (1998)
29. L. Alvarez, A. Righi, T. Guillard, S. Rols, E. Anglaret, D. Laplaze, J.-L. Sauvajol, Chem. Phys. Lett. **316**, 186 (2000)
30. J. W. Mintmire, C. T. White, Phys. Rev. Lett. **81**, 2506 (1998)
31. C. T. White, T. N. Todorov, Nature (London) **393**, 240 (1998)
32. S. J. Tans, R. M. Verschueren, C. Dekker, Nature **393**, 49 (1998)
33. P. Lambin, Private communication
34. J. Yu, K. Kalia, P. Vashishta, Europhys. Lett. **32**, 43 (1995)
35. R. Saito, T. Takeya, T. Kimura, G. Dresselhaus, M. S. Dresselhaus, Phys. Rev. B **57**, 4145 (1998)
36. O. Madelung, *Solid State Theory* (Springer, Berlin, Heidelberg 1978)
37. R. Saito, G. Dresselhaus, M. S. Dresselhaus, In *Science and Technology of Carbon Nanotubes*, K. Tanaka, T. Yamabe, K. Fukui (Eds.), (Elsevier Science Ltd., Oxford 1999) pp. 51–62
38. S. Guha, J. Menéndez, J. B. Page, G. B. Adams, Phys. Rev. B **53**, 13106 (1996)
39. M. J. Matthews, M. A. Pimenta, G. Dresselhaus, M. S. Dresselhaus, M. Endo, Phys. Rev. B **59**, R6585 (1999)
40. H. Kataura, A. Kimura, Y. Ohtsuka, S. Suzuki, Y. Maniwa, T. Hanyu, Y. Achiba, Jpn. J. Appl. Phys. **37**, L616 (1998)
41. J. Chen, M. A. Hamon, H. Hu, Y. Chen, A. M. Rao, P. C. Eklund, R. C. Haddon, Science **282**, 95 (1998)
42. H. Kataura. (unpublished)
43. S. Iijima, T. Ichihashi, Nature (London) **363**, 603 (1993)
44. D. S. Bethune, C. H. Kiang, M. S. de Vries, G. Gorman, R. Savoy, J. Vazquez, R. Beyers, Nature (London) **363**, 605 (1993)
45. M. Ichida, J. Phys. Soc. Jpn. **68**, 3131 (1999)
46. I. Umezu, M. Daigo, K. Maeda, Jpn. J. Appl. Phys. **33**, L873 (1994)
47. T. Pichler, M. Knupfer, M. S. Golden, J. Fink, A. Rinzler, R. E. Smalley, Phys. Rev. Lett. **80**, 4729 (1998)
48. J.-C. Charlier, J. P. Michenaud, Phys. Rev. Lett. **70**, 1858 (1993)
49. J.-C. Charlier, X. Gonze, J. P. Michenaud, Europhys. Lett. **29**, 43 (1995)
50. E. Richter, K. R. Subbaswamy, Phys. Rev. Lett. **79**, 2738 (1997)
51. Y. K. Kwon, S. Saito, D. Tománek, Phys. Rev. B **58**, R13314 (1998)
52. U. D. Venkateswaran, A. M. Rao, E. Richter, M. Menon, A. Rinzler, R. E. Smalley, P. C. Eklund, Phys. Rev. B **59**, 10928 (1999)
53. L. Henrard, E. Hernández, P. Bernier, A. Rubio, Phys. Rev. B **60**, R8521 (1999)
54. D. Kahn, J. P. Lu, Phys. Rev. B **60**, 6535 (1999)
55. G. Dresselhaus, M. A. Pimenta, R. Saito, J.-C. Charlier, S. D. M. Brown, P. Corio, A. Marucci, M. S. Dresselhaus, in *Science and Applications of Nanotubes*, D. Tománek, R. J. Enbody (Eds.), (Kluwer Academic, New York 2000) pp. 275–295

56. M. S. Dresselhaus, M. A. Pimenta, K. Kneipp, S. D. M. Brown, P. Corio, A. Marucci, G. Dresselhaus, in *Science and Applications of Nanotubes*, D. Tománek, R. J. Enbody (Eds.), (Kluwer Academic, New York 2000) pp. 253–274
57. A. Kasuya, M. Sugano, Y. Sasaki, T. Maeda, Y. Saito, K. Tohji, H. Takahashi, Y. Sasaki, M. Fukushima, Y. Nishina, C. Horie, Phys. Rev. B **57**, 4999 (1998)
58. M. Sugano, A. Kasuya, K. Tohji, Y. Saito, Y. Nishina, Chem. Phys. Lett. **292**, 575 (1998)
59. X. P. Tang, A. Kleinhammes, H. Shimoda, L. Fleming, K. Y. Bennoune, C. Bower, O. Zhou, Y. Wu, MRS Proc. **593**, J. Robertson, J. P. Sullivan, O. Zhou, T. B. Allen, B. F. Coll (Eds.), (2000)
60. S. D. M. Brown, P. Corio, A. Marucci, M. S. Dresselhaus, M. A. Pimenta, K. Kneipp, Phys. Rev. B **61**, R5137 (2000)
61. A. M. Rao, P. C. Eklund, S. Bandow, A. Thess, R. E. Smalley, Nature (London) **388**, 257 (1997)
62. R. S. Lee, H. J. Kim, J. E. Fischer, A. Thess, R. E. Smalley, Nature (London) **388**, 255 (1997)
63. H. Kataura, Y. Kumazawa, N. Kojima, Y. Maniwa, I. Umezu, S. Masubuchi, S. Kazama, Y. Ohtsuka, S. Suzuki, Y. Achiba, Mol. Cryst. Liquid Cryst.0 (2000)
64. Y. Maniwa, M. Hayashi, Y. Kumazawa, H. Tou, H. Kataura, H. Ago, Y. Ono, T. Yamabe, K. Tanaka, AIP Conf. Proc. **442**, 87 (1998)
65. A. Fujiwara, K. Tomiyama, H. Suematsu, M. Yumura, K. Uchida, Phys. Rev. B **60**, 13492 (1999)
66. A. Bachtold, C. Strunk, J. P. Salvetat, J. M. Bonard, L. Forró, T. Nussbaumer, C. Schönenberger, Nature **397**, 673 (1999)
67. S. Bandow, Jpn J. Appl. Phys. **36**, L1403 (1997)
68. V. Z. Mordkovich, Mol. Cryst. Liquid Cryst. (2000)
69. X. Zhao, M. Ohkohchi, M. Wangm, S. Iijima, T. Ichihashi, Y. Ando, Carbon **35**, 775 (1997)
70. X. Zhao, Y. Ando, Jpn. J. Appl. Phys. **37**, 4846 (1998)
71. H. Kataura, Y. Achiba, X. Zhao, Y. Ando, In *MRS Symp. Proc., Boston, Fall 1999*, J. Robertson, J. P. Sullivan, O. Zhou, T. B. Allen, B. F. Coll (Eds.) (Materials Research Society Press, Pittsburgh, PA 2000)
72. C. Piskoti, J. Yarger, A. Zettl, Nature **393**, 771 (1998)
73. Y. Saito, T. Yoshikawa, S. Bandow, M. Tomita, T. Hayashi, Phys. Rev. B **48**, 1907 (1993)
74. X. Blase, L. X. Benedict, E. L. Shirley, S. G. Louie, Phys. Rev. Lett. **72**, 1878 (1994)
75. S. G. Louie, chapter 6 in this volume
76. K. Kaneto, M. Tsuruta, G. Sakai, W. Y. Cho, Y. Ando, Synth. Met. **103**, 2543 (1999)
77. J. W. McClure, Phys. Rev. **108**, 612 (1957)
78. K. S. Krishnan, Nature (London) **133**, 174 (1934)

# Electron Spectroscopy Studies of Carbon Nanotubes

Jörg H. Fink[1] and Philippe Lambin[2]

[1] Institut für Festkörper- und Werkstofforschung Dresden
Postfach 270016, 01171 Dresden, Germany
j.fink@ifw-dresden.de

[2] Département de Physique, Facultés Universitaires Notre-Dame de la Paix
61 rue du Bruxelles, 5000 Namur, Belgium
philippe.lambin@scf.fundp.ac.be

**Abstract.** Electron spectroscopies play an essential role in the experimental characterization of the electronic structure of solids. After a short description of the techniques used, the paper reviews some of the most important results obtained on purified single- and multi-wall carbon nanotubes, and intercalated nanotubes as well. An analysis of the occupied and unoccupied electron states, and the plasmon structure of the nanotubes is provided. How this information can be used to characterize the samples is discussed. Whenever possible, a comparison between nanotube data and those from graphite and $C_{60}$ fullerene is made, as to draw a coherent picture in the light of recent theories on the electronic properties of these C-$sp^2$ materials.

Carbon-based $\pi$-electron system are becoming more and more important in solid state physics, chemistry and in material science. These systems comprise graphite, conjugated polymers and oligomers, fullerenes and carbon nanotubes. Many systems can be "doped" or intercalated. In this way new materials can be tailored having interesting properties. These new materials have often become model compounds in solid state physics since they show metallic and semiconducting behavior, superconductivity and magnetism. Correlation effects, which result from electron-electron interactions, and the electron-phonon interaction, are important in many of these systems. Dimensionality plays an important role, too. Thus many of the interesting questions of present solid state research are encountered again in the carbon-based $\pi$-electron systems.

Moreover, some of these materials have high technical potential. They show remarkable mechanical properties, e.g., a record-high elastic modulus. In addition, electronic devices from conjugated carbon systems are coming close to realization and commercialization. Transistors and organic light-emitting diodes based on conjugated polymers or molecules are already on the market. Industry is strongly interested in the field emission from carbon nanotubes to fabricate bright light sources and flat-panel screens. Finally, future nanoscale electronics systems can possibly be realized using carbon materials such as nanotubes.

M. S. Dresselhaus, G. Dresselhaus, Ph. Avouris (Eds.): Carbon Nanotubes,
Topics Appl. Phys. **80**, 247–272 (2001)

The exceptional properties of these materials are strongly related to their electronic structure. The mechanical properties are mainly caused by the strong covalent bonds between the tightly-bond $\sigma$-orbitals. On the other hand, the interesting electronic properties are related to the loosely bound $\pi$-electrons. It is evident that the study of the electronic structure of these materials is an important task. In this paper we review recent investigations of the electronic structure of the quasi one-dimensional Single-Wall Carbon NanoTubes (SWNTs) and Multi-Wall Carbon NanoTubes (MWNTs), based on electron spectroscopy studies. We compare these results with the better known results from the quasi two-dimensional graphite and the quasi zero-dimensional fullerene solids. We emphasize that electron spectroscopy studies of carbon nanotubes are just at an early stage, because many of these techniques are extremely surface sensitive and surfaces suitable for representing bulk properties are difficult to prepare. Therefore, most of the reliable results have been obtained by less surface sensitive methods.

## 1 Electron Spectroscopies

In this section we describe various electron spectroscopies which have been used to study the electronic structure of carbon nanotubes. The techniques are illustrated schematically in Fig. 1. For this illustration we use the electronic structure of a carbon-based $\pi$-electron system. In these compounds the valence electrons of carbon are predominantly in the $sp^2$ configuration, i.e., one $s$-electron and two $p$-electrons form the $sp^2$-hybrid which has trigonally directed $\sigma$ bonds in a plane. In the solid, these $\sigma$ orbitals form strong covalent bonds with the $\sigma$ orbitals from neighboring carbon atoms. Therefore, occupied $\sigma$ and unoccupied $\sigma^*$ bonds are formed. The third C $2p$ electron is in a $2p_z$-orbital perpendicular to the plane and forms a weaker $\pi$ bond with the $2p_z$-orbitals of neighboring C atoms. The electrons from the C $2p_z$ orbitals in this configuration are usually called $\pi$-electrons. Due to the weaker bonding, the splitting between the occupied $\pi$-bonds and the unoccupied $\pi^*$-bands is weaker. The electronic structure of the valence and conduction bands in such systems together with the C $1s$ core level are shown in Fig. 1. In addition, transitions used in the various techniques are shown in Fig. 1.

In Photoelectron Spectroscopy (PES) [1], photoelectrons are ejected from the solid by a photon with the energy $h\nu$. The kinetic energy, $E_{\text{kin}}$, and the intensity of the photoelectrons are measured. The binding energy of the ejected electron is then given by the Einstein relation $E_{\text{B}} = h\nu - E_{\text{kin}} - \Phi$, where $\Phi$ is the work function. In a first approximation, the intensity of the photoelectrons as a function of $E_{\text{B}}$ yields the density of occupied states. Detecting the angle of the ejected photoelectrons relative to the surface normal gives information on the wave vector of the electrons in the solid. Therefore, using this Angular Resolved PhotoEmission Spectroscopy (ARPES) the band structure of the occupied bands can be probed. Measuring the binding energy, $E_{\text{B}}$,

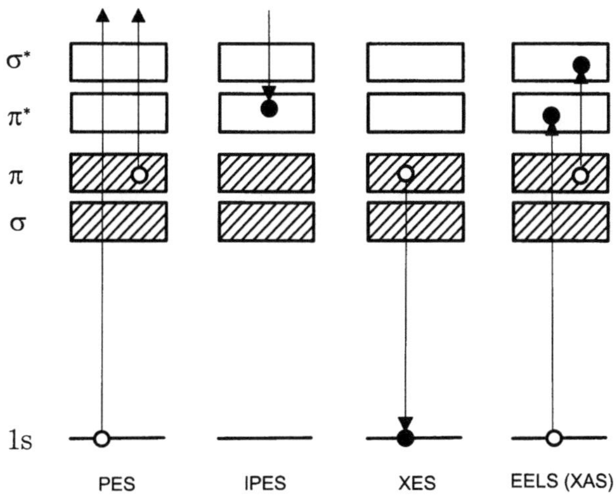


**Fig. 1.** Illustration of various electron spectroscopies. The electronic structure of carbon-based $\pi$-electron systems is sketched. PES: Photoemission, IPES: Inverse Photoemission, XES: X-ray Emission Spectroscopy, EELS: Electron Energy-Loss Spectroscopy, XAS: X-ray Absorption Spectroscopy

relative to the Fermi level of the core electrons, provides information on the chemical bonding of the C atoms. The so-called chemical shift of $E_B$ is determined by the charge on the excited atom and the Madelung potential from the surrounding atoms. Finally, one should remark that the mean free path of the photoelectron is only several Å. Therefore, this technique is extremely surface sensitive.

In the Inverse PhotoElectron Spectroscopy (IPES) [1], electrons are injected into the solid, and the intensity and the energy of the photons from the Bremsstrahlung is recorded. From these measurements, information on the density of unoccupied states can be derived. Similar to ARPES, in Angular-Resolved Inverse PhotoEmission Spectroscopy (ARIPES), the angle of the incoming electrons is varied and the band structure of the unoccupied bands can be measured. Since the mean free path of the injected electrons is again only several Å IPES and ARIPES are also surface sensitive techniques.

In X-ray Emission Spectroscopy (XES) [2], a core hole, (e.g. in the C 1$s$ shell) is created by electron bombardment or by X-ray irradiation. The fluorescent decay of the core hole by transitions from the occupied valence bands to the unoccupied core state is monitored. The intensity as a function of the energy of the fluorescent radiation gives information on the density of occupied states. Since a dipole transition to a localized core orbital is involved, only states with the appropriate symmetry character that are localized in the vicinity of the core orbital will contribute to the spectra. Consequently, this technique provides information on the local partial density of states. Starting from the core hole in the C $1s$ level, the contributions of C $2p$ states to the occupied density of states are measured.

In Inelastic soft X-ray Scattering (RIXS) [3,4], which is a technique closely related to XES, additional information can be obtained on the wave vector of the occupied states, similar to ARPES. Here, a core hole is created by selectively promoting the core electron into an unoccupied state of a chosen energy. Synchrotron radiation is used to induce excitations from the core level into various parts of the conduction bands. Under certain conditions, valence-band electrons with the same wave vector as the excited electron in the conduction band will then contribute predominantly in the consecutive X-ray emission process. As a result part of the band structure of the occupied bands can be probed selectively by variation of the excitation energy.

In Electron Energy-Loss Spectroscopy (EELS) [5], transitions from the core level into unoccupied states can be performed using the inelastic scattering of high-energy electrons. At small scattering angles, only dipole excitations are allowed. Therefore, starting from the C $1s$ level, the local partial density of unoccupied states having C $2p$ character is probed. However, since there is a core hole in the final state, excitonic effects have to be taken into account. When these effects are strong, the spectral weight at the bottom of the bands is enhanced at the expense of the spectral weight at higher energy. Similar information can be obtained in X-ray Absorption Spectroscopy (XAS) [6]. There, using synchrotron radiation, the absorption coefficient is measured near the threshold of a core excitation. In carbon-based materials, again, the local partial density of unoccupied states with C $2p$ character is probed.

In the low-energy range of the EELS spectra, excitations from occupied to unoccupied states are probed. In contrast to optical spectroscopy, the measured loss function is not dominated by absorption maxima but by the collective excitations of electrons, i.e. the plasmons [7]. The concept of plasmons can be introduced in a continuous-medium approximation, at least in the long-wavelength limit. Neglecting retardation effects, the electric field, $\boldsymbol{E}$, and displacement vector, $\boldsymbol{D}$, generated by a perturbation of the polarization of the medium, satisfy the Maxwell equations $\boldsymbol{\nabla} \times \boldsymbol{E} = 0$ and $\boldsymbol{\nabla} \cdot \boldsymbol{D} = 0$. In an infinite, homogeneous material, solutions can be sought in the form of plane waves, with frequency $\omega$ and wave vector $\boldsymbol{q}$. The above equations then become $\boldsymbol{q} \times \boldsymbol{E} = 0$ and $\boldsymbol{q} \cdot \boldsymbol{D} = 0$, with $\boldsymbol{D} = \varepsilon_0 \varepsilon(\omega, \boldsymbol{q}) \boldsymbol{E}$, where $\varepsilon$ is the dielectric function of the medium. The two possible plane-wave solutions are

$$\varepsilon(\omega, \boldsymbol{q}) = 0 \quad \text{and} \quad \boldsymbol{q} \times \boldsymbol{E} = 0 \tag{1}$$

$$1/\varepsilon(, \omega, \boldsymbol{q}) = 0 \quad \text{and} \quad \boldsymbol{q} \cdot \boldsymbol{D} = 0 \tag{2}$$

The first solution corresponds to longitudinal modes, the plasmons probed in EELS. The second modes are transverse ones and these can be excited by electromagnetic waves.

The probability per length unit that the electron loses the energy $\hbar\omega$ and momentum $\hbar\boldsymbol{q}$ in a single-scattering event is proportional to the so-called loss function [8], the imaginary part of the inverse frequency- and momentum-dependent dielectric function,

$$\mathrm{Im}[-1/\varepsilon(\omega, \boldsymbol{q})] = \varepsilon_2/(\varepsilon_1^2 + \varepsilon_2^2) . \tag{3}$$

For an anisotropic material such as graphite, $\varepsilon$ denotes $\hat{\mathbf{q}} \cdot \epsilon \cdot \hat{\mathbf{q}}$ with $\hat{\mathbf{q}}$ a unit vector in the direction of the wave-vector transfer, and $\epsilon$ the dielectric tensor. The loss function has maxima when the real part $\mathrm{Re}[\varepsilon] = \varepsilon_1$ is zero and $\mathrm{Im}[\varepsilon] = \varepsilon_2$ is small. That condition immediately shows that the longitudinal modes (1) are the ones that can be excited by a traveling electron. These modes generate an electric field in the medium that interacts with the electron through the Coulomb force $e\boldsymbol{E}$, and this force slows the particle down. The energy losses come from that interaction.

EELS has the advantage to provide information on the dielectric function of the sample over a broad frequency interval. Kramers-Kronig analysis can then be used, which allows one to determine $\mathrm{Re}[-1/\varepsilon(\omega, \boldsymbol{q})]$, the dielectric function $\varepsilon_1 + \mathrm{i}\varepsilon_2$, and the optical constants from the measured loss function. Furthermore, performing angular resolved measurements, the wave vector and consequently the wavelength of the excitations can be varied. In terms of optical spectroscopy, not only vertical but also non-vertical transitions can be excited. Because EELS is performed with high-energy electrons ($E \sim 200\,\mathrm{keV}$) in transmission through free-standing samples with a thickness of $\sim 0.1\,\mu\mathrm{m}$, this technique, like optical absorption spectroscopy, is not surface sensitive.

## 2 Graphite and $C_{60}$

As an illustration of results which can be derived using techniques described in the previous section, we present now some spectroscopic results on quasi two-dimensional graphite and the quasi zero-dimensional fullerene $C_{60}$ (for a recent review on fullerenes, see [9]). These results on graphite and $C_{60}$ will help to understand the electron spectroscopy results on the quasi one-dimensional carbon nanotubes.

PES spectra of graphite [10] are shown in Fig. 2. For $E_\mathrm{B} < 7.5$ eV, the spectral weight is mainly caused by the density of occupied $\pi$-states. Here we see a linear increase of intensity with increasing binding energy followed by a maximum at $E_\mathrm{B} \sim 3\,\mathrm{eV}$ in agreement with LDA band structure calculations [12]. The spectral weight at higher binding energies with a peak at 7.5 eV is mainly caused by the density of states of the occupied $\sigma$-bands.

In Fig. 3 we show ARPES and ARIPES data of occupied and unoccupied bands of graphite, respectively, together with LDA band-structure calculations [1]. We focus our discussion primarily on the $\pi$-bands. A simple tight-binding calculation of a single graphene sheet yields a total width of the $\pi$-bands of $6\gamma_0$ were $\gamma_0$ is the $pp\pi$ hopping integral between two carbon sites.

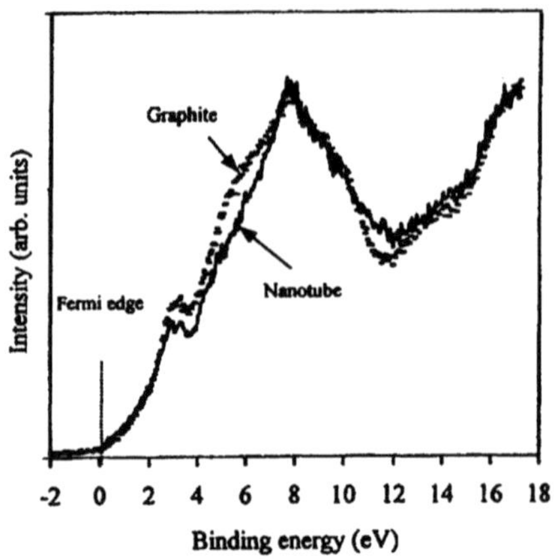


**Fig. 2.** Photoemission spectra of graphite and annealed multiwall carbon nanotubes recorded with a photon energy of 40.8 eV. (from [10])

This calculation also yields a flat band region near the $M$-point in the Brillouin zone, near $E_\mathrm{B} = \pm\gamma_0$ for both the occupied $\pi$ and unoccupied $\pi^*$ states. In a real graphite crystal, there is a small splitting of the bands due to the weak van der Waals interaction between the graphene layers. The flat band region causes a maximum (van Hove singularity) in the density of states [12], which is also detected in the PES spectrum at $E_\mathrm{B} \approx 3\,\mathrm{eV}$ (see Fig. 2). The binding energy of this maximum, the energy separation of the two flat band regions $E = 5\,\mathrm{eV} = 2\gamma_0$ and the total width of the $\pi$ bands $W = 15\,\mathrm{eV} = 6\gamma_0$ yields a hopping integral for the $\pi$ electron between C atoms of $\gamma_0 \sim 2.5\,\mathrm{eV}$. However, a somewhat larger value, 3.1 eV, is needed to reproduce the fine details of the graphite band structure [13]. The gap between the $\sigma$ and the $\sigma^*$ band is about 9 eV and the total width of the $\sigma$ bands is of the order of 40 eV.

Figure 4 displays a PES spectrum of solid $C_{60}$, which is quite different from that of graphite [14]. What is striking is the sharp and well separated features which correspond to the highly degenerate molecular levels of $C_{60}$ [15]. Since the interaction between the $C_{60}$ molecules is about 50 times smaller than the intra-ball hopping integral, the broadening of the molecular levels in the solid is rather small. This is in agreement with band structure calculations which yield a width of the molecular-level derived bands of about 0.5 eV [16]. A further broadening observed in the PES spectra is due to excitations of phonons and probably also due to a multiplet splitting. The features for $E_\mathrm{B} < 5\,\mathrm{eV}$ can be interpreted by molecular levels having predominantly $\pi$ character. At higher binding energies, also $\sigma$ molecular orbitals contribute to the spectral weight. The comparison of the PES spectra of graphite and $C_{60}$ clearly demonstrates the difference between a quasi two-dimensional and a quasi zero-dimensional system. In the former case, wide $\pi$-bands are observed, while in the latter, the density of states is dominated by only slightly broadened molecular levels.

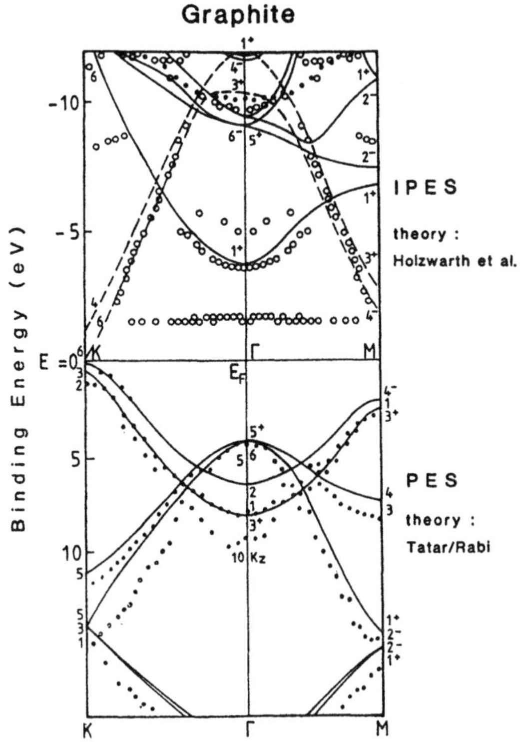


**Fig. 3.** ARPES and IPES of graphite. The IPES data have been obtained from highly oriented pyrolytic graphite. Therefore, the $\Gamma M$ and $\Gamma K$ directions are indistinguishable. *Solid line:* band-structure calculations [11]. (from [1])

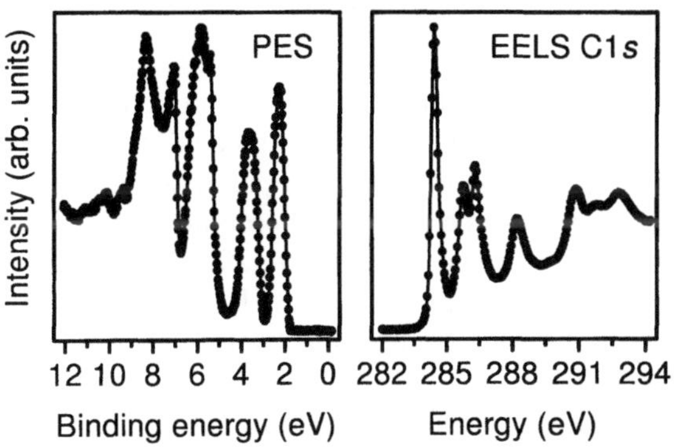


**Fig. 4.** *Left-hand panel:* Photoelectron spectrum of solid $C_{60}$. The excitation energy was $h\nu = 21.2\,\text{eV}$. *Right-hand panel:* C 1*s* excitation spectrum of solid $C_{60}$ measured with EELS (from [14])

In Fig. 5 we show XES spectra of graphite and $C_{60}$, which also should provide information on the occupied density of states [17]. The shape of the XES spectrum of graphite is rather similar to that of the PES spectrum shown in Fig. 4. For solid $C_{60}$, more pronounced features appear near 280 and 282 eV

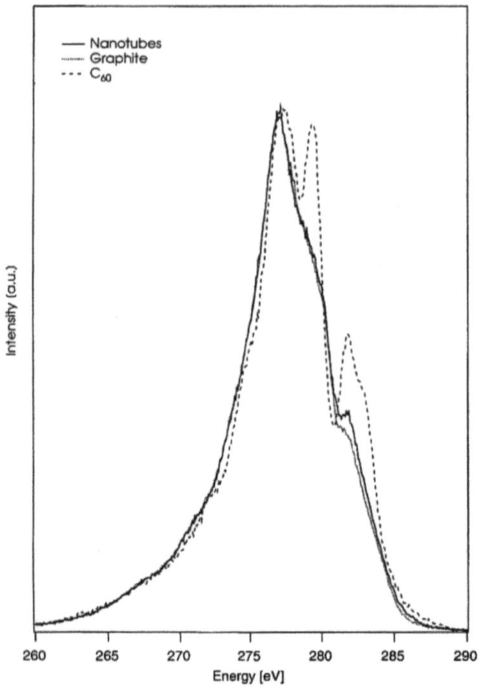


**Fig. 5.** XES of carbon nanotubes (*heavy line*), graphite (*light line*) and solid $C_{60}$ (*dotted curve*). (from [17])

due to the quasi molecular electronic structure of this material. This is also observed for the unoccupied states which are probed by transitions from the C 1*s* core level to the unoccupied states.

An EELS spectrum of graphite is presented in Fig. 6f. One maximum is observed at 285 eV, corresponding to transitions into unoccupied $\pi^*$ states [18]. The width of this resonance is considerably reduced compared to the width of the unoccupied $\pi^*$ band. This comes from an excitonic enhancement of the spectral weight at the bottom of the $\pi^*$ band [19]. Above 291 eV, core-level excitations to the unoccupied $\sigma^*$ bands take place. In $C_{60}$ (Fig. 4, right-hand panel) [14], four features corresponding to unoccupied $\pi^*$ molecular levels are present below the $\sigma^*$ onset at 291 eV.

Finally we come to valence band excitations recorded by EELS. In Fig. 7 (left figure) we show the loss function (3) of graphite together with the real and the imaginary part of the dielectric function, $\varepsilon_1(\omega, \boldsymbol{q})$ and $\varepsilon_2(\omega, \boldsymbol{q})$, derived from a Kramers–Kronig analysis [5]. The data were taken at a small wave vector parallel to the graphene sheets (small scattering angle), $q = 0.1\,\text{Å}^{-1}$, which is much smaller than the Brillouin zone. In this case the data are comparable to optical data derived previously from reflectivity measurements [20]. In this geometry, the in-plane component of the graphite dielectric tensor is probed. Its imaginary part, $\varepsilon_2$, which is related to absorption, shows a Drude-like tail at low energy due to the small concentration of free carri-

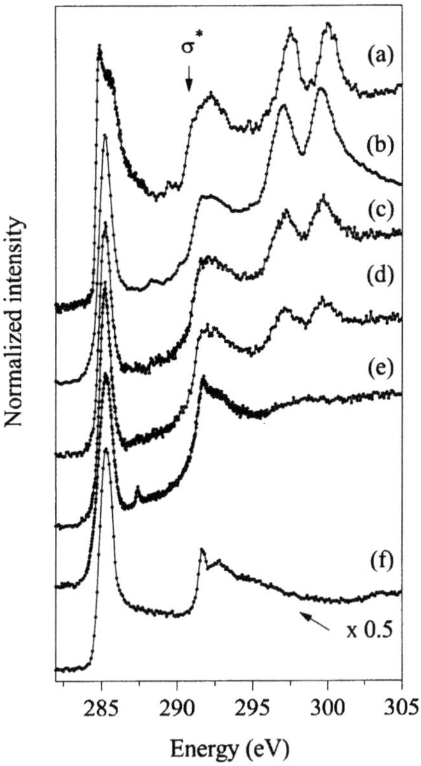


**Fig. 6.** C 1$s$ and K 2$p$ excitation spectra of: (**a**) intercalated graphite $KC_8$; intercalated SWNTs with C/K ratio of (**b**) 7$\pm$ 1; (**c**) 16 $\pm$ 2, and (**d**) 34 $\pm$ 5; (**e**) pristine SWNTs, and (**f**) graphite. (from [18])

ers (electrons and holes). The tail is followed by an oscillator resonance at 4 eV. The latter corresponds predominantly to a $\pi$–$\pi^*$ transition between the flat band regions at the $M$-point [21]. The second peak in $\varepsilon_2$ near 12 eV is predominantly caused by $\sigma$–$\sigma^*$ transitions. The $\pi$-resonance at 4 eV causes a zero-crossing of $\varepsilon_1$ near 6 eV where $\varepsilon_2$ is small and, therefore, the loss function (3) shows a maximum there, i.e. a plasmon is observed. Since this plasmon is related to a $\pi$–$\pi^*$ interband transition, we call this plasmon an interband plasmon or a $\pi$-plasmon. When turning to higher wave vectors, i.e., going from vertical transitions to non-vertical transitions, the energy of the transitions from $\pi$ to $\pi^*$ states increases and therefore the $\pi$-plasmon shows dispersions to higher energies.

The second peak in the loss function of graphite at 27 eV is caused by the zero-crossing of $\varepsilon_1$ near 25 eV. This zero-crossing is related both to the number of valence electrons and to the energy of the $\pi$–$\pi^*$ and, predominantly, of the $\sigma$–$\sigma^*$ transitions. Since this plasmon involves all the valence electrons, it is called the $\pi+\sigma$ plasmon. Using a simple Lorentzian model for the dielectric function,

$$\varepsilon(\omega) = 1 + \frac{\omega_{\mathrm{p}}^2}{\omega_0^2 - \omega^2 - \mathrm{i}\gamma\omega} \tag{4}$$

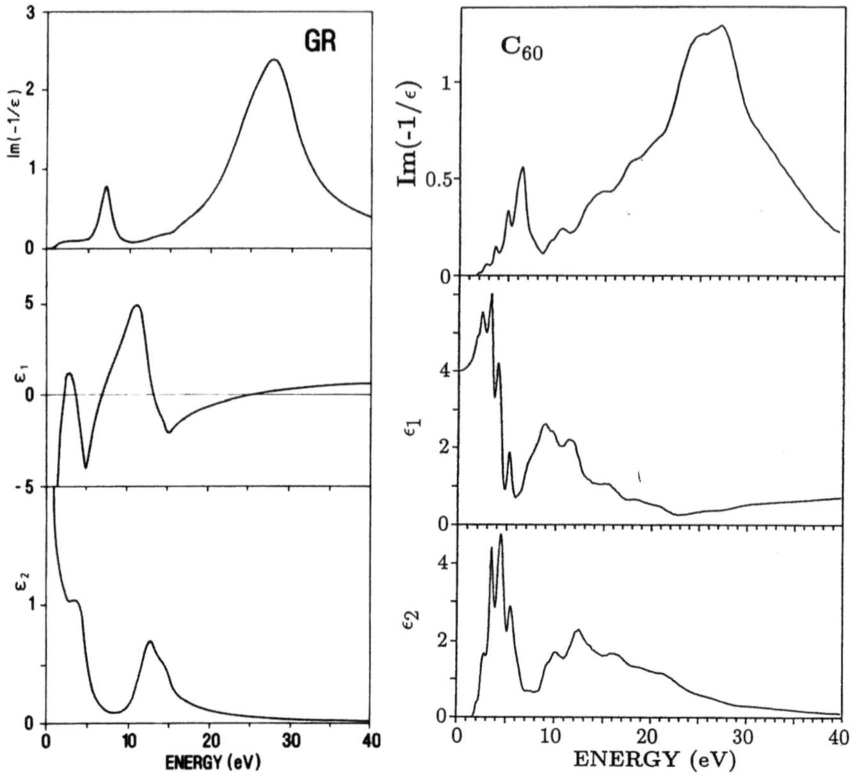


**Fig. 7.** Electron energy loss function $\text{Im}[-1/\varepsilon]$, and the real ($\varepsilon_1$) and imaginary ($\varepsilon_2$) parts of the dielectric function of graphite and solid $C_{60}$. (from [5] and [22])

yields for the plasmon energy $E'_\text{p} = \sqrt{E_0^2 + E_\text{p}^2}$ where $E_0 = \hbar\omega_0$ is the oscillator energy, $\gamma$ is a damping frequency, and $E_\text{p} = \hbar\sqrt{ne^2/m\varepsilon_0}$ is the free-electron plasmon energy, with the valence electron density $n$ and effective mass $m$. This relation clearly shows that the dispersion of the $\pi + \sigma$ plasmon as a function of the wave vector is not only determined by the free electron plasmon dispersion $E_\text{p}(q) = E_\text{p}(0) + \alpha\hbar^2 q^2/m$ with $\alpha = (3/5)E_\text{F}/E_\text{p}(0)$, but also by the wave-vector dependence of the $\sigma$ and the $\pi$ oscillators, i.e., mainly the $\sigma$–$\sigma^*$ transition.

For solid $C_{60}$ the loss function (see Fig. 7, right-hand figure) [22] shows differences when compared to graphite. There is a gap of 1.8 eV followed by several $\pi$–$\pi^*$ transitions between the well-separated molecular levels having $\pi$-electron character. Following the same argumentation as for graphite, these $\pi$ oscillators now cause several $\pi$ plasmons in the energy range 1.8 to 6 eV, although $\varepsilon_1$ does not vanish in that interval. The last peak at 5 eV has the highest intensity because $\varepsilon_2$ is smaller there. In addition, there are several $\sigma$–$\sigma^*$ transitions which cause an almost zero-crossing of $\varepsilon_1$ near 22 eV, leading to a wide maximum in the loss function, i.e. the $\pi + \sigma$-plasmon [23]. As mentioned before, the molecular orbital levels are only slightly broadened by the interaction between the $C_{60}$ molecules, which implies that the $\pi$-electrons are strongly localized on the molecules. Therefore, there is almost no wave-vector dependence of the energy of the $\pi$–$\pi^*$ transitions, and consequently no dispersion of the $\pi$-plasmons is detected [22].

# 3 Occupied States of Carbon Nanotubes

To illustrate the problems of the surface contamination encountered in PES spectroscopy of carbon nanotubes, we show in Fig. 8 wide-range X-ray induced PES [24] of a film of purified SWNTs. The dominant feature at 284.5 eV corresponds to the C 1*s* level. The binding energy is close to values observed in graphite and fullerenes. In addition, strong contaminations with O, N and Na are detected which show up by the O 1*s*, O:KLL (Auger peak), N 1*s* and Na 1*s* peaks, respectively. After annealing the sample in UHV at 1000°C, most of these contaminants have disappeared. However, small Ni and Co contamination from the catalyst remains, as indicated by the low intensity Ni 2*p* and Co 2*p* lines.

In Fig. 9 we show PES spectra of annealed SWNT films taken with the photon energies $h\nu = 21.2$ and 40.8 eV. For the higher energy photons, the PES spectrum is quite similar to the graphite spectrum shown in Fig. 2. A peak is observed at $E_B = 2.9$ eV which can be assigned to a large density of states of the $\pi$ states near $E_B = \gamma_0$ (Fig. 11). A further peak is observed at $E_B = 7.5$ eV caused by a large density of $\sigma$ states. For $h\nu = 21.2$ eV, the C 2*s* states at lower $E_B$ are more pronounced.

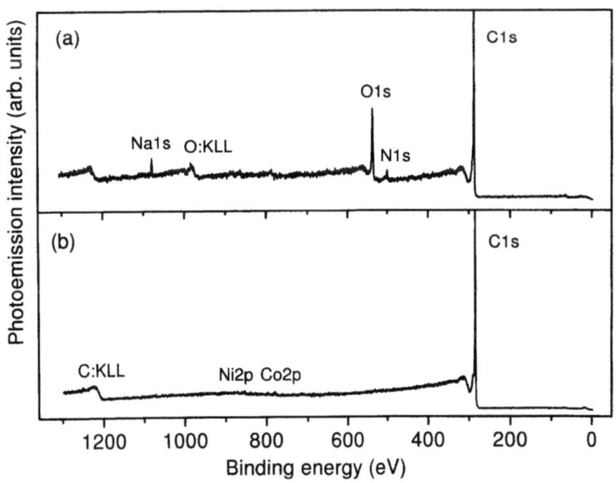


**Fig. 8.** X-ray induced photoelectron spectra of purified single-wall carbon nanotubes (**a**) without annealing and (**b**) after annealing at 1000°C (from [24])

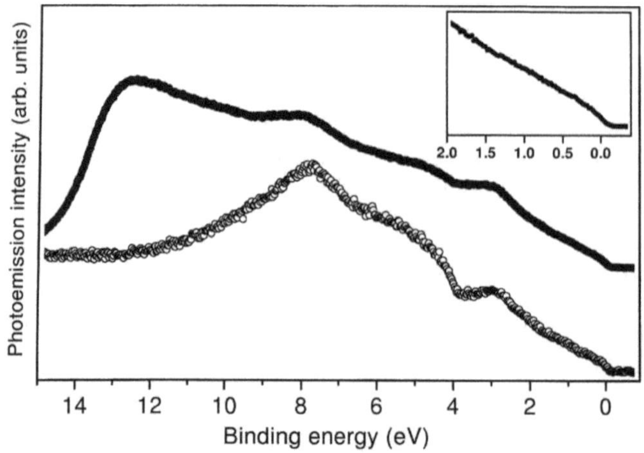


**Fig. 9.** Photoelectron spectra of annealed SWNT using a photon energy of $h\nu = 21.2\,\mathrm{eV}$ (•, shifted vertically for more clarity) and 40.8 eV (○). The *inset* gives results for low binding energies on an expanded scale (from [24])

Of particular interest is the question whether a Fermi edge can be observed from the metallic tubes. For one-dimensional systems, important deviations from Fermi liquid theory or even its breakdown is expected. Theoretical works predict a Luttinger liquid model in which the spectral weight close to the Fermi level should be suppressed [25]. Recent transport measurements indicate a Luttinger-liquid behavior in metallic SWNTs [26] and MWNTs as well [27]. The parameters derived from these measurements would lead to a considerable suppression of the spectral weight near $E_F$. In the experiment, a clear Fermi edge is observed which, however, is probably due to small metallic catalyst particles (Ni and Co) that are also detected in the core level spectra. Further information on the occupied density of states of SWNTs is obtained from XES spectra, which are shown in Fig. 5, along with the spectra for $C_{60}$ and micro-crystalline graphite for comparison [17]. The overall shape of the spectral profile is intermediate between graphite and $C_{60}$, in particular close to 282 eV. A closer inspections, however, yields better agreement between graphite and SWNTs than between $C_{60}$ and SWNTs. This observation indicates that the nanotubes are closer to graphite than to $C_{60}$ fullerite.

Some information on the band dispersion in SWNTs can be obtained from RIXS [17]. The corresponding spectra are shown in Fig. 10. In the spectra excited at 285 eV (bottom curve), the broad feature at 270 eV emission energy, labeled B, is due to emission from the lowest two occupied bands at the $K$ point, corresponding to states around 13 eV below $E_F$ in the band structure shown in Fig. 3. The reason for this assignment comes from the fact that the lowest excited states just above $E_F$, which can be reached with photons of 285 eV, are also located at the $K$ point, and because the wave vector is conserved. Similarly, the peak labeled A can be assigned to occupied $\pi$-bands

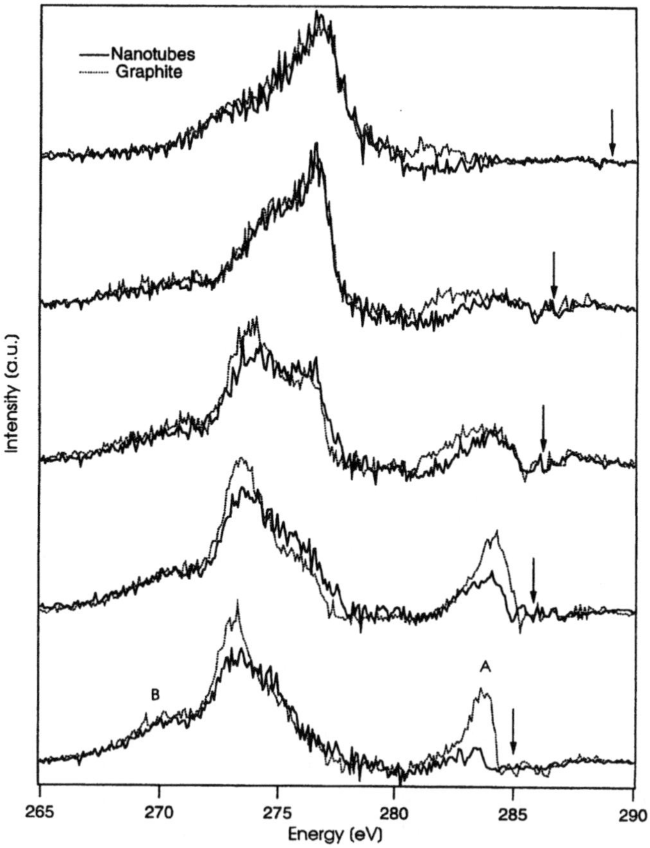


**Fig. 10.** Resonant inelastic X-ray scattering spectra of single-wall carbon nanotubes (*heavy line*) and non-oriented graphite (*light line*). The photon energies (*indicated by the arrows*) used for the excitation are 285.0, 285.8, 286.2, 286.6 and 289.0 eV (*from bottom to top*). The raw data were corrected for the incoherent scattering and diffuse reflection (from [17])

close to the $K$ point. With increasing excitation energy the feature A moves to lower energy both for graphite and SWNTs, which is expected for a dispersive $\pi$-band (Fig. 3). Regarding the structure of SWNTs, it is not surprising that the occupied density of states and the data for the $\pi$-band dispersion of SWNTs is very close to that of graphite. Each SWNT can be thought of as a single graphene layer that has been wrapped into a cylinder with a diameter of about 1.5 nm.

In Fig. 2 we show PES data for annealed MWNTs together with data for graphite. Again the difference between the two materials is rather small. Since the MWNTs consist of several graphene layers wrapped up into concentric layers, this close similarity of the electronic structure of the two systems is expected.

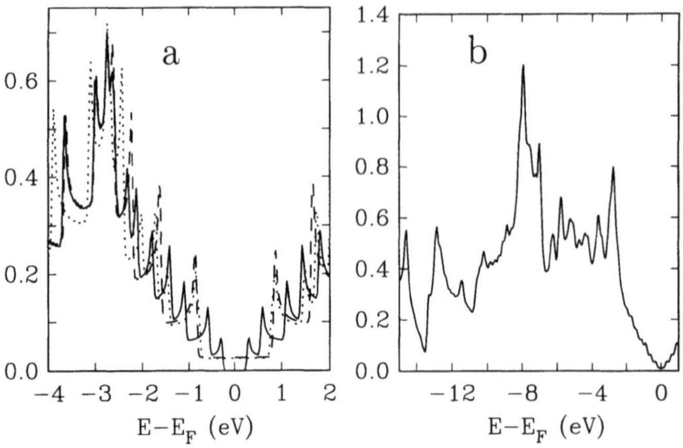


**Fig. 11.** (**a**) Calculated electronic density of states (states/eV/atom) of three different nanotubes derived from tight binding ($\gamma_0 = 2.75\,\mathrm{eV}$): (12,8) (*full curve*), (10,10) (*dashed curve*) and (16,4) (*dotted curve*). (**b**) Arithmetic average of the densities of states of the 7 nanotubes with indices (10,10), (11,9), (12,8), (13,7), (14,6), (15,5), and (16,4). These calculations involve both $\pi$ and $\sigma$ states [30]

On the other hand, a closer look at the structure would indicate differences in the electronic structure between graphite and carbon nanotubes. Firstly, there is a quantum confinement effect perpendicular to the axis of the tube. The wave function of the rolled-up graphene sheet has to satisfy periodic boundary conditions around the circumference. Hence the component of the Bloch wave vector perpendicular to the axis, $k_\perp$, can only assume discrete values. This means also that discrete energy values for the $k_\perp$ values are allowed [28]. Therefore, the electronic structure might be regarded as molecular-like in the circumferential direction. This molecular electronic structure for $\boldsymbol{k}$ perpendicular to the axis leads to van Hove singularities in the density of states [29], not encountered in that of graphite. The calculated density of states of the semiconducting (12,8) and the metallic (10,10) and (16,4) nanotubes are shown in Fig. 11. The above-mentioned van Hove singularities are clearly recognized.

Regarding the PES data shown in Fig. 9 that were recorded with an energy resolution of 0.1 eV, one does not see the expected occupied van Hove singularities that should appear near 0.3, 0.6 and 1.2 eV for the semiconducting SWNTs, and at 0.9 eV for the metallic tubes. These singularities have clearly been observed on individual nanotubes by scanning tunneling spectroscopy [31,32] and also through resonant Raman scattering [33,34]. The sample used in Fig. 9 comprised different kinds of SWNTs, including a distribution of diameters and chiralities. It is now well established from various diffraction experiments that a rope of SWNTs can mix tubes with different helicities [35], and the nanotube diameters may vary from one rope to the other within some limits [36]. Taking an average of the densities of states of different nanotubes brings them closer to graphite. This effect is clearly shown in Fig. 11b, which represents the arithmetic mean of the densities of states of 7 nanotubes with wrapping indices (10,10), (11,9) ... (15,5), (16,4) and diameters 1.4 $\pm$0.03 nm. The averaging process and the small broadening of the densities of states used in the calculations (0.1 eV) washed out most of the one-dimensional van Hove singularities of the individual nanotubes. In the experiment, the peaks are possibly broadened by phonon excitations, as is also observed in PES spectra of fullerenes. Defects and correlation effects may also broaden the spectral weights related to these singularities.

In MWNTs, which have diameters of the order of 10 nm, the van Hove singularities should appear at considerably lower energy [27]. In this case it is clear that they cannot be detected in the PES spectrum. In effect, Fig. 2 show experimental PES data of annealed MWNTs together with data of graphite. Again the difference between the two materials is rather small. Since the MWNTs consist of several graphene layers wrapped up into concentric layers, this close similarity between the electronic structures of the two systems is expected.

## 4 Unoccupied States of Carbon Nanotubes

There is only little information presently available from electron spectroscopies on the unoccupied density of states of carbon nanotubes. Only core level C 1$s$ EELS [18] and XAS [17] data on SWNTs have been reported. In Fig. 6e we show C 1$s$ excitation spectra recorded by EELS in transmission. As in graphite (see Fig. 6f) the first peak at 285 eV corresponds to transitions into unoccupied $\pi^*$-states while above 292 eV mainly unoccupied $\sigma^*$-states are detected. Since in graphite the maximum of the density of states 2.5 eV above $E_F$ is strongly shifted to lower energy due to the interaction with the core hole, the maximum appears at 1 eV above threshold. Probably the same happens in the SWNT spectra. The unoccupied van Hove singularities should appear at 0.3, 0.6 and 1.2 eV above threshold for the semiconductors, and at 0.9 eV above threshold for the metallic SWNTs. However, in this energy range the spectral weight is dominated by the peaked $\pi^*$ density of states, which is shifted to lower energies. Therefore, no details of the unoccupied density of states near the Fermi level could be detected in the EELS spectra, although the energy resolution in these experiments was 0.1 eV. The origin of the small peak at 287 eV which does not appear in all samples, is at present unclear.

## 5 Excited States on Carbon Nanotubes

In this section we review valence-band excitations recorded by momentum dependent EELS measurements on transmission [37]. In Fig. 12 we show the loss functions of purified SWNTs for various momentum transfers, $q$. The

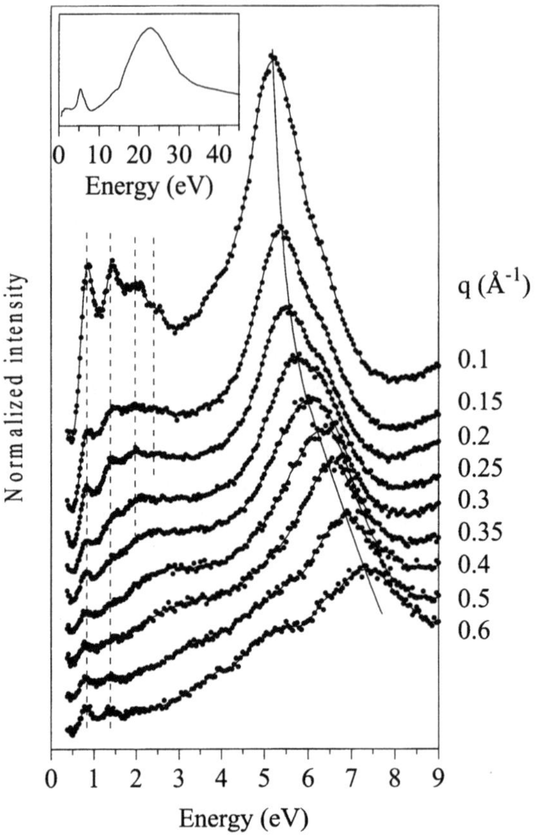


**Fig. 12.** The magnitude of the loss function of purified single-wall carbon nanotubes for various momentum transfers $q$. The inset shows the loss function over a larger energy range for $q = 0.15\,\text{Å}^{-1}$ (from [37])

wide energy-range loss function for $q = 0.15\,\text{Å}^{-1}$ is shown in the insert. Similar to graphite, a $\pi$-plasmon is detected at 5.2 eV and a $\pi+\sigma$-plasmon appears at 21.5 eV. The same plasmon energies have been reported from EELS spectra recorded using a transmission electron microscope [38]. The reason for the lower energy of the $\pi + \sigma$-plasmon compared to that of graphite partly comes from the fact that the carbon density, and therefore the electron density which determines the free electron plasmon energy, $E_\text{p}$, is smaller (see Sect. 2). But also, unlike the case of planar graphite, the wave-vector transfer $\boldsymbol{q}$ is tangential to some parts of the nanotube and normal to other parts, simply due to curvature. The anisotropy of the graphene layer is responsible for a down shift of the plasmon energy as compared to graphite. This interpretation is supported by theoretical calculations of EELS spectra based on atomic, discrete-dipole excitations in hyper-fullerenes [39]. The influence of the layer anisotropy on the plasmon excitation spectrum has clearly been demonstrated in the case of multi-shell systems [40,41].

Coming back to (1), the dielectric function of the nanotube sample is best defined for a crystalline arrangement of identical tubes, such as realized in a close-packed rope. A Lindhardt-like formula for $\varepsilon(\omega, \boldsymbol{q})$ can then be determined and the plasmons come out of the calculations as the zeros of the real part of it. Such calculations have been performed for the $\pi$ electrons described by a tight-binding Hamiltonian, while ignoring the inter-tube coupling [42,43]. When the electric field is parallel to the nanotube axis, the real part of the long-wavelength dielectric function, $\varepsilon_1(\omega)$, is shown to vanish for a special energy $\hbar\omega$ close to $2\gamma_0 \approx 6$ eV [42]. The corresponding longitudinal mode is the $\pi$-plasmon. Its excitation energy is independent on the tube diameter and chirality [44].

The momentum-dependent measurements (Fig. 12) show a strong dispersion of the $\pi$-plasmon, similar to that of graphite. As discussed in Sect. 2, this plasmon is related to a $\pi$–$\pi^*$ interband transition near 5 eV, and therefore the dispersion indicates dispersive bands of delocalized $\pi$-electrons as in graphite. This dispersion can only occur along the SWNT axis and therefore the $\pi$-plasmon at 5.2 eV is related to a collective oscillation of $\pi$-electrons along the tube. On the other hand, the low-energy peaks at 0.85, 1.45, 2.0, and 2.55 eV show no dispersion as a function of the momentum transfer, similar to the $\pi$-plasmons in solid $C_{60}$ (see the discussion in Sect. 2). This indicates that the low-energy maxima are related to excitations of *localized* electrons. It is tempting to attribute these excitations to collective excitations around the circumference of the tubes. The interband transitions which cause these peaks are then related to transitions between the van Hove singularities shown in Fig. 11a. In this view, the lowest two peaks appear as due to semiconducting tubes while that at 2.0 eV can be assigned to metallic tubes. This is also supported by regarding the data of $\varepsilon_2$ derived from Kramers-Kronig analysis (not shown). Peaks are detected at 0.65, 1.2 and 1.8 eV, which are close to the energy distance between van Hove singularities of semiconducting and metallic SWNTs in the tube diameter range relevant here. Measurements of the occupied and unoccupied density of states by scanning tunneling spectroscopy are in agreement with these results [31,32]. From the intensity of the transitions, one can immediately conclude that about two third of the tubes are semiconducting while one third is metallic. Later on, transitions between the van Hove singularities have also been observed at 0.68, 1.2 and 1.8 eV with higher energy-resolution by optical absorption measurements [45,46], which agree very well with the positions of the $\varepsilon_2(\omega)$ peaks.

The $\varepsilon_2$ data (not shown) or the optical absorption data are in a first approximation a measure of the joint density of states averaged over all existing SWNT structures existing in the sample. From band-structure calculations, it was predicted that the gap in the semiconducting samples or generally the energy distances between the van Hove singularities are inversely proportional to the diameter of the SWNTs [47,48]. This was confirmed by scanning tunneling microscopy and spectroscopy measurements [31,32,49]. Then the energies of the transitions yield information on the mean diameter of the tubes in the sample. The fact that well-defined peaks are observed in the loss function or in the optical absorption indicates a narrow diameter distribution, as otherwise the energetically different interband transitions would wash out the maxima like in Fig. 11b.

On the other hand, recent high-resolution optical measurements have detected a fine structure of the absorption lines [46]. This is illustrated in Fig. 13 where typical optical absorption data of the first three low-energy lines are shown as a function of the synthesis temperature. Peaks A and B correspond again to transitions between van Hove singularities in semiconducting tubes, while peak C corresponds to an excitation of metallic tubes. With increasing growth temperature, the mean energy of the three peaks are shifted to lower energies, i.e. the mean diameter of the tubes increases with increasing synthesis temperature. Moreover, the peaks clearly show a fine structure which is more pronounced in the high-energy peaks (C). The energy position of the vast majority of the sub peaks remained constant within the resolution

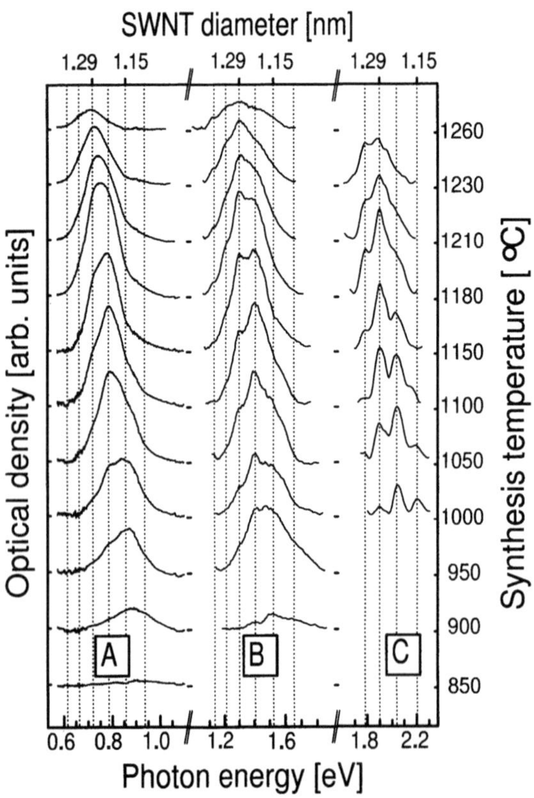


**Fig. 13.** Optical absorption peaks of purified single-wall nanotubes as a function of the synthesis temperature, displayed both on an energy and a diameter axis. The background is subtracted. The *dotted lines* indicate the groups of nanotube diameters separated by $\Delta d = 0.07\,\mathrm{nm}$ (from [46])

limit for all synthesis conditions studied. This points to the fact that the investigated material consists of nanotubes with a discrete number of diameters grouped around preferred values independent of variations of the process parameters. Furthermore, it was found that the positions of the sub peaks are equidistantly separated on the diameter scale with nearly the same values of $\Delta d$ for both semiconducting and metallic SWNTs. Such an equivalent spacing between nanotube diameters ($\Delta d \approx 0.07\,\text{nm}$), common to both semiconducting and metallic SWNTs, can be realized for nanotubes close to the $(n, n)$ armchair geometry. Provided there are no selection rules to favor special helicities, this observation indicates a preferred formation of SWNTs in the vicinity to the armchair configuration. This result certainly needs further investigations, since recent electron-diffraction experiments have revealed a uniform distribution of helicity in the ropes [35], except perhaps on rare occasions where a preferred armchair arrangement was detected [50]. If confirmed, the analysis discussed here clearly shows that the energy of the excited states, in particular the interband excitations between van Hove singularities, first detected by EELS measurements [37] yields valuable information not only on the mean diameter but also on the helicity of SWNT material.

In the following we focus again on the dispersion data of the plasmons of SWNTs which are summarized in Fig. 14, together with similar data on graphite. The dispersions of the $\pi$- and the $\pi + \sigma$-plasmons of both materials look similar. Graphite is a three-dimensional solid, since the Coulomb interplane interaction is comparable to the intraplanar interaction. As for SWNTs, the individual tubes form bundles having diameters of about 10 nm. Up to 100 SWNTs are located on an hexagonal lattice in a section perpendicular to the bundle, and the distance between the tubes is comparable to the inter-sheet distance in graphite. Such a rope can be described as an effective bulk medium in the spirit of, for instance, the Maxwell–Garnett theory [51]. With such an effective dielectric function at hand, (1) predicts a plasmon dispersion typical of a three-dimensional material. This explains why the quasi two-dimensional graphite system and the quasi one-dimensional SWNT systems show a dispersion like in three-dimensional solids.

On an isolated nanotube, by contrast, theoretical calculations [52,53] predict a plasmon frequency which vanishes with decreasing wave vector, but only for this mode that has full rotational symmetry, whereas in three-dimensions, $\hbar\omega(q)$ approaches a finite value at zero $q$. The $\pi$- and $\sigma$-plasmons in an isolated MWNT are also predicted to behave the same way, however, with a crossover from 1D (low $q$ corresponding to a wavelength larger than the nanotube diameter) to 3D (large $q$) [54]. Strictly speaking, this acoustic behavior applies to free-electron like materials, which the undoped nanotubes are not. In the tight-binding picture, the $\pi$-plasmon of a SWNT has a finite energy at $q = 0$, of the order of $2\gamma_0$ as mentioned above. This optical-like character of the nanotube plasmon is further reinforced in a bundle [55]. Here, the plasmon frequency depends on both the parallel and perpendicular

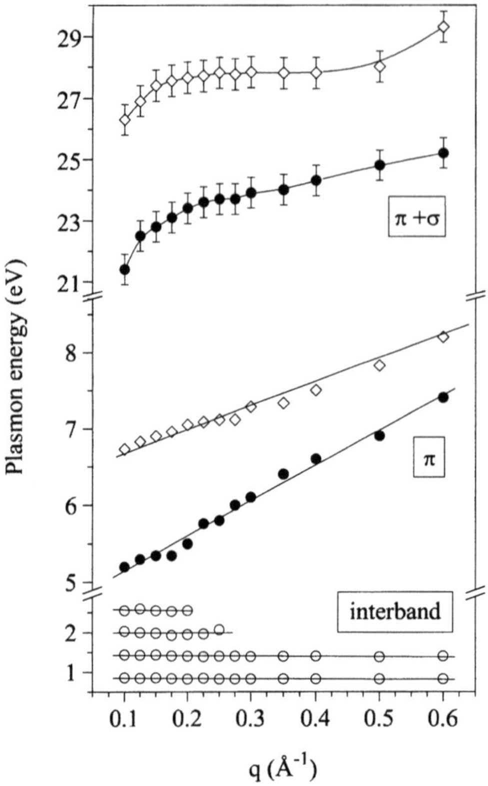


**Fig. 14.** Dispersion of the $\pi$-plasmon, the $\pi + \sigma$-plasmon (•), and of the excitations due to interband transitions between van Hove singularities (∘) of purified single-wall carbon nanotubes. For comparison, the dispersion of the $\pi$- and $\sigma$- plasmons in graphite for momentum transfers parallel to the planes is included (⋄) (from [37])

components of the transferred wave vector, and it decreases with decreasing $\sqrt{q_\perp^2 + q_\parallel^2}$, as observed experimentally.

At the end of this section we discuss studies of the plasmons of *MWNTs* with an inner diameter of 2.5–3 nm and an outer diameter of 10–12 nm, thus consisting of 10–14 layers. In Fig. 15 we show the $\pi$-plasmon for various momentum transfers [56,57]. In the inset, the loss function for $q = 0.1\,\text{Å}^{-1}$ is shown in a larger energy range. For $q = 0.1\,\text{Å}^{-1}$, the $\pi$-plasmon is near 6 eV while the $\pi+\sigma$-plasmon is near 22 eV. No low-energy excitations are observed due to the larger tube diameter and the corresponding smaller energy of the van Hove singularities. For $q > 0.1\,\text{Å}^{-1}$, the plasmon energies are intermediate between those of graphite and SWNTs. For small momentum transfers, the $\pi + \sigma$-plasmon of the MWNTs shows a decrease in energy, as in the SWNTs. Contrary to the EELS behavior in SWNTs, a strongly decreasing $\pi$-plasmon energy is observed in this momentum range, which could be explained by a transition from three-dimensional behavior to one-dimensional behavior with decreasing momentum transfer, as already discussed before.

Recently, EELS spectra of an isolated MWNT have been obtained in a scanning transmission electron microscope, in a well-defined geometry [58,59].

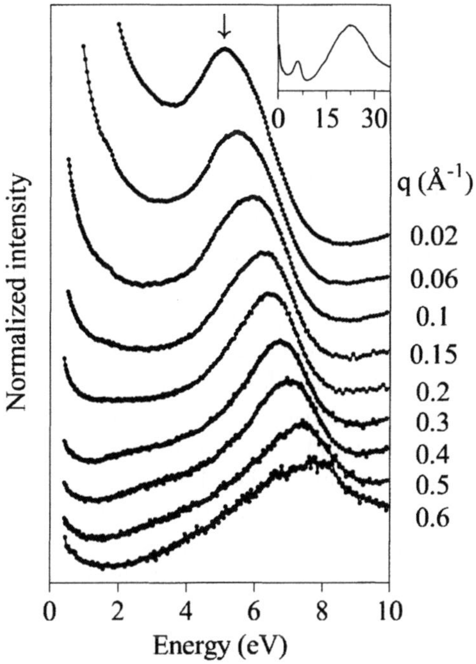


**Fig. 15.** Plot of the normalized EELS intensity vs energy showing the dispersion of the $\pi$-plasmon (*indicated by the arrow*) of annealed multi-wall carbon nanotubes for various values of momentum transfers. The *inset* shows the loss function over a wider energy range (from [56] and [57])

Here an electron beam of nanometer diameter is used in near-perpendicular orientation with respect to the nanotube axis. The EELS spectrum is recorded as a function of the impact parameter and the evolution of the loss peaks is traced against this parameter. If, in the experiments, the $\pi$-plasmon remained near 6 eV, insensitive to the impact parameter, the $\pi+\sigma$-plasmon shifted from 27 eV (graphite like) for a near-zero impact parameter to about 23 eV when the beam came closer to a nanotube edge. The interpretation of the shift is that a beam passing through the center of the nanotube probes the in-plane component of the dielectric function of graphite, whereas the off-center beam probes the out-of-plane component [44]. The real parts of the corresponding dielectric components of graphite do not vanish at the same energy, and therefore there is a shift of the plasmon position with increasing impact parameter. This is an effect of the graphite dielectric anisotropy mentioned above.

## 6 Intercalated Single-Wall Nanotubes

Right after the discovery of SWNTs it was clear that one should try to intercalate these materials, in analogy with the well-known examples of Graphite Intercalation Compounds (GICs) and of $C_{60}$. Consequently, a decrease of the resistivity by one order of magnitude has been detected when SWNTs were exposed to potassium (or bromine) vapors [60]. The intercalation of the bundles by K can be followed by electron diffraction [18]. Upon successive intercalation, the first Bragg peak, the so-called rope-lattice peak, characteristic of the nanotube triangular lattice, shifts to lower momentum. This is consistent with an expansion of the inter-nanotube spacings concomitant with intercalation in between the SWNTs in the bundle.

The maximal intercalation can be derived from the intensity of the C 1$s$ and K 2$p$ excitations recorded by EELS [18] (see Fig. 6). A comparison with data of $KC_6$ GICs also shown in Fig. 6 yields the highest concentration of C/K $\approx$ 7 for SWNTs intercalated to saturation, which is essentially a similar value as for stage I GIC $KC_8$. Like in K-intercalated graphite compounds, the filling of unoccupied $\pi^*$-states by K 4$s$ electrons cannot clearly be detected in the C 1$s$ spectra (see Fig. 6) [61], since excitonic effects yield deviations of the measured spectral weight compared to the density of states (see Sect. 2).

To discuss the changes of the conduction-band structure upon intercalation, we show in Fig. 16 the loss function of K-intercalated SWNTs for two different concentrations together with that of GIC $KC_8$ [61]. For the latter compound, besides the bound $\pi$-plasmon, a charge carrier plasmon due to the filled $\pi^*$ bands is observed at about 2.5 eV. In the case of the interca-

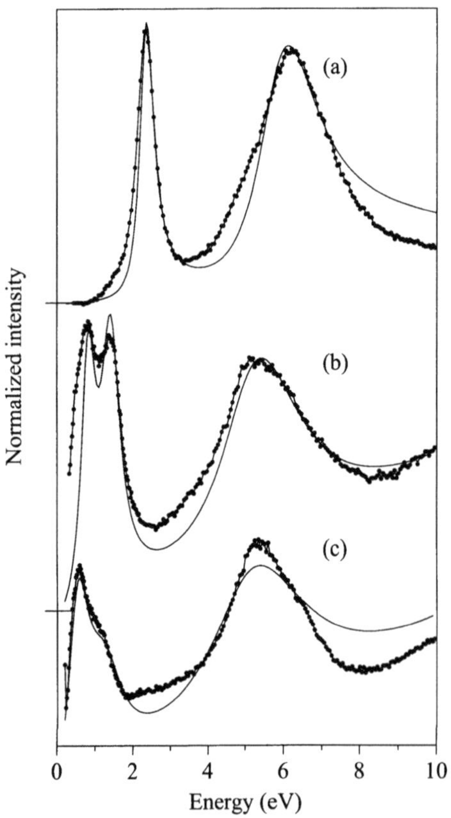


**Fig. 16.** Loss function at a momentum transfer of 0.15 Å$^{-1}$ for (**a**) GIC $KC_8$; intercalated SWNTs with C/K ratios (**b**) 7 ± 1 and (**c**) 16 ± 2. The *solid lines* represent a fit of a Drude–Lorentz model (from [18])

lated SWNTs, the low-energy features cannot be satisfactorily described by a charge carrier plasmon alone. The introduction of an additional interband excitation located at 1.2 eV is necessary. Taking into account the previous discussion on undoped SWNTs, this excitation corresponds to the second transition between the van Hove singularities of the semiconducting SWNTs. In this view, the first transition at 0.6 eV has disappeared by the filling of the lowest unoccupied van Hove singularity due to K 4*s* electrons transfered to the nanotubes, which lead to a shift of the Fermi level. A filling of these states and a related disappearance of the corresponding interband transitions upon intercalation has been also observed in optical spectroscopy [34,62,63]. Within a Drude–Lorentz model [(4) generalized to include several oscillators], the low-energy features can be explained by a charge carrier plasmon in addition to the remaining interband transitions between the van Hove singularities, and the $\pi$-plasmon. The unscreened energy of the charge carrier plasmon in $KC_7$ SWNTs is 1.85 eV and 1.2 eV in $KC_{16}$ [18]. This plasmon is responsible for the peaks observed in the loss function near 0.8 and 0.6 eV, respectively. From the data shown in Fig. 16, an effective mass of the charge carriers has been derived that is about 3.5 times that of the GIC $KC_8$. This effect could be considered to follow naturally from the back-folding of the graphene band structure, which occurs on the wrapping of a graphene sheet to form a SWNT. In addition, the damping of the charge carrier plasmon in the fully intercalated bundles of SWNTs is about twice as large as in fully intercalated graphite. This can be ascribed to a considerable disorder in the nanotube system. From the scattering rates an intrinsic dc conductivity of about 1200 S/cm for the fully intercalated SWNTs could be derived. By comparison, a value of 3000 S/cm has been reported from transport measurements in samples doped with K under similar conditions [60].

## 7 Summary

At present, there exist only a limited number of electron spectroscopy studies on carbon nanotubes. The main reason for that is the non existence of good crystalline materials. Nevertheless, electron spectroscopies have considerably contributed to the understanding of the electronic structure of these new carbon-based $\pi$-electron systems. EELS experiments performed on ropes of SWNTs clearly revealed interband transitions between van Hove singularities of the density of states. The intensity of these loss features unambiguously showed that the samples mixed metallic and semiconducting nanotubes in approximate statistical ratio 1:2. The $\pi$- and $\pi + \sigma$-plasmon structure of the nanotubes was shown to be similar to that of graphite with, however, a few differences among which the indication of quasi-one dimensional dispersion, especially in multi-wall tubes. EELS made it possible to measure the concentration of potassium in intercalated bundles. Charge transfer in potassium-doped nanotubes was also clearly demonstrated from low-energy

EELS spectra. In addition, these methods provided information on the diameter distribution and, indirectly, on the helicity of the nanotubes. Much more detailed studies can be performed in the future when well ordered, oriented and fully dense materials will become available, which even may have a unique helicity.

### Acknowledgments

We are grateful to R. Friedlein, M.S. Golden, M. Knupfer and T. Pichler (Dresden), A. Lucas, P. Rudolf and L. Henrard (Namur) for fruitful discussions.

## References

1. S. Hüfner, *Photoelectron Spectroscopy*, (Springer, Berlin, Heidelberg 1995)
2. B. K. Agarwal, *X-Ray Spectroscopy*, (Springer, Berlin, Heidelberg 1979)
3. J. Lüning, J.-E. Rubensson, C. Ellmers, S. Eisebitt, W. Eberhardt, Appl. Phys. B **56**, 13147 (1997)
4. For a recent overview, see the issue Appl. Phys. A **65**, No.2 (1997)
5. J. Fink, Adv. Electron. Electron Phys. **75**, 121 (1989)
6. D. D. Vvedensky, Topics Appl. Phys. **69**, 139 (Springer, Berlin, Heidelberg 1992)
7. H. Raether, *Excitation of Plasmons and Interband Transitions by Electrons*, Springer Tracts Modern Phys. **88** (Springer, Berlin, Heidelberg 1980)
8. Ch. Kittel, *Introduction to Solid State Physics*, 7th edn. (Wiley, New York 1996)
9. P. Rudolf, M. S. Golden, P. A. Brühwiller, J. Electr. Spectrosc. **100**, 409 (1999)
10. P. Chen, X. Wu, X. Sun, J. Liu, W. Ji, K. L. Tan, Phys. Rev. Lett. **82**, 2548 (1999)
11. R. C. Tatar, S. Rabii, Phys. Rev. B **25**, 4126 (1982)
    N. A. W. Holwarth, S. G. Louie, S. Rabii, Phys. Rev. B **26**, 5382 (1982)
12. J. C. Charlier, X. Gonze, J. P. Michenaud, Phys. Rev. Lett. **43**, 4579 (1991)
13. W. W. Toy, M. S. Dresselhaus, G. Dresselhaus, Phys. Rev. Lett. **15**, 4077 (1977)
14. M. S. Golden, M. Knupfer, J. Fink, J. F. Armbruster, T. R. Cummins, H. A. Romberg, M. Roth, M. Sing, M. Schmitt, E. Sohmen, J. Phys. Condens. Matter **7**, 8219 (1995)
15. J. H. Weaver, J. L. Martins, T. Komeda, Y. Chen, T. R. Ohno, G. H. Kroll, N. Troullier, R. E. Haufler, R. E. Smalley, Phys. Rev. Lett. **66**, 1741 (1991)
16. S. Saito, A. Oshiyama, Phys. Rev. Lett. **66**, 2637 (1991)
17. S. Eisenbitt, A. Karl, W. Eberhardt, J. E. Fischer, C. Sathe, A. Agni, J. Nordgren, Appl. Phys. A **67**, 89 (1998)
18. T. Pichler, M. Sing, M. Knupfer, M. S. Golden, J. Fink, Solid State Commun. **109**, 721 (1999)
19. E. J. Mele, J. J. Ritsko, Phys. Rev. Lett. **43**, 68 (1979)
20. E. A. Taft, H. R. Phillip, Phys. Rev. **A197**, 138 (1965)
21. E. Tossati, F. Bassani, Nuovo Cimento B **65**, 161 (1970)

22. E. Sohmen, J. Fink, W. Krätschmer, Z. Phys. B-Condensed Matter **86**, 87 (1992)
23. Y. Saito, H. Shinohara A. Ohshita, Japn. J. Appl. Phys. **30**, L1068 (1991)
24. X. Liu, H. Peisert, T. Pichler, M. Knupfer, M. S. Golden, J. Fink, unpublished
25. M. Grioni, J. Voigt, *High-resolution photoemission studies of low-dimensional systems*, in *Electron Spectroscopies Applied to Low-Dimensional Materials*, H. Stanberg, H. Huges (Eds.), (Kluwer Academic, Dordrecht 2000)
26. M. Bockrath, D. H. Cobden, J. Lu. A. G. Rinzler, R. E. Smalley, L. Balents, P. L. Mc Enen, Nature **397**, 598 (1999)
Z. Yao, H. W.Ch. Postma, L. Balents, C. Dekker, Nature **402**, 273 (1999)
27. C. Schönenberger, A. Bachtold, C. Strunk, J. P. Salvetat, L. Forró, Appl. Phys. A **69**, 283 (1999)
28. N. Hamada, S. I. Sawada, A. Oshiyama, Phys. Rev. Lett. **68**, 1579 (1992)
29. R. Saito, M. Fujita, G. Dresselhaus, M. S. Dresselhaus, Appl. Phys. Lett. **60**, 2204 (1992)
30. J. C. Charlier, Ph. Lambin, Phys. Rev. Lett. **57**, R15037 (1998)
31. J. W. G. Wildöer, L. C. Venema, A. G. Rinzler, R. E. Smalley, C. Dekker, Nature **391**, 59 (1998)
32. T. W. Odom, J. L. Huang, Ph. Kim, Ch. M. Lieber, Nature **391**, 62 (1998)
33. A. M. Rao, E. Richter, S. Bandow, B. Chase, P. C. Eklund, K. A. Williams, S. Fang, K. R. Subbaswamy, M. Menon, A. Thess, R. E. Smalley, G. Dresselhaus, M. S. Dresselhaus, Science **275**, 187 (1997)
34. R. Saito H. Kataura, chapter in this volume
35. L. Henrard, A. Loiseau, C. Journet, P. Bernier, Eur. Phys. J. B **13**, 661 (2000)
36. S. Rols, E. Anglaret, J. L. Sauvajol, G. Coddens, A. J. Dianoux, Appl. Phys. A **69**, 591 (1999)
37. T. Pichler, M. Knupfer, M. S. Golden, J. Fink, A. Rinzler, R. E. Smalley, Phys. Rev. Lett. **80**, 4729 (1998)
38. R. Kuzno, M. Teranchi, M. Tanaka, Y. Saito, Jpn. J. Appl. Phys. **33**, L1316 (1994)
39. L. Henrard Ph. Lambin, J. Phys. B. **29**, 5127 (1996)
40. A. A. Lucas, L. Henrard, Ph. Lambin, Phys. Rev. Lett. **49**, 2888 (1994)
41. T. Stöckli, J. M. Bonard, A. Chatelain, Z. L. Wang, P. Stadelmann, Phys. Rev. Lett. **57**, 15599 (1998)
42. M. F. Lin, K. W. K. Shung, Phys. Rev. Lett. **50**, 17744 (1994)
43. S. Tasaki, K. Maekawa, T. Yamabe, Phys. Rev. Lett. **57**, 9301 (1998)
44. O. Stéphan, P. M. Ajayan, C. Colliex, F. Cyroy-Lackmann, E. Sandre, Phys. Rev. Lett. **53**, 13824 (1996)
45. H. Kataura, Y. Kumazawa, Y. Haniwa, I. Umezu, S. Suzuki, Y. Ohtsuka, Y. Achiba, Synth. Met. **103**, 2555 (1999)
46. O. Jost, A. A. Gorbunov, W. Pompe, T. Pichler, R. Friedlein, M. Knupfer, M. Reibold, H.-D. Bauer, L. Dunsch, M. S. Golden, J. Fink, Appl. Phys. Lett. **75**, 2217 (1999)
47. M. S. Dresselhaus, G. Dresselhaus, P. C. Eklund, *Science of Fullerenes and Carbon Nanotubes* (Academic, San Diego 1996)
48. R. Saito, G. Dresselhaus, M. S. Dresselhaus, Phys. Rev. Lett. **61**, 2981 (2000)
49. C. H. Olk, J. P. Heremans, J. Mater. Res. **9**, 259 (1994)
50. L. C. Qin, S. Iijima, H. Kataura, Y. Maniwa, S. Suzuki, Y. Achiba, Chem. Phys. Lett. **268**, 101 (1997)

51. F. J. Garcia-Vidal, J. M. Pitarke, J. B. Pendry, Phys. Rev. Lett. **78**, 4289 (1997)
52. M. F. Lin, K. W. K. Shung, Phys. Rev. Lett. **47**, 6617 (1993)
53. S. Das Sarma, E. H. Hwang, Phys. Rev. Lett. **54**, 1936 (1996)
54. C. Yannouleas, E. G. Bogachek, U. Landman, Phys. Rev. Lett. **53**, 10225 (1996)
55. M. F. Lin, D. S. Chuu, Phys. Rev. Lett. **57**, 10183 (1998)
56. R. Friedlein, T. Pichler, M. Knupfer, M. S. Golden, K, Mukhopadhyay, T. Sugai, H. Shinohara, J. Fink, in *Electronic Properties of Novel Materials - Science and Technology of Molecular Nanostructures*, AIP Conf. Proc. **486**, 351 (1999)
57. R. Friedlein, PhD thesis, TU Dresden (2000)
58. L. Henrard, O. Stéphan, C. Colliex, Synt. Met. **103**, 2502 (1999)
59. M. Kociak, L. Henrard, O. Stéphan, K. Suenaga, C. Colliex, Phys. Rev. Lett. **61**, 13936 (2000)
60. R. S. Lee, H. J. Kim, J. E. Fischer, A. Thess, R. E. Smalley, Nature **388**, 255 (1997)
61. J. J. Ritsko, Phys. Rev. Lett. **25**, 6452 (1982)
62. P. Petit, C. Mathis, C. Journet, P. Bernier, Chem. Phys. Lett. **305**, 370 (1999)
63. S. Kazaoui, N. Minami, R. Jacquemin, H. Kataura, Y. Achiba, Phys. Rev. Lett. **60**, 13339 (1999)

# Phonons and Thermal Properties of Carbon Nanotubes

James Hone

Department of Physics, University of Pennsylvania
Philadelphia, PA 19104-6317, USA
hone@caltech.edu

**Abstract.** The thermal properties of carbon nanotubes display a wide range of behaviors which are related both to their graphitic nature and their unique structure and size. The specific heat of individual nanotubes should be similar to that of two-dimensional graphene at high temperatures, with the effects of phonon quantization becoming apparent at lower temperatures. Inter-tube coupling in SWNT ropes, and interlayer coupling in MWNTs, should cause their low-temperature specific heat to resemble that of three-dimensional graphite. Experimental data on SWNTs show relatively weak inter-tube coupling, and are in good agreement with theoretical models. The specific heat of MWNTs has not been examined theoretically in detail. Experimental results on MWNTs show a temperature dependent specific heat which is consistent with weak inter-layer coupling, although different measurements show slightly different temperature dependences. The thermal conductivity of both SWNTs and MWNTs should reflect the on-tube phonon structure, regardless of tube-tube coupling. Measurements of the thermal conductivity of bulk samples show graphite-like behavior for MWNTs but quite different behavior for SWNTs, specifically a linear temperature dependence at low $T$ which is consistent with one-dimensional phonons. The room-temperature thermal conductivity of highly aligned SWNT samples is over 200 W/mK, and the thermal conductivity of individual nanotubes is likely to be higher still.

## 1 Specific Heat

Because nanotubes are derived from graphene sheets, we first examine the specific heat $C$ of a single such sheet, and how $C$ changes when many such sheets are combined to form solid graphite. We then in Sect. 1.2 consider the specific heat of an isolated nanotube [1], and the effects of bundling tubes into crystalline ropes and multi-walled tubes (Sect. 1.3). Theoretical models are then compared to experimental results (Sect. 1.4).

### 1.1 Specific Heat of 2-D Graphene and 3-D Graphite

In general, the specific heat $C$ consists of phonon $C_{\mathrm{ph}}$ and electron $C_{\mathrm{e}}$ contributions, but for 3-D graphite, graphene and carbon nanotubes, the dominant contribution to the specific heat comes from the phonons. The phonon contribution is obtained by integrating over the phonon density of states with

M. S. Dresselhaus, G. Dresselhaus, Ph. Avouris (Eds.): Carbon Nanotubes,
Topics Appl. Phys. **80**, 273–287 (2001)

a convolution factor that reflects the energy and occupation of each phonon state:

$$C_{\mathrm{ph}} = \int_0^{\omega_{\max}} k_{\mathrm{B}} \left(\frac{\hbar\omega}{k_{\mathrm{B}}T}\right)^2 \frac{\mathrm{e}^{\left(\frac{\hbar\omega}{k_{\mathrm{B}}T}\right)} \rho(\omega)\, d\omega}{\left(\mathrm{e}^{\frac{\hbar\omega}{k_{\mathrm{B}}T}} - 1\right)^2}, \tag{1}$$

where $\rho(\omega)$ is the phonon density of states and $\omega_{\max}$ is the highest phonon energy of the material. For nonzero temperatures, the convolution factor is 1 at $\omega = 0$, and decreases smoothly to a value of $\sim 0.1$ at $\hbar\omega = k_{\mathrm{B}}T/6$, so that the specific heat rises with $T$ as more phonon states are occupied. Because $\rho(\omega)$ is in general a complicated function of $\omega$, the specific heat, at least at moderate temperatures, cannot be calculated analytically.

At low temperature ($T \ll \Theta_{\mathrm{D}}$), however, the temperature dependence of the specific heat is in general much simpler. In this regime, the upper bound in (1) can be taken as infinity, and $\rho(\omega)$ is dominated by acoustic phonon modes, i.e., those with $\omega \to 0$ as $k \to 0$. If we consider a single acoustic mode in $d$ dimensions that obeys a dispersion relation $\omega \propto k^{\alpha}$, then from (1) it follows that:

$$C_{\mathrm{ph}} \propto T^{(d/\alpha)} \qquad (T \ll \Theta_{\mathrm{D}}). \tag{2}$$

Thus the low-temperature specific heat contains information about both the dimensionality of the system and the phonon dispersion.

A single graphene sheet is a 2-D system with three acoustic modes, two having a very high sound velocity and linear dispersion [a longitudinal (LA) mode, with $v$=24 km/s, and an in-plane transverse (TA) mode, with $v$=18 km/s] and a third out-of-plane transverse (ZA) mode that is described by a parabolic dispersion relation, $\omega = \delta k^2$, with $\delta \sim 6 \times 10^{-7}\,\mathrm{m}^2/s$ [2,3]. From (2), we see that the specific heat from the in-plane modes should display a $T^2$ temperature dependence, while that of the out-of plane mode should be linear in $T$. Equation (1) can be evaluated separately for each mode; the contribution from the ZA mode dominates that of the in-plane modes below room temperature.

The phonon contributions to the specific heat can be compared to the expected electronic specific heat of a graphene layer. The unusual linear $k$ dependence of the electronic structure $E(k)$ of a single graphene sheet at $E_{\mathrm{F}}$ (see Fig. 2 of [4]) produces a low-temperature electronic specific heat that is quadratic in temperature, rather than the linear dependence found for typical metals [5]. *Benedict* et al. [1] show that, for the in-plane modes in a graphene sheet,

$$\frac{C_{\mathrm{ph}}}{C_{\mathrm{e}}} \approx \left(\frac{v_F}{v}\right)^2 \approx 10^4 . \tag{3}$$

The specific heat of the out-of plane mode is even higher, and thus phonons dominate the specific heat, even at low $T$, and all the way to $T = 0$.

Combining weakly interacting graphene sheets in a correlated stacking arrangement to form solid graphite introduces dispersion along the $c$-axis, as the system becomes three-dimensional. Since the $c$-axis phonons have very low frequencies, thermal energies of ~50 K are sufficient to occupy all ZA phonon states, so that for $T > 50\,\mathrm{K}$ the specific heat of 3-D graphite is essentially the same as that of 2-D graphene. The crossover between 2-D and interplanar coupled behavior is identified as a maximum in a plot of $C_{\mathrm{ph}}$ vs. $T^2$ [6,7]. We will see below that this type of dimensional crossover also exists in bundles of SWNTs. The electronic specific heat is also significantly changed in going from 2-D graphene to 3-D graphite: bulk graphite has a small but nonzero density of states at the Fermi energy $N(E_{\mathrm{F}})$ due to $c$-axis dispersion of the electronic states. Therefore 3-D graphite displays a small linear $C_{\mathrm{e}}(T)$, while $C_{\mathrm{ph}}$ has no such term. For 3-D graphite, the phonon contribution remains dominant above ~1 K [8,9].

## 1.2 Specific Heat of Nanotubes

Figure 1 shows the low-energy phonon dispersion relations for an isolated (10,10) nanotube. Rolling a graphene sheet into a nanotube has two major effects on the phonon dispersion. First, the two-dimensional band-structure

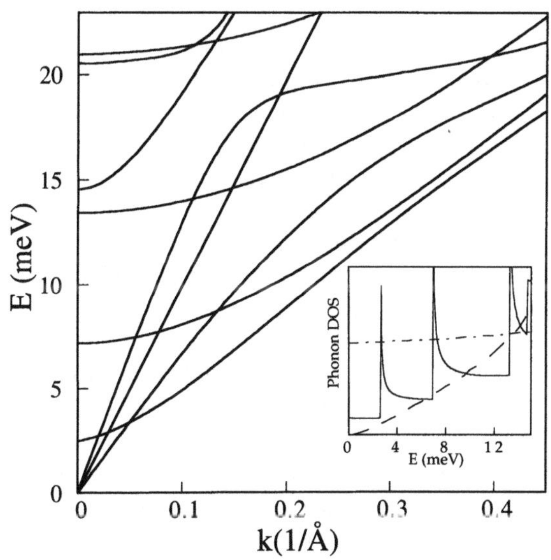


**Fig. 1.** Low-energy phonon dispersion relations for a (10,10) nanotube. There are four acoustic modes: two degenerate TA modes ($v = 9\,\mathrm{km/s}$), a 'twist' mode ($v = 15\,\mathrm{km/s}$), and one TA mode ($v = 24\,\mathrm{km/s}$) [3]. The *inset* shows the low-energy phonon density of states of the nanotube (*solid line*) and that of graphite (*dashed line*) and graphene (*dot-dashed line*). The nanotube phonon DOS is constant below 2.5 meV, then increases stepwise as higher subbands enter; there is a 1-D singularity at each subband edge

of the sheet is collapsed onto one dimension; because of the periodic boundary conditions on the tube, the circumferential wavevector is quantized and discrete 'subbands' develop. From a zone-folding picture, the splitting between the subbands at the $\Gamma$ point is of order [1]

$$\Delta E = k_\mathrm{B}\Theta_\mathrm{subband} \approx \frac{\hbar v}{R}, \tag{4}$$

where $R$ is the radius of the nanotube and $v$ is the band velocity of the relevant graphene mode. The second effect of rolling the graphene sheet is to rearrange the low-energy acoustic modes. For the nanotube there are now four, rather than three, acoustic modes: an LA mode, corresponding to motion of the atoms along the tube axis, two degenerate TA modes, corresponding to atomic displacements perpendicular to the nanotube axis, and a 'twist' mode, corresponding to a torsion of the tube around its axis. The LA mode is exactly analogous to the LA mode in graphene. The TA modes in a SWNT, on the other hand, are a combination of the in-plane and out-of-plane TA modes in graphene, while the twist mode is directly analogous to the in-plane TA mode. These modes all show linear dispersion (there is no nanotube analogue to the ZA mode) and high phonon velocities: $v_\mathrm{LA} = 24\,\mathrm{km/s}$, $v_\mathrm{TA} = 9\,\mathrm{km/s}$, and $v_\mathrm{twist} = 15\,\mathrm{km/s}$ for a (10,10) tube [3]. Because all of the acoustic modes have a high velocity, the splitting given in (4) corresponds to quite high temperatures, on the order of 100 K for a 1.4 nm-diameter tube. In the calculated band structure for a (10,10) tube, the lowest subband enters at ~2.5 meV (30 K), somewhat lower in energy than the estimate given by (4).

The inset to Fig. 1 shows the low-energy phonon density of states $\rho(\omega)$ of a (10,10) nanotube (solid line), with $\rho(\omega)$ of graphene (dot-dashed line) and graphite (dashed line) shown for comparison. In contrast to 2-D graphene and 3-D graphite, which show a smoothly-varying $\rho(\omega)$, the 1-D nanotube has a step-like $\rho(\omega)$, which has 1-D singularities at the subband edges. The markedly different phonon density of states in carbon nanotubes results in measurably different thermal properties at low temperature.

At moderate temperatures, many of the phonon subbands of the nanotube will be occupied, and the specific heat will be similar to that of 2-D graphene. At low temperatures, however, both the quantized phonon structure and the stiffening of the acoustic modes will cause the specific heat of a nanotube to differ from that of graphene. In the low $T$ regime, only the acoustic bands will be populated, and thus the specific heat will be that of a 1-D system with a linear $\omega(k)$. In this limit, $T \ll \hbar v / k_\mathrm{B} R$, (1) can be evaluated analytically, yielding a linear $T$ dependence for the specific heat [1]:

$$C_\mathrm{ph} = \frac{3 k_\mathrm{B}^2 T}{\pi \hbar v \rho_\mathrm{m}} \times \frac{\pi^2}{3}, \tag{5}$$

where $\rho_\mathrm{m}$ is the mass per unit length, $v$ is the acoustic phonon velocity, and $R$ is the nanotube radius. Thus the circumferential quantization of the nanotube phonons should be observable as a linear $C(T)$ dependence at the lowest temperatures, with a transition to a steeper temperature dependence above the thermal energy for the first quantized state.

Turning to the electron contribution, a metallic SWNT is a one-dimensional metal with a non-zero density of states at the Fermi level. The electronic specific heat will be linear in temperature [1]:

$$C_{\mathrm{e}} = \frac{4\pi k_{\mathrm{B}}^2 T}{3\hbar v_{\mathrm{F}} \rho_{\mathrm{m}}}\,, \tag{6}$$

for $T \ll \hbar v_{\mathrm{F}}/k_{\mathrm{B}}R$, where $v_{\mathrm{F}}$ is the Fermi velocity and $\rho_{\mathrm{m}}$ is again the mass per unit length. The ratio between the phonon and the electron contributions to the specific heat is [1]

$$\frac{C_{\mathrm{ph}}}{C_{\mathrm{e}}} \approx \frac{v_{\mathrm{F}}}{v} \approx 10^2\,, \tag{7}$$

so that even for a metallic SWNT, phonons should dominate the specific heat all the way down to $T = 0$. The electronic specific heat of a semiconducting tube should vanish roughly exponentially as $T \to 0$ [10], and so $C_{\mathrm{e}}$ will be even smaller than that of a metallic tube. However, if such a tube were doped so that the Fermi level lies near a band edge, its electronic specific heat could be significantly enhanced.

### 1.3 Specific Heat of SWNT Ropes and MWNTs

As was mentioned above, stacking graphene sheets into 3-D graphite causes phonon dispersion in the $c$ direction, which significantly reduces the low-$T$ specific heat. A similar effect should occur in both SWNT ropes and MWNTs. In a SWNT rope, phonons will propagate both along individual tubes and between parallel tubes in the hexagonal lattice, leading to dispersion in both the longitudinal (on-tube) and transverse (inter-tube) directions. The solid lines in Fig. 2 show the calculated dispersion of the acoustic phonon modes in an infinite hexagonal lattice of carbon nanotubes with 1.4 nm diameter [6]. The phonon bands disperse steeply along the tube axis and more weakly in the transverse direction. In addition, the 'twist' mode becomes an optical mode because of the presence of a nonzero shear modulus between neighboring tubes. The net effect of this dispersion is a significant reduction in the specific heat at low temperatures compared to an isolated tube (Fig. 3). The dashed lines show the dispersion relations for the higher-order subbands of the tube. In this model, the characteristic energy $k_{\mathrm{B}}\Theta_{\mathrm{D}}^{\perp}$ of the inter-tube modes is ($\sim 5$ meV), which is larger than the subband splitting energy, so that 3-D dispersion should obscure the effects of phonon quantization. We will address the experimentally-measured inter-tube coupling below.

The phonon dispersion of MWNTs has not yet been addressed theoretically. Strong phonon coupling between the layers of a MWNT should cause roughly graphite-like behavior. However, due to the lack of strict registry

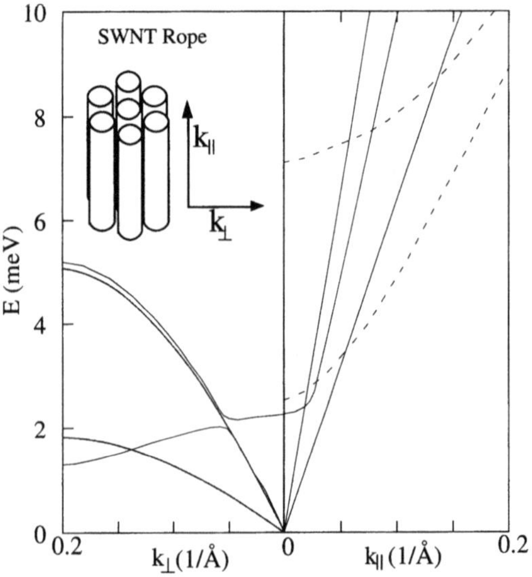


**Fig. 2.** Calculated dispersion of the acoustic phonon modes in an infinite rope of 1.4 nm diameter SWNTs [6]. The phonon velocity is high in the longitudinal direction and lower in the transverse direction. The first two higher-order subbands (*dashed lines*) are shown for comparison

between the layers in a MWNT, the interlayer coupling could conceivably be much weaker than in graphite, especially for the twist and LA modes, which do not involve radial motion. The larger size of MWNTs, compared to SWNTs, implies a significantly smaller subband splitting energy (4), so that the thermal effects of phonon quantization should be measurable only well below 1 K.

## 1.4 Measured Specific Heat of SWNTs and MWNTs

The various curves in Fig. 3 show the calculated phonon specific heat for isolated (10,10) SWNTs, a SWNT rope crystal, graphene, and graphite. The phonon contribution for a (10,10) SWNT was calculated by computing the phonon density of states using the theoretically derived dispersion curves [3] and then numerically evaluating (1); $C(T)$ of graphene and graphite was calculated using the model of *Al-Jishi* and *Dresselhaus* [12]. Because of the high phonon density of states of a 2-D graphene layer at low energy due to the quadratic ZA mode, 2-D graphene has a high specific heat at low $T$. The low temperature specific heat for an isolated SWNT is however significantly lower than that for a graphene sheet, reflecting the stiffening of the acoustic modes due to the cylindrical shape of the SWNTs. Below $\sim 5$ K, the predicted $C(T)$ is due only to the linear acoustic modes, and $C(T)$ is linear in $T$, a behavior which is characteristic of a 1-D system. In these curves, we can see clearly the effects of the interlayer (in graphite) and inter-tube (in SWNT ropes) dispersion on the specific heat. Below $\sim$50 K, the phonon specific heat of graphite and SWNT ropes is significantly below that of graphene or isolated (10,10) SWNTs. The measured specific heat of graphite [8,9] matches the phonon

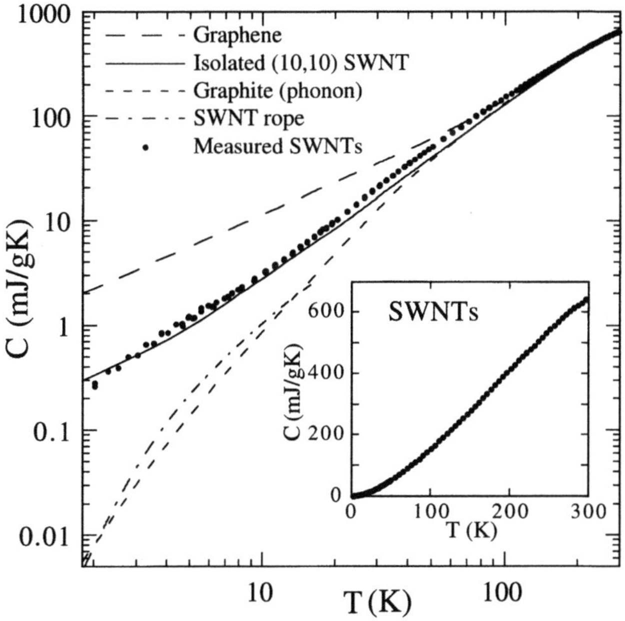


**Fig. 3.** Measured specific heat of SWNTs, compared to predictions for the specific heat of isolated (10,10) tubes, SWNT ropes, a graphene layer and graphite [11]

contribution above 5 K, below which temperature the electronic contribution is also important.

The filled points in Fig. 3 represent the measured specific heat of SWNTs [11]. The measured $C(T)$ agrees well with the predicted curve for individual (10,10) nanotubes. $C(T)$ for the nanotubes is significantly smaller than that of graphene below ~50 K, confirming the relative stiffness of the nanotubes to bending. On the other hand, the measured specific heat is larger than that expected for SWNT ropes. This suggests that the tube-tube coupling in a rope is significantly weaker than theoretical estimates [6,11].

Figure 4 highlights the low-temperature behavior of the specific heat. The experimental data, represented by the filled points, show a linear slope below 8 K, but the slope does not extrapolate to zero at $T = 0$, as would be expected for perfectly isolated SWNTs. This departure from ideal behavior is most likely due to a weak transverse coupling between neighboring tubes. The measured data can be fit using a two-band model, shown in the inset. The dashed line in Fig. 4 represents the contribution from a single (four-fold degenerate) acoustic mode, which has a high on-tube Debye temperature $\Theta_{\mathrm{D}}^{\|}$ and a much smaller inter-tube Debye temperature $\Theta_{\mathrm{D}}^{\perp}$. The dot-dashed line represents the contribution from the first (doubly-degenerate) subband, with minimum energy $k_{\mathrm{B}}\Theta_{\mathrm{subband}}$. Because $\Theta_{\mathrm{subband}} > \Theta_{\mathrm{D}}^{\perp}$, the subband is assumed to be essentially one-dimensional. The solid line represents the sum

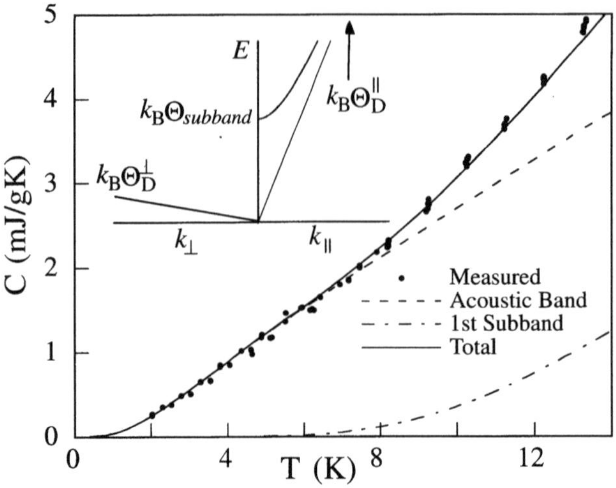


**Fig. 4.** Measured specific heat of SWNTs compared to a two-band model with transverse dispersion (*inset*). The fitting parameters used are $\Theta_D^{\parallel} = 960\,\mathrm{K}$; $\Theta_D^{\perp} = 50\,\mathrm{K}$; and $\Theta_{\mathrm{subband}} = 13\,\mathrm{K}$ [11]

of the two contributions, and fits the data quite well. The derived value for the on-tube Debye temperature is $\Theta_D^{\parallel} = 960\,\mathrm{K}$ (80 meV), which is slightly lower than the value of 1200 K (100 meV) which can be derived from the calculated phonon band structure. The transverse Debye temperature $\Theta_D^{\perp}$ is 13 K (1.1 meV), considerably smaller than the expected value for crystalline ropes (5 meV) or graphite (10 meV). Finally, the derived value of $\Theta_{\mathrm{subband}}$ is 50 K (4.3 meV), which is larger than the value of 30 K (2.5 meV) given by the calculated band structure [3].

Figure 5 shows the two reported measurements of the specific heat of MWNTs, along with the theoretical curves for graphene, isolated nanotubes, and graphite. *Yi* et al. [13] used a self-heating technique to measure the specific heat of MWNTs of 20–30 nm diameter produced by a CVD technique, and they find a linear behavior from 10 K to 300 K. This linear behavior agrees well with the calculated specific heat of graphene below 100 K, but is lower than all of the theoretical curves in the 200–300 K range. Agreement with the graphene specific heat, rather than that for graphite, indicates a relatively weak inter-layer coupling in these tubes. Because the specific heat of graphene at low $T$ is dominated by the quadratic ZA mode, the authors postulate that this mode must also be present in their samples. As was discussed above (Sect. 1.1,1.2), such a band should should not exist in nanotubes. However, the phonon structure of large-diameter nanotubes, whose properties should approach that of graphene, has not been carefully studied. *Mizel* et al. report a direct measurement of the specific heat of arc-produced MWNTs [6]. The specific heat of their sample follows the theoretical curve for an isolated nanotube, again indicating a weak inter-layer coupling, but shows no evidence of a graphene-like quadratic phonon mode. At present the origin of the discrepancy between the two measurements is unknown, although the sample morphologies may be different.

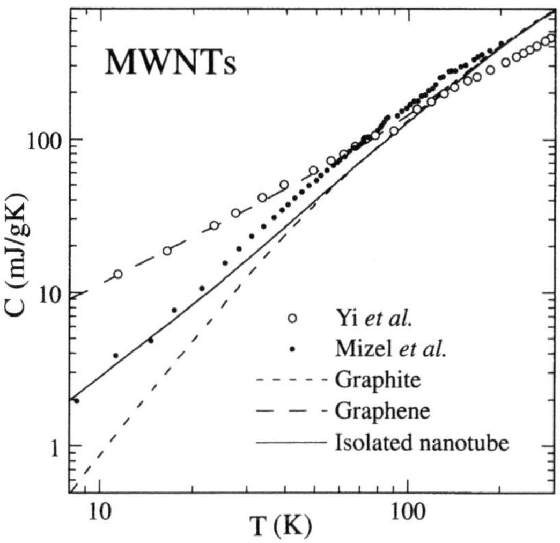


**Fig. 5.** Measured specific heat of MWNTs [6,13], compared to the calculated phonon specific heat of graphene, graphite, and isolated nanotubes

## 2 Thermal Conductivity

Carbon-based materials (diamond and in-plane graphite) display the highest measured thermal conductivity of any known material at moderate temperatures [14]. In graphite, the thermal conductivity is generally dominated by phonons, and limited by the small crystallite size within a sample. Thus the apparent long-range crystallinity of nanotubes has led to speculation [15] that the longitudinal thermal conductivity of nanotubes could possibly exceed the in-plane thermal conductivity of graphite. Thermal conductivity also provides another tool (besides the specific heat) for probing the interesting low-energy phonon structure of nanotubes. Furthermore, nanotubes, as low-dimensional materials, could have interesting high-temperature properties as well [16]. In this section, we will first discuss the phonon and electronic contributions to the thermal conductivity in graphite. Then we will examine the thermal conductivity of multi-walled and single-walled nanotubes.

The diagonal term of the phonon thermal conductivity tensor can be written as:

$$\kappa_{zz} = \sum C {v_z}^2 \tau , \tag{8}$$

where $C$, $v$, and $\tau$ are the specific heat, group velocity, and relaxation time of a given phonon state, and the sum is over all phonon states. While the phonon thermal conductivity cannot be measured directly, the electronic contribution $\kappa_e$ can generally be determined from the electrical conductivity by the Wiedemann–Franz law:

$$\frac{\kappa_e}{\sigma T} \approx L_0, \tag{9}$$

where the Lorenz number $L_0 = 2.45 \times 10^{-8}\,(\mathrm{V/K})^2$. Thus it is in principle straightforward to separate the electronic and lattice contributions to $\kappa(T)$. In graphite, phonons dominate the specific heat above ~20 K [17], while in MWNTs and SWNTs, the phonon contribution dominates at all temperatures.

In highly crystalline materials and at high temperatures ($T > \Theta_\mathrm{D}/10$), the dominant contribution to the inelastic phonon relaxation time $\tau$ is phonon-phonon Umklapp scattering. At low temperatures, however, Umklapp scattering disappears and inelastic phonon scattering is generally due to fixed sample boundaries or defects, yielding a constant $\tau$. Thus at low temperature ($T < \Theta_\mathrm{D}/10$), the temperature dependence of the phonon thermal conductivity is similar to that of the specific heat. However, in an anisotropic material, the weighting of each state by the factor $v^2\tau$ becomes important. The thermal conductivity is most sensitive to the states with the highest band velocity and scattering time. In graphite, for instance, the *ab*-plane thermal conductivity can be closely approximated by ignoring the inter-planar coupling [17]. From this argument, we would expect that the temperature-dependence of the thermal conductivity of SWNT ropes and MWNTs should be close to that of their constituent tubes. However, bundling individual tubes into ropes or MWNTs may introduce inter-tube scattering, which could perturb somewhat both the magnitude and the temperature dependence of the thermal conductivity.

## 2.1 Thermal Conductivity of MWNTs

In highly graphitic fibers, $\kappa(T)$ follows a $T^{2.3}$ temperature dependence until $\sim 100\,K$, then begins to decrease with increasing $T$ above ~150 K [18]. This decrease in $\kappa(T)$ above 100 K is due to the onset of phonon-phonon Umklapp scattering, which grows more effective with increasing temperature as higher-energy phonons are populated. In less graphitic fibers, the magnitude of $\kappa$ is significantly lower, and the Umklapp peak in $\kappa(T)$ is not seen, because grain-boundary scattering dominates $\kappa(T)$ to higher temperatures.

Figure 6 shows the thermal conductivity of CVD-grown MWNTs, on a linear scale, from 4 K to 300 K [13]. Because of the large diameter of these tubes, we expect them to act essentially as 2-D phonon materials. Indeed, at low temperature ($T < 100\,\mathrm{K}$), $\kappa(T)$ increases as $\sim T^2$, similar to the $T^{2.3}$ behavior in graphite. The room-temperature thermal conductivity is small, comparable to the less-graphitic carbon fibers, and the MWNTs do not show a maximum in $\kappa(T)$ due to Umklapp scattering; both properties are consistent with a small crystallite size.

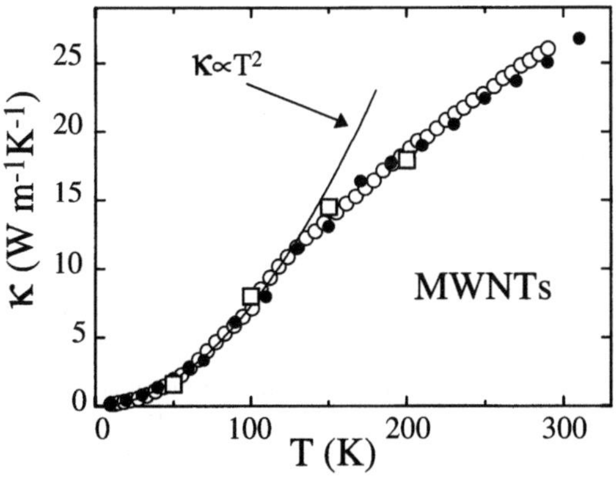


**Fig. 6.** Measured thermal conductivity of MWNTs [13] from 4 K to 300 K

## 2.2 Thermal Conductivity of SWNTs

Figure 7 represents the measured $\kappa(T)$ of a bulk sample of laser-vaporization produced SWNTs, with ~1.4-nm diameter [20]. The different temperature-dependence of $\kappa(T)$ reflects the much smaller size of SWNTs compared to MWNTs. $\kappa(T)$ increases with increasing $T$ from 8 K to 300 K, although a gradual decrease in the slope above 250 K may indicate the onset of Umklapp scattering. Most striking is a change in slope near 35 K: below this temperature, $\kappa(T)$ is linear in $T$ and extrapolates to zero at $T$=0. We will discuss the low-temperature behavior in detail below.

Although the temperature dependence of the thermal conductivity is the same for all 1.4 nm diameter SWNT samples, the magnitude of $\kappa(T)$ is sensitive to sample geometry. In disordered 'mat' samples, the the room-temperature thermal conductivity is ~35 W/mK. However, in samples consisting of aligned SWNTs, the room-temperature thermal conductivity is

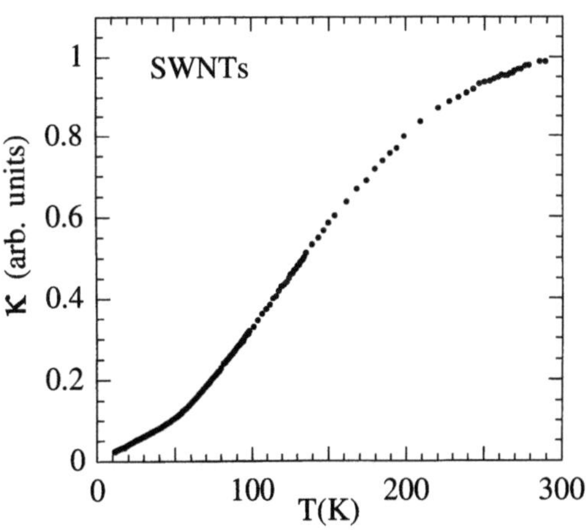


**Fig. 7.** Thermal conductivity of a bulk sample of SWNTs [19]

above 200 W/mK [21], within an order of magnitude of the room-temperature thermal conductivity of highly crystalline graphite. Because even such an aligned sample contains many rope-rope junctions, it is likely that a single tube, or a rope of continuous tubes, will have significantly higher thermal conductivity than the bulk samples.

Simultaneous measurement of the electrical and thermal conductance of bulk SWNT samples yields a Lorenz ratio $\kappa/\sigma T$ which is more than two orders of magnitude greater than the value for electrons at all temperatures. Thus the thermal conductivity is dominated by phonons, as expected.

Figure 8 highlights the low-$T$ behavior of the thermal conductivity of SWNTs [19]. As discussed above, the linear $T$ dependence of $\kappa(T)$ likely reflects the one-dimensional band-structure of individual SWNTs, with linear acoustic bands contributing to thermal transport at the lowest temperatures and optical subbands entering at higher temperatures. $\kappa(T)$ can be modeled using a simplified two-band model (shown in the inset to Fig. 8), considering a single acoustic band and one subband. In a simple zone-folding picture, the acoustic band has a dispersion $\omega = vk$ and the first subband has dispersion $\omega^2 = v^2k^2 + \omega_0^2$, where $\omega_0 = v/R$. The thermal conductivity from each band can then be estimated using (8) and assuming a constant scattering time $\tau$. Thus $\tau$ provides an overall scaling factor, and $v$ sets the energy scale $\hbar\omega_0$ of the splitting between the two bands.

Figure 8 shows the measured $\kappa(T)$ of SWNTs, compared to the results of the two-band model discussed above, with $v$ chosen to be 20 km/s, which is between that of the 'twist' ($v$ = 15 km/s) and LA ($v$ = 24 km/s) modes. The top dashed line represents $\kappa(T)$ of the acoustic band: it is linear in $T$, as expected for a 1-D phonon band with linear dispersion and constant $\tau$. The lower dashed line represents the contribution from the optical subband;

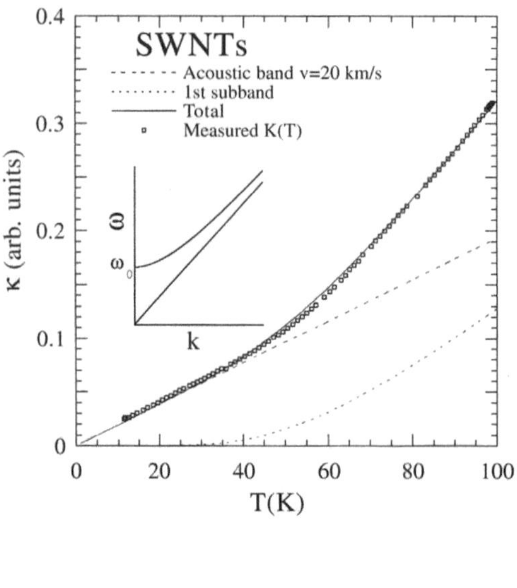


**Fig. 8.** Measured low-temperature thermal conductivity of SWNTs, compared to a two-band model [19]

it is frozen out at low temperatures, and begins to contribute near 35 K. The solid line is the sum of the two contributions, and is quite successful in fitting the experimental data below ~100 K. The phonon energies, and temperature scale of the observed linear behavior, are higher in the thermal conductivity measurements than in the heat capacity measurements (Sect. 1.4). This may be due to the preferential weighting of higher-velocity modes in the thermal conductivity, although more detailed modeling is needed to resolve this issue. The measured linear slope can be used to calculate the scattering time, or, equivalently, a scattering length. A room-temperature thermal conductivity of 200 W/mK (as is seen in the bulk aligned samples) implies a phonon scattering length of 30 nm, although this value is likely to be higher for single tubes.

We have seen above that the small size of nanotubes causes phonon quantization which can be observed both in the heat capacity and in the thermal conductivity at low temperatures. The restricted geometry of the tubes may also affect the thermal conductivity at high temperature since Umklapp scattering should be suppressed in one dimensional system because of the unavailability of states into which to scatter [16,22]. Extension of these measurements to higher temperatures, as well as additional theoretical modeling, should prove interesting.

## References


1. L. X. Benedict, S. G. Louie, M. L. Cohen, Solid State Commun. **100**, 177–180 (1996)
2. D. Sanchez-Portal, E. Artacho, J. M. Soler, A. Rubio, P. Ordejón, Phys. Rev. B **59**, 12678–12688 (1999)
3. R. Saito, G. Dresselhaus, M. S. Dresselhaus, *Physical Properties of Carbon Nanotubes* (Imperial College Press, London 1998)
4. R. Saito, H. Kataura, chapter in this volume
5. C. Kittel, in *Introduction to Solid State Physics*, 6th edn. (Wiley, New York, 1986)
6. A. Mizel, L. X. Benedict, M. L. Cohen, S. G. Louie, A. Zettl, N. K. Budra, W. P. Beyermann, Phys. Rev. B **60**, 3264 (1999)
7. R. Nicklow, N. Wakabayashi, H. G. Smith, Phys. Rev. B **5**, 4951 (1972)
8. W. DeSorbo, G. E. Nichols, J. Phys. Chem. Solids **6**, 352 (1958)
9. M. G. Alexander, D. P. Goshorn, D. Guérard, P. Lagrange, M. El Makrini, D. G. Onn, Synth. Met. **2**, 203 (1980)
10. Neil W. Ashcroft, N. David Mermin, *Solid State Physics* (Harcourt Brace , New York 1976)
11. J. Hone, B. Batlogg, Z. Benes, A. T. Johnson, J. E. Fischer, Science **289**, 1730 (2000)
12. R. Al-Jishi, Lattice Dynamics of Graphite Intercalation Compounds, PhD thesis, Massachusetts Institute of Technology (1982)
13. W. Yi, L. Lu, Zhang Dian-lin, Z. W. Pan, S. S. Xie, Phys. Rev. B **59**, R9015 (1999)

14. G. W. C. Kaye, T. H. Laby, *Tables of Physical and Chemical Constants*, 16th edn. (Longman, London 1995)
15. R. S. Ruoff, D. C. Lorents, Carbon **33**, 925 (1995)
16. D. T. Morelli, J. Heremans, M. Sakamoto, C. Uher, Phys. Rev. Lett. **57**, 869 (1986)
17. B. T. Kelly, in *Physics of Graphite* (Applied Science, London 1981)
18. J. Heremans, C. P. Beetz, Jr., Phys. Rev. B **32**, 1981 (1985)
19. J. Hone, M. Whitney, C. Piskoti, A. Zettl, Phys. Rev. B **59**, R2514 (1999)
20. A. Thess, R. Lee, P. Nikolaev, H. Dai, P. Petit, J. Robert, C. Xu, Y. H. Lee, S. G. Kim, A. G. Rinzler, D. T. Colbert, G. E. Scuseria, D. Tománek, J. E. Fischer, R. E. Smalley, Science **273**, 483–487 (1996)
21. J. Hone, M. C. Llaguno, N. M. Nemes, A. J. Johnson, J. E. Fischer, D. A. Walters, M. J. Casavant, J. Schmidt, R. E. Snalley, Appl. Phys. Lett. **77**, 666 (2000)
22. R. E. Peierls, in *Quantum Theory of Solids* (Oxford Univ. Press, London 1955)

# Mechanical Properties of Carbon Nanotubes

Boris I. Yakobson[1] and Phaedon Avouris[2]

[1] Center for Nanoscale Science and Technology and Department of Mechanical Engineering and Materials Science, Rice University, Houston, TX, 77251–1892, USA
biy@rice.edu
[2] IBM T.J. Watson Research Center Yorktown Heights, NY 10598, USA
avouris@us.ibm.com

**Abstract.** This paper presents an overview of the mechanical properties of carbon nanotubes, starting from the linear elastic parameters, nonlinear elastic instabilities and buckling, and the inelastic relaxation, yield strength and fracture mechanisms. A summary of experimental findings is followed by more detailed discussion of theoretical and computational models for the entire range of the deformation amplitudes. Non-covalent forces (supra-molecular interactions) between the nanotubes and with the substrates are also discussed, due to their significance in potential applications.

It is noteworthy that the term *resilient* was first applied not to nanotubes but to smaller fullerene cages, when Whetten *et al.* studied the high-energy collisions of $C_{60}$, $C_{70}$, and $C_{84}$ bouncing from a solid wall of H-terminated diamond [6]. They observed no fragmentation or any irreversible atomic rearrangement in the bouncing back cages, which was somewhat surprising and indicated the ability of fullerenes to sustain great elastic distortion. The very same property of resilience becomes more significant in the case of carbon nanotubes, since their elongated shape, with the aspect ratio close to a thousand, makes the mechanical properties especially interesting and important due to potential structural applications.

## 1 Mechanical Properties and Mesoscopic Duality of Nanotubes

The utility of nanotubes as the strongest or stiffest elements in nanoscale devices or composite materials remains a powerful motivation for the research in this area. While the jury is still out on practical realization of these applications, an additional incentive comes from the fundamental materials physics. There is a certain duality in the nanotubes. On one hand they have molecular size and morphology. At the same time possessing sufficient translational

M. S. Dresselhaus, G. Dresselhaus, Ph. Avouris (Eds.): Carbon Nanotubes,
Topics Appl. Phys. **80**, 287–329 (2001)

symmetry to perform as very small (nano-) crystals, with a well defined primitive cell, surface, possibility of transport, etc. Moreover, in many respects they can be studied as well defined engineering structures and many properties can be discussed in traditional terms of moduli, stiffness or compliance, geometric size and shape. The mesoscopic dimensions (a nanometer scale diameter) combined with the regular, almost translation-invariant morphology along their micrometer scale lengths (unlike other polymers, usually coiled), make nanotubes a unique and attractive object of study, including the study of mechanical properties and fracture in particular.

Indeed, fracture of materials is a complex phenomenon whose theory generally requires a multiscale description involving microscopic, mesoscopic and macroscopic modeling. Numerous traditional approaches are based on a macroscopic continuum picture that provides an appropriate model except at the region of actual failure where a detailed atomistic description (involving real chemical bond breaking) is needed. Nanotubes, due to their relative simplicity and atomically precise morphology, offer us the opportunity to address the validity of different macroscopic and microscopic models of fracture and mechanical response. Contrary to crystalline solids where the structure and evolution of ever-present surfaces, grain-boundaries, and dislocations under applied stress determine the plasticity and fracture of the material, nanotubes possess simpler structure while still showing rich mechanical behavior within elastic or inelastic brittle or ductile domains. This second, theoretical-heuristic value of nanotube research supplements their importance due to anticipated practical applications. A morphological similarity of fullerenes and nanotubes to their macroscopic counterparts, geodesic domes and towers, compels one to test the laws and intuition of macro-mechanics in the scale ten orders of magnitude smaller.

In the following, Sect. 2 provides a background for the discussion of nanotubes: basic concepts from materials mechanics and definitions of the main properties. We then present briefly the experimental techniques used to measure these properties and the results obtained (Sect. 3). Theoretical models, computational techniques, and results for the elastic constants, presented in Sect. 4, are compared wherever possible with the experimental data. In theoretical discussion we proceed from linear elastic moduli to the nonlinear elastic behavior, buckling instabilities and shell model, to compressive/bending strength, and finally to the yield and failure mechanisms in tensile load. After the linear elasticity, Sect. 4.1, we outline the non-linear buckling instabilities, Sect. 4.2. Going to even further deformations, in Sect. 4.3 we discuss irreversible changes in nanotubes, which are responsible for their inelastic relaxation and failure. Fast molecular tension tests (Sect. 4.3) are followed by the theoretical analysis of relaxation and failure (Sect. 4.4), based on intramolecular dislocation failure concept and combined with the computer simulation evidence. We discuss the mechanical deformation of the nanotubes caused by their attraction to each other (supramolecular interactions) and/or to, the substrates, Sect. 5.1. Closely related issues of manipulation of the tubes position and shape, and their self-organization into ropes and rings, caused by the seemingly weak van der Waals forces, are presented in the Sects. 5.2,5.3. Finally, a brief summary of mechanical properties is included in Sect. 6.

## 2 Mechanics of the Small: Common Definitions

Nanotubes are often discussed in terms of their materials applications, which makes it tempting to define "materials properties" of a nanotube itself. However, there is an inevitable ambiguity due to lack of translational invariance in the transverse directions of a singular nanotube, which is therefore not a material, but rather a structural member.

A definition of elastic moduli for a solid implies a spatial uniformity of the material, at least in an average, statistical sense. This is required for an accurate definition of any intensive characteristic, and generally fails in the nanometer scale. A single nanotube possesses no translational invariance in the radial direction, since a hollow center and a sequence of coaxial layers are well distinguished, with the interlayer spacing, $c$, comparable with the nanotube radius, $R$. It is essentially an engineering *structure*, and a definition of any material-like characteristics for a nanotube, while heuristically appealing, must always be accompanied with the specific additional assumptions involved (e.g. the definition of a cross-section area). Without it confusion can easily cripple the results or comparisons. The standard starting point for defining the elastic moduli as $1/V\ \partial^2 E/\partial\varepsilon^2$ (where $E$ is total energy as a function of uniform strain $\varepsilon$) is not a reliable foothold for molecular structures. For nanotubes, this definition only works for a strain $\varepsilon$ in the axial direction; any other deformation (e.g. uniform lateral compression) induces *non-uniform* strain of the constituent layers, which renders the previous expression misleading. Furthermore, for the hollow fullerene nanotubes, the volume $V$ is not well defined. For a given length of a nanotube $L$, the cross section area $A$ can be chosen in several relatively arbitrary ways, thus making both volume $V = LA$ and consequently the moduli ambiguous. To eliminate this problem, the intrinsic elastic energy of nanotube is better characterized by the energy change not per volume but per area $S$ of the constituent graphitic layer (or layers), $C = 1/S\ \partial^2 E/\partial\varepsilon^2$. The two-dimensional spatial uniformity of the graphite layer ensures that $S = lL$, and thus the value of $C$, is unambiguous. Here $l$ is the total circumferential length of the graphite layers in the cross section of the nanotube. Unlike more common material moduli, $C$ has dimensionality of surface tension, N/m, and can be defined in terms of measurable characteristics of nanotube,

$$C = (1/L)\partial^2 E/\partial\varepsilon^2 / \int dl\ . \tag{1}$$

The partial derivative at zero strain in all dimensions except along $\varepsilon$ yields an analog of the elastic stiffness $C_{11}$ in graphite, while a free-boundary (no lateral traction on the nanotube) would correspond to the Young's modulus $Y = S_{11}^{-1}$ ($S_{11}$ being the elastic compliance). In the latter case, the nanotube Young's modulus can be recovered and used,

$$Y = C \textstyle\int dl/A \ , \quad \text{or} \quad Y = C/h \ , \tag{2}$$

but the non-unique choice of cross-section $A$ or a thickness $h$ must be kept in mind. For the bending stiffness $K$ correspondingly, one has ($\kappa$ being a beam curvature),

$$K \equiv (1/L)\partial^2 E/\partial\kappa^2 = C \int y^2 dl, \tag{3}$$

where the integration on the right hand side goes over the cross-section length of all the constituent layers, and $y$ is the distance from the neutral surface. Note again, that this allows us to completely avoid the ambiguity of the monoatomic layer "thickness", and to relate only physically measurable quantities like the nanotube energy $E$, the elongation $\varepsilon$ or a curvature $\kappa$. If one adopts a particular convention for the graphene thickness $h$ (or equivalently, the cross section of nanotube), the usual Young's modulus can be recovered, $Y = C/h$. For instance, for a bulk graphite $h = c = 0.335$ nm, $C = 342$ N/m and $Y =$ 1.02 GPa, respectively. This choice works reasonably well for large diameter multiwall tubes (macro-limit), but can cause significant errors in evaluating the axial and especially bending stiffness for narrow and, in particular, single-wall nanotubes.

*Strength* and particularly *tensile strength* of a solid material, similarly to the elastic constants, must ultimately depend on the strength of its interatomic forces/bonds. However, this relationship is far less direct than in the case of linear-elastic characteristics; it is greatly affected by the particular arrangement of atoms in a periodic but imperfect lattice. Even scarce imperfections in this arrangement play a critical role in the material nonlinear response to a large force, that is, plastic yield or brittle failure. Without it, it would be reasonable to think that a piece of material would break at $Y/8$–$Y/15$ stress, that is about 10% strain [3]. However, all single-phase solids have much lower $\sigma_Y$ values, around $Y/10^4$, due to the presence of dislocations, stacking-faults , grain boundaries, voids, point defects, etc. The stress induces motion of the pre-existing defects, or a nucleation of the new ones in an almost perfect solid, and makes the deformation irreversible and permanent. The level of strain where this begins to occur at a noticeable rate determines the *yield strain* $\varepsilon_Y$ or *yield stress* $\sigma_Y$ . In the case of tension this threshold reflects truly the strength of chemical-bonds, and is expected to be high for C–C based material.

A possible way to strengthen some materials is by introducing extrinsic obstacles that hinder or block the motion of dislocations [32]. There is a limit to the magnitude of strengthening that a material may benefit from, as too many obstacles will freeze (pin) the dislocations and make the solid brittle. A single-phase material with immobile dislocations or no dislocations at all breaks in a brittle fashion, with little work required. The reason is that it is energetically more favorable for a small crack to grow and propagate. Energy dissipation due to crack propagation represents materials toughness, that is a work required to advance the crack by a unit area, $G > 2\gamma$ (which can be just above the doubled surface energy $\gamma$ for a brittle material, but is several orders of magnitude greater for a ductile material like copper). Since the $c$-edge dislocations in graphite are known to have very low mobility, and are the so called *sessile* type [36], we must expect that nanotubes *per se* are brittle, unless the temperature is extremely high. Their expected high strength does not mean significant toughness, and as soon as the yield point is reached, an individual nanotube will fail quickly and with little dissipation of energy. However, in a large microstructured material, the pull-out and relative shear *between* the tubes and the matrix can dissipate a lot of energy, making the overall material (composite) toughness improved. Although detailed data is not available yet, these differences are important to keep in mind.

*Compression strength* is another important mechanical parameter, but its nature is completely different from the strength in tension. Usually it does not involve any bond reorganization in the atomic lattice, but is due to the buckling on the surface of a fiber or the outermost layer of nanotube. The standard measurement [37] involves the so called "loop test" where tightening of the loop abruptly changes its aspect ratio from 1.34 (elastic) to higher values when kinks develop on the compressive side of the loop. In nanotube studies, this is often called *bending strength*, and the tests are performed using an atomic force microscope (AFM) tip [74], but essentially in both cases one deals with the same intrinsic instability of a laminated structure under compression [62].

These concepts, similarly to linear elastic characteristics, should be applied to carbon and composite nanotubes with care. At the current stage of this research, nanotubes are either assumed to be structurally perfect or to contain few defects, which are also defined with atomic precision (the traditional approach of the physical chemists, for whom a molecule is a well-defined unit). A proper averaging of the "molecular" response to external forces, in order to derive meaningful material characteristics, represents a formidable task for theory. Our quantitative understanding of inelastic mechanical behavior of carbon, BN and other inorganic nanotubes is just beginning to emerge, and will be important for the assessment of their engineering potential, as well as a tractable example of the physics of fracture.

# 3 Experimental Observations

There is a growing body of experimental evidence indicating that carbon nanotubes (both MWNT and SWNT) have indeed extraordinary mechanical properties. However, the technical difficulties involved in the manipulation of these nano-scale structures make the direct determination of their mechanical properties a rather challenging task.

### 3.1 Measurements of the Young's modulus

Nevertheless, a number of experimental measurements of the Young's modulus of nanotubes have been reported.

The first such study [71] correlated the amplitude of the thermal vibrations of the free ends of anchored nanotubes as a function of temperature with the Young's modulus. Regarding a MWNT as a hollow cylinder with a given wall thickness, one can obtain a relation between the amplitude of the tip oscillations (in the limit of small deflections), and the Young's modulus. In fact, considering the nanotube as a cylinder with the high elastic constant $c_{11} = 1.06$ TPa and the corresponding Young's modulus 1.02 TPa of graphite and using the standard beam deflection formula one can calculate the bending of the nanotube under applied external force. In this case, the deflection of a cantilever beam of length $L$ with a force $F$ exerted at its free end is given by $\delta = FL^3/(3YI)$, where $I$ is the moment of inertia. The basic idea behind the technique of measuring free-standing room-temperature vibrations in a TEM, is to consider the limit of small amplitudes in the motion of a vibrating cantilever, governed by the well known fourth-order wave equation, $y_{tt} = -(YI/\varrho A)y_{xxxx}$, where $A$ is the cross sectional area, and $\varrho$ is the density of the rod material. For a clamped rod the boundary conditions are such that the function and its first derivative are zero at the origin and the second and third derivative are zero at the end of the rod. Thermal nanotube vibrations are essentially elastic relaxed phonons in equilibrium with the environment; therefore the amplitude of vibration changes stochastically with time. This stochastic driven oscillator model is solved in [38] to more accurately analyze the experimental results in terms of the Gaussian vibrational-profile with a standard deviation given by

$$\sigma^2 = \sum_{n=0}^{\infty} \sigma_n^2 = 0.4243 \frac{L^3 kT}{Y(D_\mathrm{o}^4 - D_\mathrm{i}^4)}, \tag{4}$$

with $D_\mathrm{o}$ and $D_\mathrm{i}$ the outer and inner radii, $T$ the temperature and $\sigma_n$ the standard deviation. An important assumption is that the nanotube is uniform along its length. Therefore, the method works best on the straight, clean nanotubes. Then, by plotting the mean-square vibration amplitude as a function of temperature one can get the value of the Young's modulus.

This technique was first used in [71] to measure the Young's modulus of carbon nanotubes. The amplitude of those oscillations was defined by means of careful TEM observations of a number of nanotubes. The authors obtained an average value of 1.8 TPa for the Young's modulus, though there was significant scatter in the data (from 0.4 to 4.15 TPa for individual tubes). Although this number is subject to large error bars, it is nevertheless indicative of the exceptional axial stiffness of these materials. More recently studies

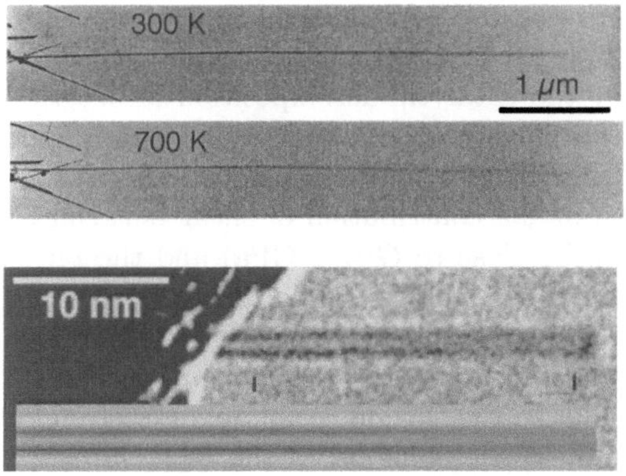


**Fig. 1.** *Top panel:* bright field TEM images of free-standing multi-wall carbon nanotubes showing the blurring of the tips due to thermal vibration, from 300 to 600 K. Detailed measurement of the vibration amplitude is used to estimate the stiffness of the nanotube beam [71]. *Bottom panel:* micrograph of single-wall nanotube at room temperature, with the inserted simulated image corresponding to the best-squares fit adjusting the tube length $L$, diameter $d$ and vibration amplitude (in this example, $L = 36.8$ nm, $d = 1.5$ nm, $\sigma = 0.33$ nm, and $Y = 1.33 \pm 0.2$ TPa) [38]

on SWNT's using the same technique have been reported, Fig. 1 [38]. A larger sample of nanotubes was used, and a somewhat smaller average value was obtained, $Y = 1.25 - 0.35/+0.45$ TPa, around the expected value for graphite along the basal plane. The technique has also been used in [14] to estimate the Young's modulus for BN nanotubes. The results indicate that these composite tubes are also exceptionally stiff, having a value of $Y$ around 1.22 TPa, very close to the value obtained for carbon nanotubes.

Another way to probe the mechanical properties of nanotubes is to use the tip of an AFM (atomic force microscope) to bend anchored CNT's while simultaneously recording the force exerted by the tube as a function of the displacement from its equilibrium position. This allows one to extract the Young's modulus of the nanotube, and based on such measurements [74] have reported a mean value of $1.28 \pm 0.5$ TPa with no dependence on tube diameter for MWNT, in agreement with the previous experimental results. Also [60] used a similar idea, which consists of depositing MWNT's or SWNT's bundled in ropes on a polished aluminum ultra-filtration membrane. Many tubes are then found to lie across the holes present in the membrane, with a fraction of their length suspended. Attractive interactions between the nanotubes and the membrane clamp the tubes to the substrate. The tip of an AFM is then used to exert a load on the suspended length of the nanotube, measuring at the same time the nanotube deflection. To minimize the uncertainty of the applied force, they calibrated the spring constant of each AFM tip (usually 0.1 N/m) by measuring its resonant frequency. The slope of the deflection versus force curve gives directly the Young's modulus for a known length and tube radius. In this way, the mean value of the Young's modulus obtained for arc-grown carbon nanotubes was $0.81 \pm 0.41$ TPa. (The same study applied to disordered nanotubes obtained by the catalytic decomposition of acetylene gave values between 10 to 50 GPa. This result is likely due to the higher density of structural defects present in these nanotubes.) In the case of ropes, the analysis allows the separation of the contribution of shear between the constituent SWNT's (evaluated to be close to $G = 1$ GPa) and the tensile modulus, close to 1 TPa for the individual tubes. A similar procedure has also been used [48] with an AFM to record the profile of a MWNT lying across an electrode array. The attractive substrate-nanotube force was approximated by a van der Waals attraction similar to the carbon–graphite interaction but taking into account the different dielectric constant of the $SiO_2$ substrate; the Poisson ratio of 0.16 is taken from *ab initio* calculations. With these approximations the Young modulus of the MWNT was estimated to be in the order of 1 TPa, in good accordance with the other experimental results.

An alternative method of measuring the elastic bending modulus of nanotubes as a function of diameter has been presented by *Poncharal* et al. [52]. The new technique was based on a resonant electrostatic deflection of a multiwall carbon nanotube under an external ac-field. The idea was to apply a time-dependent voltage to the nanotube adjusting the frequency of the source to resonantly excite the vibration of the bending modes of the nanotube, and to relate the frequencies of these modes directly to the Young modulus of the sample. For small diameter tubes this modulus is about 1 TPa, in good agreement with the other reports. However, this modulus is shown to decrease by one order of magnitude when the nanotube diameter increases (from 8 to 40 nm). This decrease must be related to the emergence of a different bending mode for the nanotube. In fact, this corresponds to a wave-like distortion of the inner side of the bent nanotube. This is clearly shown in Fig. 2. The amplitude of the wave-like distortion increases uniformly from essentially zero for layers close to the nanotube center to about 2–3 nm for the outer layers without any evidence of discontinuity or defects. The non-linear behavior is discussed in more detail in the next section and has been observed in a static rather than dynamic version by many authors in different contexts [19,34,41,58].

Although the experimental data on elastic modulus are not very uniform, overall the results correspond to the values of in-plane rigidity (2) $C = 340 - 440$ N/m, that is to the values $Y = 1.0 - 1.3$ GPa for multiwall tubules, and to $Y = 4C/d = (1.36 - 1.76)$ TPa nm/d for SWNT's of diameter $d$.

## 3.2 Evidence of Nonlinear Mechanics and Resilience of Nanotubes

Large amplitude deformations, beyond the Hookean behavior, reveal nonlinear properties of nanotubes, unusual for other molecules or for the graphite fibers. Both experimental evidence and theory-simulations suggest the ability

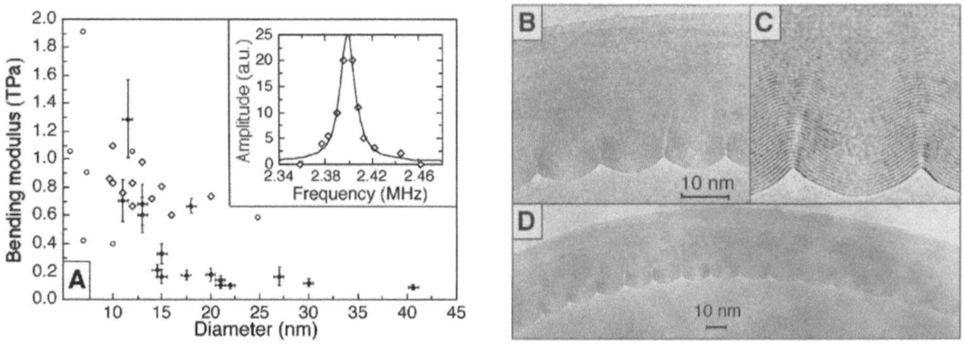


**Fig. 2.** A: bending modulus $Y$ for MWNT as a function of diameter measured by the resonant response of the nanotube to an alternating applied potential (the *inset* shows the Lorentzian line-shape of the resonance). The dramatic drop in Y value is attributed to the onset of a wave-like distortion for thicker nanotubes. D: high-resolution TEM of a bent nanotube with a curvature radius of 400 nm exhibiting a wave-like distortion. B,C: the magnified views of a portion of D [52]

of nanotubes to significantly change their shape, accommodating to external forces without irreversible atomic rearrangements. They develop kinks or ripples (multiwalled tubes) in compression and bending, flatten into deflated ribbons under torsion, and still can reversibly restore their original shape. This resilience is unexpected for a graphite-like material, although folding of the mono-atomic graphitic sheets has been observed [22]. It must be attributed to the small dimension of the tubules, which leaves no room for the stress-concentrators — micro-cracks or dislocation failure piles (cf. Sect. 4.4), making a macroscopic material prone to failure. A variety of experimental evidence confirms that nanotubes can sustain significant nonlinear elastic deformations. However, observations in the nonlinear domain rarely could directly yield a measurement of the threshold stress or the force magnitudes. The facts are mostly limited to geometrical data obtained with high-resolution imaging.

An early observation of noticeable flattening of the walls in a close contact of two MWNT has been attributed to van der Walls forces pressing the cylinders to each other [59]. Similarly, a crystal-array [68] of parallel nanotubes will flatten at the lines of contact between them so as to maximize the attractive van der Waals intertube interaction (see Sect. 5.1). Collapsed forms of the nanotube ("nanoribbons"), also caused by van der Waals attraction, have been observed in experiment (Fig. 3d), and their stability can be explained by the competition between the van der Waals and elastic energies (see Sect. 5.1).

Graphically more striking evidence of resilience is provided by bent structures [19,34], Fig. 4. The bending seems fully reversible up to very large bending angles, despite the occurrence of kinks and highly strained tubule regions

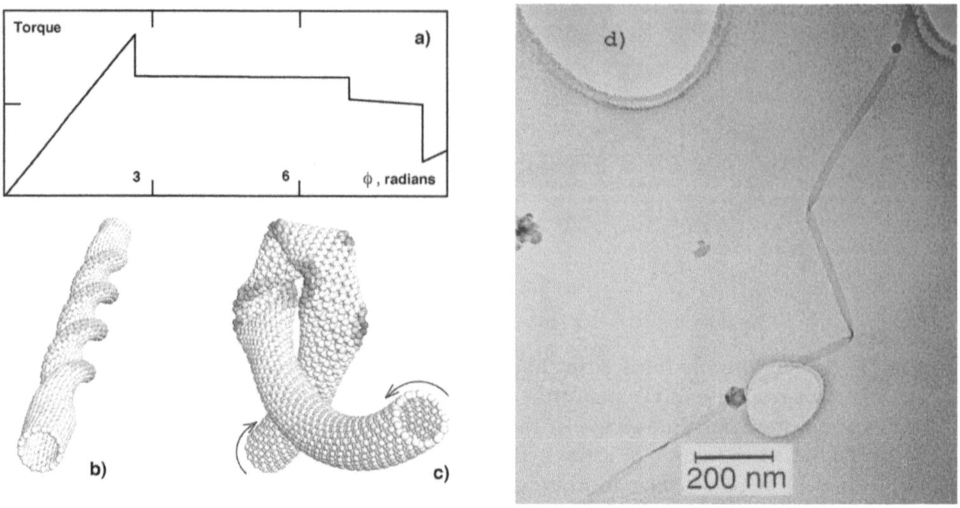


**Fig. 3.** Simulation of torsion and collapse [76]. The strain energy of a 25 nm long (13, 0) tube as a function of torsion angle $f$ (**a**). At the first bifurcation the cylinder flattens into a straight spiral (**b**) and then the entire helix buckles sideways, and coils in a forced tertiary structure (**c**). Collapsed tube (**d**) as observed in experiment [13]

in simulations, which are in excellent morphological agreement with the experimental images [34]. This apparent flexibility stems from the ability of the $sp^2$ network to rehybridize when deformed out of plane, the degree of $sp^2$–$sp^3$ rehybridization being proportional to the local curvature [27]. The accumulated evidence thus suggests that the strength of the carbon–carbon bond does not guarantee resistance to radial, normal to the graphene plane deformations. In fact, the graphitic sheets of the nanotubes, or of a plane graphite [33] though difficult to stretch are easy to bend and to deform.

A measurement with the Atomic Force Microscope (AFM) tip detects the "failure" of a multiwall tubule in bending [74], which essentially represents nonlinear buckling on the compressive side of the bent tube. The measured local stress is 15–28 GPa, very close to the calculated value [62,79]. Buckling and rippling of the outermost layers in a dynamic resonant bending has been directly observed and is responsible for the apparent softening of MWNT of larger diameters. A variety of largely and reversibly distorted (estimated up to 15% of local strain) configurations of the nanotubes has been achieved with AFM tip [23,30]. The ability of nanotubes to "survive the crash" during the impact with the sample/substrate reported in [17] also documents their ability to reversibly undergo large nonlinear deformations.

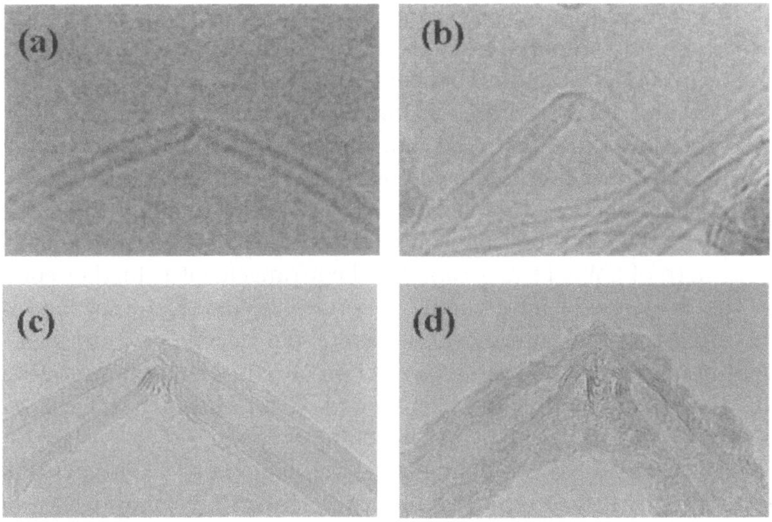


**Fig. 4.** HREM images of bent nanotubes under mechanical duress. (**a**) and (**b**) single kinks in the middle of SWNT with diameters of 0.8 and 1.2 nm, respectively. (**c**) and (**d**) MWNT of about 8nm diameter showing a single and a two-kink complex, respectively [34]

### 3.3 Attempts of Strength Measurements

Reports on measurements of carbon nanotube strength are scarce, and remain the subject of continuing effort. A nanotube is too small to be pulled apart with standard tension devices, and too strong for tiny "optical tweezers", for example. The proper instruments are still to be built, or experimentalists should wait until longer nanotubes are grown.

A bending strength of the MWNT has been reliably measured with the AFM tip [74], but this kind of failure is due to buckling of graphene layers, not the C–C bond rearrangement. Accordingly, the detected strength, up to 28.5 GPa, is two times lower than 53.4 GPa observed for non-laminated SiC nanorods in the same series of experiments. Another group [23] estimates the maximum sustained tensile strain on the outside surface of a bent tubule as large as 16%, which (with any of the commonly accepted values of the Young's modulus) corresponds to 100–150 GPa stress. On the other hand, some residual deformation that follows such large strain can be an evidence of the beginning of yield and the 5/7-defects nucleation. A detailed study of the failure via buckling and collapse of matrix-embedded carbon nanotube must be mentioned here [41], although again these compressive failure mechanisms are essentially different from the bond-breaking yield processes in tension (as discussed in Sects. 4.3,4.4).

Actual tensile load can be applied to the nanotube immersed in matrix materials, provided the adhesion is sufficiently good. Such experiments, with stress-induced fragmentation of carbon nanotube in a polymer matrix has been reported, and an estimated strength of the tubes is 45 GPa, based on a simple isostrain model of the carbon nanotube-matrix. It has also to be remembered that the authors [72] interpret the contrast bands in HRTEM images as the locations of failure, although the imaging of the carbon nanotube through the polymer film limits the resolution in these experiments.

While a singular single-wall nanotube is an extremely difficult object for mechanical tests due to its small molecular dimensions, the measurement of the "true" strength of SWNTs in a rope-bundle arrangement is further complicated by the weakness of inter-tubular lateral adhesion. External load is likely to be applied to the outermost tubules in the bundle, and its transfer and distribution across the rope cross-section obscures the interpretation of the data. Low shear moduli in the ropes (1 GPa) indeed has been reported [60].

Recently, a suspended SWNT bundle-rope was exposed to a sideways pull by the AFM tip [73]. It was reported to sustain reversibly many cycles of elastic elongation up to 6%. If this elongation is actually transferred directly to the individual constituent tubules, the corresponding tensile strength then is above 45 GPa. This number is in agreement with that for multiwalled tubes mentioned above [72], although the details of strain distribution can not be revealed in this experiment.

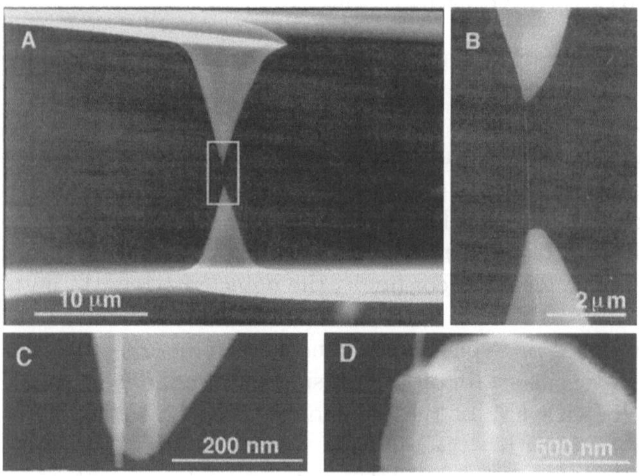


**Fig. 5.** A: SEM image of two oppositely aligned AFM tips holding a MWCNT which is attached at both ends on the AFM silicon tip surface by electron beam deposition of carbonaceous material. The lower AFM tip in the image is on a soft cantilever whose deflection is used to determine the applied force on the MWCNT. B–D: Large magnification SEM image of the indicated region in (A) and the weld of the MWCNT on the top AFM tip [84]

A direct tensile, rather than sideways, pull of a multiwall tube or a rope has a clear advantage due to simpler load distribution, and an important step in this direction has been recently reported [84]. In this work tensile-load experiments (Fig. 5) are performed for MWNTs reporting tensile strengths in the range of 11 to 63 GPa with no apparent dependence on the outer shell diameter. The nanotube broke in the outermost layer ("sword in sheath" failure) and the analysis of the stress-strain curves (Fig. 6) indicates a Young's modulus for this layer between 270 and 950 GPa. Moreover, the measured strain at failure can be as high as 12% change in length. These high breaking strain values also agree with the evidence of stability of highly stressed graphene shells in irradiated fullerene onions [5].

In spite of significant progress in experiments on the strength of nanotubes that have yielded important results, a direct and reliable measurement remains an important challenge for nanotechnology and materials physics.

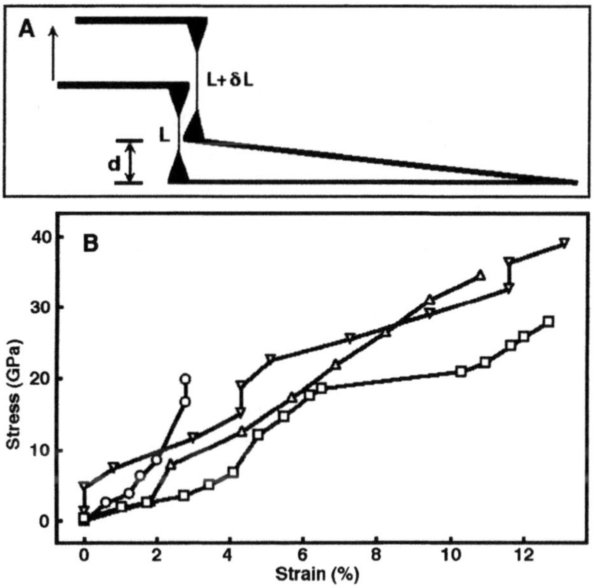


**Fig. 6. A:** A schematic explaining the principle of the tensile-loading experiment. **B:** Plot of stress versus strain curves for individual MWCNTs [84]

# 4 Theoretical and Computational Models

## 4.1 Theoretical Results on Elastic Constants of Nanotubes

An early theoretical report based on an empirical Keating force model for a finite, capped (5,5) tube [49] could be used to estimate a Young's modulus about 5 TPa (five times stiffer than iridium). This seemingly high value is likely due to the small length and cross-section of the chosen tube (only 400 atoms and diameter $d = 0.7$ nm). In a study of structural instabilities of SWNT at large deformations (see Sect. 4.2) the Young's modulus that had to be assigned to *the wall* was 5 TPa, in order to fit the results of molecular dynamics simulations to the continuum elasticity theory [75,76]. From the point of view of elasticity theory, the definition of the Young's modulus involves the specification of the value of the thickness $h$ of the tube wall. In this sense, the large value of $Y$ obtained in [75,76] is consistent with a value of $h = 0.07$ nm for the thickness of the graphene plane. It is smaller than the value used in other work [28,42,54] that simply took the value of the graphite interlayer spacing of $h = 0.34$ nm. All these results agree in the values of inherent stiffness of the graphene layer $Yh = C$, (2), which is close to the value for graphite, $C = Yh = 342$ N/m. Further, the effective moduli of a material uniformly distributed within the entire single wall nanotube cross section will be $Y_t = 4C/d$ or $Y_b = 8C/d$, that is different for axial tension or bending, thus emphasizing the arbitrariness of a "uniform material" substitution.

The moduli $C$ for a SWNT can be extracted from the second derivative of the *ab initio* strain energy with respect to the axial strain, $d^2E/d\varepsilon^2$. Recent calculations [61] show an average value of 56 eV, and a very small variation between tubes with different radii and chirality, always within the limit of accuracy of the calculation. We therefore can conclude that the effect of curvature and chirality on the elastic properties of the graphene shell is small. Also, the results clearly show that there are no appreciable differences between this elastic constant as obtained for nanotubes and for a single graphene sheet. The *ab initio* results are also in good agreement with those obtained in [54] using Tersoff-Brenner potentials, around 59 eV/atom, with very little dependence on radius and/or chirality.

Tight-binding calculations of the stiffness of SWNTs also demonstrate that the Young modulus depends little on the tube diameter and chirality [28], in agreement with the first principles calculations mentioned above. It is predicted that carbon nanotubes have the highest modulus of all the different types of composite tubes considered: BN, $BC_3$, $BC_2N$, $C_3N_4$, CN [29]. Those results for the C and BN nanotubes are reproduced in the left panel of Fig. 7. The Young's modulus approaches the graphite limit for diameters of the order of 1.2 nm. The computed value of $C$ for the wider carbon nanotubes is 430 N/m; that corresponds to 1.26 TPa Young's modulus (with $h = 0.34$ nm), in rather good agreement with the value of 1.28 TPa reported for multi-wall nanotubes [74]. Although this result is for MWNT, the similarity between SWNT is not surprising as the intra-wall C–C bonds mainly determine the moduli. From these results one can estimate the Young's modulus for two relevant geometries: (i) *multiwall* tubes, with the normal area calculated using the interlayer spacing h approximately equal to the one of graphite, and (ii) *nanorope* or *bundle* configuration of SWNTs, where the tubes form a hexagonal closed packed lattice, with a lattice constant of $(d + 0.34$ nm$)$. The results for these two cases are presented in the right panel of Fig. 7. The MWNT

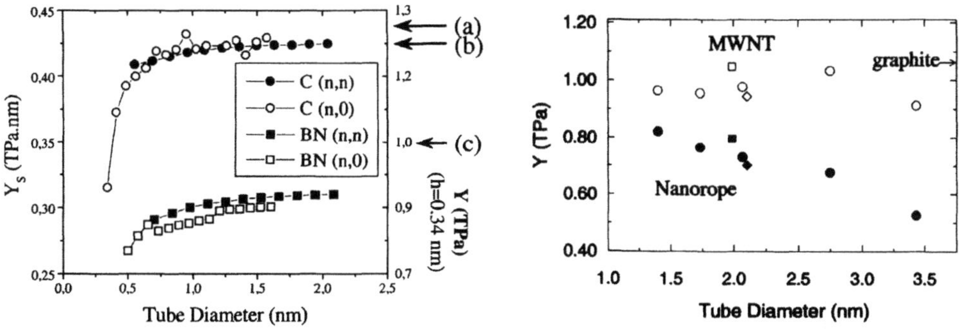


**Fig. 7.** *Left panel:* Young modulus for armchair and zig-zag C- and BN- nanotubes. The values are given in the proper units of TPa · nm for SWNTs (*left axis*), and converted to TPa (*right axis*) by taking a value for the graphene thickness of 0.34 nm. The experimental values for carbon nanotubes are shown on the *right-hand-side:* (**a**) 1.28 TPa [74]; (**b**) 1.25 TPa [38]; (**c**) 1 TPa for MWNT [48]. *Right panel:* Young's modulus versus tube diameter in different arrangements. *Open symbols* correspond to the multi-wall geometry (10 layer tube), and *solid symbols* for the SWNT crystalline-rope configuration. In the MWNT geometry the value of the Young's modulus does not depend on the specific number of layers (adapted from [61])

geometry gives a value that is very close to the graphitic one. The rope geometry shows a decrease of the Young's modulus with the increasing tube diameter, simply proportional to the decreasing mass-density. The computed values for MWNT and SWNT ropes are within the range of the reported experimental data, (Sect. 3.1).

Values of the Poisson ratio vary in different model computations within the range 0.15–0.28, around the value 0.19 for graphite. Since these values always enter the energy of the tube in combination with unity (5), the deviations from 0.19 are not, overall, very significant. More important is the value of another modulus, associated with the tube curvature rather than in-plane stretching. Fig. 8 shows the elastic energy of carbon and the newer composite BN and $BC_3$ SWNT. The energy is smaller for the composite than for the carbon tubules. This fact can be related to a small value of the elastic constants in the composite tubes as compared to graphite. From the results of Fig. 8 we clearly see that the strain energy of C, BN and $BC_3$ nanotubes follows the $D'/d^2$ law expected from linear elasticity theory, cf. (5). This dependence is satisfied quite accurately, even for tubes as narrow as $(4,4)$. For carbon armchair tubes the constant in the strain energy equation has a value of $D' = 0.08\,eV\,\mathrm{nm}^2/\,\mathrm{atom}$ (and up to 0.09 for other chiral tubes) [61]. Previous calculations using Tersoff and Tersoff-Brenner potentials [54] predict the same dependence and give a value of $D' \sim 0.06\,\mathrm{eV\,nm}^2/\,\mathrm{atom}$ and $D' \sim 0.046\,\mathrm{eV\,nm}^2/\,\mathrm{atom}$. The latter corresponds to the value $D = 0.85\,\mathrm{eV}$ in the energy per area as in (5), since the area per atom is $0.0262\,\mathrm{nm}^2$. We note in

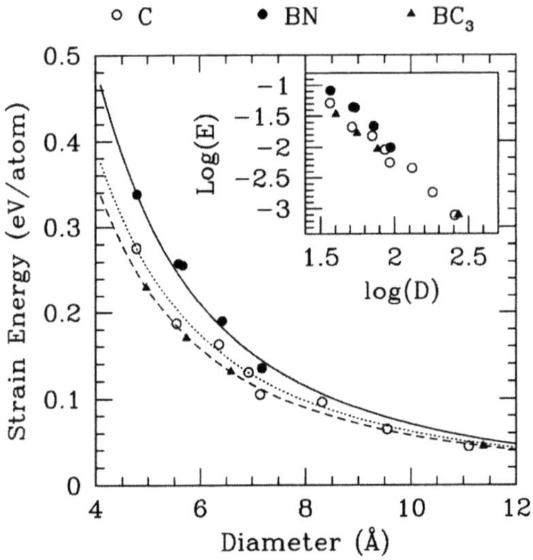


**Fig. 8.** Ab initio results for the total strain energy per atom as a function of the tubule diameter, $d$, for C- (*solid circle*), $BC_3$- (*solid triangle*) and BN-(*open circle*) tubules. The data points are fitted to the classical elastic function $1/d^2$. The *inset* shows in a log plot more clearly the $1/d^2$ dependence of the strain energy for all these tubes. We note that the elasticity picture holds down to sub-nanometer scale. The three calculations for $BC_3$ tubes correspond to the $(3,0)$, $(2,2)$ and $(4,0)$ tubes (adapted from [7,46,47,56])

Fig. 8 that the armchair $(n,n)$ tubes are energetically more stable as compared to other chiralities with the same radius. This difference is, however, very small and decreases as the tube diameter increases. This is expected, since in the limit of large radii the same graphene value is attained, regardless of chirality. It is to some extent surprising that the predictions from elasticity theory are so similar to those of the detailed *ab initio* calculations. In [1] a complementary explanation based on microscopic arguments is provided. In a very simplified model the energetics of many different fullerene structures depend on a single structural parameter: the *planarity* $\phi_\pi$, which is the angle formed by the $\pi$-orbitals of neighbor atoms. Assuming that the change in total energy is mainly due to the change in the nearest neighbor hopping interaction between these orbitals, and that this change is proportional to $\cos(\phi_\pi)$, the $d^{-2}$ behavior is obtained. By using non-self-consistent first-principles calculations they have obtained a value of $D' = 0.085\,\mathrm{eV\,nm^2/\,atom}$, similar to the self-consistent value given above.

### 4.2 Nonlinear Elastic Deformations and Shell Model

Calculations of the elastic properties of carbon nanotubes confirm that they are extremely rigid in the axial direction (high tensile) and more readily distort in the perpendicular direction (radial deformations), due to their high aspect ratio. The detailed studies, stimulated first by experimental reports of visible kinks in the molecules, lead us to conclude that, in spite of their molecular size, nanotubes obey very well the laws of continuum shell theory [2,39,70].

One of the outstanding features of fullerenes is their hollow structure, built of atoms densely packed along a closed surface that defines the overall shape. This also manifests itself in dynamic properties of molecules, which greatly resemble the macroscopic objects of continuum elasticity known as *shells.* Macroscopic shells and rods have long been of interest: the first study dates back to Euler, who discovered the elastic instability. A rod subject to longitudinal compression remains straight but shortens by some fraction $\varepsilon$, proportional to the force, until a critical value (Euler force) is reached. It then becomes unstable and buckles sideways at $\varepsilon > \varepsilon_{cr}$, while the force almost does not vary. For hollow tubules there is also a possibility of local buckling in addition to buckling as a whole. Therefore, more than one bifurcation can be observed, thus causing an overall nonlinear response of nanotubes to the large deforming forces (note that local mechanics of the constituent shells may well still remain within the elastic domain).

In application to fullerenes, the theory of shells now serves a useful guide [16,25,63,75,76,78], but its relevance for a covalent-bonded system of only a few atoms in diameter was far from being obvious. MD simulations seem better suited for objects that small. Perhaps the first MD-type simulation indicating the macroscopic scaling of the tubular motion emerged in the study of nonlinear resonance [65]. Soon results of detailed MD simulations for a nanotube under axial compression allowed one to introduce concepts of elasticity of shells and to adapt them to nanotubes [75,76]. MD results for other modes of load have also been compared with those suggested by the continuum model and, even more importantly, with experimental evidence [34] (see Fig. 4 in Sect. 3.2).

Figure 9 shows a simulated nanotube exposed to *axial compression.* The atomic interaction was modeled by the Tersoff-Brenner potential, which reproduces the lattice constants, binding energies, and the elastic constants of graphite and diamond. The end atoms were shifted along the axis by small steps and the whole tube was relaxed by the conjugate-gradient method while keeping the ends constrained. At small strains the total energy (Fig. 9a) grows as $E(\varepsilon) = (\frac{1}{2})E'' \cdot \varepsilon^2$, where $E'' = 59$ eV/atom. The presence of four singularities at higher strains was quite a striking feature, and the patterns (b)–(e) illustrate the corresponding morphological changes. The shading indicates strain energy per atom, equally spaced from below 0.5 eV (brightest) to above 1.5 eV (darkest). The sequence of singularities in $E(\varepsilon)$ corresponds to a loss of molecular symmetry from $D_{\infty h}$ to $S_4$, $D_{2h}$, $C_{2h}$ and $C_1$. This evolution of the molecular structure can be described within the framework of continuum elasticity.

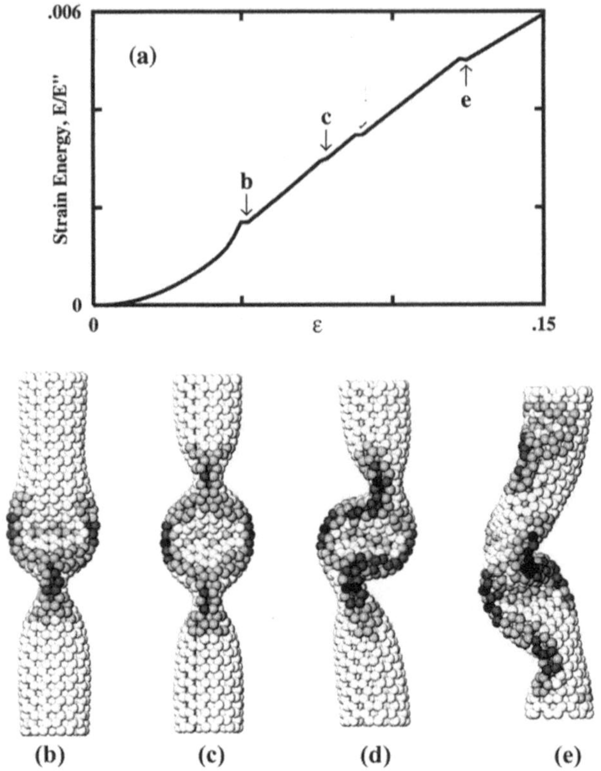


**Fig. 9.** Simulation of a (7, 7) nanotube exposed to axial compression, $L = 6\,\mathrm{nm}$. The strain energy (**a**) displays four singularities corresponding to shape changes. At $\varepsilon_c = 0.05$ the cylinder buckles into the pattern (**b**), displaying two identical flattenings, "fins", perpendicular to each other. Further increase of $\varepsilon$ enhances this pattern gradually until at $\varepsilon_2 = 0.076$ the tube switches to a three-fin pattern (**c**), which still possesses a straight axis. In a buckling sideways at $\varepsilon_3 = 0.09$ the flattenings serve as hinges, and only a plane of symmetry is preserved (**d**). At $\varepsilon_4 = 0.13$ an entirely squashed asymmetric configuration forms (**e**) (from [75])

The intrinsic symmetry of a graphite sheet is hexagonal, and the elastic properties of two-dimensional hexagonal structures are isotropic. A curved sheet can also be approximated by a uniform shell with only two elastic parameters: flexural rigidity $D$, and its resistance to an in-plane stretching, the in-plane stiffness $C$. The energy of a shell is given by a surface integral of the quadratic form of local deformation,

$$E = \frac{1}{2}\int\!\!\int \{D[(\kappa_x + \kappa_y)^2 - 2(1-\nu)(\kappa_x\kappa_y - \kappa_{xy}^2)] + \frac{C}{(1-\nu^2)}[(\varepsilon_x + \varepsilon_y)^2 - 2(1-\nu)(\varepsilon_x\varepsilon_y - \varepsilon_{xy}^2)]\}dS\,, \tag{5}$$

where $\kappa$ is the curvature variation, $\varepsilon$ is the in-plane strain, and $x$ and $y$ are local coordinates). In order to adapt this formalism to a graphitic tubule, the values of $D$ and $C$ are identified by comparison with the detailed *ab initio* and semi-empirical studies of nanotube energetics at small strains [1,54]. Indeed, the second derivative of total energy with respect to axial strain corresponds to the in-plane rigidity $C$ (cf. Sect. 3.1). Similarly, the strain energy as a function of tube diameter $d$ corresponds to $2D/d^2$ in (5). Using the data of [54], one obtains $C = 59$ eV/atom $= 360$ J/m$^2$, and $D = 0.88$ eV. The Poisson ratio $\nu = 0.19$ was extracted from a reduction of the diameter of a tube stretched in simulations. A similar value is obtained from experimental elastic constants of single crystal graphite [36]. One can make a further step towards a more tangible picture of a tube as having wall thickness $h$ and Young's modulus $Y_s$. Using the standard relations $D = Yh^3/12(1-\nu^2)$ and $C = Y_s h$, one finds $Y_s = 5.5$ TPa and $h = 0.067$ nm. With these parameters, linear stability analysis [39,70] allows one to assess the nanotube behavior under strain.

To illustrate the efficiency of the shell model, consider briefly the case of imposed axial compression. A trial perturbation of a cylinder has a form of Fourier harmonics, with $M$ azimuthal lobes and $N$ half-waves along the tube (Fig. 10, inset), i.e. sines and cosines of arguments $2My/d$ and $N\pi x/L$. At a critical level of the imposed strain, $\varepsilon_c(M, N)$, the energy variation (4.1) vanishes for this shape disturbance. The cylinder becomes unstable and lowers its energy by assuming an $(M, N)$-pattern. For tubes of $d = 1$ nm with the shell parameters identified above, the critical strain is shown in Fig. 10. According to these plots, for a tube with $L > 10$ nm the bifurcation is first attained for $M = 1$, $N = 1$. The tube preserves its circular cross section and

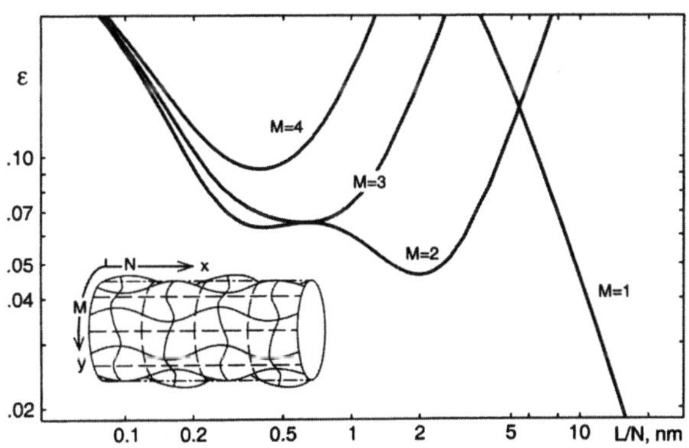


**Fig. 10.** The critical strain levels for a continuous, 1 nm wide shell-tube as a function of its scaled length $L/N$. A buckling pattern $(M, N)$ is defined by the number of half-waves $2M$ and $N$ in $y$ and $x$ directions, respectively, e.g., a (4, 4)-pattern is shown in the *inset*. The effective moduli and thickness are fit to graphene (from [75])

buckles sideways as a whole; the critical strain is close to that for a simple rod,

$$\varepsilon_c = 1/2(\pi d/L)^2\,, \tag{6}$$

or four times less for a tube with hinged (unclamped) ends. For a shorter tube the situation is different. The lowest critical strain occurs for $M = 2$ (and $N \geq 1$, see Fig. 10), with a few separated flattenings in directions perpendicular to each other, while the axis remains straight. For such a local buckling, in contrast to (6), the critical strain depends little on length and estimates to $\varepsilon_c = 4\sqrt{D/C}\; d^{-1} = (2/\sqrt{3})(1-\nu^2)^{-1/2}\; hd^{-1}$ in the so-called *Lorenz limit.* For a nanotube one finds,

$$\varepsilon_c = 0.077\ \mathrm{nm}/d\,. \tag{7}$$

Specifically, for the 1 nm wide tube of length $L = 6\,\mathrm{nm}$, the lowest critical strains occur for the $M = 2$ and $N = 2$ or 3 (Fig. 10), and are close to the value obtained in MD simulations, (Fig. 9a). This is in accord with the two- and three-fin patterns seen in Figs. 9b,c. Higher singularities cannot be quantified by the linear analysis, but they look like a sideways beam buckling, which at this stage becomes a non-uniform object.

Axially compressed tubes of greater length and/or tubes simulated with hinged ends (equivalent to a doubled length) first buckle sideways as a whole at a strain consistent with (6). After that the compression at the ends results in bending and a local buckling inward. This illustrates the importance of the "beam-bending" mode, the softest for a long molecule and most likely to attain significant amplitudes due to either thermal vibrations or environmental forces. In simulations of *bending,* a torque rather than force is applied at the ends and the bending angle $\theta$ increases stepwise. While a notch in the energy plot can be mistaken for numerical noise, its derivative $dE/d\theta$ drops significantly, which unambiguously shows an increase in tube compliance — a signature of a buckling event. In bending, only one side of a tube is compressed and thus can buckle. Assuming that it buckles when its local strain, $\varepsilon = K \cdot (d/2)$, where $K$ is the local curvature, is close to that in axial compression, (7), we estimate the critical curvature as

$$K_c = 0.155\ \mathrm{nm}/d^2\,. \tag{8}$$

This is in excellent agreement (within 4%) with extensive simulations of single wall tubes of various diameters, helicities and lengths [34]. Due to the end effects, the average curvature is less than the local one and the simulated segment buckles somewhat earlier than at $\theta_c = K_c L$, which is accurate for longer tubes.

In simulations of *torsion*, the increase of azimuthal angle $\phi$ between the tube ends results in energy and morphology changes shown in Fig. 3. In the continuum model, the analysis based on (5) is similar to that outlined above, except that it involves skew harmonics of arguments like $N\pi x/L \pm 2My/d$. For overall beam-buckling ($M = 1$),

$$\phi_c = 2(1 + \nu)\pi \tag{9}$$

and for the cylinder-helix flattening ($M = 2$),

$$\phi_c = 0.055 \text{ nm}^{3/2}\ L/d^{5/2}\,. \tag{10}$$

The latter should occur first for $L < 136\ d^{5/2}$ nm, which is true for all tubes we simulated. However, in simulations it occurs later than predicted by (10). The ends, kept circular in simulation, which is physically justifiable, by a presence of rigid caps on normally closed ends of a molecule, deter the through flattening necessary for the helix to form (unlike the local flattening in the case of an axial load).

In the above discussion, the specific values of the parameters $C$ and $D$ (or $Y$ and $h$) are chosen to achieve the best correspondence between the elastic-shell and the MD simulation *within the same study*, performed with the Tersoff-Brenner potential. Independent studies of nanotube dynamics under compression generally agree very well with the above description, although they reveal reasonable deviations in the parameter values [16,25]. More accurate and realistic values can be derived from the TB or the *ab initio* calculations [1,7,57] of the elastic shell, and can be summarized in the somewhat "softer but thicker" shell [76]. Based on a most recent study [28] one obtains effective shell parameters $C = 415$ J/m$^2$ and $D = 1.6$ eV $= 2.6 \times 10^{-19}$ J, that is correspondingly $Y_s = 4.6$ TPa and $h = 0.09$ nm, cf. Sect. 4.1.

Simulations of nanotubes under mechanical duress lead to shapes very different in appearance. At the same time there are robust traits in common: a deformation, proportional to the force within Hooke's law, eventually leads to a collapse of the cylinder and an abrupt change in pattern, or a sequence of such events. The presence of a snap-through buckling of nanotubes allows for a possibility of "shape memory", when in an unloading cycle the switch between patterns occurs at a somewhat lower level of strain. A small hysteresis observed in simulations is practically eliminated by thermal motion at any finite temperature. However, this hysteresis is greatly enhanced by the presence of van der Waals attraction which causes the tube walls to "stick"-flatten together after the collapse, Fig. 3d [13]. The simulations at even a low temperature (e.g. 50 K) shows strongly enhanced thermal vibrations in the vicinity of every pattern switch, while before and after the transition only barely noticeable ripples are seen. Physically, this indicates softening of the system, when one of the eigenvalues tends to zero in the vicinity of the bifurcation.

While several reports focus on a nonlinear dynamics of an open-end SWNT, when the terminal ring atoms are displaced gradually in simulation, a more realistic interaction of a cap-closed SWNT with the (diamond or graphite) substrates has been studied recently [25]. An inward cap collapse and/or sideways sliding of the nanotube tip along the substrate are observed, in addition to the buckling of the tubule itself. Furthermore, an interaction of a small (four SWNT) bundle and a double-wall tubule with the substrates has been also reported [26].

An atomistic modeling of multi-layer tubes remains expensive. It makes extrapolation of the continuum model tempting, but involves an interlayer van der Waals interaction. The flexural rigidity scales as $\sim h^3$ in case of a coherent, and as $\sim h$ for an incoherent stack of layers sliding with respect to each other when the tube is deformed; this affects the mechanical properties and still has to be investigated.

Direct simulations of the tubules under hydrostatic pressure have not been reported to the best of our knowledge. In this scale anisotropic lateral forces in a molecular crystal packing are more plausible than a uniform pressure. An ability of a shell-tubule to bifurcate in a flattened form makes it an example of a two-level system, which manifests in the phase-transition behavior of SWNT crystal, as was first described in [68] and is now indicated by several experimental reports. While the faceting in the triangular crystal packing results in a partial wall flattening, a singular tubule under hydrostatic pressure can collapse completely. One can resort to continuum elasticity and estimate a pressure leading to an inward buckling as $p_{\mathrm{c}} = 2Y(h/d)^3$, that is thousands of atmospheres for a nanometer tube. However, it drops fast with the diameter and is assisted by a flattening effects of twisting or bending and by van der Waals attraction between the opposite walls [13]. Such collapse cannot occur simultaneously throughout the significant SWNT length, but rather propagates at a certain speed depending on the ambient over-pressure $u \propto \sqrt{(p - p_{\mathrm{c}})}$. This pressure dependence [76] is similar to the observations on macroscopic objects like underwater pipelines [50].

## 4.3 Atomistics of High Strain-Rate Failure

The simulations of compression, torsion, and tension described above (Sect. 4.2) do not show any bond breaking or atoms switching their positions, in spite of the very large local strain in the nanotubes. This supports the study of *axial tension*, where no shape transformations occur up to an extreme dilation. How strong in tension is a carbon nanotube? Since the tensile load does not lead to any shell-type instabilities, it is transferred more directly to the chemical bond network. The inherent strength of the carbon-carbon bond indicates that the tensile strength of carbon nanotubes might exceed that of other known fibers. Experimental measurements remain complex (Sect. 3.3) due to the small size of the grown single tubes. In the meantime, some tests are being done in computer modeling, especially well suited to the fast strain rate [75,76,77,78]. Indeed, a simulation of an object with thousand atoms even using a classical potential interaction between atoms is usually limited to picoseconds up to nanoseconds of real physical time. This is sufficiently long by molecular standards, as is orders of magnitude greater than the periods of intramolecular vibrations or intermolecular collision times. However, it is still much less than a normal test-time for a material, or an engineering structure. Therefore a standard MD simulation addresses a "molecular strength" of the CNT, leaving the true mechanisms of material behavior to the more subtle considerations (Sect. 4.4).

In MD simulation, the high-strain-rate test proceeds in a very peculiar manner. Fast stretching simply elongates the hexagons in the tube wall, until at the critical point an atomic disorder suddenly nucleates: one or a few C–C bonds break almost simultaneously, and the resulting "hole" in a tube wall becomes a crack precursor (see Fig. 11a). The fracture propagates very quickly along the circumference of the tube. A further stage of fracture displays an interesting feature, the formation of two or more distinct chains of atoms, $\ldots = C = C = C = \ldots$, spanning the two tube fragments, Fig. 11b. The vigorous motion (substantially above the thermal level) results in frequent collisions between the chains; they coalesce, and soon only one such chain survives. A further increase of the distance between the tube ends does not break this chain, which elongates by increasing the number of carbon atoms that pop out from both sides into the necklace. This scenario is similar to the monatomic chain unraveling suggested in field-emission exper-

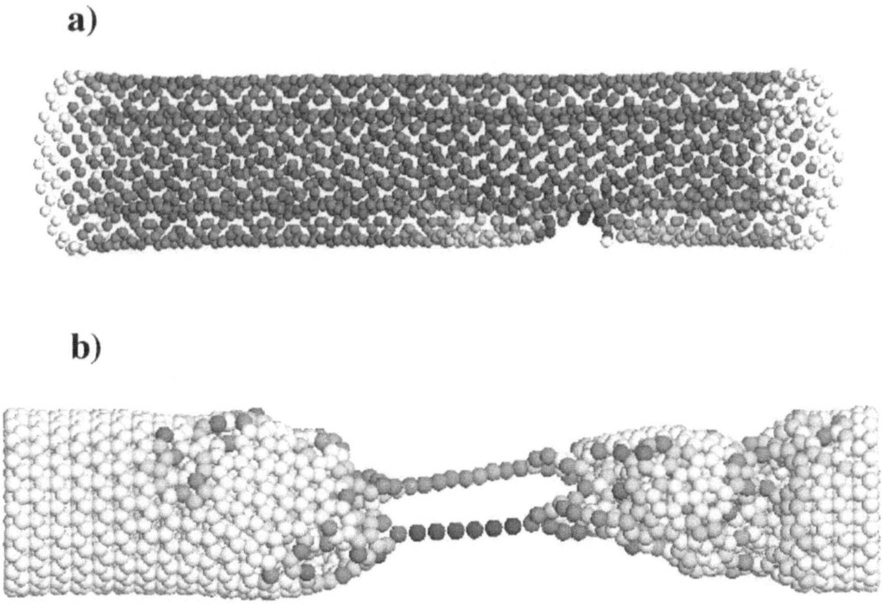


**Fig. 11.** High strain rate tension of a two-wall tube begins from the outermost layer, nucleating a crack precursor (**a**), where the atomic size is reduced to make the internal layer visible. Eventually it leads to the formation of monatomic chains (**b**) (from [77])

iments [53], where the electrostatic force unravels the tube like the sleeve of a sweater. Notably, the breaking strain in such fast-snap simulations is about 30%, and varies with temperature and the strain rate. (For a rope of nanotubes this translates to a more than 150 GPa breaking stress.) This high breaking strain value is consistent with the stability limit (inflection point on the energy curve) of 28% for symmetric low-temperature expansion of graphene sheet [64], and with some evidence of stability of highly stresses graphene shells in irradiated fullerene onions [5].

## 4.4 Yield Strength and Relaxation Mechanisms in Nanotubes

Fast strain rate (in the range of 100 MHz) simulations correspond to the elongation of the tubule at percents of the speed of sound. In contrast to such "molecular tension test", materials engineering is more concerned with the static or slow tension conditions, when the sample is loaded during significantly longer time. Fracture, of course, is a kinetic process where time is an important parameter. Even a small tension, as any non-hydrostatic stress, makes a nanotube thermodynamically meta-stable and a generation of defects energetically favorable. In order to study a slow strength-determining relaxation process, preceding the fast fracture, one should either perform extensive simulations at exceedingly elevated temperature [9,10], or apply dislocation failure theory [79,81]. It has been shown that in a crystal lattice such as the wall of a CNT, a yield to deformation must begin with a homogeneous nucleation of a slip by the shear stress present. The non-basal edge dislocations emerging in such a slip have a well-defined core, a pentagon-heptagon pair, 5/7. Therefore, the prime dipole is equivalent to the Stone–Wales (SW) defect [20] (Fig. 12). The nucleation of this prime dislocation dipole "unlocks" the nanotube for further relaxation: either brittle cleavage or a plastic flow. Remarkably, the latter corresponds to a motion of dislocations along the helical paths (glide "planes") within the nanotube wall. This causes a stepwise (quantized) necking, when the domains of different chiral symmetry and, therefore, different electronic structure are formed, thus coupling the mechanical and electrical properties [79,80]. It has further been shown [10,51,62,79,80,81,85] that the energetics of such nucleation explicitly depend on nanotube helicity.

Below, we deduce [79,81], starting with dislocation theory, the atomistics of mechanical relaxation under extreme tension. Locally, the wall of a nanotube differs little from a single graphene sheet, a two-dimensional crystal of carbon. When a uniaxial tension $\sigma$ (N/m — for the two-dimensional wall it is convenient to use force per unit length of its circumference) is applied it can be represented as a sum of expansion (locally isotropic within the wall) and a shear of a magnitude $\sigma/2$ (directed at $\pm 45°$ with respect to tension). Generally, in a macroscopic crystal the shear stress relaxes by a movement of *dislocations*, the edges of the atomic extra-planes. Burgers vector $\boldsymbol{b}$ quantifies the mismatch in the lattice due to a dislocation [32]. Its glide requires only

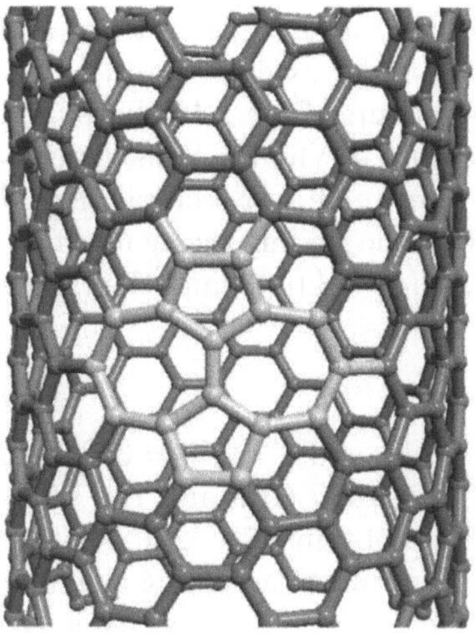

**Fig. 12.** Stone–Wales (SW) dipole embedded in a nanotube hexagonal wall [67]

local atomic rearrangements and presents the easiest way for strain release, provided there is sufficient thermal agitation. In an initially *perfect* lattice such as the wall of a nanotube, a yield to a great axial tension begins with a homogeneous *nucleation* of a slip, when a dipole of dislocations (a tiny loop in three-dimensional case) first has to form. The formation and further glide are driven by the reduction of the applied-stress energy, as characterized by the elastic Peach-Koehler force on a dislocation failure. The force component along $\boldsymbol{b}$ is proportional to the shear in this direction and thus depends on the angle between the Burgers vector and the circumference of the tube,

$$f_{\mathrm{b}} = -\frac{1}{2}\sigma|b|\sin 2\theta\,, \tag{11}$$

The max $|f_{\mathrm{b}}|$ is attained on two $\pm 45°$ lines, which mark the directions of a slip in an isotropic material under tension.

The graphene wall of the nanotube is not isotropic; its hexagonal symmetry governs the three glide planes — the three lines of closest zigzag atomic packing, oriented at 120° to each other (corresponding to the $\{10\bar{1}\,l\}$ set of planes in three-dimensional graphite). At non-zero shear these directions are prone to slip. The corresponding c-axis edge dislocations involved in such a slip are indeed known in graphite [21,36]. The six possible Burgers vectors $1/3a\langle 2\bar{1}\bar{1}\ 0\rangle$ have a magnitude $b = a = 0.246$ nm (lattice constant), and the dislocation core is identified as a 5/7 pentagon-heptagon pair in the honeycomb lattice of hexagons. Therefore, the primary nucleated dipole must have a 5/7/7/5 configuration (a 5/7 attached to an inverted 7/5 core). This configuration is obtained in the perfect lattice (or a nanotube wall) by a 90° rotation of a single C–C bond, well known in fullerene science as a Stone–Wales diatomic interchange [20]. One is led to conclude that the SW transformation is equivalent to the smallest slip in a hexagonal lattice and must play a key role in the nanotube relaxation under external force.

The preferred glide is the closest to the maximum-shear ±45° lines, and depends on how the graphene strip is rolled-up into a cylinder. This depends on nanotube helicity specified by the chiral indices $(c_1, c_2)$ or a chiral angle $\theta$ indicating how far the circumference departs from the leading zigzag motif $\boldsymbol{a}_1$. The max $|f_\mathrm{b}|$ is attained for the dislocations with $\boldsymbol{b} = \pm(0,1)$ and their glide reduces the strain energy,

$$E_\mathrm{g} = -|f_\mathrm{b} a| = -Ca^2/2 \cdot \sin(2\theta + 60°)\varepsilon \,, \tag{12}$$

per one displacement, $a$. Here $\varepsilon$ is the applied strain, and $C = Yh = 342$ N/m can be derived from the Young modulus of $Y = 1020$ GPa and the interlayer spacing $h = 0.335$ nm in graphite; one then obtains $Ca^2/2 = 64.5$ eV. Equation (12) allows one to compare different nanotubes (assuming a similar amount of pre-existing dislocations); the more energetically favorable is the glide in a tube, the earlier it must yield to applied strain.

In a pristine nanotube-molecule, the 5/7 dislocations have first to emerge as a dipole, by a prime SW transformation. Topologically, the SW defect is equivalent to either one of the two dipoles, each formed by an $\sim a/2$ slip. Applying (11) to each of the slips one finds,

$$E_\mathrm{sw} = E_o - A\varepsilon - B\sin(2\theta + 30°)\varepsilon \,. \tag{13}$$

The first two terms, the zero-strain formation energy and possible isotropic dilation, do not depend on nanotube symmetry. The symmetry-dependent third term, which can also be derived as a leading term in the Fourier series, describes the essential effect: SW rotation gains more energy in the armchair ($\theta = 30°$) nanotube, making it the weakest, most inclined to SW nucleation of the dislocations, in contrast to the zigzag ($\theta = 0$) where the nucleation is least favorable.

Consider, for example, a $(c, c)$ armchair nanotube as a typical representative (we will also see below that this armchair type can undergo a more general scenario of relaxation.) The initial stress-induced SW rotation creates a geometry that can be viewed as either a dislocation dipole or a tiny crack along the equator. Once "unlocked", the SW defect can ease further relaxation. At this stage, both brittle (dislocation pile-up and crack extension), or plastic (separation of dislocations and their glide away from each other) routes are possible, the former usually at larger stress and the latter at higher temperatures [9,10,79,80,81].

Formally, both routes correspond to a further sequence of SW switches. The 90° rotation of the bonds at the "crack tip" (Fig. 13, left column) will result in a 7/8/7 flaw and then 7/8/8/7 etc. This further strains the bonds-partitions between the larger polygons, leading eventually to their breakage, with the formation of greater openings like 7/14/7 etc. If the crack, represented by this sequence, surpasses the critical Griffith size, it cleaves the tubule.

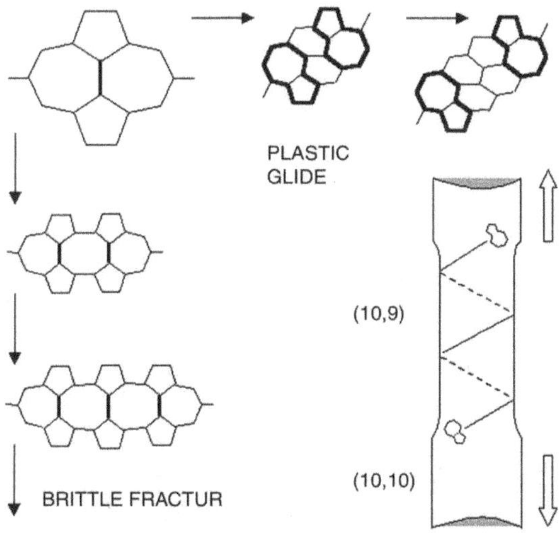


**Fig. 13.** SW transformations of an equatorially oriented bond into a vertical position creates a nucleus of relaxation (*top left corner*). It evolves further as either a crack (brittle fracture route, *left column*) or as a couple of dislocations gliding away along the spiral slip plane (plastic yield, *top row*). In both cases only SW rotations are required as elementary steps. The stepwise change of the nanotube diameter reflects the change of chirality (*bottom right image*) causing the corresponding variations of electrical properties [81]

In a more interesting distinct alternative, the SW rotation of another bond (Fig. 13, top row) divides the 5/7 and 7/5, as they become two dislocation cores separated by a single row of hexagons. A next similar SW switch results in a double-row separated pair of the 5/7's, and so on. This corresponds, at very high temperatures, to a plastic flow *inside* the nanotube-molecule, when the 5/7 and 7/5 twins glide away from each other driven by the elastic forces, thus reducing the total strain energy [cf. (12)]. One remarkable feature of such glide is due to mere cylindrical geometry: the glide "planes" in case of nanotubes are actually spirals, and the slow thermally-activated Brownian walk of the dislocations proceeds along these well-defined trajectories. Similarly, their extra-planes are just the rows of atoms also curved into the helices.

A nanotube with a 5/7 defect in its wall loses axial symmetry and has a bent equilibrium shape; the calculations show [12] the junction angles $< 15°$. Interestingly then, an exposure of an even achiral nanotube to the axially symmetric tension generates two 5/7 dislocations, and when the tension is removed, the tube "freezes" in an asymmetric configuration, S-shaped or C-shaped, depending on the distance of glide, that is time of exposure. Of course the symmetry is conserved statistically, since many different shapes form under identical conditions.

When the dislocations sweep a noticeable distance, they leave behind a tube segment changed strictly following the topological rules of dislocation theory. By considering a planar development of the tube segment containing a 5/7, for the new chirality vector $\boldsymbol{c}'$ one finds,

$$(c_1', c_2') = (c_1, c_2) - (b_1, b_2)\,, \tag{14}$$

with the corresponding reduction of diameter, $d$. While the dislocations of the first dipole glide away, a generation of another dipole results, as shown above, in further narrowing and proportional elongation under stress, thus forming a neck. The orientation of a generated dislocation dipole is determined every time by the Burgers vector closest to the lines of maximum shear ($\pm 45°$ cross at the end-point of the current circumference-vector $\boldsymbol{c}$). The evolution of a $(c, c)$ tube will be: $(c, c) \to (c, c-1) \to (c, c-2) \to \ldots (c, 0) \to [(c-1, 1)$ or $(c, -1)] \to (c-1, 0) \to [(c-2, 1)$ or $(c-1, -1)] \to (c-2, 0) \to [(c-3, 1)$ or $(c-2, -1)] \to (c-3, 0)$ etc. It abandons the armchair $(c, c)$ type entirely, but then oscillates in the vicinity of to be zigzag (c,0) kind, which appears a peculiar attractor. Correspondingly, the diameter for a $(10, 10)$ tube changes stepwise, $d = 1.36$, 1.29, 1.22, 1.16 nm, etc., the local stress grows in proportion and this quantized necking can be terminated by a cleave at late stages. Interestingly, such plastic flow is accompanied by the change of electronic structure of the emerging domains, governed by the vector $(c_1, c_2)$. The armchair tubes are metallic, others are semiconducting with the different band gap values. The 5/7 pair separating two domains of different chirality has been discussed as a pure-carbon heterojunction [11,12]. It is argued to cause the current rectification detected in a nanotube nanodevice [15] and can be used to modify, in a controlled way, the electronic structure of the tube. Here we see how this electronic heterogeneity can arise from a mechanical relaxation at high temperature: if the initial tube was armchair-metallic, the plastic dilation transforms it into a semiconducting type irreversibly.

*Computer simulations* have provided a compelling evidence of the mechanisms discussed above. By carefully tuning the tension in the tubule and gradually elevating its temperature, with extensive periods of MD annealing, the first stages of the mechanical yield of CNT have been observed [9,10]. In simulation of tensile load the novel patterns in plasticity and breakage, just described above, clearly emerge.

Classical MD simulations have been carried out for tubes of various geometries with diameters up to 13 nm. Such simulations, although limited by the physical assumptions used in deriving the interatomic potential, are still invaluable tools in investigating very large systems in the time scales that are characteristic of fracture and plasticity phenomena. Systems containing up to 5000 atoms have been studied for simulation times of the order of nanoseconds. The ability of the classical potential to correctly reproduce the energetics of the nanotube systems has been verified through comparisons with TB and *ab initio* simulations [9,10].

Beyond a critical value of the tension, an armchair nanotube under axial tension releases its excess strain via spontaneous formation of a SW defect through the rotation of a C-C bond producing two pentagons and two heptagons, 5/7/7/5 (Fig. 14). Further, the calculations [9,10] show the energy of the defect formation, and the activation barrier, to decrease approximately linearly with the applied tension; for (10,10) tube the formation energy can be approximated as $E_{\mathrm{sw}}(\mathrm{eV}) = 2.3 - 40\varepsilon$. The appearance of a SW defect represents the nucleation of a (degenerate) dislocation loop in the planar hexagonal network of the graphite sheet. The configuration 5/7/7/5 of this primary dipole is a 5/7 core attached to an inverted 7/5 core, and each 5/7 defect can indeed further behave as a single edge dislocation in the graphitic plane. Once nucleated, the dislocation loop can split in simulations into two

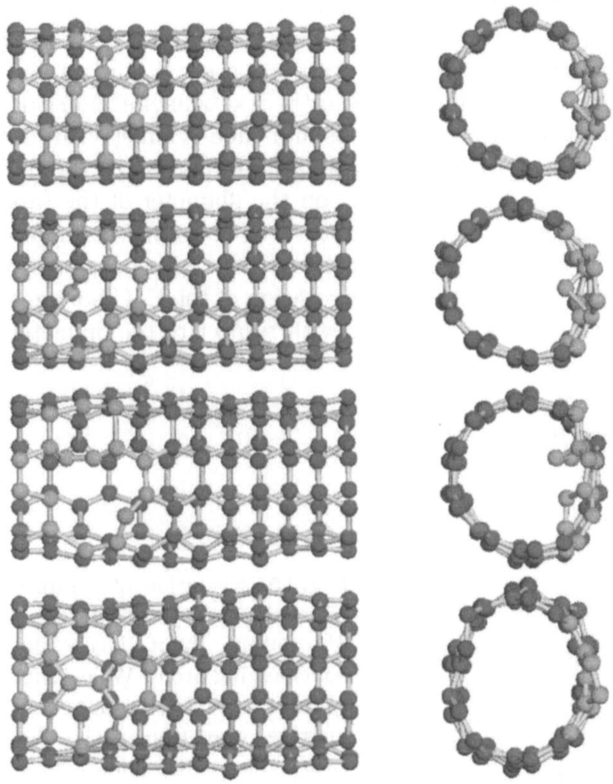

**Fig. 14.** Kinetic mechanism of 5/7/7/5 defect formation from an *ab-initio* quantum mechanical molecular dynamics simulation for the (5, 5) tube at 1800 K [10]. The atoms that take part in the Stone–Wales transformation are in lighter gray. The four snapshots show the various stages of the defect formation, from *top to bottom:* system in the ideal configurations ($t = 0$ ps); breaking of the first bond ($t = 0.10$ ps); breaking of the second bond ($t = 0.15$ ps); the defect is formed ($t = 0.20$ ps)

dislocation cores, $5/7/7/5 \leftrightarrow 5/7+7/5$, which are then seen to glide through successive SW bond rotations. This corresponds to a plastic flow of dislocations and gives rise to possible ductile behavior. The thermally activated migration of the cores proceeds along the well-defined trajectories (Fig. 15) and leaves behind a tube segment changed according to the rules of dislocation theory, (14). The tube thus abandons the armchair symmetry $(c, c)$ and undergoes a visible reduction of the diameter, a first step of the possible quantized necking in "intramolecular plasticity" [79,80,81].

The study, based on the extensive use of classical, tight-binding and ab initio MD simulations [10], shows that the different orientations of the carbon bonds with respect to the strain axis (in tubes of different symmetry) lead to different scenarios. Ductile or brittle behaviors can be observed in nanotubes of different indices under the same external conditions. Furthermore, the behavior of nanotubes under large tensile strain strongly depends on their symmetry and diameter. Several modes of behavior are identified, and a map of their ductile-vs-brittle behavior has been proposed. While graphite is brittle, carbon nanotubes can exhibit plastic or brittle behavior under deformation, depending on the external conditions and tube symmetry. In the case of a zig-zag nanotube (longitudinal tension), the formation of the SW defect is strongly dependent on curvature, i.e., on the diameter of the tube and gives rise to a wide variety of behaviors in the brittle-vs-ductile map of stress response of carbon nanotubes [10]. In particular, the formation energy of the off-axis 5/7/7/5 defect (obtained via the rotation of the C–C bond oriented 120° to the tube axis) shows a crossover with respect to the diameter.

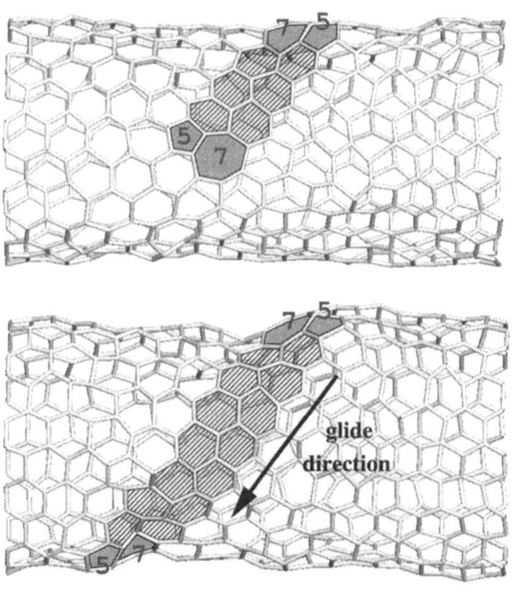


**Fig. 15.** Evolution of a (10,10) nanotube at $T = 3000$ K, strain 3 % within about 2.5 ns time. An emerging Stone–Wales defect splits into two 5/7 cores which migrate away from each other, each step of this motion being a single-bond rotation. The shaded area indicates the migration path of the 5/7 edge dislocation failure [9] and the resulting nanotube segment is reduced to the (10,9) in accord with (14) [80,81]

It is negative for $(c, 0)$ tubes with $c < 14$ ($d < 1.1$ nm). The effect is clearly due to the variation in curvature, which in the small-diameter tubes makes the process energetically advantageous. Therefore, above a critical value of the curvature a plastic behavior is possible and the tubes can be ductile.

Overall, after the nucleation of a first 5/7/7/5 defect in the hexagonal network either brittle cleavage or plastic flow are possible, depending on tube symmetry, applied tension and temperature. Under high strain and low temperature conditions, all tubes are brittle. If, on the contrary, external conditions favor plastic flow, such as low strain and high temperature, tubes of diameter less than approximately 1.1 nm show a completely ductile behavior, while larger tubes are moderately or completely brittle depending on their symmetry.

# 5 Supramolecular Interactions

Most of the theoretical discussions of the structure and properties of carbon nanotubes involve free unsupported nanotubes. However, in almost all experimental situations the nanotubes are supported on a solid substrate with which they interact. Similarly, nanotubes in close proximity to each other will interact and tend to associate and form larger aggregates [69,82].

## 5.1 Nanotube–Substrate and Nanotube–Nanotube Interactions: Binding and Distortions

These nanotube–substrate interactions can be physical or chemical. So far, however, only physical interactions have been explored. The large polarizability of carbon nanotubes (see article by S. Louie in this volume) implies that these physical interactions (primarily van der Waals forces) are significant. One very important consequence of the strong adhesive forces with which carbon nanotubes bind to a substrate is the deformation of the atomic structure of the nanotube itself. An experimental demonstration of this effect is given in Fig. 16, which shows non-contact AFM images of two pairs of overlapping multi-wall nanotubes deposited on an inert H-passivated silicon surface. The nanotubes are clearly distorted in the overlap regions with the upper nanotubes bending around the lower ones [30,31]. These distortions arise from the tendency of the upper CNTs to increase their area of contact with the substrate so as to increase their adhesion energy. Countcracting this tendency is the rise in strain energy produced from the increased curvature of the upper tubes and the distortion of the lower tube. The total energy of the system can be expressed as an integral of the strain energy $U(\kappa)$ and the adhesion energy $V(z)$ over the entire tube profile: $E = \int\{U(\kappa) + V[z(x)]dx\}$. Here, $\kappa$ is the local tube curvature and $V[z(x)]$ the nanotube-substrate interaction potential at a distance $z$ above the surface. Using the experimental value of Young's modulus for MWNTs [71,74] and by fitting to the experimentally observed nanotube profile, one can estimate the binding energy from the observed distortion. For example, for a 100 Å diameter MWNT a binding energy of about 0.8 eV/Å is obtained. Therefore, van der Waals binding energies, which for individual atoms or molecules are weak (typically 0.1 eV), can be quite strong for mesoscopic systems such as the CNTs. High binding energies imply that strong forces are exerted by nanotubes on underlying surface features such as steps, defects, or other nanotubes. For example, the force leading to the compression of the lower tubes in Fig. 16a is estimated to be as high as 30 nN. The effect of these forces can be observed as a reduced inter-tube electrical resistance in crossed tube configurations similar to those shown in Fig. 16 [24].

The axial distortions of CNTs observed in AFM images are also found in molecular dynamics and molecular mechanics simulations. Molecular mechanics represents a simple alternative to the Born-Oppenheimer approximation-based electronic structure calculations. In this case, nuclear motion is studied assuming a fixed electron distribution associated with each atom. The molecular system is described in terms of a collection of spheres representing the atoms, which are connected with springs to their neighbors. The motion of the atoms is described classically using appropriate potential energy functions. The advantage of the approach is that very large systems (many thousands of atoms) can be easily simulated. Figure 17a,b show the results of such simulations involving two single-walled (10,10) CNTs crossing each other over a graphite slab [31]. In addition to their axial distortion, the two nanotubes develop a distorted, non-circular cross-section in the overlap region. Further results on the radial distortions of single-walled nanotubes due to van der Waals interactions with a graphite surface are shown in Fig. 17c. The adhesion forces tend to flatten the bottom of the tubes so as to increase the area of contact. At the same time, there is an increase in the curvature of the tube and therefore a rise in strain energy $E_S$. The resulting overall shape is again dictated by the optimization of these two opposing trends. Small diameter

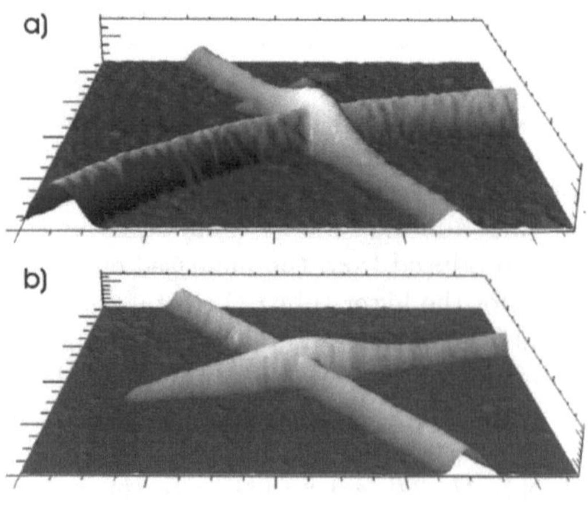


**Fig. 16.** AFM non-contact mode images of two overlapping multi-wall nanotubes. The *upper* tubes are seen to wrap around the *lower* ones which are slightly compressed. The size of image (**a**) is 330 nm × 330 nm and that of (**b**) is 500 nm × 500 nm [4]

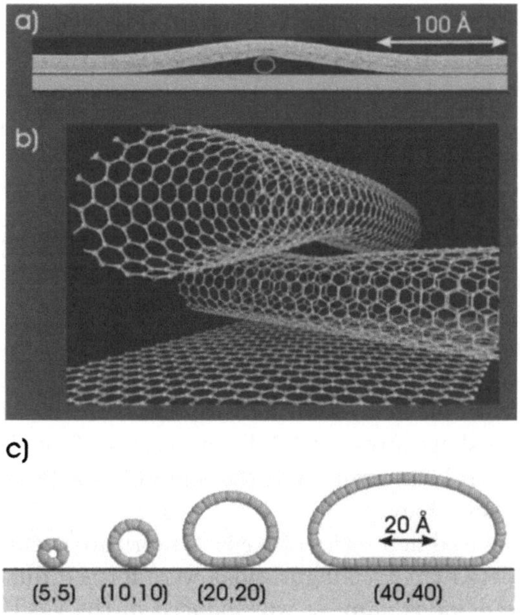


**Fig. 17.** Molecular mechanics calculations on the axial and radial deformation of single-wall carbon nanotubes. (**a**) Axial deformation resulting from the crossing of two (10,10) nanotubes. (**b**) Perspective close up of the same crossed tubes showing that both tubes are deformed near the contact region. (**c**) Computed radial deformations of single-wall nanotubes adsorbed on graphite [4]

tubes that already have a small radius of curvature $R_C$ resist further distortion ($E_S \propto R_C^{-2}$), while large tubes flatten out and increase considerably their binding energy [by 115% in the case of the (40,40) tube]. In the case of MWNTs, we find that as the number of carbon shells increases, the overall gain in adhesion energy due to distortion decreases as a result of the rapidly increasing strain energy [31].

The AFM results and the molecular mechanics calculations indicate that carbon nanotubes in general tend to adjust their structure to follow the surface morphology of the substrate. One can define a critical radius of surface curvature $R_{CRT}$ above which the nanotube can follow the surface structure or roughness. Given that the strain energy varies more strongly with tube diameter ($\propto d^4$) than the adhesion energy ($\propto d$), the critical radius is a function of the NT diameter. For example, $R_{CRT}$ is about $(12d)^{-1}$ for a CNT with a $d = 1.3\,\text{nm}$, while it is about $(50d)^{-1}$ for a CNT with $d = 10\,nm$.

### 5.2 Manipulation of the Position and Shape of Carbon Nanotubes

A key difference between the mechanical properties of CNTs and carbon fibers is the extraordinary flexibility and resistance to fracture of the former. Furthermore, the strong adhesion of the CNTs to their substrate can stabilize highly strained configurations. Deformed, bent and buckled nanotubes were clearly observed early in TEM images [34]. One can also mechanically manipulate and deform the CNTs using an AFM tip and then study the properties of the deformed structures using the same instrument [23,30]. For this purpose one uses the AFM in the so-called contact mode with normal forces of the order of 10–50 nN [30]. It was found that most MWNTs can sustain multiple bendings and unbending without any observable permanent damage. Bending of MWNTs induces buckling, observed in the form of raised points along the CNT, due to the collapsing of shells. When the bending curvature is small a series of regularly spaced buckles appear on the inside wall of the nanotube [23]. This phenomenon is analogous to axial bifurcations predicted by a continuum mechanics treatment of the bending of tubes [39].

In studies of electrical or other properties of individual CNTs it is highly desirable to be able to manipulate them and place them in particular positions of the experimental setup, such as on metal electrodes in conductance studies, or in order to build prototype electronic devices structures. Again the AFM can be used for this purpose. The shear stress of CNTs on most surfaces is high, so that not only can one control the position of the nanotubes at even elevated temperatures, but also their shape.

In Fig. 18, a MWNT is manipulated in a series of steps to fabricate a simple device [4]. While highly distorted CNT configurations were formed during the manipulation process, no obvious damage was induced in the CNT. The same conclusion was reached by molecular dynamics modeling of the bending of CNTs [34]. The ability to prepare locally highly strained configurations stabilized by the interaction with the substrate, and the well known dependence of chemical reactivity on bond strain suggest that manipulation may be used to produce strained sites and make them susceptible to local chemistry. Furthermore, bending or twisting CNTs changes their electrical properties [35,55] and, in principle, this can be used to modify the electrical behavior of CNTs through mechanical deformation.

### 5.3 Self-Organization of Carbon Nanotubes: Nanotube Ropes, Rings, and Ribbons

Van der Waals forces play an important role not only in the interaction of the nanotubes with the substrate but also in their mutual interaction [68]. The different shells of a MWNT interact primarily by van der Waals forces; single-walled tubes form ropes for the same reason [69]. In these ropes the nanotubes form a regular triangular lattice. Calculations have shown that the binding forces in a rope are substantial. For example, the binding energy of 1.4 nm diameter SWNTs is estimated to be about 0.48 eV/nm, and rises to 1.8 eV/nm for 3 nm diameter tubes [68]. The same study showed that the nanotubes may be flattened at the contact areas to increase adhesion [68]. Aggregation of single-walled tubes in ropes is also expected to affect their electronic structure. When a rope is formed from metallic $(10,10)$ nanotubes a pseudogap of the order of 0.1 eV is predicted to open up in the density of states due to the breaking of mirror symmetry in the rope [18].

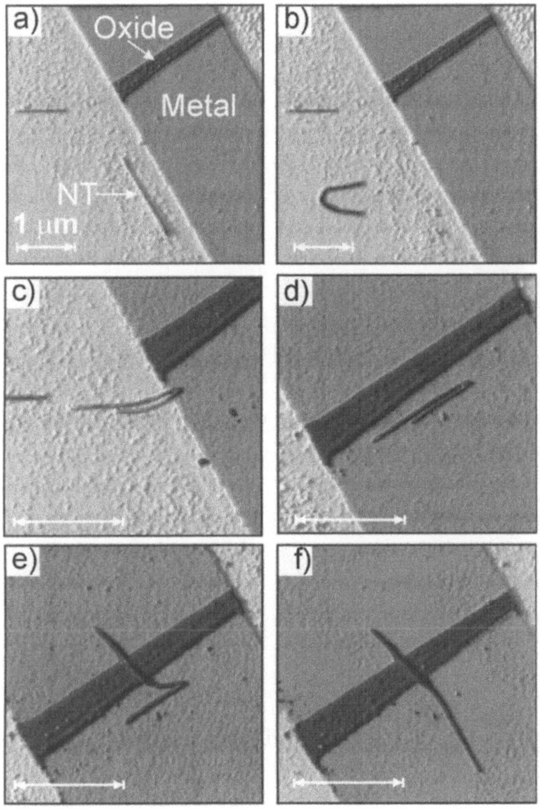


**Fig. 18.** AFM manipulation of a single multi-wall carbon nanotube such that electrical transport through it can be studied. Initially, the nanotube is located on the insulating ($SiO_2$) part of the sample. In a stepwise fashion (not all steps are shown) it is dragged up the 80 Å high metal thin film wire and finally is stretched across the oxide barrier [4]

A different manifestation of van der Waals interactions involves the self-interaction between two segments of the same single-wall CNT to produce a closed ring (loop) [44,45]. Nanotube rings were first observed in trace amounts in the products of laser ablation of graphite and were assigned a toroidal structure [40]. More recently, rings of SWNTs were synthesized with large yields (up to 50%) from straight nanotube segments, Fig. 19. These rings were shown to be coils not tori [45].

The formation of coils by CNTs is particularly intriguing. While coils of biomolecules and polymers are well known structures, they are stabilized by a number of interactions that include hydrogen bonds and ionic interactions [8]. On the other hand, the formation of nanotube coils is surprising, given the high flexural rigidity of CNTs and the fact that CNT coils can only be stabilized by van der Waals forces. However, estimates based on continuum mechanics show that in fact it is easy to compensate for the strain energy induced by the coiling process through the strong adhesion between tube segments in the coil. Figure 20 shows how a given length of nanotube $l$ should be divided between the perimeter of the coil, $2\pi R$, that defines the strain energy and the interaction length, $l_i = l - 2\pi R$, that contributes to

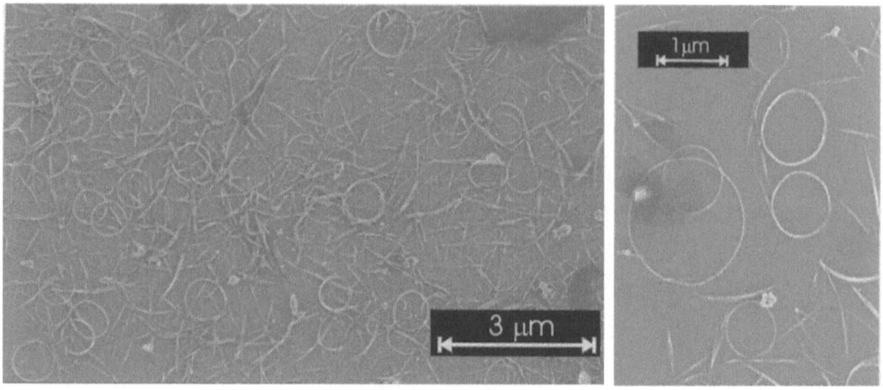


**Fig. 19.** Scanning electron microscope images of rings of single-wall nanotubes dispersed on hydrogen-passivated silicon substrates [45]

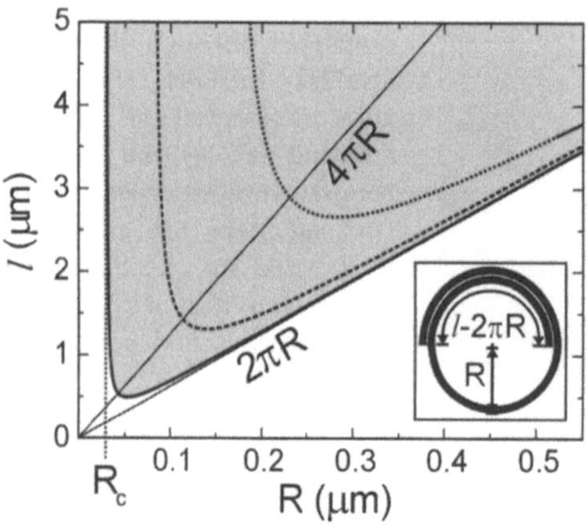


**Fig. 20.** Thermodynamic stability limits for rings formed by coiling single wall nanotubes with radii of 0.7 nm (plain line), 1.5 nm (dashed line), and 4.0 nm (dotted line) calculated using a continuum elastic model [45]

the adhesion (see the schematic in the inset) so that a stable structure is formed [45]. From this figure it is clear that the critical radius $R_C$ for forming rings is small, especially for small radius CNTs such as the (10,10) tube ($r = 0.7\,\text{nm}$).

The coiling process is kinetically controlled. The reason is easy to understand; to form a coil the two ends of the tube have to come first very close to each other before any stabilization (adhesion) begins to take place. This bending involves a large amount of strain energy $E_S \propto R^{-2}$, and the activation energy for coiling will be of the order of this strain energy (i.e. several eV). Similar arguments hold if, instead of a single SWNT, one starts with a SWNT rope. Experimentally, the coiling process is driven by exposure to ultrasound [44]. Ultrasonic irradiation can provide the energy for thermal activation [66], however, it is unrealistic to assume that the huge energy needed is supplied in the form of heat energy. It is far more likely that mechanical processes associated with cavitation, i.e. the formation and collapse of small bubbles in the aqueous solvent medium that are generated by the ultrasonic waves, are responsible for tube bending [66]. The nanotubes may act as nucleation centers for bubble formation so that a hydrophobic nanotube trapped at the bubble-liquid interface is mechanically bent when the bubble collapses. Once formed, a nanotube "proto-ring" can grow thicker by the attachment of other segments of SWNTs or ropes. The synthesis of nanotube rings opens the door for the fabrication of more complex nanotube-based structures relying on a combination of mechanical manipulation and self-adhesion forces.

Finally, we note that opposite sections of the carbon atom shell of a nanotube also attract each other by van der Waals forces, and under certain conditions this attraction energy ($E_{\mathrm{vdW}}$) may lead to the collapse of the nanotube to a ribbon-like structure. Indeed, such structures are often observed in TEM [13] and AFM images [43] of nanotubes (primarily multi-wall tubes). The elastic curvature energy per unit length of a tube is proportional to $1/R$ ($R$, radii of the tubes). However, for a fully collapsed single-wall tubule, the energy contains the higher curvature energy due to the edges, independent of the initial radius, and a negative (attractive) van der Waals contribution, $\varepsilon_{\mathrm{vdW}} \sim 0.03-0.04\,\mathrm{eV/\,atom}$, that is proportional to $R$ per unit length. Collapse occurs when the latter term prevails above a certain critical tube radii $R_{\mathrm{c}}$ that increases with increasing number $N$ of shells of the nanotube. For example: $R_{\mathrm{c}}(N=1) \sim 8d_{\mathrm{vdW}}$ and $R_{\mathrm{c}}(N=8) \sim 19d_{\mathrm{vdW}}$ [13]. The thickness of the collapsed strip-ribbon is obviously $(2N-1)d_{\mathrm{vdW}}$. Any torsional strain imposed on a tube by the experimental environment favors flattening [55,75,76] and facilitates the collapse. The twisting and collapse of a nanotube brings important changes to its electrical properties. For example, a metallic armchair nanotube opens up a gap and becomes a semiconductor as shown in Fig. 21.

## 6 Summary: Nanomechanics at a Glance

In summary, it seems useful to highlight the 'nanomechanics at a glance', based on the knowledge accumulated up-to-date, and omitting technical details and uncertainties. Carbon nanotubes demonstrate very high stiffness to an axial load or a bending of small amplitude, which translates to the record-high efficient linear-elastic moduli. At larger strains, the nanotubes (especially, the single-walled type) are prone to buckling, kink forming and collapse, due to the hollow shell-like structure. These abrupt changes (bifurcations) manifest themselves as singularities in the non-linear stress-strain curve, but are reversible and involve no bond-breaking or atomic rearrangements. This resilience corresponds, quantitatively, to a very small sub-angstrom effective thickness of the constituent graphitic shells. Irreversible yield of nanotubes begins at extremely high deformation (from several to dozens percent of in-plane strain, depending on the strain rate) and high

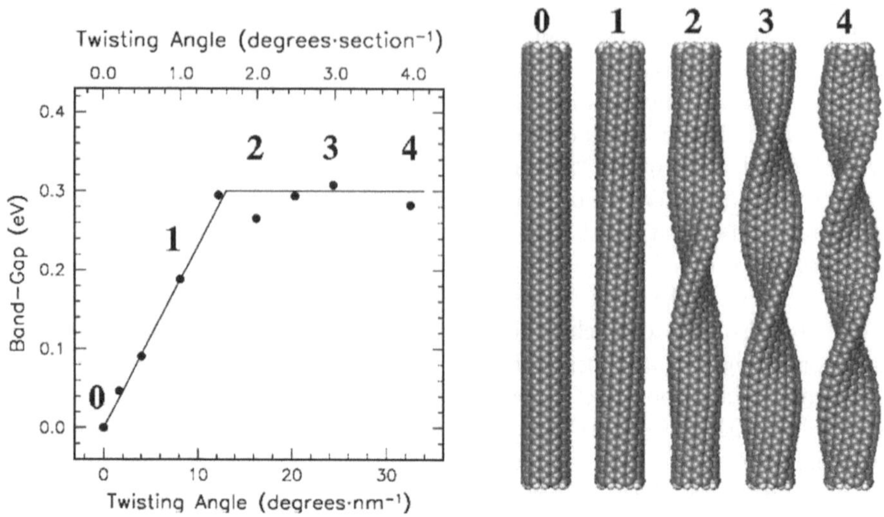


**Fig. 21.** *Right:* Relaxed structures of a (6,6) nanotube computed using molecular mechanics as a function of the twisting angle. *Left:* Computed band-gap energy using extended Huckel theory as a function of the twisting angle [55]

temperature. The atomic relaxation begins with the edge dislocation dipole nucleation, which (in case of carbon) involves a diatomic interchange, i.e. a ninety-degree bond rotation. A sequence of similar diatomic steps ultimately leads to failure of the nanotube filament. The failure threshold (yield strength) turns out to depend explicitly on nanotube helicity, which demonstrates again the profound role of symmetry for the physical properties, either electrical conductivity or mechanical strength. Finally, the manifestation of mechanical strength in the multiwalled or bundled nanotubes (ropes) is obscured by the poor load transfer from the exterior to the core of such larger structure. This must lead to lower apparent strength and even lower linear moduli, as they become limited by the weak lateral interaction between the tubules rather than by their intrinsic carbon bond network. The ultimate strength of nanotubes and their ensembles is an issue that requires the modeling of inherently mesoscopic phenomena, such as plasticity and fracture, on a microscopic, atomistic level, and constitutes a challenge from the theoretical as well as experimental points of view.

## Acknowledgements

B.I.Y. acknowledges support from the U.S. AFOSR/AFRL and from the NASA Ames Center.

## References


1. G. Adams, O. Sankey, J. Page, M. O'Keeffe, D. Drabold, Science **256**, 1792 (1992)
2. H. G. Allen, P. S. Bulson: *Background to Buckling* (McGraw-Hill, London 1980) p. 582
3. M. F. Ashby, Acta. Met. **37**, 1273 (1989)
4. Ph. Avouris, Hertel, T., Martel, R., Schmidt, T., Shea, H. R., Walkup, R. E., Appl. Surf. Sci. **141**, 201 (1999)
5. F. Banhart, P. M. Ajayan, Nature **382**, 433 (1996)
6. R. D. Beck, P. S. John, M. M. Alvarez, F. Diederich, R. L. Whetten, J. Phys. Chem. **95**, 8402 (1991)
7. X. Blase, A. Rubio, S. G. Louie, M. L. Cohen, Europhys. Lett. **28**, 335 (1994)
8. J. W. Bryson, S. F. Betz, H. S. Lu, D. J. Suich, H. X. Zhou, K. T. O'Neil, W. F. DeGrado, Science **270**, 935 (1995)
9. M. Buongiorno-Nardelli, B. I. Yakobson, J. Bernholc, Phys. Rev. Lett. **81**, 4656 (1998)
10. M. Buongiorno-Nardelli, B. I. Yakobson, J. Bernholc, Phys. Rev. B **57**, 4277 (1998)
11. J.-C. Charlier, T. W. Ebbesen, P. Lambin, Phys. Rev. B **53**, 11108 (1996)
12. L. Chico, V. H. Crespi, L. X. Benedict, S. G. Loui, M. L. Cohen, Phys. Rev. Lett. **76**, 971 (1996)
13. N. G. Chopra, L. X. Benedict, V. H. Crespi, M. L. Cohen, S. G. Louie, A. Zettl, Nature **377**, 135 (1995)
14. N. G. Chopra, A. Zettl, Solid State Commun. **105**, 297 (1998)
15. P. G. Collins, A. Zettl, H. Bando, A. Thess, R. E. Smalley, Science **278**, 100 (1997)
16. C. F. Cornwell, L. T. Wille, Solid State Commun. **101**, 555 (1997)
17. H. Dai, J. H. Hafner, A. G. Rinzler, D. T. Colbert, R. E. Smalley, Nature **384**, 147 (1996)
18. P. Delaney, H. J. Choi, J. Ihm, S. G. Louie, M. L. Cohen, Nature (London), **391**, 466 (1998)
19. J. F. Despres, E. Daguerre, K. Lafdi, Carbon **33**, 87 (1995)
20. M. S. Dresselhaus, G. Dresselhaus, P. C. Eklund, *Science of Fullerenes and Carbon Nanotubes* (Academic, San Diego 1996) p. 965
21. M. S. Dresselhaus, G. Dresselhaus, K. Sugihara, I. L. Spain, H. A. Goldberg, *Graphite Fibers and Filaments* (Springer, Berlin, Heidelberg 1988) p. 382
22. T. W. Ebbesen, H. Hiura, Adv. Mater. **7**, 582 (1995)
23. M. R. Falvo, G. J. Clary, R. M. Taylor, V. Chi, F. P. Brooks, S. Washburn, R. Superfine, Nature **389**, 582 (1997)
24. M. S. Fuhrer, J. Nygard, L. Shih, M. Forero, Y.-G. Yoon, M. S. C. Mazzoni, H. J. Choi, J. Ihm, S. G. Louie, A. Zettl, P. L. McEuen, Science, **288**, 494 (2000)
25. A. Garg, J. Han, S. B. Sinnott, Phys. Rev. Lett. **81**, 2260 (1998)
26. A. Garg, S. B. Sinnott, Phys. Rev. B **60**, 13786 (1999)
27. R. C. Haddon, Science **261**, 1545 (1993)
28. E. Hernández, C. Goze, P. Bernier A. Rubio, Phys. Rev. Lett. **80**, 4502 (1998)
29. E. Hernández, C. Goze, P. Bernier, A. Rubio, Appl. Phys. A **68**, 287 (1999)
30. T. Hertel, R. Martel, Ph. Avouris, J. Phys. Chem. B **102**, 910 (1998)
31. T. Hertel, R. E. Walkup, Ph. Avouris, Phys. Rev. B **58**, 13870 (1998)

32. J. P. Hirth, J. Lothe, *Theory of Dislocations* (Wiley, New York 1982) p. 857
33. H. Hiura, T. W. Ebbesen, J. Fujita, K. Tanigaki, T. Takada, Nature **367**, 148 (1994)
34. S. Iijima, C. Brabec, A. Maiti, J. Bernholc, J. Chem. Phys. **104**, 2089 (1996)
35. C. L. Kane, E. J. Mele, Phys. Rev. Lett. **78**, 1932 (1997)
36. B. T. Kelly, *Physics of Graphite* (Applied Science Publishers, London 1981) p. 478
37. V. V. Kozey, H. Jiang, V. R. Mehta, S. Kumar, J. Mater. Res. **10**, 1044 (1995)
38. A. Krishnan, E. Dujardin, T. W. Ebbesen, P. N. Yanilos, M. M. J. Treacy, Phys. Rev. B **58**, 14013 (1998)
39. L. D. Landau, E. M. Lifshitz: *Theory of Elasticity* (Pergamon, Oxford 1986)
40. J. Liu, H. Dai, J. H. Hafner, D. T. Colbert, R. E. Smalley, S. J. Tans, C. Dekker, Nature (London) **385**, 780 (1997)
41. O. Lourie, D. M. Cox, H. D. Wagner, Phys. Rev. Lett. **81**, 1638 (1998)
42. J. P. Lu, Phys. Rev. Lett. **79**, 1297 (1997)
43. R. Martel, T. Schmidt, H. R: Shea, T. Hertel, Ph. Avouris, Appl. Phys. Lett. **73**, 2447 (1998)
44. R. Martel, H. R. Shea, Ph. Avouris, Nature (London) **398**, 582 (1999)
45. R. Martel, H. R. Shea, Ph. Avouris, J. Phys. Chem. B **103**, 7551 (1999)
46. Y. Miyamoto, A. Rubio, S. G. Louie, M. L. Cohen, Phys. Rev. B **50**, 4976 (1994)
47. Y. Miyamoto, A. Rubio, S. G. Louie, M. L. Cohen, Phys. Rev. B **50**, 18360 (1994)
48. J. Muster, M. Burghard, S. Roth, G. S. Dusberg, E. Hernandez, A. Rubio, J. Vac. Sci. Technol. **16**, 2796 (1998)
49. G. Overney, W. Zhong, D. Tománek, Z. Phys. D **27**, 93 (1993)
50. A. C. Palmer, J. H. Martin, Nature **254**, 46 (1975)
51. D. Pierson, C. Richardson, B. I. Yakobson, 6th Foresight Conference on Molecular Nanotechnology, Santa Clara, CA, http:// www.foresight.org /Conferences (1998)
52. P. Poncharal, Z. L. Wang, D. Ugarte, W. A. de Heer, Science **283**, 1513 (1999)
53. A. G. Rinzler, J. H. Hafner, P. Nikolaev, L. Lou, S. G. Kim, D. Tománek, P. Nordlander, D. T. Colbert, R. E. Smalley, Science **269**, 1550 (1995)
54. D. H. Robertson, D. W. Brenner, J. W. Mintmire, Phys. Rev. B **45**, 12592 (1992)
55. A. Rochefort, Ph. Avouris, F. Lesage, D. R. Salahub, Phys. Rev. B **60**, 13824 (1999)
56. A. Rubio, J. L. Corkill, M. L. Cohen, Phys. Rev. B **49**, 5081 (1994)
57. A. Rubio, Cond. Matter News **6**, 6 (1997)
58. R. S. Ruoff, D. C. Lorents, Bull. APS **40**, 173 (1995)
59. R. S. Ruoff, J. Tersoff, D. C. Lorents, S. Subramoney, B. Chan, Nature **364**, 514 (1993)
60. J. P. Salvetat, G. A. D. Briggs, J. M. Bonard, R. R. Bacsa, A. J. Kulik, T. Stöckli, N. A. Burnham, L. Forro, Phys. Rev. Lett. **82**, 944 (1999)
61. D. Sánchez-Portal, E. Artacho, J. M. Soler, A. Rubio, P. Ordejón, Phys. Rev. B **59**, 12678 (1999)
62. R. E. Smalley, B. I. Yakobson, Solid State Commun. **107**, 597 (1998)
63. D. Srivastava, M. Menon, K. Cho, Phys. Rev. Lett. **83**, 2973 (1999)
64. P. Stumm, R. S. Ruoff, B. I. Yakobson (unpublished, 2000)

65. B. G. Sumpter, D. W. Noid, J. Chem. Phys. **102**, 6619 (1995)
66. K. S. Suslick, (Ed.), *Ultrasound: Its Chemical Physical and Biological Effects* (VCH, Weinheim 1988)
67. M. Terrones, H. Terrones, Fullerene Sci. Technol. **4**, 517 (1996)
68. J. Tersoff, R. S. Ruoff, Phys. Rev. Lett. **73**, 676 (1994)
69. A. Thess, R. Lee, P. Nikolaev, H. Dai, P. Petit, J. Robert, C. Xu, Y. H. Lee, S. G. Kim, A. G. Rinzler, D. T. Colbert, G. E. Scuseria, D. Tománek, J. E. Fischer, R. E. Smalley, Science **273**, 483 (1996)
70. S. P. Timoshenko, J. M. Gere, *Theory of Elastic Stability* (McGraw-Hill, New York 1998) p. 541
71. M. M. J. Treacy, Ebbesen, T. W., J. M. Gibson, Nature **381**, 678 (1996)
72. H. D. Wagner, O. Lourie, Y. Feldman, R. Tenne, Appl. Phys. Lett. **72**, 188 (1998)
73. D. A. Walters, Ericson, L. M., Casavant, M. J., Liu, J., Colbert, D. T., Smith, K. A., Smalley, R. E., Appl. Phys. Lett. **74**, 3803 (1999)
74. E. W. Wong, P. E. Sheehan, C. M. Lieber, Science **277**, 1971 (1997)
75. B. I. Yakobson, C. J. Brabec, J. Bernholc, Phys. Rev. Lett. **76**, 2511 (1996)
76. B. I. Yakobson, C. J. Brabec, J. Bernholc, J. Computer-Aided Materials Design **3**, 173 (1996)
77. B. I. Yakobson, M. P. Campbell, C. J. Brabec, J. Bernholc, Comput. Mater. Sci. **8**, 341 (1997)
78. B. I. Yakobson, R. E. Smalley, American Scientist **85**, 324 (1997)
79. B. I. Yakobson, *Fullerenes* (Electrochem. Soc., Paris; ECS, Pennington 1997) p. 549
80. B. I. Yakobson, Physical Property Modification of Nanotubes, U. S. Patent Application 60/064,539 (1997)
81. B. I. Yakobson, Appl. Phys. Lett. **72**, 918 (1998)
82. B. I. Yakobson, G. Samsonidze, G. G. Samsonidze, *Carbon* **38**, 1675 (2000)
83. M. Yu, B. I. Yakobson, R. S. Ruoff, *J. Phys. Chem. B Lett.* **104**, 8764 (2000)
84. Yu, M., O. Lourie, M. Dyer, K. Moloni, T. Kelly, R. S. Ruoff, Science **287**, 637 (2000)
85. P. Zhang, P. E. Lammert, V. H. Crespi, Phys. Rev. Lett. **81**, 5346 (1998)

# Physical Properties of Multi-wall Nanotubes

László Forró[1] and Christian Schönenberger[2]

[1] Department of Physics, Ecole Polytechnique Fédérale de Lausanne
1015 Lausanne, Switzerland
forro@igahpse.epfl.ch

[2] Department of Physics and Astronomy, University of Basel
4056 Basel, Switzerland
christian.schoenenberger@unibas.ch

**Abstract.** After a short presentation on the preparation and structural properties of Multi-Wall carbon NanoTubes (MWNTs), their outstanding electronic, magnetic, mechanical and field emitting properties are reviewed. The manifestation of mesoscopic transport properties in MWNTs is illustrated through the Aharonov–Bohm effect, universal conductance fluctuations, the weak localization effect and its power-law temperature/field dependences. Measurements of the Young's modulus of individual nanotubes show the high strength of tubes having well-graphitized walls. Electron Spin Resonance (ESR) measurements indicate the low-dimensional character of the electronic states even for relatively large diameter tubes. The conducting nature of the tubes, together with their large curvature tip structure, make them excellent electron and light emitters suitable for applications.

With the discovery of Multi-Wall carbon NanoTubes (MWNTs) by *Iijima* in 1991, a new era has started in the physics and chemistry of carbon nanostructures [1]. After the synthesis of Single Wall carbon NanoTubes (SWNTs) in 1993 by *Bethune* and coworkers [2] and by *Iijima* et al. [3], the main stream of carbon research shifted towards the SWNTs, especially through the development of an efficient synthesis method for their large scale production by *Smalley* and colleagues [4]. Nevertheless, MWNTs present several complementary attractive features with respect to SWNTs, both for basic science and for applications. For example, one advantage of MWNTs is that they can be grown without magnetic catalytic particles, which are certainly disturbing for magnetic, and probably for transport measurements, as well. The larger diameter of the MWNTs enables us to study quantum interference phenomena, such as the Aharonov–Bohm effect, in magnetic fields accessible in the laboratory, while study of the same phenomenon would require 600 T fields in the case of SWNTs. The Russian-doll structure allows better mechanical stability and higher rigidity for the MWNTs which is needed for scanning probe tip applications. Even for making nanotube composites, for which the first step is the chemical functionalization of the tube walls, the multi-wall configuration is more advantageous, since efficient load transfer can be achieved without damaging the stiffness of the internal tubes.

The paper is organized as follows. First we will briefly review the production methods for MWNTs. Since the control of sample quality is a condition

M. S. Dresselhaus, G. Dresselhaus, Ph. Avouris (Eds.): Carbon Nanotubes,
Topics Appl. Phys. **80**, 329–390 (2001)

*sine qua non* for all studies, we present the purification and the structural characterization of the nanotubes. Different methods for filling the hollow interior of the nanotubes with a variety of elements is described. The longest section of the chapter is devoted to the transport properties of MWNTs, since these properties are really spectacular, and most of the mesoscopic transport phenomena can be studied on this system. The magnetic properties are investigated with the Electron Spin Resonance (ESR) technique and the fingerprints of low-dimensionality are shown. The field emission properties of MWNTs, which allowed the first application of these structures in flat panel displays, are studied for individual nanotubes. Light emission follows the discussion on electron emission. This phenomenon is presented not only in the field emission configuration, but also by using an STM tip for electron injection. The mechanical properties of MWNTs (as compared to SWNT ropes) are also discussed. A review of the role of the defects, deformations, and mechanical manipulation of MWNTs is also presented.

# 1 Production Methods and Purification

Carbon nanotubes are produced in several different ways. One method relies on a carbon arc where a current on the order of 70 to 100 A passes through a graphite rod (which serves as the anode) to a graphite cathode in a He atmosphere [1,5,6]. In the arc process, a rod-shaped deposit forms on the cathode as shown in Fig. 1. This deposit has a hard grayish outer shell, which houses a soft deep black material. This material consists of fine fibers (typically 1 mm in length and a fraction of a mm in diameter), which under High Resolution Transmission Electron Microscopy (HRTEM) are revealed to be dense bundles of MWNTs [6]. Production rates are reasonably high: several hundred mg of raw material is produced in about 10 min. The material is quite heterogeneous and consists of MWNTs with a rather large dispersion in their

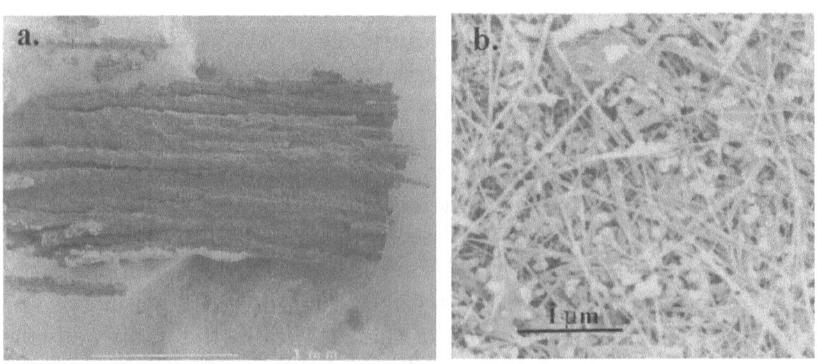


**Fig. 1.** Scanning Electron Microscope image of the (**a**) rod-shaped deposit on the cathode and of (**b**) a zoomed part of a deposit, which shows the heterogeneous character of the as-grown nanotube soot [8]

outer and inner diameters. The MWNTs are typically 10 nm in diameter and are on the order of 1 μm long. The tubes tend to be very straight (when not stressed, of course). Besides MWNTs, this method also produces rather large amounts of graphitic material in the form of multilayered fullerenes (or carbon onions) and amorphous carbon which covers the nanotubes. Purification steps are often employed [7,8] with varying degrees of success.

For the sake of completeness, we mention that a variation of this synthesis method can be used to produce SWNTs (with diameters of the order of 1 nm), by replacing the graphite anode with a hollow graphite tube which is filled with graphite powder containing powdered transition metal (i.e., Fe, Co, Ni,) and/or rare earth metal (Y, Lu, Gd) catalysts, and other metals as well (i.e., Li, B, Si, Sn, Te, Pb,...) [2,12]. This method was originally discovered by *Bethune* and co-workers [2] and by *Iijima* and co-workers [3]. A method which holds much promise was subsequently introduced by *Smalley* and coworkers [4], and involves the laser ablation in a He atmosphere of a graphite target which is impregnated with a transition metal catalyst. The material which is produced by this process is very rich in long bundles of SWNTs. Subsequent purification steps allow this material to be refined (through the elimination of other graphitic particles) to "bucky paper", which is a dense mat of purified SWNTs, and this bucky paper is used for many physical properties measurements.

Other, less exotic, nanotube production methods are based on the thermal decomposition of hydrocarbons in the presence of a catalyst (for a review, see *Fonseca* et al. [13]). These methods [14] are most closely related to traditional carbon fiber production schemes, which have been known for a long time [15,16]. There have been several adaptations to nanotube production, involving a hydrocarbon (i.e., acetylene, methane) and very small catalyst particles (i.e., Co, Cu, Fe) supported on a substrate (silica, zeolite) [17,18,19]. At elevated temperatures, the hydrocarbon decomposes on the catalyst, whereby long nanotubes are spun from the particle (Fig. 2). These methods produce an impressive variety of nanotubes [20], many of which are regularly coiled, forming long helices [21]. This helix-forming feature indicates the presence of defects (non-hexagonal units) in the graphitic structure [22]. The method is relatively simple to apply and produces nanotubes in large quantities. Thus, vapor phase growth is of interest, since large quantities of nanotubes are necessary for many applications, such as making composite materials.

A further advantage of the gas phase growth method is that it enables the deposition of carbon nanotubes on pre-designed lithographic structures [23], producing ordered arrays, a dense forests of aligned nanotube films on surfaces (Fig. 2) which can be used in applications such as thin-screen technology, electron guns, etc.

No matter which type of carbon nanotubes we are dealing with, the first step in the study of carbon nanotubes is technological: their purification. Purification steps, that are commonly performed, may consist of controlled ox-

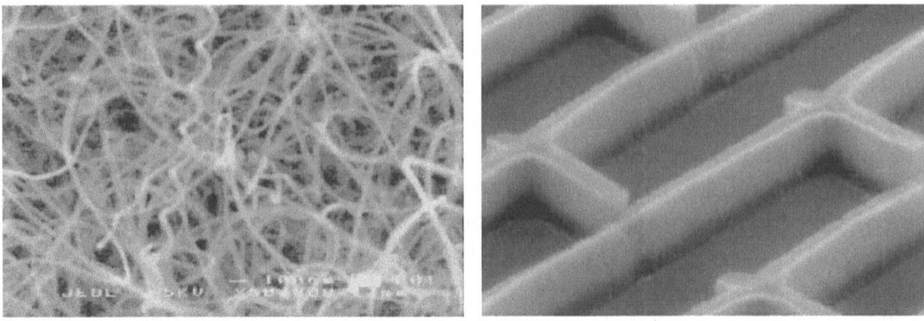

**Fig. 2.** On the *left*, a Scanning Electron Micrograph (SEM) of MWNTs prepared by catalytic decomposition of acetylene, and, on the *right*, an ordered array of nanotubes deposited on pre-designed lithographic structures [23]

idation, chemical treatment, filtration and other procedures, each with their advantages and disadvantages. The very first purification method consisted of burning the raw soot in air. Since the non-nanotube graphitic structures are less resistive to oxidation, they will burn first. This method resulted in a relatively pure nanotube soot, but nevertheless with damaged tube walls and tips, and resulted in losing about 99 wt% of the material [24]. It is possible to save a much larger percentage of tubes by using surfactants and then successive filtrations [25]. A liquid-phase separation of nanotubes and nanoparticles was performed by filtering a kinetically stable colloidal dispersion consisting of the carbon material in a water/surfactant solution, thus allowing extraction of the nanotubes from the suspension, while leaving the nanoparticles in the filtrate. Further purification was accomplished by size-selection through controlled flocculation of the dispersion. Final separation yields as high as 90% in weight were obtained without any damage to the tube tips or tube walls following the flocculation separation process. The result of this separation is seen in Fig. 3. An interesting side-product of this purification method is a large quantity of polyhedral, onion-like carbon particles, which are interesting nanostructures in their own right.

## 2 Crystallographic Structure

### 2.1 Crystallographic Structure as Seen by HRTEM

For fundamental studies and for some applications, one is looking for "perfect" nanotube structures, which means well-graphitized nanotube walls, with a level of defects as low as possible. High-Resolution Transmission Electron Microscopy (HRTEM) provides valuable information about the nanotube quality prepared by different synthesis methods. The MWNTs grown by the arc discharge method have generally the best structures, presumably due to the high temperature of the synthesis process. It is likely that during

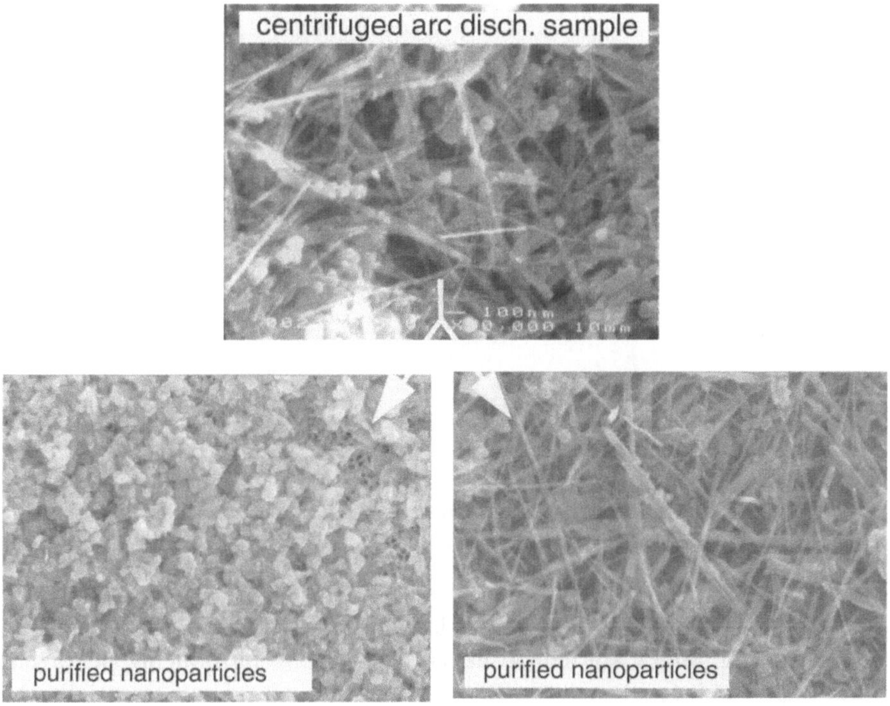


**Fig. 3.** Scanning Electron Micrographs (SEM) of a MWNT deposit (*top*) and MWNTs after purification (*bottom right*) and as a 'side-product' of onion-like carbon nanoparticles (*bottom left*) [7,8]

the growth process, most of the defects are annealed. In the HRTEM image these rolled-up graphene sheets appear as straight lines (Fig. 4). Each nested graphite cylinder has a diameter which is given by two integers $m$ and $n$, the coefficients of the lattice vectors $\boldsymbol{a}_1$ and $\boldsymbol{a}_2$ in the chiral vector $C_h = n\boldsymbol{a}_1 + m\boldsymbol{a}_2$ which describes the way the graphene sheet is rolled-up, and relates to the helicity of the nanotube [26,27], and is described elsewhere in this volume [28,29]. The diameter $d_\mathrm{t}$ of the $(n, m)$ nanotube is given by

$$d_\mathrm{t} = a(m^2 + mn + n^2)^{1/2}/\pi\,. \tag{1}$$

The interlayer spacing in MWNTs (0.34 nm) is close to that of turbostratic graphite [16,30]. Since nanotubes are composed of nearly coaxial cylindrical layers, each with different helicities, the adjacent layers are generally non-commensurate, i.e., the stacking cannot be classified as AA or AB as in graphite [16,30]. The consequence of this interplanar stacking disorder is a decreased electronic coupling between the layers relative to graphite. MWNTs also have structural features which are different from the well-ordered graphitic side-walls: the tip, the internal closures, and the internal tips within the central part of the tube, called the "bamboo" structures

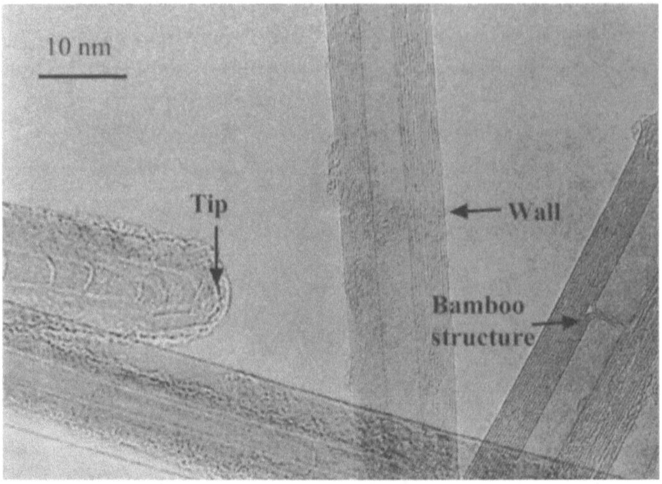


**Fig. 4.** HRTEM images of MWNTs grown by the arc-discharge method, showing the well graphitized wall, the tip and the "bamboo" structure [9,10,11]

(Fig. 4). The structure of the tips is closely related to that of the icosahedral fullerenes [31], where the curvature is mediated by introducing pentagons (and higher polygons) into the structure, while maintaining essentially the $sp^2$ electronic structure (i.e., 3 bonds) at each carbon atom site. A variety of tip structures are commonly observed [32], and an usual feature is that the tubules within the MWNT often close in pairs at their tips. The tips play an especially important role in the electronic and field emission properties of nanotubes. There has been considerable debate concerning the structure of MWNTs, i.e., whether in fact the tubules close on themselves or if nanotubes sometimes are scrolled [21] (i.e., whether one or more of the sheets of graphite that are rolled have two edges). This interpretation followed analysis of HRTEM and electron diffraction studies of nanotubes which revealed that the number of different chiral angles which are observed in a MWNT are usually less than the number of tubules. Anomalous layer spacings are also observed (at least for catalytically grown tubes). This would be the case if one sheet produced several cylindrical layers in the tube. However, the consensus viewpoint from many HRTEM studies appears to favor the Russian doll model, since there has been little evidence for the edges which should result from graphene scrolls. However, nanotubes morphologies may vary considerably with the production method. This is especially true for catalytically grown nanotubes. If the growth conditions are not adjusted properly, then poorly graphitized walls result, and tubes with so-called coffee-cup structures form (Fig. 5).

There can be other types of defects which are more difficult to visualize like point defects in planar graphite, consisting of vacant sites (Schottky defects) and displacements of atoms to interstitial positions (Frenkel defects).

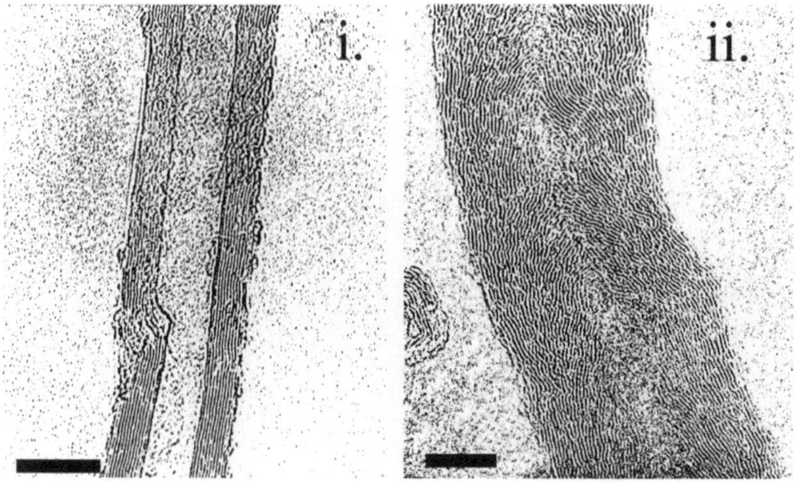


**Fig. 5.** HRTEM images of MWNTs produced by the decomposition of acetylene using a Co/silica catalyst, (i) at 720°C and (ii) at 900°C. All scale bars are 10 nm [7]

These are quite costly in energy due to the strength of the broken bonds, but these defects can be introduced by high energy electron irradiation [33]. Ultrasonic treatment may also produce defects in MWNTs [34] and SWNTs [35], which become more susceptible to chemical attack after being subjected to high intensity ultrasound treatment. Defective structures also result after severe chemical attack (i.e., in hot nitric acid and oxidation at elevated temperature) which has the effect of destroying the tips and hence provides a method to open the tubes.

## 2.2 Filling of Carbon Nanotubes

The hollow interior of carbon nanotubes naturally presents us with an opportunity to fill the nanotube core with atoms, molecules, or metallic wires, thereby forming nanocomposites with new electronic or magnetic properties. In early approaches, electrodes composed of carbon impregnated with the filling material were used [36,37]. This resulted in a large variety of filled graphitic structures, metallic carbides, but with relatively low yield. *Loiseau* and collaborators [38] realized that a minute quantity of sulfur present in the graphite/metal mixture helps to wet the surface of the cavity, so that the tubes fill up over micron distances with a high yield [20]. Nanowires of various elements (or their sulfides) were formed by this method, including transition metals (Cr, Ni, Co, Fe), rare earth metals (Gd, Sm, Dy) and covalent elements (Ge, Se, Sb). Figure 6 displays a nanotube with a crystalline wire of Ge inside the core [38,20].

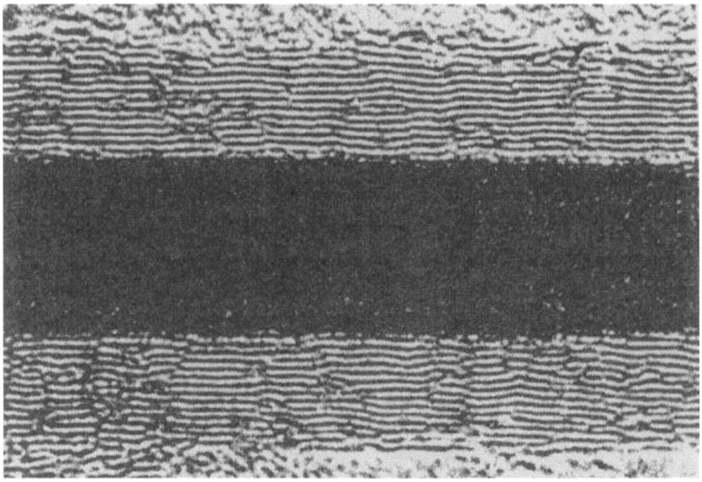

**Fig. 6.** HRTEM image of a MWNT filled with crystalline Ge. [38,20]

Other approaches to filling the hollow core involved the exploitation of nanotube capillarity [39]. This procedure for filling nanotubes may involve a chemical method, using wet chemistry [40]; or (b) a physical method, where capillarity forces induce the filling by a molten material [41,42,43]. In the wet-chemistry method, the nanotubes are refluxed in a nitric acid bath in order to open their tips. When a metal salt is simultaneously used in the bath, it is possible to obtain oxide or pure metal particles by a subsequent annealing process. Also, the metal/salt solvent bath can also be used on previously opened tubes. However, the drawback of this method is that it requires chemical manipulation over a long time to open nanotubes containing various tip structures, "bamboo" closures and various numbers of SWNT constituent layers. Furthermore, the reduction of the metal salts inside the tubes does not give a continuous wire, but rather yields metallic blobs. This can be observed in Fig. 7 in the case of a nanotube filled with silver [43].

In the physical method, the nanotubes are filled by first opening them, for example, by heating them in air to oxidize the nanotube tips preferentially, and the opened tubes are then submerged in molten material [41,42,43]. Studies of filling nanotubes with liquid metals have shown that only certain metals will enter the nanotubes and that the metal surface tension is the determining factor [37,43]. Further studies along these lines have demonstrated that the capillary action is related to the diameter of the inner cavity of the nanotube. This was explained in terms of the polarizability of the cavity wall, which is related to the radius of curvature, through the pyramidalization angle (bond-bending angle) and the polarizability of the filling material [43]. The filled nanotubes with different metals offer exciting structures for future electronic studies. In the following sections, we focus on a review of electrical and magnetic measurements that have been performed on arc-discharge

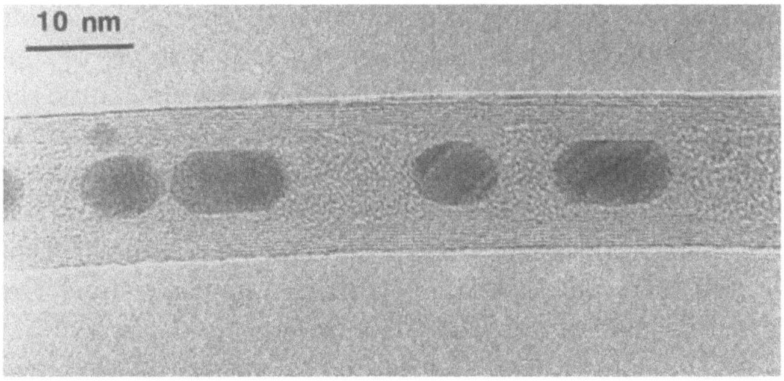


**Fig. 7.** HRTEM of a carbon nanotube whose inner cavity is filled with silver particles (lattice fringes correspond to [111] planes of Ag, with a lattice spacing of 0.23 nm). The tube cavity is first filled with molten silver nitrate that is subsequently reduced to metal by electron irradiation, in situ, in the electron microscope [43]

grown MWNTs, which show good structural order and their purity is well controlled.

## 3 Electronic Properties of Multi-wall Carbon Nanotubes

The electronic and electrical properties of low-dimensional conductors are an exciting area of research. Very rich phase diagrams have been predicted. Since a small set of 1-D modes is sufficient to describe the low-energy electronic properties of carbon nanotubes, they are considered prototype 1-D molecular conductors. This is particularly true for Single-Wall NanoTubes (SWNTs). Many interactions are especially strong in 1-D. The Coulomb interaction, for example, cannot efficiently be screened, leading to a strongly correlated electron gas, called a Luttinger liquid, whose low-energy excitations are long-range density waves. In Luttinger liquids a pseudo-gap develops for the conventional quasi-particles. Because MWNTs consist of several coaxially arranged SWNTs, one may expect that MWNTs do not qualify as 1-D conductors. However, there is now convincing evidence that Luttinger-liquid-like features are present in MWNTs too [44]. Moreover, the electric current introduced into a MWNT from the outside is confined to a large extent to the outermost SWNT [45,46]. In this respect, studying MWNTs is somewhat similar to studying transport in large diameter SWNTs. This is interesting because SWNTs of diameters similar to the outer diameter of a typical MWNT (20 nm) have not yet been observed. Large diameter nanotubes have specific advantages: they favor low-ohmic contact and allow investigation of quantum-interference phenomena in a magnetic field, such as the Aharonov–Bohm effect. This is not possible in SWNTs, because a huge magnetic field of order 1000 T would then be required. This section focuses on the electrical and electronic properties of MWNTs with an emphasis on electrical measurements. Though quite a variety of experimental results are available today, very fundamental questions remain unanswered, partly because of conflicting results. For example, at present we do not know with certainty the electronic ground state of carbon nanotubes. Do quasi-particles in the sense of Fermi liquid theory exist or is the ground state a correlated many-body state (charge-density wave)? May it be possible that superconducting or ferromagnetic fluctuations are present, and affect the ground state? To what extent are carbon nanotubes (SWNTs and MWNTs) ballistic quantum wires? Because it is impossible to discuss all available data in depth, we focus on a few representative measurements by which the present picture of the electronic properties of MWNTs can be illustrated. The selections made by the authors is biased by their own research. The presentation is therefore neither complete, nor balanced.

This section is structured as follows. Section 3.1, written in a form accessible to any scientist with a basic background in condensed matter physics, provides additional introductory material to help the reader understand the chapter, and more detailed reviews can be found in [47,48]. Section 3.2 briefly reviews recent work regarding electrical transport in MWNTs (see also [44]). This section has been added for those interested in entering this field of research. Next, we mention in Sect. 3.3 important results regarding transport in films, assemblies and ropes of MWNTs. We emphasize the measured positive Hall coefficient and briefly mention recent results on the thermopower. Most of the text is concerned with measurements on single MWNTs (Sect. 3.4). This section is further divided into parts on: 1) electrical transport in zero magnetic field (temperature and gate-dependence of the resistance), 2) magnetotransport (magnetic-field-dependence), and finally, 3) spectroscopy, highlighting density of state features in carbon nanotubes. The chapter ends in Sect. 3.9 with a discussion of the most pressing issues.

## 3.1 Multi-wall Nanotubes in Relation to Graphene Sheets

A MWNT is composed of a set of coaxially arranged SWNTs of different radii [1]. The distance between nearest-neighbor shells corresponds, within a good approximation, to the van-der-Waals distance between adjacent carbon sheets in graphite (3.4 Å) [16]. The outer diameter of MWNTs depends on the growth process. It is typically of order 20 nm for arc-discharge grown tubes, but can attain values exceeding 100 nm for some CVD-grown MWNTs. (The term nanofiber is reserved for multi-wall nanotubes in the diameter range 10–100 nm [16].) Transmission electron microscopy has shown that these large diameter tubes contain a considerable amount of defects (Fig. 5), though examples of MWNTs with lower defect densities have been demonstrated [16]. Carbon shells are often not continuous throughout the tube. For this reason, we will focus only on electrical measurements obtained from arc-discharge grown MWNTs which have been demonstrated to be highly graphitized [7,9].

Nanotubes are a special class of fullerene molecules, namely those that are quantum wires with 1-dimensional (1-D) translation symmetry. SWNTs are the simplest of these objects. A SWNT can be constructed from a slice of graphene (a single planar sheet of the honeycomb lattice of graphite) rolled into a seamless cylinder, i.e., with all carbon atoms covalently bound to three neighbor carbons. A large number of such foldings are possible. In fact, each graphene lattice vector (called a wrapping or chiral vector) defines a unique nanotube [47]. Regarding electronic properties, SWNTs can be further classified as being either semiconductors or metals [49,50,51]. If all wrapping vectors would be realized with equal probability, 1/3 of the SWNTs would be metallic and 2/3 semiconducting. Assuming that neighboring shells do not interact, the electronic properties of MWNTs would be similar to a set of independent SWNTs with radii in the range of a few nanometers to $\approx 10$ nm. Though the coupling strength between adjacent shells might be low, it possibly cannot be neglected altogether [52]. For this reason, a much richer band structure might be expected. Graphite is, in fact, an excellent example for demonstrating the importance of the weak inter-plane coupling for ground-state properties. Graphite is a semimetal, and the finite Density Of States (DOS) at the Fermi energy can be traced back to the three-dimensional crystal structure with finite interplanar coupling which leads to a small band overlap of 40 meV [16,47]. MWNTs are interesting because they allow us to study the transition from the SWNT single molecule to the behavior of the macroscopic crystal (graphite). MWNTs are mesoscopic, in-between the single-wall nanotube molecules and planar graphite. In this respect, MWNTs play the same role as carbon onions play to fullerene research [16,29,53]. An immediate question arises now: are the electronic properties of MWNTs closer to graphite or do MWNTs rather behave as a set of independent SWNTs? In order to be able to discuss the electrical measurements of MWNTs, it is useful to briefly highlight the important electronic properties of the two reference systems: graphite and ideal SWNTs.

Starting with graphite, we will first review the band structure of a single layer of graphite, called a graphene sheet. The carbon atoms in a graphene sheet form a planar hexagonal lattice, the honeycomb lattice. Each carbon atom is covalently bound to three neighbor carbons by $sp^2$ molecular orbitals. The fourth valence electron, an atomic $p_z$ state, hybridizes with all other $p_z$ orbitals to form a delocalized $\pi$-band. It is important to recognize that the unit-cell of graphene contains two carbon atoms. Hence, the $\pi$-band has to accommodate two electrons per unit cell. In the case of an even number of valence electrons, a material can in general be either a metal or a semiconductor. The system that is realized depends on the specific form of the overlap integral. Figure 8 shows the result of a simple tight binding calculation [54]. It is seen that the bonding $\pi$-band is always energetically below the antibonding $\pi^\star$ band for all wavevectors, except at the corner points of the Brillouin zone boundary (the $K$ points), where the band splitting is zero

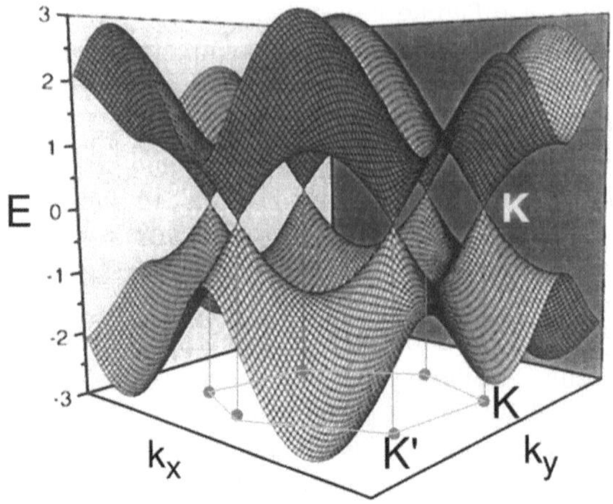


**Fig. 8.** Tight-binding dispersion relations of the $\pi$- (*lower half*) and $\pi^*$ (*upper half*) bands of graphene. The energy is expressed in units of the nearest-neighbor C–C overlap integral $\gamma_0 \simeq 3.0\,\mathrm{eV}$ and measured relative to the Fermi energy $E_F$ for neutral (undoped) graphene [55]

by the symmetry of the honeycomb lattice [47]. With regard to band overlap, graphene is very unusual. Theoretically, all bonding states will be filled up to the corner points, which coincide with the Fermi energy $E_F$. Since the DOS is zero at $E_F$, graphene is not a true metal. However, it also not a true semiconductor, because there is no excitation gap at the $K$ points. Graphene can thus be characterized as a zero-gap semiconductor or as a zero-DOS metal.

The insight into the band structure of graphene allows us to calculate the Drude resistivity. In the vicinity of $E_F$, the electron states can be approximated by $E = \pm\hbar v_F|\boldsymbol{k}|$, where $\boldsymbol{k}$ is the 2-D-wavevector measured with respect to the two independent Brillouin corner points [56]. Energies are measured relative to the charge neutrality level, and the Fermi velocity is taken to be $v_F = 10^6\,\mathrm{m/s}$ [54,57,58,59]. For the electron DOS, we obtain $n_{2D}(E) = 2E/\pi(\hbar v_F)^2$. Using the Einstein equation $\sigma_{2D} = e^2 n_{2D} D$, which relates the conductivity $\sigma_{2D}$ to the diffusion coefficient $D = v_F l_e/2$ and the electron density, we obtain for the energy-dependent conductivity $\sigma_{2D}(E)$

$$\sigma_{2D}(E) = \left(\frac{2e^2}{h}\right)\frac{E}{\hbar v_F} l_e\,, \tag{2}$$

where $l_e$ is the carrier scattering length. In order to be able to calculate the equilibrium sheet conductivity $\sigma_{2D}$ at a finite temperature ($T$), the exact position of the Fermi level needs to be known. Without any doping, $E_F = 0$. For a real graphene surface a finite doping is unavoidable due to adsorbates causing partial charge transfer. In this more realistic situation, the Fermi level will be either below (hole doping) or above (electron doping) the neutrality point. Two cases have to be distinguished: the high- and low-temperature limits, defined by $kT \gg |E_F|$ and $kT \ll |E_F|$, respectively. For the latter (either strongly doped, or sufficient low temperature) $\sigma_{2D}$ is given by (2) with $E$ replaced by $|E_F|$. Note, that the mean-free path is in general $T$-dependent, due to electron-phonon scattering, for example, thus $l_e$ increases with decreasing $T$ and approaches a finite zero-temperature value, only determined by static disorder. The equation therefore predicts true metallic behavior for the temperature-dependent resistance, i.e., the electric resistance should decrease with decreasing temperature due to the gradual suppression of electron–phonon scattering. In the high-temperature limit ($kT \gg |E_F|$, i.e., low doping or ideal graphene) $\sigma_{2D}$ is given (up to a numerical factor of order 1) by (2) with $E$ replaced by $kT$. That $\sigma_{2D} \propto T$ reflects the fact that the carrier density is proportional to $T$. Also this equation predicts a metallic temperature-dependence of the resistance provided that $l_e \propto T^{-p}$ with $p \geq 1$. This condition is always met, except for highly disordered samples for which $l_e$ is a constant over a substantial temperature range. The resistance of such 'dirty' samples should display a $R(T) \propto T^{-1}$ dependence. Unfortunately, there are no measurements for the electric sheet resistance of graphene, since graphene is a theoretical object. In contrast, there is an extensive source of literature for graphite (including graphitic fibers and turbostratic graphite) [15,16]. Graphite consists of a regular A–B–A–B stacking of graphene planes. There is a weak interplane coupling of order $\Delta \sim 10$ meV, which results in semimetallic behavior and a band overlap of 40 meV. Most importantly, graphite develops a finite DOS at $E_F$ and is therefore a metal. The sheet conductivity is approximately given by (2) with $E$ replaced by $\Delta$ (if $kT \ll \Delta$). We therefore expect a metallic temperature-dependent resistivity for crystalline graphite in the temperature range $kT \ll \Delta$. Figure 9 shows the measured temperature-dependence of the electric resistivity $\rho(T)$ for different forms of carbon fibers [15]. The upper curves correspond to disordered pregraphitic carbons while the ones with lower $\rho$ correspond to highly graphitized samples. Because we will be discussing arc-discharge grown MWNTs in this chapter, arc-grown graphitic fibers (graphite whiskers [16,60]) provide a good reference material. Both arc-grown and single crystalline graphite display a metallic $\rho(T)$, i.e., the resistivity decreases with decreasing temperature $T$. From the typical value of $5 \times 10^{-6}\ \Omega\,\mathrm{m}$ (100 K) for the resistivity, a sheet resistance $R_\square$ of $R_\square = 1.5\,\mathrm{k}\Omega$ is deduced. Taking (2) with $E$ replaced by $\Delta \sim 10\,\mathrm{meV}$, we obtain a scattering length of $l_e = 600\,\mathrm{nm}$. Because the zero-temperature resistivity of single-crystalline graphite can be more than an order of magnitude lower, large mean-free paths for scattering $l_e > 1\,\mu\mathrm{m}$ are possible. Based on this consideration we can expect that carbon nanotubes may be 1-D ballistic conductors, provided that nanotubes of structural quality similar to single-crystalline graphite can be obtained. For the uppermost curve in Fig. 9, corresponding to highly disordered carbon fibers, the mean-free path is only of order 1.5 nm. In such disordered materials the graphene planes have a high density of structural defects as well as random stacking

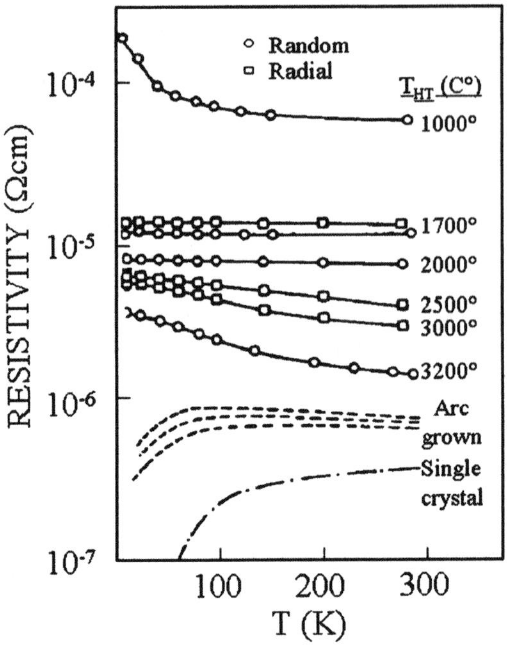


**Fig. 9.** Electrical resistivity $\rho$ as a function of temperature $T$ for different form of graphitic fibers taken from [15]. The higher up the curves, the larger the disorder. The curve with the lowest resistivity corresponds to crystalline graphite

(turbostratic graphite) [16], and the interplane coupling is reduced to $\Delta \approx 0$. The properties of disordered carbon fibers are therefore very well approximated by a set of independent sheets of 'dirty' graphene. For the comparison with MWNTs we should memorize the following two observations: 1) disordered graphite is characterized by a resistance upturn at low temperatures (localized behavior) with a typical sheet resistance of $R_{\square} \approx 100\,\mathrm{k}\,\Omega$, and 2) crystalline graphite is characterized by a metallic $R(T)$ with $R_{\square} \approx 1\,\mathrm{k}\,\Omega$ [61].

Next, we will briefly discuss the electronic properties of the other reference compound, that is of a perfect single-wall carbon nanotube. A SWNT is obtained from a slice of graphene wrapped into a seamless cylinder. The periodic boundary condition around the nanotube circumference causes quantization of the transverse wavevector component. Let us denote the wavevector along the tube direction by $k$ and the transverse component by $k_{\perp}$. The allowed $k_{\perp}$ are spaced by $2/d_{\mathrm{t}}$, where $d_{\mathrm{t}}$ is the tube diameter. The 1-D band structure can now easily be constructed using for example the 2-D tight-binding band structure $E(k_x, k_y)$ of graphene, shown in Fig. 8. Each $k_{\perp}$ within the first Brillouin zone gives rise to a 1-D subband (1-D dispersion relation) by expressing $E$ as a function of $k$ for the given $k_{\perp}$. Hence, a set of subbands $E_{k_{\perp}}(k)$ is obtained. A large diameter tube will have many such subbands, while a small diameter tube has only a few. It turns out that both metallic and semiconducting nanotubes are possible. A nanotube is metallic if and only if the $K$ points belong to the set of allowed $k$-vectors. In the $\boldsymbol{k} \cdot \boldsymbol{p}$ scheme, the eigenfunctions in the vicinity of the $K$ points are approximated by a product of one of two fast oscillating graphene wavefunctions $\Psi_K(\boldsymbol{r})$, changing sign on the scale of interatomic distances, and a slowly varying envelope function $F(\boldsymbol{r})$ [58]. The periodic boundary condition leads to the following condition for $\Psi_K$ [62]:

$$\Psi_K(\boldsymbol{r}+\boldsymbol{L}) = \Psi_K(\boldsymbol{r}) \exp(\mathrm{i}2\pi\nu/3)\,, \tag{3}$$

where $\boldsymbol{L}$ measures the length around the tube circumference and $\nu = 0, \pm 1$, depending on the wrapping or chiral vector [55]. Taking for the slowly varying function $F$ a plane wave (free electrons) and noting that $F$ should cancel the phase factor in (3), we obtain the condition

$$k_\perp = \frac{2}{d_\mathrm{t}}(n - \nu/3) \quad n = 0, \pm 1, \pm 2, \ldots \tag{4}$$

for the allowed transverse wavevectors. The approximate 1-D dispersion relations, valid in the vicinity of $E_\mathrm{F}$, can now be derived using $E(\boldsymbol{k}) = \pm \hbar v_\mathrm{F} |\boldsymbol{k}|$. Explicitly:

$$E_n(k) = \pm E_0 \sqrt{(n-\nu/3)^2 + (k d_\mathrm{t}/\pi)^2} \tag{5}$$

with $E_0 = 2\hbar v_\mathrm{F}/d_\mathrm{t}$. These approximate dispersion relations are shown in Fig. 10. As can be seen, there are metallic ($\nu = 0$) and semiconducting ($\nu = \pm 1$) nanotubes. The bandgap of a semiconducting tube is $E_\mathrm{g}(d_\mathrm{t}) = 2E_0/3$, which is inversely proportional to the tube diameter. We obtain 0.68 eV for a tube with $d = 1.3$ nm. Since MWNTs have rather larger diameters of order 20 nm, the corresponding bandgaps are only $E_\mathrm{g} \approx 44$ meV. We mention, that the set of 1-D subbands in Fig. 10 needs to be duplicated because the $K$ and $K'$ points which are equivalent in the absence of a magnetic field are both in the first Brillouin zone [47,55]. Furthermore, the crossing at the Fermi energy between the bonding and antibonding states need not necessarily be at $k = 0$ [47]. The Fermi level crossing for zig-zag $(n, 0)$ tubes, which can be metallic or semiconducting, is at $k = 0$. In contrast, the level crossings are found at $k = \pm 2\pi/(3a_0)$ for armchair $(n, n)$ tubes. Armchair tubes are always metallic. For all metallic nanotubes, the Fermi energy intersects two 1-D branches with positive velocity (right-movers) and two with negative velocity (left-movers) [47]. This corresponds to two 1-D modes (not counting spin) leading to a quantized conductance of $G = 4e^2/h$ for an ideal nanotube. Figure 10 also shows the corresponding DOS, which are characterized by sharp features (van Hove singularities) at the onset of subbands. The observations of these singularities in, for example, tunneling spectroscopy, is a direct proof for the presence of a 1-D band structure. This has indeed been observed in SWNTs using scanning-tunneling spectroscopy at low temperatures [63,64]. Since MWNTs consist of tubes with larger diameters ($d_\mathrm{t}$), we

(a)

(b)

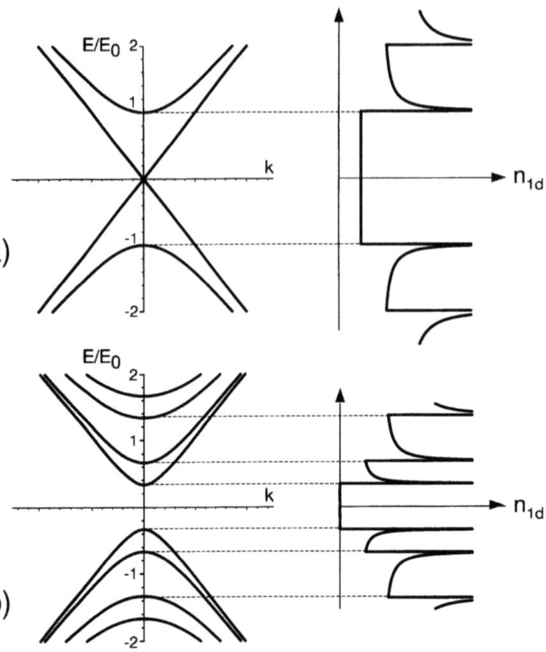


**Fig. 10.** Approximate dispersion relations (*left*) for the 1-dimensional electronic energy bands and the corresponding density of state (*right*) of metallic (**a**), and semiconducting (**b**), single-wall carbon nanotubes close to the Fermi energy. The energy scale is in units of $E_0 = 2\hbar v_F/d_t$

need to understand how the van Hove singularities develop if $d_t$ approaches values comparable to the mean-free path $l_e$. The picture sketched above is valid if $l_e \gg d_t$ (1-D-ballistic transport). In contrast, if $l_e \ll d_t$ transport is 2-D-diffusive and the density of states should closely resemble that for graphene without any singularities. If, on the other hand, $l_e$ is of the same order as $d_t$, transport is neither fully 2-D-diffusive, nor 1-D-ballistic. In this regime, the characteristic subband features in the DOS are still present, albeit considerably broadened. One expects to see broadened peaks in the DOS with a mean level spacing of the 1-D subbands given by

$$\overline{\Delta E_{sb}} = E_0/2 = \frac{\hbar v_F}{d_t} \sim 33\,\text{meV}\ (d_t = 20\,\text{nm})\,, \tag{6}$$

where $E_0 = 2\hbar v_F/d_t$.

For a metallic nanotube, two subbands are occupied at $E_F$. Consequently, the electric conductance $G$ is predicted to be $G = 4e^2/h$, provided the complete wire is ballistic. We not only require $l_e \gg d_t$, but in addition ballistic transport requires that $l_e \gg L$, where $L$ is the length of the electrically contacted nanotube segment. More precisely, ballistic transport requires that backscattering must be absent altogether, taking into account both the nanotube *together* with the electric contacts. This condition is very hard to satisfy in the laboratory [44]. Important for MWNTs is the question of how the band structure might change due to inter-tube coupling. This question has only recently been addressed [65,66,67]. However, the main consequence of the low-energy properties can simply be guessed. It is convenient to consider only two tubes which couple only weakly to each other. If one is semiconducting and the other is metallic, it is obvious that the low-energy properties are just determined by the metallic one, with no modifications in the DOS around zero energy. If the two tubes are metallic, the situation is expected to be more complicated. However, in practice it is most likely that the two tubes have different chiralities. For example, one tube could be of the armchair type and the other a zig-zag one, to mention an extreme case. Because the zero-energy bands cross at different $k$ points for the two tubes, hybridization is very weak around $E_F$ and the total DOS is just the sum of the two. Strong band structure modifications are only expected for tubes of similar chiralities [66]. We should keep in mind, however, that this picture is not valid, if the tubes are relatively strongly doped such that $E_F$ is shifted either into the valence or conduction bands. For a comparison of results obtained for SWNTs with MWNT studies, the following three aspects of SWNTs should be emphasized: 1) an *ideal* SWNT is either a metal or a semiconductor with a considerable band gap, 2) the band structure consists of a set of 1-D-subbands, leading to van Hove singularities in the density of states, and 3), the electrical conductance $G$ is quantized in units of $4e^2/h$, provided that backscattering is absent altogether.

### 3.2 Electrical Transport in MWNTs: A Brief Review

A remarkable variety of physical phenomena have been observed in electrical transport. The first signature of quantum effects was found in the magnetoresistance (MR) of MWNTs. *Song* et al. studied bundles of MWNTs [68], while *Langer* et al. was able to measure the MR of a single MWNT for the first time [69]. In both cases a negative MR was observed at low temperatures indicative of weak localization [70,71,72,73]. From these MR experiments, the phase-coherence length $l_\phi$ was found to be small, amounting to $< 20\,\text{nm}$ at 0.3 K, in strong contrast with the ballistic transport theoretically expected for a perfect nanotube. However, evidence for much larger coherence lengths in SWNTs was provided by the observation of zero-dimensional states in single-electron tunneling experiments [74,75], and in other experiments discussed below.

Recently, a pronounced Aharonov–Bohm resistance oscillation has been observed in MWNTs [46]. This experiment has provided compelling evidence that $l_\phi$ can exceed the circumference of the tube, so that large coherence lengths are possible for MWNTs too [76]. Since the magnetic-flux modulated resistance is in agreement with an Aharonov–Bohm flux of $h/2e$, this oscillatory effect is supposed to be caused by conventional weak-localization for which the backscattering of electrons is essential. In essence, as in the work of *Langer* et al. [69], 2-D-diffusive transport could explain the main observation reasonably well.

The Aharonov–Bohm experiment provides a convincing proof that the electric current flows in the outermost (metallic) graphene tube, at least at low temperatures $T < 70\,\mathrm{K}$. Presumably, this is a consequence of the way in which the nanotubes are contacted. In general, electrodes are evaporated over a MWNT, and the electrodes therefore contact the outermost tube preferentially. Since it is essentially only the outermost tube that carries the current, large diameter single graphene cylinders can now be investigated. Recently, proximity-induced superconductivity was found in weak links formed by a bundle of SWNTs in contact with two superconducting banks [77]. Furthermore, spin transport also has been considered. Here, a MWNT was contacted by two bulk ferromagnets (e.g., Co) and the electrical resistance was measured as a function of the relative orientation of the bulk magnetization in the two ferromagnetic contacts [78]. A resistance change of order 10% was found, from which the authors estimate the spin-flip scattering length to be $> 130\,\mathrm{nm}$. All the striking results mentioned above were obtained by contacting a single nanotube with the aid of micro- and nano-structured technologies.

Alternative approaches for contacting nanotubes have been developed as well. For example, *Dai* et al. [79] and *Thess* et al. [80] have measured the voltage drop along nanotubes using movable tips, and *Kasumov* et al. have developed a pulsed-laser deposition method [81]. Furthermore, scanning-probe manipulation schemes were developed [82,83,84], and recently is has been shown that SWNTs can directly be synthesized to bridge pre-patterned structures [85]. Still another elegant method allowing one to contact a single MWNT electrically has been used by *Frank* et al. [45], whereby a single nanotube extending out of a MWNT fiber (a macro-bundle) is contacted by gently immersing the fiber into a liquid metal (e.g., mercury). Immersing and pulling out the nanotube repeatedly is claimed to have a cleaning effect. In particular, graphitic particles are removed from the tubes. After some repeated immersions, an almost universal conductance step behavior is observed, with steps close to the quantized value $G_0 = 2e^2/h$. From these experiments, the researchers conclude that transport in MWNTs is ballistic over distances on the order of $> 1\,\mu\mathrm{m}$. This is a very striking result because the experiments were conducted at room temperature. At present, it is not clear why the conductance is close to $G_0$ instead of the theoretically expected value of $2G_0$.

## 3.3 Hall-Effect and Thermopower in Assembled Nanotubes

Already in 1994, *Song* et al. studied the electronic transport properties of macro-bundles of MWNTs with diameters tens of μm [68]. The authors measured the temperature-dependence of the resistance $R$, its change in magnetic field $B$ and the Hall effect. Above 60 K, the MagnetoResistance (MR) is positive, i.e., $R$ increases with magnetic field; they however also found that a pronounced negative MR peak develops at low temperatures. This negative MR-dependence is suggestive of Weak Localization (WL). WL is an interference correction to the Drude resistance of a metal. WL primarily lowers the diffusion coefficient $D$ due to constructive interference of mutually time-reversed quasiclassical electron trajectories in zero magnetic field. Because only trajectories of lengths shorter than the phase-coherence length $l_\phi$ can participate, sufficiently low temperatures are usually required. For $T < 10$ K, the conductivity $\sigma$ was found to show a $\ln(T)$ dependence. This is in agreement with 2-D-WL theory, so that $l_\phi < d_t$, where $d_t$ is the diameter of a typical MWNT within the bundle. This relatively short coherence length suggests a small mean-free path for scattering from static disorder, implying that the MWNTs are fairly disordered. This is further supported by the temperature dependence of the measured conductance $G = R^{-1}$. This temperature dependence of the conductance, as well as of the Hall coefficient, are shown in Fig. 11, where $G$ is shown to be proportional to $T$ over a remarkably large temperature range of 50–300 K. Below $\approx 50$ K the $T$-dependence weakens to finally turn into a logarithmic dependence, characteristic of 2-D-WL. From the theoretical models, which have been developed for graphite in Sect. 3.1, a linear $G(T)$ is predicted for the conductance of disordered graphene. In this case, $l_e$ is mostly determined by scattering from static disorder, and therefore $l_e$ is only weakly $T$-dependent. This yields $G \propto T$ because of the energy-dependent DOS of graphene. The cross-over at 50 K suggests metallic behavior which can be either due to intertube coupling (like in graphite) or due to doping, thus shifting $E_F$ away from the neutrality point. Because the change of the Hall coefficient $R_H$ from $\approx 0$ (high $T$) to a pronounced positive value (low $T$) coincides with the 'flattening' in

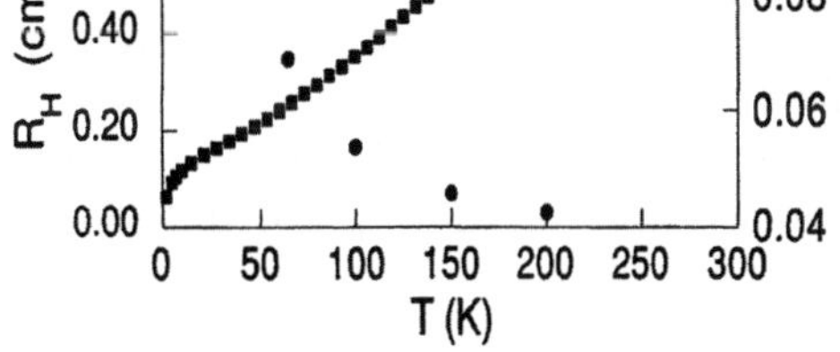


**Fig. 11.** The Hall coefficient $R_H$ (*left scale*) and the conductance (*right scale*) vs temperature $T$ (taken from *Song* et al. [68])

$G(T)$, the nanotubes in Fig. 11 are most likely slightly (hole) doped. From the cross-over temperature ($\approx 50\,\mathrm{K}$), the Fermi energy is estimated to be $\approx -4\,\mathrm{meV}$. Using the theoretical DOS for graphene, the sheet doping level is found to be $n_{2D} = 1.5 \times 10^9\,\mathrm{cm}^{-2}$, which has to be compared to the bulk doping level of $n_{3D} = 6 \times 10^{18}\,\mathrm{cm}^{-3}$ obtained from $R_H$. If the material would be homogeneous and densely packed then $n_{2D} = n_{3D} d_{I_p}$, where $d_{I_p} = 3.4\,\mathrm{Å}$ is the interplane distance of graphite. Under these assumptions, we obtain $n_{2D} = 2 \times 10^{11}\,\mathrm{cm}^{-3}$ from $R_H$, which is more than two orders of magnitude larger than the previous estimate. Though appearing to be inconsistent, this disagreement can easily be resolved. The apparent inconsistency is caused by the assumption of complete filling. Moreover, we know today that most of the electric current in transport measurements is confined to the outermost metallic SWNT at low temperature [45,46]. Hence, the volume fraction cannot exceed $d_{I_p}/d_t \approx 50$, thus resolving this apparent inconsistency.

From fits of the low-$T$ conductance to WL theory, Song et al. could give estimates for several important parameters. For example, they obtained $R_\square \approx 6\,\mathrm{k}\Omega$ for the sheet resistance, $D \approx 50\,\mathrm{cm}^2/\mathrm{s}$ for the diffusion coefficient and $l_e \approx 5\,\mathrm{nm}$ for the mean-free path. The main results of this work, a positive $R_H$ suggesting hole doping and interference corrections at low $T$, have both been confirmed later by *Baumgartner* et al. with measurements on oriented MWNT films [86].

We have added the following discussion, to emphasize the question of doping in carbon nanotubes. This has recently received quite some attention, in particular, regarding research on SWNTs. The Hall coefficient clearly suggests hole doping (of unknown origin). Much more evidence for hole doping is emerging now from other work. The thermoelectric power $S$ has been found to be positive and approximately linear in $T$ at low $T$, both for SWNT and MWNT ropes [87,88,89]. The linear temperature dependence of the thermopower suggest metallic behavior. The substantial positive value of $S$ of order 40–60 $\mu\mathrm{V/K}$ at 300 K initially posed a serious problem, since $S$ should vanish, if $E_F$ lies at the charge neutrality-point, as theoretically expected (for graphite $S \approx -4\,\mu\mathrm{V/K}$). However, these findings of a substantial positive $S$ are in agreement with the above-mentioned positive Hall coefficient, suggesting unintentional hole doping. If we assume that $E_F$ is considerably shifted into the valence band, then the large magnitude of $S$ can easily be understood. Assuming, for simplicity, an energy-independent scattering length, but taking the DOS to be that of graphene, then $S$ is interestingly exactly given by the standard textbook formula for free electrons, i.e., $S = -\pi^2 k^2 T/(3eE_F)$. From the measured value of $S$ one derives $-0.14\,\mathrm{eV}$ for $E_F$ which amounts to a doping level of $n_{2D} = 5 \times 10^{10}\,\mathrm{cm}^{-2}$. This level may appear to be large, but if we express the doping in terms of the elementary charge ($e$) per unit tube-length, there is on average only one $e$ per 300 nm (assuming a SWNT with $d_t = 2\,\mathrm{nm}$). Most recently, it has been shown that the magnitude and even the sign of $S$ can be changed by annealing the nanotubes in vacuum.

It is believed that such a treatment removes oxygen, which may act as the dopant [90,91,92]. Nanotubes might be more sensitive to enviromental conditions than initially believed. Clearly more work is needed to understand the nature of the doping and of the doping level.

## 3.4 Electrical Transport Measurement Techniques for Single MWNTs

Electric measurements on single MWNTs have been performed in three ways: (1) metallic leads are attached to a single tube supported on a piece of Si wafer with the aid of microfabrication technology [46,69,81,93]; (2) the end of a macrobundle of MWNTs, fixed on an moveable manipulator, is steered above a beaker containing a liquid metal (e.g., mercury) and the MWNTs are then gently lowered into the liquid metal. According to this method, a single MWNT makes contact to the liquid metal first, enabling conductance measurements to be made on a single nanotube [45], as shown in Fig. 12, and (3) a scanning-tunneling microscope can be used to measure the local electronic density of states by measuring the bias-dependent differential conductance while the tip is positioned above a single MWNT.

Because lithography is now widely used for contacting nanotubes, we discuss the lithographic approach in detail in the following. A droplet of a dispersion of nanotubes (NTs) is used to spread the NTs onto a piece of thermally oxidized Si. Then, a PMMA resist layer is spun over the sample. An array of electrodes, each consisting of two or more contact fingers together with their bonding pads, is exposed by electron-beam lithography. After development, a metallic film (mostly Au) is evaporated over the structure and then lifted off. The sample is now inspected by either Scanning-Electron Microscopy (SEM) or Scanning-Force Microscopy (SFM) and the structures that have one single

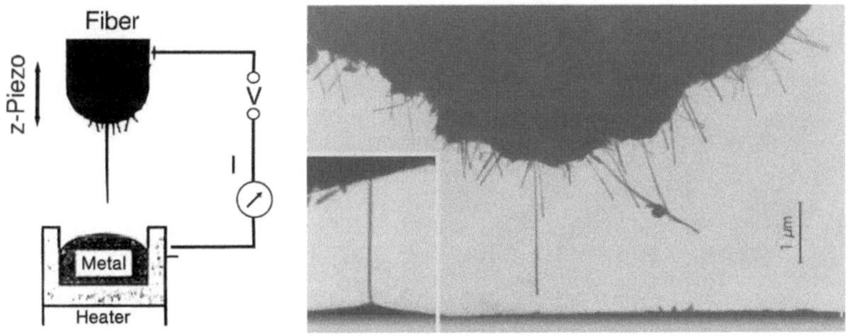


(a)

**Fig. 12.** (**a**) Schematic diagram of a contacting apparatus. The fiber is moved into the liquid metal, while the conductance $G$ is measured. Taken from *Frank* et al. [45]. (**b**) Transmission electron micrograph of the end of a nanotube fiber with one single MWNT extending far out from the fiber [45] nanotube lying under the electrodes are selected for electric measurements. An example of a single MWNT contacted by four Au fingers is shown in Fig. 13. Since the success of this contacting scheme works by chance, it is obvious that the yield is low. There are many structures which have either no or several NTs (NanoTubes) contacted in parallel. Since a large array of more than 100 structures can readily be fabricated, however, this scheme has turned out to be very convenient.

Alternatively, it is also possible to first structure a regular pattern of alignment marks on the substrate. After adsorbing the NTs, the sample is first imaged with SEM or SFM in order to locate suitable NTs. Having noted the coordinates of the designated nanotubes aided by the alignment marks, the electrodes can be structured directly onto the respective NTs with high precision. This improves the yield at the cost of an additional lithography step. Let us emphasize that the Au electrodes are evaporated directly onto the nanotubes. The reverse scheme is also possible. Here, electrode structures are made first and the NTs are adsorbed thereafter. In this scheme (nanotube over the contacts), the contact resistances are typically found to be large ($> 1\,\mathrm{M}\Omega$). It was only with the aid of local electron exposure directly onto the NT–Au contacts that this resistance could be lowered to acceptable values [95]. In contrast, the contact resistance $R_c$ can be surprisingly small in the former scheme (nanotube under the contacts). $R_c$ is of order $0.1 \ldots 20\,\mathrm{k}\Omega$ with an average of $4\,\mathrm{k}\Omega$ [44]. $R_c$ has been determined by comparing the 2-terminal ($R_{2t}$) with the 4-terminal ($R_{4t}$) resistance according to $R_c = (R_{2t} - R_{4t})/2$.

An 'ideal' contact is defined to have no backscattering and to inject electrons in all modes equally. Electrons incident from the NT to a contact will then be adsorbed by the contact with unit probability. Because the contact couples to both right and left propagating modes equally, Ohm's law should

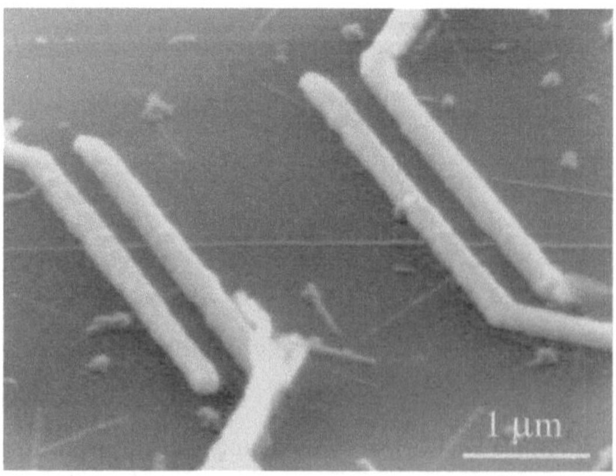


**Fig. 13.** SEM of a single multi-wall nanotube electrically contacted by four Au fingers from above. The separation between the inner two contacts is 2μm [94]

be valid in this limit. It is important to realize that for ideal contacts, $R_{4t}$ cannot contain any nonlocal contributions, i.e., contributions to the resistance that arise from a NT segment not located in between the inner two contacts. Any sign of non-locality points to the presence of non-ideal contacts. Conceptually, it is very hard to image that the Au-electrodes provide an ideal contact to a metallic nanotube, because the electronic properties change abruptly from something which is more like a semimetal (well within the nanotube) to a high-carrier density metal well within the metal contact.

In order to confirm the ballistic 1-D nature of an ideal metallic nanotube in transport measurements, ideal contacts (no backscattering) are required. Only then, will the conductance be quantized and equal to $G = 4e^2/h$ [96]. Since the metallic electrodes are extended and it is very likely that the coupling from the metal to the tube is not uniform, metal-tube tunneling occurs at many different spots simultaneously. Unfortunately, models for such contacts are not yet available. The interpretation of the measured resistance would be much easier, if we were able to intentionally choose the strength of the coupling at the contacts. If this were possible, one could fabricate 4-terminal devices with low-ohmic contacts for the outer two contacts (placed as far apart as possible) and very weak-coupling (non-perturbing) contacts for the inner two. $R_{4t}$ of such a device would then give the real intrinsic nanotube resistance caused by scattering in the tube. In particular, $R_{4t} = 0$ for the ballistic ideal metallic NT. Furthermore, a high-ohmic contact (a true tunneling contact) can be used for tunneling spectroscopy from which the density of states in the tube can be derived.

Until now, lithographically fabricated contacts are only accidentally high-ohmic. However, an elegant non-invasive measurement of the electrostatic potential is possible with scanning-probe microscopy in which a tip is used to probe the potential along a biased nanotube [97]. In addition to allowing for more than just two contacts, electrostatic gates can be added to supported NTs enabling one to tune the carrier concentration (Fermi energy). A metal electrode (not in direct contact with the NT) or the substrate itself can serve as a gate. In the later case, which is preferred by most researchers, a degenerately-doped Si substrate is used and the isolation between the gate and the NT is provided by a thin oxide layer with a thickness of 0.1-1 µm.

## 3.5 Electrical Measurements on Single MWNTs

Figure 14 shows the characteristic dependence of the equilibrium electric resistance $R$ on the temperature $T$ for a single MWNT with low-ohmic contacts, where it is shown that $R(T)$ increases with decreasing $T$. This decrease is rather moderate and distinctly different to what was found in macrobundles and films, discussed in Sect. 3.3, see Fig. 11, where it was found that $R^{-1} = G(T) \propto T$. In contrast, $R(T)$ in Fig. 14 cannot be expressed as a simple power-law, so that if we would try to write $G \propto T^{\alpha}$,

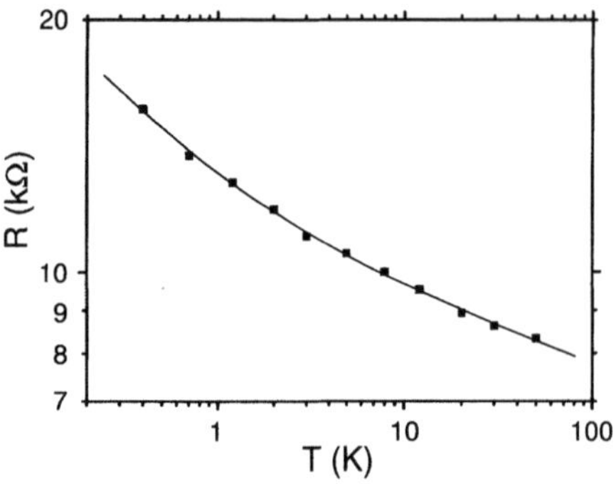


**Fig. 14.** Four-terminal equilibrium electric reistance $R$ vs $T$ of a single arc-grown MWNT with a contact separation of 300 nm [98]

then $\alpha$ would range from 0.07 (100 K) to 0.19 (0.4 K). Hence, the $T$ dependence for a single MWNT is weaker as compared to macro-bundles. According to our previous discussion (Sect. 3, 3.1), the non-metallic looking behavior of Fig. 14 could be taken as a sign that MWNTs are similar to highly disordered graphite (Fig. 9). This is, however, not true. Taking $R(300K) = 6\,\mathrm{k}\Omega$, the sheet resistance and the resistivity are found to be $R_\square = 1.3\,\mathrm{k}\Omega$ and $\rho = 0.4\,\mu\Omega\,\mathrm{m}$, respectively. Here, we have used again the fact that $R$ is determined by the outermost metallic SWNT [46]. Only arc-grown graphitic fibers or single-crystalline graphite have shown such low resitivities. But contrary to MWNTs, these 'clean' materials display a metallic $R(T)$-dependence: the resistance decreases with decreasing temperature. Based on $R_\square$ we conclude that single arc-grown MWNT are highly graphitized with a low degree of disorder, comparable to that of single-crystalline graphite, and, that the resistance increase must be a characteristic feature for this tubular one-dimensional form of graphite. The low-$T$ resistance of MWNTs differs also from SWNTs. Measurements on SWNTs almost always display Coulomb Blockade (CB) behavior below $\approx 10\,\mathrm{K}$ [74,75]. In this regime of single-electron tunneling, the whole NT acts as an island weakly coupled to the environment, i.e., with contact resistances larger than the resistance quantum $R_0 = h/2e^2 = 12.9\,\mathrm{k}\Omega$ [99]. For low applied voltages, $R$ is exponentially suppressed once $kT \ll E_c$, where $E_c = e^2/2C$ is the single-electron charging energy of the NT with capacitance $C$ with respect to the environment. A typical value is $E_c \approx 1\,\mathrm{meV}$ for a $1\,\mu\mathrm{m}$ long nanotube segment. Figure 14 is representative for all measured MWNTs with low-ohmic contacts. Coulomb blockade is not observed. According to theory, 2/3 of the tubes should be semiconducting with a substantial band gap (even for MWNTs) of order $E_g \approx 40\,\mathrm{meV}$ for a 20 nm diameter tube. This band gap should give rise to an exponential temperature-dependence, according to $R(T) \propto \exp(2E_g/kT)$ which has never been observed. MWNTs have many tubes in parallel, from which it is clear that the metallic ones will take over at low temperatures.

Because the electrodes are in contact with the outermost shell, the outermost metallic NT dominates. If the outermost tube happens to have a gap, a larger contact resistance might be expected because the electrons would have to tunnel from the electrode through the semiconducting tube into the first metallic NT. This also has not been observed until now. Of all MWNTs contacted from above (i.e., with metal evaporated over the NT), appreciably high-ohmic contacts are rare (only 10 %) [44]. A possible explantion could be that the NTs are doped, either intrinsically or due to charge transfer from the metallic contact.

A metallic gate allows study of the dependence of the nanotube conductance on the carrier concentration. Gate sweeps are also important for providing a convincing proof of CB. If $G(T)$ is determined by CB at low $T$, then $G$ should be periodic in the gate voltage $V_g$ with a period given by $e/C$. Figure 15 shows two such gate-sweeps for a single MWNT at 0.4 K, one for zero magnetic field, the other in a transverse field of $B = 2\,\mathrm{T}$. Here $G$ is strongly modulated by $V_g$ with rms-fluctuations of order $0.3G_0$ around an average value of $0.8G_0$. Coulomb blockade would result in a periodic modulation with a period in $V_g$ estimated to be 50 mV. But the fluctuations in Fig. 15 occur rather on a scale of 1 V. One might argue that the NT splits into a sequence of smaller islands at low $T$, which would give rise to a fluctuation pattern with a larger characteristic voltage scale. This picture is quite similar to strong localization for which an exponential resistance increase would be expected at low $T$, which is not observed. We therefore conclude that CB is not relevant for MWNTs with low-ohmic contacts down to 300 mK.

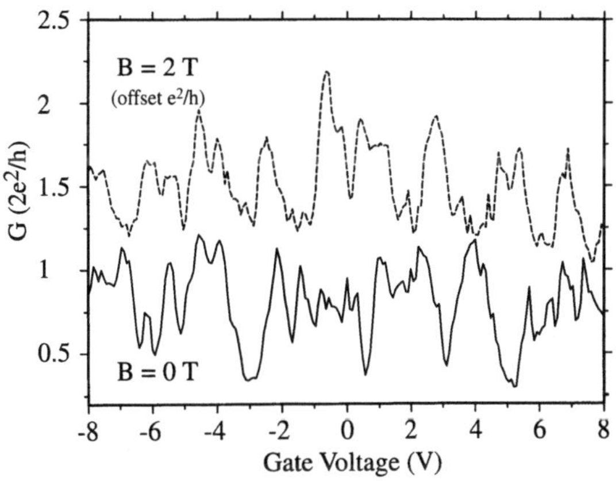


**Fig. 15.** Conductance of a 300 nm single MWNT segment at $T = 0.4\,\mathrm{K}$ as a function of gate voltage applied to the degenerately doped Si substrate. The nanotube is spaced from the back-gate by a Si-oxide layer 400 nm in thickness. Lower (*upper*) curve is measured in zero (2 T) magnetic field. The *upper curve* is offset vertically by $e^2/h$ for clarity [98]

It is remarkable that $G > G_0$ for certain $V_g$ in Fig. 15. The observed pattern reflects the quantum states in this nanotube segment approaching the 0D limit. The pattern is aperiodic because of random scattering, and the NT behaves like a chaotic cavity. The fluctuation is suggestive of so-called Universial-Conductance Fluctuations (UCF) [100,101,102].

The upper curve in Fig. 15 demonstrates that the pattern is completely changed in a magnetic field of 2 T. Because the two curves show no correlations, the so-called correlation field is estimated for be $B_c < 2$ T. This allows us to estimate the phase-coherence length $l_\phi$ which must be $> 100$ nm. Hence, this 300 nm NT segment is certainly 1-D with respect to quantum coherence and even close to 0D (quantum dot). Let us estimate the expected mean level spacing $\Delta E_{0D}$ for an ideal NT quantum dot of length $L$. If spin-degeneracy is removed, then $\Delta E_{0D} = hv_F/(4L)$, amounting to 3 meV for $L = 300$ nm. One now needs to know the leverage factor $\Delta V_g/\Delta E_F$ which we estimate to be 100. Hence, 0D features should show up in the gate-dependence on a voltage scale of order 0.3 V, which is indeed the case.

These measurements also allow us to estimate the mean-free path $l_e$. Because strong localization is not observed, we conclude that the localization length is $l_{loc} > l_\phi$. With $l_{loc} \sim N l_e$, where $N \approx 4$ is the number of modes, we see that $l_e > 20$ nm, so that the mean-free path is of the same order as the diameter of the MWNTs. Hence, there is elastic scattering leading to the interference patterns at low temperature, but this scattering is not strong enough to localize the electron states in the MWNT. This result, that MWNTs are not free of elastic scattering (there is some disorder), has to be contrasted with work of *Frank* et al. [45]. Figure 16 shows the measured conductance $G$ of a fiber of MWNTs while lowering this fiber continuously into a liquid metal (see also Fig. 12). In Fig. 16 we see that $G$ increases in steps of magnitude close to the quantized conductance $G_0 = 2e^2/h$. Each step is due to an additional MWNT coming into contact with the liquid. The nearly equal conductance of $\approx G_0$ for each MWNT, that makes contact with the liquid Ga, has been taken as evidence for quantum ballistic transport. This effect is very striking, since the experiments in Fig. 16 were done at room temperature. Why the measured $G$ is only one-half of the expected quantized conductance for an ideal metallic NT is not understood at present [103]. The measured $G \approx G_0$ is also in disagreement with other transport measurements on nominally similar MWNTs. For example, the NT in Fig. 14 has a room temperature resistance of only 6 k$\Omega$ which corresponds to $\approx 4e^2/h$. Values for the two-terminal conductance of up to $4G_0$ have been found at room temperature [44]. This is not unexpected because higher subbands contribute to the conductance at room temperature, too. Recall, that the mean-level spacing is only 33 meV for $d_t = 20$ nm. The result of Frank et al. can be summarized as follows: the measured resistance is a contact resistance, which is quantized and the intrinsic NT resistance, determined from the depth-dependence, is

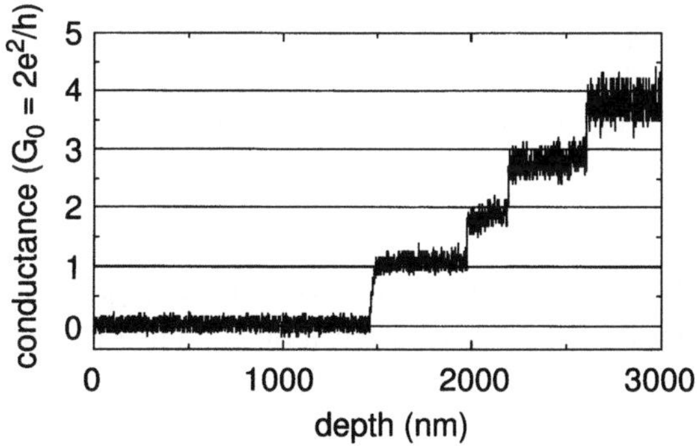


**Fig. 16.** Conductance $G$ at room temperature measured in the apparatus shown in Fig. 12 as a function of depth of immersion of the nanotube fiber into liquid gallium. As the nanotube fiber is dipped into the liquid metal, the conductance increases in steps of $G_0 = 2e^2/h$. The steps correspond to different nanotubes coming successively into contact with the liquid [104]

small. The latter is in contrast to recently published length-dependent intrinsic resistances of $4\,\mathrm{k\Omega/\mu m}$ [44] and $\approx 10\,\mathrm{k\Omega/\mu m}$ [97], respectively.

To support ballistic transport over micrometer distances, *Frank* et al. came up with another interesting experimental observation. Large electric currents of order 1 mA can be driven through MWNTs without destroying them. Based on the electric power and bulk heat conductivity for graphite, the NT is expected to evaporate due to the large temperature rise. But no melting is observed. This observation of large current densities may however not be taken as a proof for ballistic transport. Rather, it shows that dissipation is largely absent (which is an exciting fact by itself). In fact, large currents of the same magnitude can be passed through lithographically contacted MWNTs without destroying them, although these MWNTs have been proven not to be quantum-ballistic [44]. An interpretation of this phenomena is difficult because it occurs in the nonlinear transport regime for applied voltages much larger than the subband separation [105]

### 3.6 Magnetotransport

We first discuss the MagnetoResistance (MR) in a parallel magnetic field $B$. This case is very appealing because the effect of $B$ on the wavefunction can easily be described. The magnetic flux $\phi$ through the nanotube gives rise to a Aharonov–Bohm phase modifying the boundary condition of the transverse wavevector $k_\perp$ (4) into:

$$k_\perp = \frac{2}{d_\mathrm{t}}\,(n + \phi/\phi_0 - \nu/3) \quad n = 0, \pm1, \pm2, \dots\,, \tag{7}$$

$\phi_0$ is the magnetic flux quantum $h/e$ and $\nu = 0, \pm 1$ [62,106]. Similarly, the index $n$ in the approximate 1-D dispersion relation of (5) is changed into $n + \phi/\phi_0$. It then follows that the cut-off energy $E_{n,\nu}(\phi)$ for the subbands is:

$$E_{n,\nu}(\phi) = E_0 \left| n + \frac{\phi}{\phi_0} - \frac{\nu}{3} \right| . \tag{8}$$

This relation is shown in Fig. 17 for $\nu = 0$. The band structure is periodic in parallel magnetic field with the fundamental period given by $\phi_0$. The drawing corresponds to a metallic tube, for which the cutoff energy is zero for $\phi = 0$. Because $\nu$ and the scaled flux appear as a sum in (8), a metallic tube is turned into a semiconducting one depending on the magnetic flux $\phi/\phi_0$ and vice versa. At half a flux quantum $\phi/2\phi_0$ the band-gap reaches its maximum value of $E_0$, corresponding to 65 meV for a SWNT with $d_t = 20$ nm. If, as claimed before, a single nanotube dominates transport in a MWNT, a periodic metal-insulator transition should be observed as a function of parallel field at low temperatures.

The dependence of the electric resistance of MWNTs in a parallel magnetic field has been studied by *Bachtold* et al. [46] and Fig. 18 shows results of a typical (MR) measurement. If we adhere to the notion that the measured resistance in MWNTs is due to the outermost *metallic* NT, $R$ is expected to increase because of the appearance of a gap. However, on applying a parallel magnetic field $B$, the resistance rather sharply *decreases*. It is therefore clear that this effect must have another origin than the band structure modulation that we have just discussed. This decrease is associated with the phenomenon of WL [70,71,72,73].

Weak localization originates from the quantum-mechanical treatment of backscattering which contains interference terms, which add up construc-

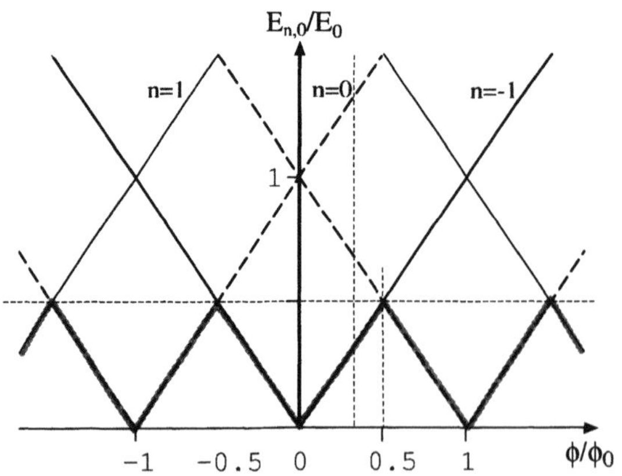


**Fig. 17.** Cutoff energy $E_{n,0}$ of 1-D-subbands for a (metallic) SWNT as a function of magnetic flux $\phi$ through the tube. The band structure is periodic in parallel magnetic field with a period given by the magnetic flux quantum $\phi_0 = h/e$. Both $E_{n,0}$ and $\phi$ and plotted in dimensionless units

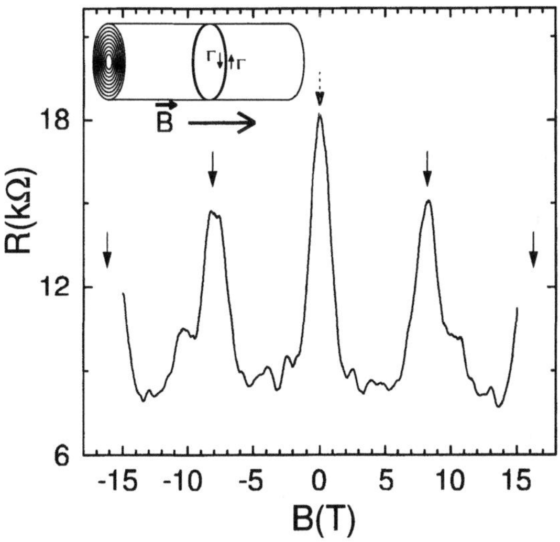


**Fig. 18.** Electrical resistance $R$ as a function of parallel magnetic field $B$ (see *inset* showing electron orbit). Arrows denote the resistance maxima corresponding to multiples of $h/2e$ in magnetic flux through the nanotube taking the outer diameter [44]

tively in zero magnetic field. Backscattering is thereby enhanced, leading to a resistance larger than the classical Drude resistance. Because the interference terms cancel in a magnetic field of sufficient strength, WL results in a negative MR. However, for the specific geometry of a cylinder (or ring), the WL contribution is periodic in the magnetic flux through the cylinder, with *half* the AB (Aharonov–Bohm) period $h/2e$ [107]. Indeed, in Fig. 18 the resistance has a second maximum at $B = 8.2\,\mathrm{T}$. From this field value, a diameter of $d_\mathrm{t} = 18\,\mathrm{nm}$ is obtained for this MWNT. As was demonstrated by *Bachtold* et al. [46], the MR agrees with the *Altshuler, Aharonov* and *Spivak* (AAS) theory [108], only if the current is assumed to flow through one of outermost cylinders with a diameter corresponding to the independently measured outer diameter of the NT. It is therefore most likely that only one cylinder actually participates in transport. The conclusion that only *one* graphene cylinder carries the current can only unambiguously be drawn from the analysis of the low-temperature MR data ($T < 20\,\mathrm{K}$). We emphasize that it is not possible to relate the resistance maxima at $\pm 8.2\,\mathrm{T}$ to a magnetic flux of $h/e$, because a tube diameter would then result which is larger than the actually measured outer diameter. The observation of a pronounced $h/2e$ resistance peak proves that backscattering is present. The NTs therefore do *not* exhibit ballistic transport.

In a parallel magnetic field, resistance maxima should occur periodically. The onset of the second resistance peak, which is expected at $B = 16.4\,\mathrm{T}$ is clearly seen in Fig. 18. From this measurement, a phase coherence length of $l_\phi \approx 200\,\mathrm{nm}$ is estimated in good agreement with the estimate given before [44].

Next, we will consider the MR in a perpendicular field, for which the MR has been extensively studied [44,68,69,76,109,110,111]. A negative MR is found for a transverse $B$ field, in agreement with weak-localization theory [70,71,72,73]. Further support for the importance of interference contributions to the transport properties comes from the observation of non-local effects in multi-terminal devices [44],

A typical MR measurement is shown in Fig. 19. On applying a transverse magnetic field, the resistance decreases (negative MR), in agreement with WL theory. Aperiodic fluctuations are observed superimposed on the negative MR background, and these fluctuations are assigned to universal conductance fluctuations (UCFs). This MR can only be fitted with 1D-WL theory, i.e., $l_\phi > \pi d_\mathrm{t}$ ($d_\mathrm{t} = 23\,\mathrm{nm}$). Best fits to theory are shown in Fig. 19 by the dashed lines, and a very good agreement is found. As a cross-check, UCF amplitudes deduced using $l_\phi$ agree with the observation. An example is given by the vertical bar in Fig. 19 which corresponds to the expected UCF amplitude at 2.5 K. Let us estimate the phase coherence length $l_\phi$ at 2.5 K. The resistance peak at zero field corresponds the a conductance change of $\Delta G = -2.2 \times 10^{-5}\,\mathrm{S}$. From $\Delta G \approx -(2e^2/h)l_\phi/L$ we obtain $l_\phi \approx 500\,\mathrm{nm}$. This particular example shows a rather large $l_\phi$ and correspondingly a surprisingly large diffusion coefficient $D$ and mean-free path $l_\mathrm{e}$. These later parameters can be obatined from the $l_\phi$ vs $T$ relation for dephasing by quasi-

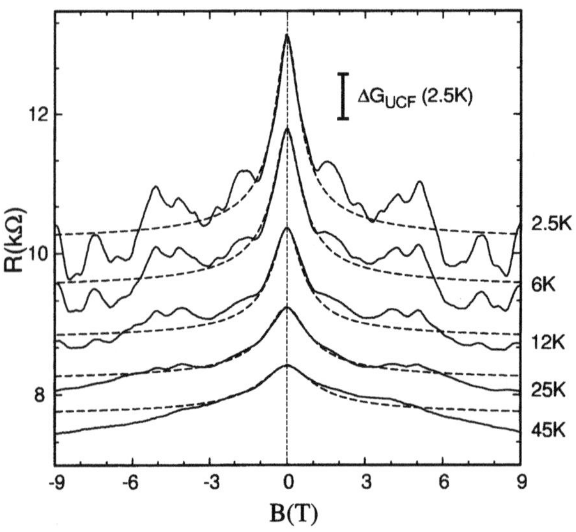


**Fig. 19.** Four-terminal MR of a MWNT in a perpendicular magnetic field for different temperatures. The voltage probes are separated by 1.9μm. *Dashed curves* show fits using one-dimensional weak-localization (1D-WL) theory. Note, that the curves are not displaced vertically for clarity [44]

elastic electron-electron scattering (Nyquist noise dephasing [112]) which is the dominant source of phase-randomization at low temperature [44]. One obtains $D = 450\text{–}900\,\mathrm{cm}^2/\mathrm{s}$ and $l_e = 90\text{–}180\,\mathrm{nm}$. Again, this mean-free path is in agreement with the condition that $l_{loc} > l_\phi$ from which $l_e > 100\,\mathrm{nm}$ follows.

Another finding in Fig. 19 worth mentioning is that weak localization results in an increase of the electric resistance at low temperatures, but this is not the only contribution to the resistance increase. For large magnetic fields $\delta G_{WL} \to 0$ but the resistance is still seen to be strongly temperature dependent. Here we note that the curves in Fig. 19 are not displaced for clarity, but rather a temperature-dependent background resistance is observed and this effect is usually associated with electron-electron interaction. While WL primarily enters as a correction to the diffusion coefficient, the carrier-carrier interaction suppresses the single-particle DOS. For a diffusive wire for which the coherence length (the thermal length) is larger than the width but smaller than the length, theory predicts a temperature dependence for the conductance correction $\delta G_{ee} \propto T^{-1/2}$ [113,114,115]. Knowing $\delta G_{WL}(T)$, one can then plot $G(T) - \delta G_{WL}(T)$ as a function of $\sqrt{T}$.

It turns out, however, that this relation does not hold [70,71,72,73]. The interference and interaction corrections $\delta G_{WL}$ and $\delta G_{ee}$ are derived by WL perturbation theory and are therefore assumed to be small. But exactly this assumption is not valid here. One therefore needs to treat the interaction exactly.

This is indeed possible in 1-D, using bozonization techniques. Carbon nanotubes are predicted to become so-called Luttinger Liquids (LL) in which a pseudo-gap in the quasi-particle DOS opens at low energy [116,117,118,119,120]. A perfect ballistic LL cannot be distinguished from the perfect Fermi liquid quantum wire in an equilibrium transport measurement whereby $G$ is just quantized to $2e^2/h$ (for one mode). However, with backscattering, $G$ is renormalized. In the so-called strong backscattering limit (low $T$ limit) $G \propto T^\alpha$ with a weaker $T$-dependence at higher temperatures. For a SWNT $\alpha \approx 0.3\text{–}0.8$, while $\alpha$ can be reduced to lower values in MWNTs due to the enhanced screening by multiple shells [121]. Though $R(T)$ in Fig. 14 does not follow a power law, the temperature dependence of $R$ is not inconsistent with LL theory. This dependence is expected for a LL in the weak-backscattering limit. LL theory includes the electron–electron interaction. In addition, scattering has been incorporated into the theory. However, the dependence on magnetic field has not yet been treated. It would be very interesting to see whether such a 'complete' theory could explain the measured MR, which can surprisingly be fitted quite well with theoretical results from WL perturbation theory based on Fermi-liquid quasi-particles [44,46,69,76]

### 3.7 Spectroscopy on Contacted MWNTs

Figure 20 shows a differential current–voltage characteristic ($dI/dV$) measured on a single MWNT, using an inner contact which by chance was high-ohmic. This particular tunneling contact had a contact resistance of $300\,\mathrm{k\Omega}$, whereas the other contacts had resistances $\ll 10\,\mathrm{k\Omega}$. The measured spectrum agrees surprisingly well with predicted spectra based on simple tight-binding calculations for a metallic NT [106,122,123]. Firstly, there is a substantial DOS at the Fermi energy, i.e., at $V = 0$, so that the NT is metallic. Secondly, the almost symmetric peak structure, appearing as a pseudo-gap is caused by the additional 1D-subbands in the valence band ($V < 0$ ) and conduction band ($V > 0$) with threshold energies of order $\approx 50\,\mathrm{meV}$. At the onset of the subbands, van Hove singularities are expected. The spectrum in Fig. 20 agrees remarkably well with scanning-tunneling measurements of *Wildöer* et al. for SWNTs [63,64]. But because of the difference in tube diameter, the energy scales are quite different. These subbands should be spaced by $2E_0 = 4\hbar v_F/d_t = 150\,\mathrm{meV}$ for the MWNT in Fig. 20 which has $d_t = 17\,\mathrm{nm}$. The measured spacing of $110\,\mathrm{mV}$ is in reasonable agreement with this estimate. The observation of van Hove features in $dI/dV$ demonstrates that the mean-free path $l_e$ cannot be much shorter than the nanotube circumference. If $l_e < \pi d_t$, all 1-D band structure features would be expected to be washed out. The observed spectrum in Fig. 20 nicely demonstrates that the unusual

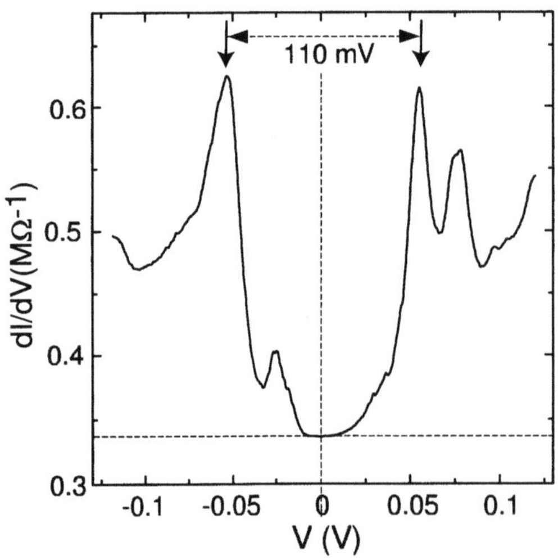


**Fig. 20.** Differential (tunneling) conductance $dI/dV$ measured on a single MWNT using a high-ohmic contact ($300\,\mathrm{k\Omega}$) at $T = 4.2\,\mathrm{K}$ [44]. This spectrum qualitatively confirms the DOS expected for a metallic nanotube in which the wave vector is quantized around the tube circumference leading to 1D-subbands. Positive (negative) voltages correspond to empty (occupied) nanotube states

band structure effects of NTs are also found for MWNTs. In most cases, however, the measured spectral features are not as sharp.

Furthermore, the prevailing $dI/dV$ spectra display a pronounced zero-bias anomaly on a smaller energy scale of 1–10 meV. Such a tunneling spectra is shown in Fig. 21. These spectra are highly structured. The observed peaks are associated with broadened van Hove singularities due to the 1-D-band structure. This assignment is strongly supported by the observed peak-shifts in a parallel magnetic field $B$. The peaks are seen to move up and down with $B$ in accordance with the Aharonov–Bohm effect. According to Fig. 17, the total peak shift amounts to $E_0/2$, corresponding to 39 meV for this MWNT ($d_t = 17$ nm). The measured, shifts are somewhat smaller, $\approx 22$ meV. With the exception of small voltages, i.e., for $|V| < 25$ meV, the mean peak spacing is $\approx 25$ meV, in agreement with the maximum peak shift. In addition to the peaks, there is a pronounced zero-bias anomaly (ZBA). At $V = 0$ the

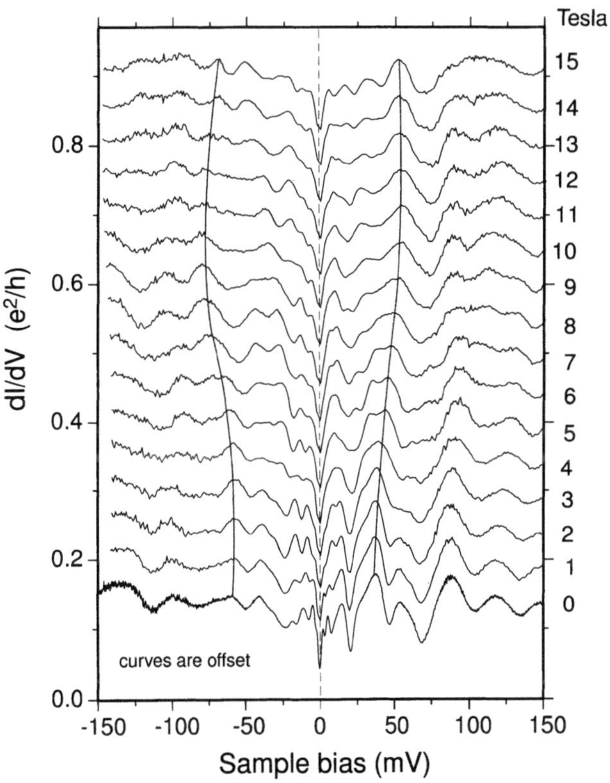


**Fig. 21.** Tunneling $dI/dV$ spectra measured on a single MWNT using a high-ohmic contact for different magnetic fields $B$ applied parallel to the tube [124]. Note, that the spectral features shift in a magnetic field (lines are a guide to the eyes for two well discernible peaks), and that there is a pronounced anomaly at $V = 0$. All curves are vertically displaced for clarity [124]

conductance is strongly suppressed, independent of $B$. This suggests that a pseudo-gap opens for low-energy quasi-particles. This finding must be related to the measured resistance increase at low temperature, as shown in Figs. 14 and 19.

Figure 22 shows a detailed analysis according to *Bockrath* et al. [125] of a zero-bias tunneling anomaly measured on a MWNT for different temperatures. The ZBA anomaly exhibits power-law scaling, i.e., $dI/dV \propto V^{\alpha}$ if $eV \gg kT$ and $dI/dV \propto T^{\alpha}$ if $eV \ll kT$. Such a dependence is in agreement with Luttinger liquid models [126]. Similar anomalies have recently been observed by *Bockrath* et al. for SWNTs [125]. Their measurement and analysis provide the first demonstration of LL behavior in carbon NTs.

The LL theory describes the interaction with a single parameter $g$ [126]. The non-interacting Fermi-liquid case corresponds to $g = 1$ and $0 < g < 1$ is valid for a LL with repulsive interaction. The parameter $g$ is determined by the ratio of the single-electron charging energy to the single-particle level spacing and has been estimated to be $g = 0.2$–$0.3$ for SWNTs [120]. Because a MWNT consists of several shells, one might expect that the inner shells strongly screen the long-range Coulomb interaction leading to $g \to 1$, i.e., to an effectively non-interacting Fermi liquid. However, it has recently been shown that $g$ is only weakly modified and theoretically only scales with $\sqrt{N}$ where $N$ is the number of shells participating in screening [121]. Although there about 20 shells, the effective $N$ is expected to be smaller and is only of order 1. LL behavior is therefore expected for MWNTs, too. Regarding the exponent $\alpha$, one distinguishes 'bulk' from 'end' tunneling. The measurement in Fig. 22 corresponds to bulk tunneling for which $\alpha = (g^{-1} + g - 2)/8$ [120].

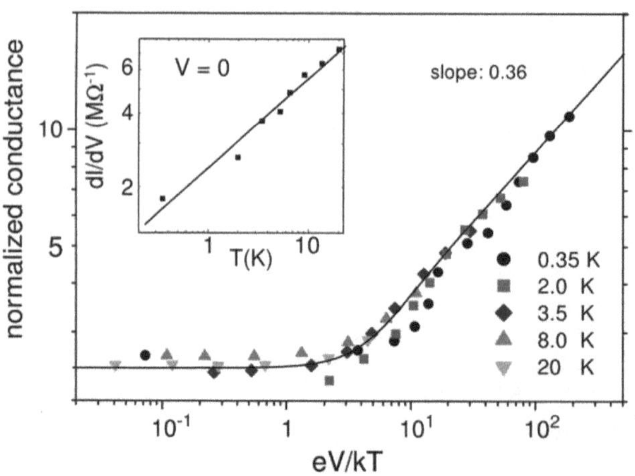


**Fig. 22.** Normalized differential tunneling conductance $dI/dV \cdot T^{-\alpha}$ vs scaled voltage $eV/kT$ measured on a single MWNT using a high-ohmic contact. The *inset* displays the equilibrium conductance as a function of temperature $T$ on a log-log plot. The *straight line* corresponds to $\alpha = 0.36$ [98]

From the experiment $\alpha \approx 0.36$, which relates to an interaction parameter of $g = 0.21$. The same value was obtained by *Bockrath* et al. for SWNTs [125]. Currently, more evidence for LL liquid behavior in nanotubes (including MWNTs) is appearing [98,127,128,129].

### 3.8 Spectroscopy Using Scanning Tunneling Probes

The individual tubes in a MWNT have helicities, described by the $(n, m)$ indexes, and the electronic structure calculations [28,47] for the SWNTs are also valid for the individual nanotubes in MWNTs. In principle, one can find metallic, or semimetallic nanotubes, if $n - m = 0$, modulo 3 is satisfied, or else the individual tubes are semiconducting, if $n - m \neq 3q$, $q = 0, 1, \ldots$. There seems to be some commensurability effect in play between the different layers, since the number of chiral angles observed in HRTEM is lower than the number of cylinders forming the MWNT [21]. Theoretically it has been predicted than even in the case of metallic layers, the nested nature of the tubes of different chiralities may introduce gaps or pseudo-gaps in the density of states in a similar way, as happens in the case of SWNTs organized in a bundle [130]. This is in contrast with graphite, where the interlayer interaction between the zero-gap semiconducting graphene layers creates a finite density of states at the Fermi level, making graphite metallic.

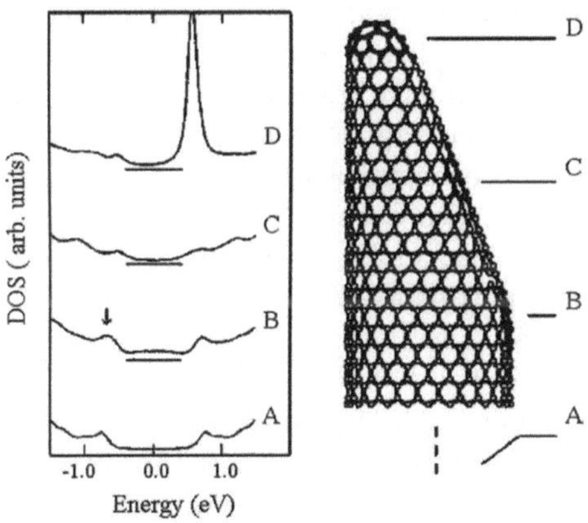


**Fig. 23.** Measured carbon nanotube differential conductances, proportional to the local density of states, along various positions near the tip (scans at B,C,D locations on the nanotube) and far from the tip (scan A) of a multi-walled carbon nanotube. The sharp peak near the Fermi level of the uppermost trace demonstrates the existence of a localized tip state [131]

The electronic structure can be measured very reliably by STM. This technique was used to image the chirality and to measure the local density of states for SWNTs [63,64]. The lateral confinement of the electronic states due to the tubular structure causes symmetric singularities in the DOS, which also help to identify the tube chirality, since there is a reliable correspondence between the peak positions, diameter and chiral indexes. Carroll and coworkers have measured the Local Density Of States (LDOS) of multi-wall carbon nanotubes for different tube diameters scanning along the tube axis towards the tips (Fig. 23) [131]. Scanning far from the ends, the measurements show the outer layer of the tubes indicate a conducting character consistent with a graphitic-like density of states at low bias and symmetric singularities with respect to zero bias, due to 1-D quantum confinement. At the nanotube tip, the STM spectra show a localized state, whose position and width varies depending on the nanotube tip structure. Calculations have shown that the strength and the position of these states with respect to the Fermi level depend sensitively on the relative positions of the pentagons and their degree of confinement at the tube ends.

## 3.9 Discussion of the Main Issues

From the previous discussion, four basic issues emerge, which need further attention in the future: (1) Why does transport appear to be ballistic in one experiment and diffusive in another? (2) What is the origin of scattering? (3) What causes hole doping and what is the energy of the doping level in single MWNTs? (4) What is the ground state of carbon nanotubes?

Let us briefly discuss these questions. Electric transport measurements on single MWNTs give conflicting results. While *Frank* et al. [45] find ballistic transport, there is convincing evidence for diffusive transport from other experiments [44,46,68,69,97]. It is important to emphasize that the estimates for the mean-free path $l_e$ from the latter experiments vary substantially, from $l_e \approx 1\,\text{nm}$ to $l_e > 100\,\text{nm}$. Taken this fact into account, the first question should be rephased to read: what is the reason for the *large variation* of scattering lengths that are observed in experiments on single MWNTs? This question immediately leads to the second question. What is the origin of the scattering? Are MWNTs defective, is the band structure modified by intertube hybridization, are there inclusions, or is the scattering caused by adsorbates? What is the role of the substrate on which the nanotubes resides in experiments using micro-fabrication? Arc-discharge grown MWNTs appear to be essentially defect-free in TEM micrographs. Furthermore, there is no evidence for bond defects from electron-spin resonance.

From the experiments and the discussion in this section the electronic intertube coupling is expected to be relatively weak. It is therefore believed that scattering is related (at least partly) to the issue of doping. MWNTs (and SWNTs) are found to be hole-doped in experiments on films and macro-bundles. Though the origin of the unintentional doping is not known yet, the apparent doping level can be changed in SWNTS if the nanotubes are heated to only 200 °C in high-vacuum [91]. This suggests that (part) of the doping might be caused by species present in air (for example by oxygen). Until now, the doping-level has not been specified in any experiment on a single MWNT. We need to find a way to quantify the doping level in the future.

The fourth question, finally, is a very intricate one: What is the ground-state of carbon nanotubes? A lot of measurements (e.g., magnetotransport) can be described by either non-interacting or only weakly interacting quasi-particles using theories like weak-localization based on the Fermi Liquid (FL) hypothesis. On the other hand, clear deviations from FL behavior has been observed too. The suppression of the quasi-particle density of states, observed in tunneling spectroscopy on single MWNTs (and SWNTs), suggests that nanotubes may develop a LL state. The ground-state question has become an even more exciting issue, because signatures for intrinsic superconductivity have recently appeared [132].

## 4 Magnetic Properties

The magnetic properties of carbon nanotubes can be effectively monitored by the Electron Spin Resonance (ESR) technique which has been intensively used to study the electronic properties of graphitic and conjugated materials [133]. This method has several advantages: it has a very high sensitivity, it responds only to the paramagnetic signal (unless static measurements are made, which are usually dominated by the large diamagnetic response), and it can distinguish between different spin species, e.g., localized and conduction electron spins. Three different quantities are determined by ESR: (1) the $g$-factor, which depends on the chemical environment of the spins via spin-orbit coupling (and also the hyperfine interaction); (2) the linewidth, which is governed by the spin relaxation mechanism; and (3) the intensity of the signal, which is proportional to the static susceptibility.

For MWNTs we are mainly interested in the ESR response of the conduction electrons. The conduction electron spin $g$-factor is determined by the spin–orbit splitting of the energy levels in the presence of a magnetic field [134,135]. In the case of degenerate bands (as in graphite at the $K$ point), theory predicts that spin–orbit coupling, which removes the degeneracy, induces a large $g$-shift which varies inversely with temperature. This is exactly what is observed in graphite when the magnetic field is perpendicular to the plane, allowing large orbital currents, thereby increasing the spin–orbit coupling. When the magnetic field is parallel to the planes, orbital currents are suppressed, and a small $g$-shift, nearly independent of temperature, is observed. When the Fermi level is shifted away from the $K$ point by doping, the $K$ point degeneracy and the $g$-shift anisotropy disappear. (Despite this understanding of the variation of the $g$-shift, a rigorous theory is still missing in graphite due to the complicated band structure near the point of degeneracy.)

Using this qualitative description for graphite, one can speculate about the $g$-factor for carbon nanotubes. When the magnetic field is parallel to the tube axis, nearly the same value is obtained for the $g$-factor as is found in graphite when the field is in the plane, except at a field for which the cyclotron radius equals the geometrical radius of the nanotubes. Such fields (typically 1 T and greater) are higher than those used in $X$-band spectrometers (~9 GHz).

When the magnetic field is perpendicular to the tube axis, orbital currents cannot completely close, as in a plane, and we can therefore expect a smaller $g$-shift than in graphite. A decrease of this $g$-shift with decreasing tube diameter is then expected. Some of these predictions are indeed realized in MWNTs [136]. The average observed $g$-value in MWNTs is 2.012, as compared to 2.018 in graphite, and the $g$-factor anisotropy is lower in MWNTs than in graphite.

Spin relaxation in metals and semimetals depends also on the spin–orbit coupling. More precisely it is the modulation of the spin–orbit coupling by lattice vibrations that causes spin relaxation. It can be shown, that in the framework of the *Elliott* theory [135], and when the spin relaxation time is proportional to the momentum relaxation time, and is governed by phonon scattering, the linewidth increases with increasing temperature. Opposite to this expectation, in graphite the linewidth increases when the temperature decreases. This behavior in graphite is attributed to motional narrowing over the $g$ value distribution. Because of the semimetallic nature of graphite, the density of states at the Fermi level is low, and hence the spin susceptibility is low, and is in the $10^{-8}$ emu/g range at room temperature [133]. We have seen that the $g$-factor in MWNTs, a local property, is very close to that of graphite. In the following subsections we will see how the spin relaxation and spin susceptibility are modified in MWNTs relative to graphite. Changing the dimensionality from 2-D to 1-D is expected to significantly modify the principal ESR characteristics. First, we have to understand what kind of ESR signal should be expected in a 1-D system.

## 4.1 Spin Relaxation in Quasi-1-D Systems

Extensive studies of the Conduction Electron Spin Resonance (CESR) linewidth ($\Delta H$) in isotropic metals have shown that the dominant process in the spin-lattice relaxation time ($T_1$) is the spin-flip scattering of conduction electrons by acoustic phonons. The same scattering process gives the momentum relaxation time $\tau_R$ measured by the electrical resistivity. Elliott has derived [134,135] a relation between the two relaxation times:

$$T_1 \sim \tau_R/(\Delta g)^2 , \tag{9}$$

where $\Delta g$ is the difference in $g$-factor between graphite and the free electron value, which reflects the strength of the spin–orbit coupling constant. Since $(\Delta g)^2$ is of the order of $10^{-4}$–$10^{-6}$, (9) indicates that one in every $10^4$–$10^6$

scattering events contributing to the resistivity gives a spin-flip, and hence the CESR linewidth can be written as

$$\Delta H = 1/\gamma T_1 \sim (\Delta g)^2/\tau_R \,. \tag{10}$$

For the majority of isotropic metals, $T_1$ is too short to give an observable CESR linewidth at room temperature. *Beuneu* and *Monod* [137] have verified the Elliott relation for most of the metals. It is well-known that the Elliott relation is violated for quasi-one-dimensional metals, such as TTF-TCNQ [138], where the linewidth does not show strong correlation with the spin-orbit coupling. Materials with similar $\Delta g$ and conductivity values as TTF-TCNQ, have quite different CESR linewidths and their $T_1$ values lie well below the universal curve established by *Beuneu* and *Monod* [137]. It is presently believed that the source of this disagreement lies in the low-dimensionality of these compounds [139]. In a strictly 1-D system the Fermi surface consists of two points at $k_F$ and $-k_F$. The possible scattering events are thus limited to those with $q \sim 0$ (forward) and $q \sim 2k_F$ (backward) momentum changes. The former process has a very small weight (even including 3-D phonons), while the latter process, reversing spin and momentum, is forbidden by time reversal symmetry. Hence to account for the observed finite ESR linewidth of quasi 1-D conductors, one has to include inter-chain scattering, because every process which could make the system resemble an isotropic metal, will broaden the line.

## 4.2 Spin Relaxation in Carbon Nanotubes

Isolated SWNTs, without inter-chain interactions, seem to be the ultimate 1-D conductors. Application of the ideas developed in Sect. 4.1 should result in a very narrow and very saturable ESR line, since most of the spin-relaxation channels are switched off. Surprisingly, no conduction ESR signal has been observed in SWNTs until now. One reason for the absence of an ESR line for the conduction electrons in SWNTs might be the strong electron-electron correlation, giving rise to the so-called Luttinger Liquid (LL) character of the 1-D electronic system. The ground state of a LL would be anti-ferromagnetic, which is ESR silent at resonance fields close to that expected for non-interacting electrons. However, there might be a less attractive reason for the absence of a CESR signal in SWNTs, and this is the presence of magnetic particles, which serve as catalysts for the SWNT synthesis, and afterwards remain in the actual SWNT samples as impurities. Because of the long carrier mean free path, and because of the even longer spin diffusion length, even a small concentration of magnetic particles can relax the conduction electrons very efficiently, causing the line to be unobservably broad. Although the MWNTs are less 1-D in comparison to SWNTs, (i.e., the MWNTs should rather be approximated by a quasi-1-D system, because of their multilayered structure and larger diameters), we believe that some signs of low-dimensionality are still present regarding the spin relaxation behavior.

Evidence for low dimensional behavior can be recognized in Fig. 24, where the ESR signal is plotted for MWNT soot with different MWNT densities: the upper signal is recorded for a dense thick MWNT film, while the lower one is the signal of MWNTs dispersed in paraffin. The ESR linewidth for the upper trace is 32 G, while for the lower trace, it is only 8 G. This demonstrates that when the phase space is increased by bringing the individual nanotubes into better contact with each other, the spin-relaxation increases like in a quasi-1-D conductor when hydrostatic pressure is applied [140]. This behavior is specific to nanotubes and it is contrary to what happens in graphitized carbon blacks. Indeed, with carbon black particles or fine graphite powder, increasing the contact between grains increases the motional averaging of the $g$-factor anisotropy which decreases the linewidth. We believe that the different behavior of the nanotubes is the fingerprint of their mesoscopic nature.

The temperature dependence of the ESR linewidth, shown in Fig. 25, is also unusual. In graphite $\Delta H$ increases with decreasing temperature, and passes through a broad maximum. This "anomalous" temperature dependence of $\Delta H$ in graphite was ascribed to the same phenomena that is responsible for the observed temperature dependence of the $g$-factor. In the case of highly purified MWNTs, this close correspondence between the two quantities breaks down as $T$ falls below 100 K. Whether the decoupling of the two quantities at low temperatures is once again the sign of low-dimensional

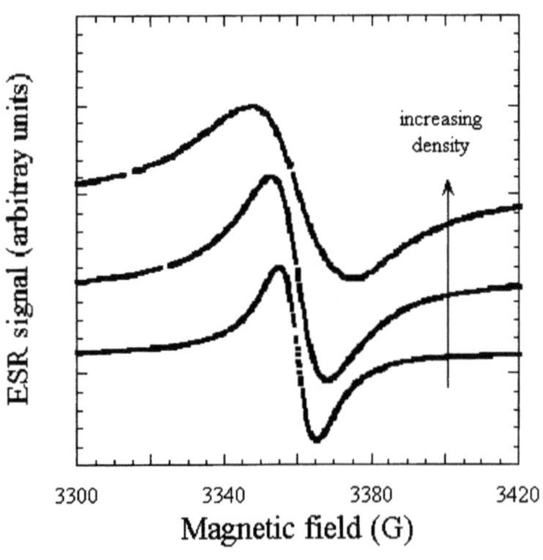


**Fig. 24.** The CESR signal of MWNTs for samples with different MWNT densities measured at 300 K as a function of the magnetic field. The intensities are not to scale. The upper signal is recorded for a dense MWNT thick film, while the lower one is the signal of MWNTs dispersed in paraffin. The narrowing linewidth with decreasing MWNT density is a fingerprint of low-dimensional spin-relaxation phenomena [141]

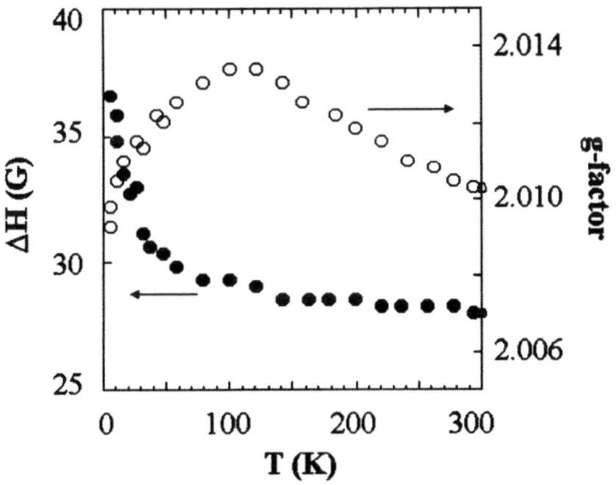


**Fig. 25.** The temperature dependence of the CESR linewidth and the $g$-factor for a highly purified thick film of MWNTs. The distinct temperature dependence for these two quantities might be also characteristic for single wall carbon nanotubes [141]

effects would require that detailed calculations of the spin and momentum relaxation be carried out for MWNTs.

The spin susceptibility $\chi_s$ for MWNTs is derived from the ESR signal strength by numerical integration. Above 40 K (Fig. 26), we find a temperature independent (Pauli) susceptibility $\chi_s = 7 \times 10^{-9}$ emu/g. In comparison, for powdered graphite, we find $\chi_s = 2 \times 10^{-8}$ emu/g, consistent with reported values between 1–4$\times 10^{-8}$ emu/g [133,142]. The Pauli behavior indicates that the aligned nanotubes are metallic or semi-metallic. For free electrons, the Pauli susceptibility is given by $\chi_s = \mu_B^2 N(E_F)$ where $\mu_B$ is the Bohr magneton and $N(E_F)$ is the density of states at the Fermi level. The measured susceptibility for MWNTs (Fig. 26) gives $N(E_F) = 2.5 \times 10^{-3}$ states/eV/atom. This density of states is comparable to that of graphite [142]. Estimating the carrier concentration $n$ given by $n = E_F N(E_F)$ and taking the Fermi energy $E_F$ for the MWNTs to be equal to that of graphite (200 K) gives $n \sim 4 \times 10^{18}\,\text{cm}^{-3}$. Although this estimate is very crude, it is, however, consistent with Hall measurements for the same sample [143], which gives an upper limit of $n = 10^{19}\,\text{cm}^{-3}$. This low carrier density is consistent with the MWNTs being semimetallic.

ESR measurements are also suitable for characterizing the defects in carbon nanotubes by their appearance in the Curie-tail (Fig. 26). It is suspected that arc-grown MWNTs contain a high defect density that modifies their electronic properties [143]. Among these defects, vacancies and interstitials can generate paramagnetic centers that can be detected by ESR. Pentagon-heptagon pair defects can also be present in the lattice. The low temperature upturn of $\chi_s(T)$ closely follows a Curie law (solid line, Fig. 26), and is thus a signature of localized spins, arising either from the localization of the carriers

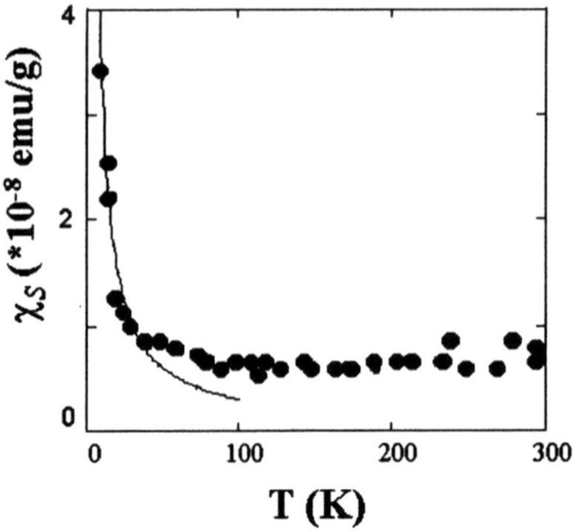


**Fig. 26.** The temperature dependence of the ESR susceptibility of highly purified MWNTs [141]

or from impurities. In either case, the Curie-tail corresponds to $1.3 \times 10^{-5}$ spins/C atom. This means that arc-grown MWNTs contain a very low density of paramagnetic defects. No substantial modifications were observed in the Curie-tail after annealing the sample at 2800°C.

Summarizing the magnetic properties seen by ESR, the MWNTs show a mesoscopic nature in their spin relaxation. The expected manifestation of strong electron-electron correlations by an enhanced spin susceptibility is absent. Instead, MWNTs exhibit a $\chi_s(T)$ dependence similar to that of graphite. MWNTs prepared by the arc-discharge method are very pure. This fact does not apply to catalytically grown MWNT samples, which have a higher defect density, and the observation of an ESR signal is precluded by the magnetic impurities left behind in these samples even after purification of the soot.

## 5 Field and Light Emission

It was very early realized that carbon nanotubes are efficient field emitters [37]. Field emission from individual MWNTs was demonstrated by attaching a nanotube to a conducting wire and observing the current after applying a negative potential to the wire [144]. Also high electric currents were found to produce electrons emission from carbon nanotube films above which an extracting grid was placed [145]. From the latter studies, it was concluded that carbon nanotube films could be used as field emission guns for technical applications, such as flat panel displays [145,146]. Five years after the initial concept was demonstrated, Samsung Advanced Technology

Institute is already manufacturing a prototype of a color display, ready to commercialize this concept [37].

When studying the field emission properties of MWNTs, it was noticed that together with electrons [37], light is emitted as well [147]. This light emission occurred in the visible part of the spectrum, and could sometimes be seen with the naked eye. Since MWNTs are excellent scanning probe tips, their light emitting property should find application in Scanning Near-field Optical Microscopy (SNOM). In our opinion the electron and light emission properties are very important features of carbon nanotubes. The electron emitting characteristics of thin films for large scales applications are treated by *Ajayan* and *Zhou* [37]. Here we present the essential characteristics of the coupled electron and light emission properties of individual MWNTs.

## 5.1 Field Emission

Field emission results from the tunneling of electrons from a metal into vacuum under application of a strong electric field. The tunneling mechanism is described in the WKB approximation for emission from metal surfaces which leads to the well known Fowler–Nordheim equation:

$$I = \alpha E_{\mathrm{eff}}^2 \exp(-\beta / E_{\mathrm{eff}}) , \tag{11}$$

where $\alpha$ is a constant related to the geometry, $E_{\mathrm{eff}}$ is the effective field at the emitter tip, and $\beta$ is a constant which is proportional to the work function. The electric field at the nanotube tip depends on the applied nanotube to grid voltage, the nanotube radius, and the nanotube length. $E_{\mathrm{eff}}$ can be as much as 1200 times higher than $E_0$, which is the grid to cathode voltage divided by the grid to cathode distance, so that field emission is readily achieved even for relatively low applied potentials.

The field emission of individual nanotubes was studied by Bonard and coworkers [7,148] in order to extract the parameters and fine details of the electron emission process. Their experimental configuration is shown in Fig. 27, which displays a typical I–V curve for a single MWNT (in this case an opened MWNT). Most single MWNT emitters, closed as well as opened MWNTs, are capable of emitting over an incredibly large current range. In this study, the maximal current that was drawn from one nanotube was 0.2 mA, and MWNTs can reach 0.1 mA routinely and repeatedly. This represents a tremendous current density for such a small object, and is actually quite close to the theoretical limit where the tube should be destroyed by resistive heating [144].

The field emission study of individual tubes shows also the importance of the tip structure (Fig. 28). Keeping the same parameters for opened and closed MWNT, the turn-on voltage for the opened MWNT is considerably higher, which suggests that the localized states at the dome-structured tip play an important role in the field emission process, so that nanotubes cannot

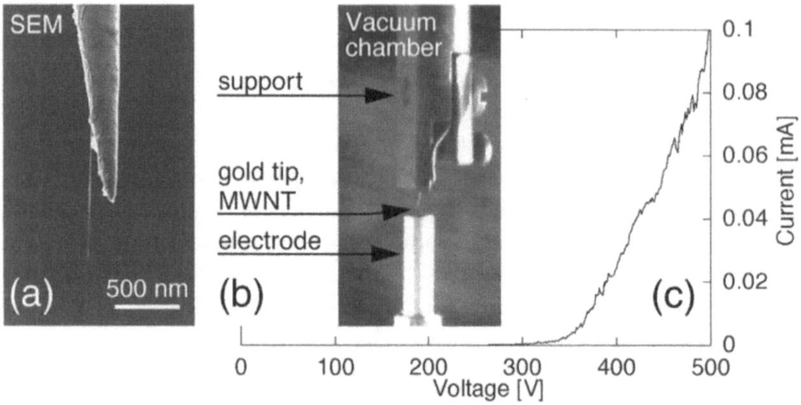


**Fig. 27.** (**a**) A single MWNT mounted on the tip of an etched gold wire. (**b**) Optical micrograph of the experimental set-up for field emission: the gold wire is fixed on a support, and placed 1 mm above the cylindrical counter-electrode. (**c**) I–V characteristics for a single opened MWNT [8,148]

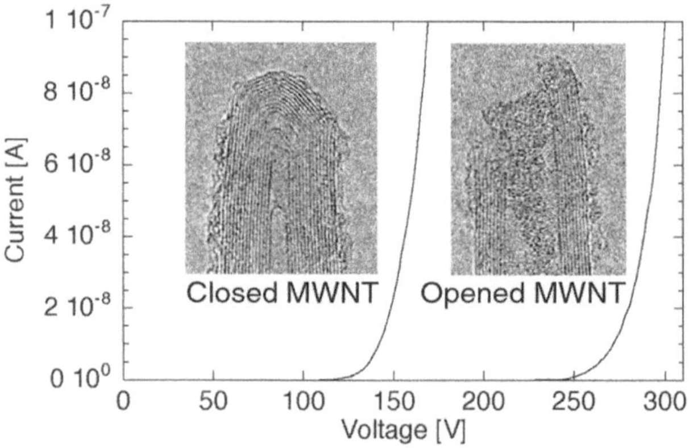


**Fig. 28.** I–V characteristics for a single closed and opened MWNT, and the corresponding TEM micrographs of typical NT tips[7]

be considered as ordinary metallic emitters. This conclusion is supported by the systematic deviations from the Fowler–Nordheim model (11) observed at high emission currents. Furthermore, the energy spread of the emission from nanotubes is very narrow (about 0.2 eV). This energy spread is typically half that of metallic emitters. These observations strongly suggest that the electrons are not emitted from a metallic continuum as in usual metallic emitters, but rather from well-defined energy levels, corresponding to localized states at the tip. Actually, the fit of the energy spread profile of the emitted current gives quite a narrow band of levels rather than a discrete energy level, but the shape of the intensity profile leaves no doubt that the emission does not happen from a metallic continuum. The broadening of the localized state might be due to interaction with the continuum [145].

In fact, theoretical calculations and STM measurements on SWNTs and MWNTs show that there is a distinctive difference between the electronic properties of the tip and the cylindrical part of the tube [131]. For MWNTs, the tube body is essentially graphitic, whereas SWNTs display a characteristic DOS [63,149] that reflects their one-dimensional character. In contrast, the local density of states at the tip exhibits sharp localized states that are correlated with the presence of pentagons, and most of the emitted current comes from occupied states just below the Fermi level.

The major factors that determine the field emission properties of the tube are the tip radius and the position of these occupied levels with respect to the Fermi level, which depends primarily on the tip geometry [131,150] (i.e., the tube chirality and diameter as well as the presence of defects). Indeed, only tubes with a band state just below or just above the Fermi level are good candidates for field emission. It is worth noting that the presence of such localized states greatly influences the emission behavior. At and above room temperature, the body of the MWNTs behave essentially as graphitic cylinders. This means that the carrier density at the Fermi level is very low, i.e., on the order of $5 \times 10^{18}$ cm$^{-3}$, which is 3 orders of magnitude lower than for a metal. Simulations show that the local density of states at the tip [131] reaches values at least 30 times higher than in the cylindrical part of the tube. Therefore, the field emission current would be far lower without these localized states for a geometrically identical tip, since the emission depends directly on this carrier density. The tip structure also strongly influences the energy and intensity of the localized states, which could explain the superiority of closed, well-ordered tips relative to opened or disordered MWNTs (Fig. 28). Another complementary explanation for this observation is that the coupling of the tip states to the metallic body states is probably far better for closed MWNTs, leading to an increased electron supply and thus to higher emitted currents.

## 5.2 Light Emission Coupled to Electron Emission

A rather unexpected behavior linked to the electron field emission was the observation of light emission as described above [147]. This light emission occurred in the visible part of the spectrum, and could sometimes even be seen with the naked eye. This light emission is not the luminescence due to resistive heating, which was observed by *Rinzler* et al. on opened nanotubes [144], and was attributed to an unraveling of carbon chains at the tip of the tube.

A typical experimental configuration for the electron/field emission is shown in Figs. 29a and b. The optical luminescence was induced by the electron field emission, since it was not detected without an applied potential (and thus emitted current). Furthermore, the emitted light intensity followed closely the variations in emitted current, as can be seen in Fig. 29c, where

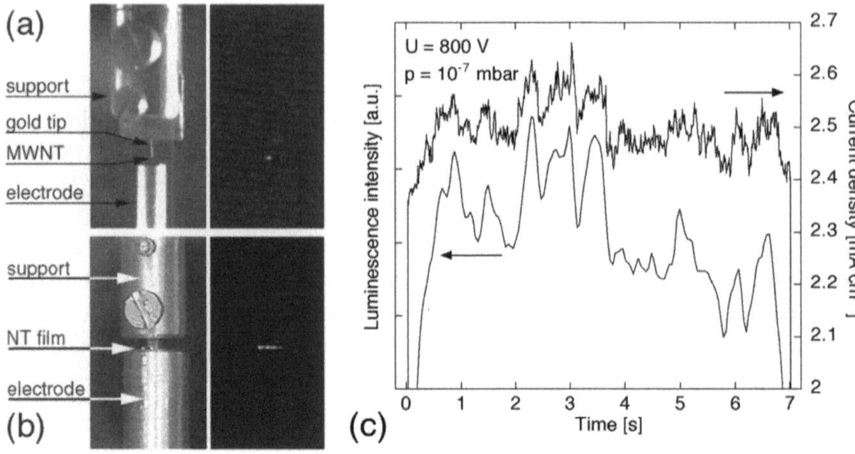


**Fig. 29.** Optical images of the experimental set-up for observing the field emission and the associated observed luminescence during field emission from an individual MWNT (**a**), and a MWNT film (**b**). The simultaneously emitted light intensity and current density as a function of time for a MWNT film (**c**) [147]

the emitted current and emitted light intensity were simultaneously measured [147].

The results in Fig. 29c demonstrate that the light emission is directly coupled to the field emission. The narrowness of the luminescence line (Fig. 30) and the very small shift of the photon energy of the luminescence peak as the emission current varies show that the luminescence is not coming from blackbody radiation or from current-induced heating effects, but rather that the photons are emitted following transitions between well-defined energy levels. Actually, the dependence of luminescence intensity $I_p$ on the emitted current $I_e$ can be reproduced by a simple two-level model [147], where the density of states at the nanotube tip is simply represented by a two level system, with the dominant emission level at energy $E_1$ below or just above the Fermi energy, and a deep level at $E_2 < E_1$. When an electron is emitted from the deep $E_2$ level, it is replaced either by an electron from the tube body, or by an electron from level $E_1$ dropping down to fill the empty state at $E_2$ thereby provoking the emission of a photon. From the Fowler–Nordheim model, the transition probability $D(E)$ can be evaluated for each level, and in the framework of this two-level model, we can write $I_e \sim D(E_1)$, and $I_p \sim D(E_2)$. It appears that $I_p$ varies as a power of $I_e$ with an exponent that depends on the energy separation between the levels [151], and this exponent is $\approx$ 1.51–1.65 for the photon emission energies observed here (typically 1.8 eV, as seen in Fig. 30). This prediction regarding the dependence of $I_p$ on $I_e$ compares well with the experimental observations.

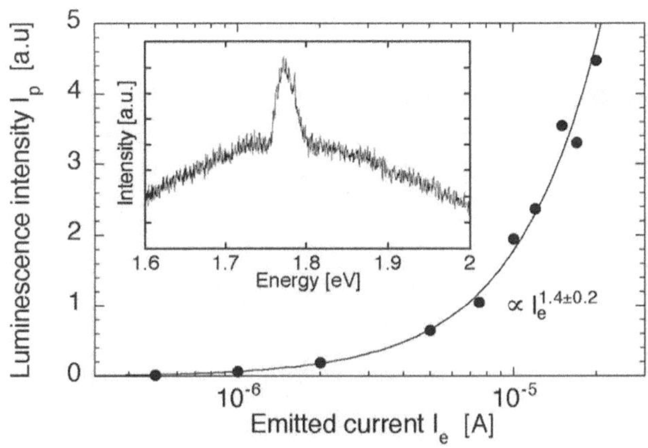


**Fig. 30.** Variation of the total emitted intensity as a function of emitted current. The *solid line* is a power law fit of the experimental data for $I_p$ vs $I_e$ on a semilog plot, which yields an exponent of $\alpha = 1.4 \pm 0.2$ for $I_p \sim I_e^{\alpha}$. *Inset:* Spectra of the field emission-induced luminescence for an individual MWNT at an emission current of 20 A [8]

### 5.3 Light Emission Induced by STM

Another type of light emission was observed in MWNTs by *Coratger* et al. [152] who observed light emission by injecting electrons by Scanning Tunneling Microscopy (STM) into MWNTs. The photon yield varied from tube to tube, but it was constant all along the tube. This is illustrated in Fig. 31, where a low resolution STM image shows a bundle of MWNTs, and while the STM topological image was taken, the photon map was recorded. The photon image shows that the emission yield depends probably on the diameter and chirality of the scanned tube, but not on the position of the STM tip along the tube. Simultaneous topographic STM images and photon spectroscopy scans show emitted photon wavelengths in the 600–1000 nm range (Fig. 32), in good agreement with the field and light emission data described above. The emission characteristics are independent of the substrate and of the material of the STM tip.

The STM induced-light emission in MWNTs could not be explained by models previously developed for light emission from metal and semiconductor surfaces, based on a radiative plasmon recombination mechanism, or from small fullerene molecules like $C_{60}$ [53,153]. The origin of the experimental observation that the light wavelength and intensity are independent of the STM tip position might be explained as follows. It was suggested [152] that the large mean free path of the electrons injected into the MWNTs is associated with an efficient resonant coupling of the localized tip states to a continuum of states. The following scenario is proposed to explain this light emission behavior. Electrons are first injected into the extended states of the outer shell. Large numbers of these electrons flow towards the substrate, but a

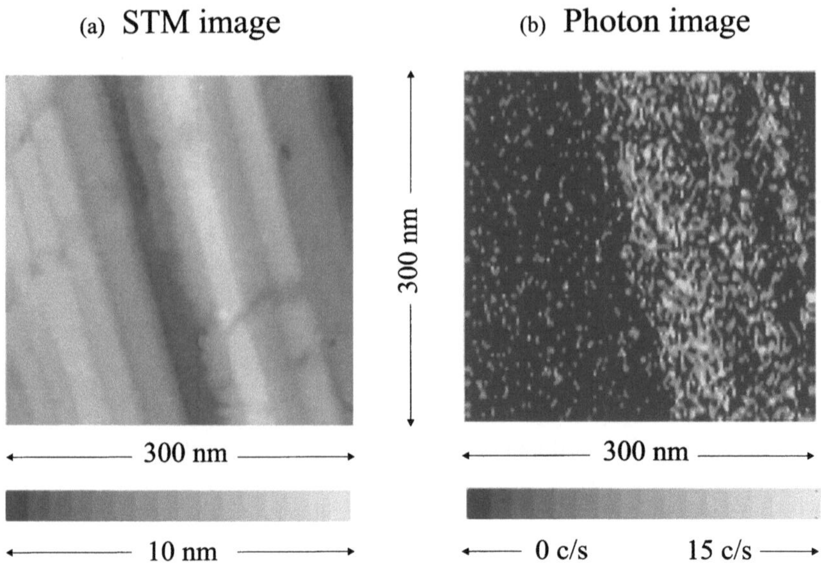


**Fig. 31.** (**a**) STM topography ($V_{\text{sample}}$=2.1 V, $I_i$ = 50 nA) and (**b**) photon mapping of a MWNT bundle recorded simultaneously. Only three nanotubes in the MWNT bundle show significant light emission. The maximum rate of emitted photons is 15 counts per second per tube [152]

fraction of the electrons spreads all over the tube in a quasi-ballistic way, like in a large molecule. Some of these electrons are trapped for a short time in localized states at the nanotube tip, where they are resonantly coupled with extended states, and occasionally decay into a neighboring localized state situated approximately 2 eV below the Fermi energy by emitting a photon. The population difference between the two quasi-localized states comes from the V-shape band structure of the graphene layer so that states higher in energy are more populated than those situated near the Fermi energy. Indeed, recent transport experiments show that the coherence length is exceptionally large in carbon nanotubes even at room temperature [154]. Differences in the emitted light characteristics probably reflect the large variety of possible geometric structures in the nanotube tip, and such experiments could perhaps offer a characteristic probe for the localized tip states in combination with Scanning Tunneling Spectroscopy.

## 6 Mechanical Properties

It is becoming clear from recent experiments [154,155,156,157,158,159,160] [161] that Carbon NanoTubes (CNTs) are fulfilling their promise to be the ultimate high strength fibers for use in materials applications. There are many outstanding problems to be overcome before composite materials, which exploit the exceptional mechanical properties of the individual nanotubes, can

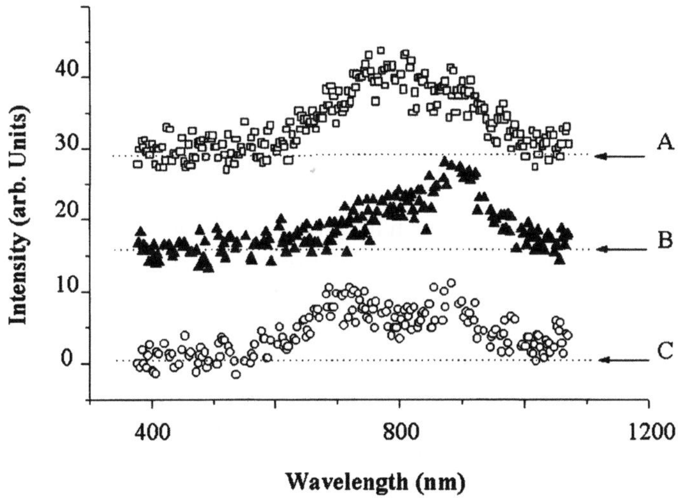


**Fig. 32.** Emission spectra of three different MWNTs acquired under similar experimental conditions. *Dotted lines* highlight position of the background. Spectrum A is one of the most recently observed and shows a peak at about 800 nm ($V_{\text{sample}}$=2 V, $I$ = 50 nA). Spectra B and C were obtained on two different nanotubes with the same Pt-Ir tip during another experiment ($V_{\text{sample}}$=2.05 V, $I$ = 35 nA). In spectrum B, a peak at 900 nm is found, while in spectrum C, another peak centered at about 710 nm is also revealed. For the three spectra, the exposure time is 5 min [152]

be fabricated. Arc-discharge methods are unlikely to produce sufficient quantities of nanotubes for such applications. Therefore, catalytically grown tubes are preferred, but these generally contain more disorder in the graphene walls and consequently they have lower moduli than the arc-grown nanotubes. Catalytically grown nanotubes, however, have the advantage that the amount of disorder (and therefore their materials properties) can be controlled through the catalysis conditions, as mentioned before. As well as optimizing the materials properties of the individual tubes for any given application, the tubes must be bonded to a surrounding matrix in an efficient way to enable load transfer from the matrix to the tubes. In addition, efficient load bearing within the tubes themselves needs to be accomplished, since, for MWNTs, experiments have indicated that only the outer graphitic shell can support stress when the tubes are dispersed in an epoxy matrix [162]. In this section, a few basic measurements are presented on individual MWNTs with different levels of disorder. The very promising SWNT ropes also present several problems, which are reviewed by showing mechanical measurements on ropes of different diameter.

## 6.1 Young's Modulus of MWNTs

The AFM technique for mechanical properties measurements developed by Salvetat and collaborators enabled characterization of the moduli of SWNT bundles [158] and MWNTs, both arc-grown and catalytically grown [159], to be carried out [158,159]. Briefly, the method involves depositing CNTs from a suspension in a liquid onto well-polished alumina ultra-filtration membranes with a pore size of about 200 nm (Whatman anodisc). By chance, CNTs occasionally span the pores and these can be subjected to mechanical testing on a nanometer length scale. The attractive interaction between the nanotube and the substrate acts to clamp the tubes to the substrate. Contact mode AFM measurements under ambient conditions are used to collect images of the suspended CNTs at various loading forces. Figure 33 shows an AFM image of a SWNT bundle suspended across a pore and a schematic representation of the mechanical test set-up. The maximum deflection of the CNT into the pore as a function of the loading force can be used to ascertain whether the behavior is elastic. If the expected linear behavior is observed, the Young's

a.

b.

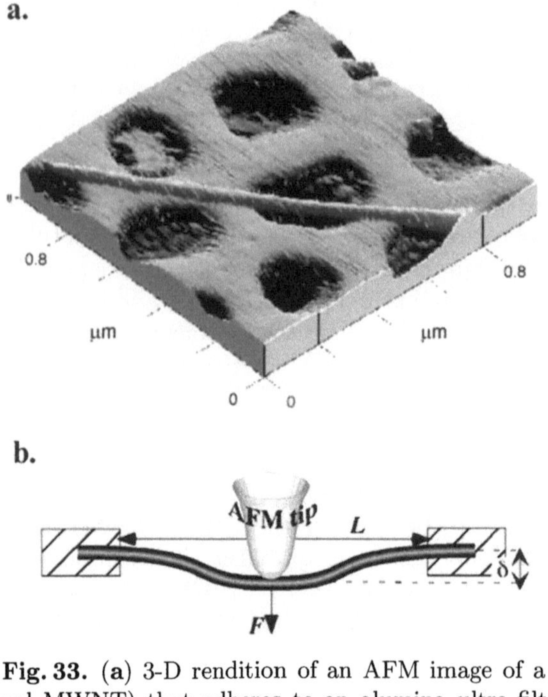


**Fig. 33.** (**a**) 3-D rendition of an AFM image of a SWNT bundle (or an individual MWNT) that adheres to an alumina ultra-filtration membrane, leading to a clamped beam configuration for mechanical testing. (**b**) Schematic representation of the measurement technique. The AFM applies a load, $F$, to the portion of nanotube with a suspended length of $L$ and the maximum deflection $\delta$ at the center of the beam is directly measured from the topographic image, along with $L$ and the diameter of the tube (measured as the height of the tube above the membrane) [158]

modulus ($E$) can be extracted using a continuum mechanics model for a clamped beam configuration. The suspended length of the CNT, its deflection as a function of load and its diameter can all be determined from the images, thereby enabling the modulus to be deduced. One great advantage of the AFM technique employed by *Salvetat* et al. [158] is its simplicity. There is no need, for example, to use complex lithographic techniques for suspending and the clamping tubes [156]. The surface forces between the CNTs and the alumina membrane are sufficiently high to maintain the clamped beam condition for the majority of MWNTs that were tested. In addition, the nanotubes are never exposed to electron radiation during the measurement, which would be the case for TEM studies [155,160]. Electron radiation will induce defects, if the energy of the electrons is high enough, and thereby alter the material properties of the CNTs. The relative ease of sample preparation for this AFM method enables a high measurement throughput, thereby allowing measurement of a variety of CNTs synthesized under different conditions and to compare the results systematically. For MWNTs grown by the arc-discharge method, it was found that the average value of the elastic modulus (or Young's modulus) $E$ is $810 \pm 410\,\mathrm{GPa}$, which is consistent with the in-plane elastic constant of graphite, $c_{11} = 1.06\,\mathrm{TPa}$ [15]. The authors did not find a significant correlation between the elastic modulus and the diameter of the tube [159]. Furthermore, no apparent difference was found in the elastic modulus between annealed and unannealed nanotubes. This suggests that point defects, if present at all, do not alter the mechanical properties of MWNTs.

## 6.2 Disorder Effect

AFM measurements of the mechanical properties of arc-grown MWNTs and MWNTs catalytically grown at different temperatures have been compared, and the results show that the Young's modulus for the catalytically grown MWNTs are lower than for arc-grown MWNTs [7]. These data are summarized in Fig. 34, with a sketch correlating the elastic modulus with the amount of order/disorder within the nanotube walls. As one might expect, the Young's modulus of the MWNTs decreases as the disorder within the walls increases. Arc-grown MWNTs, which contain very few defects, have a modulus comparable with the high values that are measured for an individual SWNT [158]. The moduli of catalytic MWNTs can vary, depending on their structure. Those, which have a highly defective kind of stacked coffee cup structure (see Fig. 5, part ii), have a very low modulus. The other catalytic MWNTs, grown at a lower temperature, showed a higher degree of order within the tubes and consequently had slightly higher moduli. The uncertainty of the measured values in this case was large. This could be due to greater uncertainties in the measurement technique, since the catalytic MWNTs were usually curved, making the continuum beam approximation less valid.

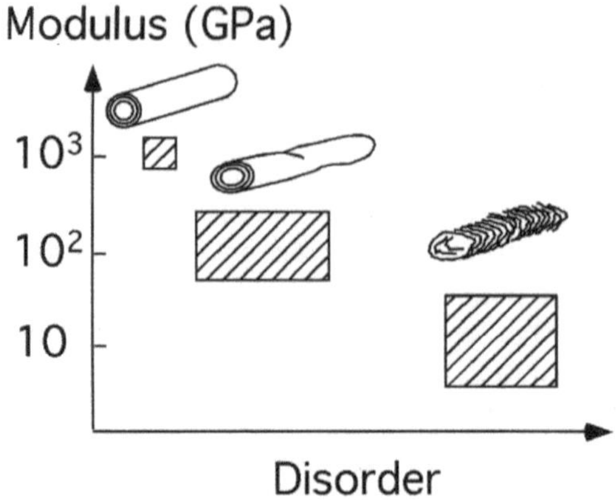


**Fig. 34.** Correlation between the measured Young's modulus of MWNTs with the amount of disorder present within the graphitic walls. Ranges of measured moduli for three different types of MWNTs are plotted against an arbitrary scale of increasing disorder. The sketch represents arc-discharge grown, and catalytically grown MWNTs at 720°C and at 900°C, respectively [163]

### 6.3 Comparison with SWNTs

Although the elastic moduli for Single-Wall NanoTube (SWNT) bundles (also known as ropes) were expected to be higher than for MWNTs, it has been demonstrated that shearing effects due to the weak intertube cohesion gives rise to significantly reduced moduli compared to individual SWNTs [158]. This is due to the fact that the SWNTs are held together in the rope by the same van-der-Waals forces that are acting between the 2D graphene layers, and these weak van der Waals forces make turbostratic graphite [16] a very good lubricant. The reduced bending modulus of these SWNT bundles is a function of the rope diameter because the magnitude of the shear modulus varies as the ratio of the length to the rope diameter. An individual SWNT has an elastic modulus of about 1 TPa, but this falls to around 100 GPa for bundles 15 to 20 nm in diameter, as shown in Fig. 35.

In order to use nanotubes as a reinforcement in composites, there are two main challenges to address: (1) to establish strong bonding between the CNTs and the surrounding matrix, and (2) to create cross-links between the shells of MWNTs and also between the individual SWNTs in SWNT bundles, so that loads can be homogeneously distributed throughout the CNTs. To exploit the excellent mechanical properties of the individual SWNTs, both of these goals should be achieved without compromising the mechanical properties of the individual SWNTs too drastically. Efforts are in progress to address these problems using post production modification of CNTs via chemical means and controlled irradiation [7].

To produce cross-links between the shells in SWNTs of SWNT bundles, the $sp^2$ carbon bonding must be disrupted to $sp^3$ bonding, so that dangling bonds are available for cross-linking. Since the $sp^2$ bonding is the essence of the CNT strength, this must not be disrupted to such a degree that the properties of the individual SWNTs in SWNT bundles are degraded. A "gentle" way to produce cross-linking is by controlled electron irradiation. In [7] the SWNT bundles were exposed to 2.5 MeV electrons with a total radiation

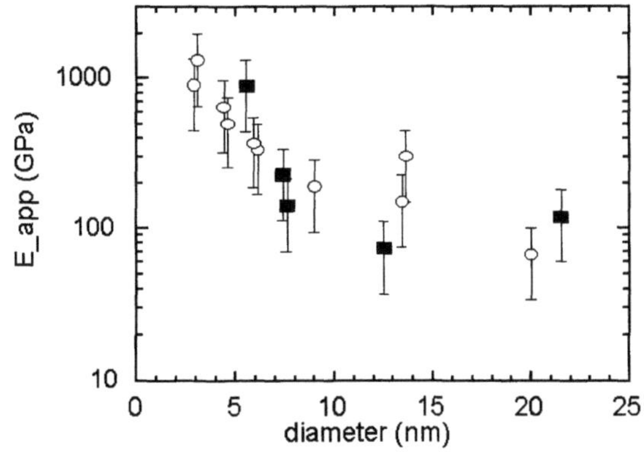


**Fig. 35.** Dependence of the apparent Young's modulus ($E_{\mathrm{app}}$) on the diameter of SWNT bundles measured using AFM. The untreated bundles are represented by *open circles* and the irradiated bundles by *filled squares*. The diameter of the individual SWNTs is in the 1.4 nm range [163]

dose of 11 Curie/cm$^2$ after synthesis. A theoretical estimation of the number of displacements that this energetic electron dose produces suggests that the irradiation will create about 1 defect per 360 carbon atoms [164]. AFM measurements of the elastic modulus of irradiated SWNT bundles are shown in Fig. 35, along with similar measurements of non-irradiated SWNT bundles. Within the errors of the measurement technique, no increase in the elastic modulus of the bundles could be clearly identified. The Young's modulus was still found to decrease with increasing bundle diameter in a similar way as that found for the non-irradiated bundles. However, the irradiation treatment does not appear to have compromised the strength of the individual SWNTs either, since the smaller diameter bundles have quite high elastic moduli. In addition, it was noticed that the irradiated bundles were more difficult to disperse in ethanol, and the morphology of the sample under AFM examination showed that the irradiated bundles exhibited a higher degree of aggregation. Taken together, these data suggest that the radiation treatment produced more bonding between the tubes, but this additional bonding was not sufficient to produce enough cross-links within the bundles to reduce shearing effects and produce bundles with higher Young's moduli. Future efforts will concentrate on optimizing the chemistry and irradiation doses to improve the mechanical properties of SWNT bundles and MWNTs.

### 6.4 Deformation of MWNTs

There is some contention about whether the elastic modulus of MWNTs varies as a function of nanotube diameter. *Poncharal* et al. [160] have suggested the formation of a rippling mode on the surface of bent MWNTs with diameters greater than about 15 nm, leading to a reduction in the measured modulus. However, a strong dependence of the measured modulus on the diameter was not observed in previous AFM measurements [159]. It is conceivable that the measurement of the modulus is force dependent and the transition to the rippling mode is not reached with the loading forces used in the AFM experiments. The TEM data in Fig. 36 obtained at large loading force show a rippling with a period of 15 nm on the compressed side of a statically bent MWNT, but the bent MWNTs in typical TEM experiments showing such phenomena have rather high curvatures [160] (much higher than typical curvatures seen in the AFM experiments). Interestingly, the rippling effect has also been observed on the compressed sides of catalytically grown MWNTs [7]. Figure 37 shows a high resolution AFM image of a catalytic MWNT (grown at 720°C) lying across a pore, the edge of which can be seen on the left of the image. The image clearly shows rippling only on the right-hand side of the tube, the direction in which the CNT is bent, and the ripple in this case has a period of roughly 16 nm. These ripples are not perpendicular to the tube axis but are inclined at approximately 30°, making the CNT left-handed. This rippling could arise from the surface forces, which constrain the CNTs on the membrane. However, rippling can also arise when the tubes are loaded in the AFM. Rippling on the upper, compressed side of these MWNTs has also been observed as the imaging force is increased. Nonlinear behavior in the loading/unloading characteristics of the catalytic tubes was frequently noticed. This also contributes to the uncertainties in measuring moduli on these kinds of MWNTs by the AFM method. It is conceivable that the onset of rippling will occur at lower curvatures, i.e., lower forces, in tubes with a higher amount of disorder.

Because of the Russian-doll structure of the MWNTs and because of the high strength of the $sp^2$ bonds, one would think that a nanotube shell cannot be easily pealed off from a MWNT. However, it is quite possible that a single

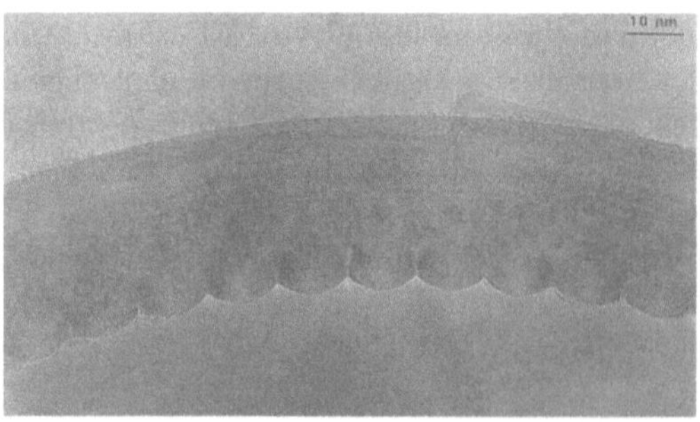


**Fig. 36.** High-resolution TEM image of a bent nanotube grown by the arc discharge method (radius of curvature 400 nm), showing the characteristic wave-like ripple distortion. The amplitude of the ripples increases continuously from the center of the tube to the outer layers of the inner arc of the bend [160]

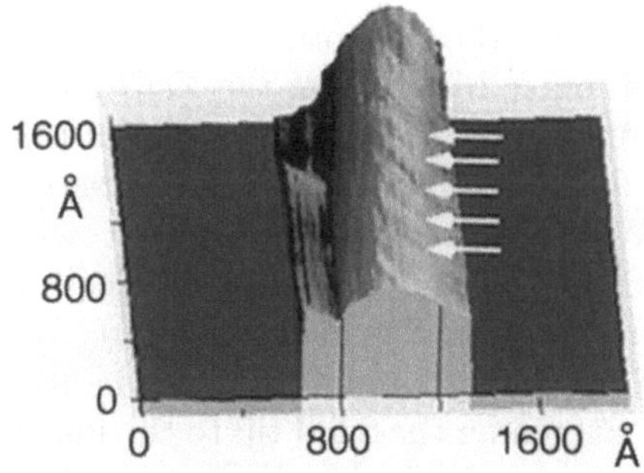


**Fig. 37.** 3-D rendering of a high-resolution AFM image of a catalytic MWNT, grown at 720°C, and suspended on an alumina ultra-filtration membrane (the 0, 254, 508 Å refer to the $z$ axis). Rippling is observed on the inner side of the natural curvature of the MWNT, with a periodicity of about 16 nm (shown by the *white arrows*), inclined at 30° to the tube axis [7]

tube can be stripped off under axial stress. This has been observed in the experiments of *Ruoff* and coworkers [165] by fixing the ends of a MWNT to two AFM tips and measuring the elongation of the nanotube, while pulling until the tube broke. The tensile strength was found to be at least an order of magnitude lower than expected, assuming a homogeneous stress distribution over all carbon shells of the multi-wall tube. But because the support only made contact to the outside of the MWNT, it is the outermost shell which is mainly stressed. *Ruoff* and coworkers [165] found that above the tensile limit, the outermost tubes rupture first, followed by a sudden and large elongation. From TEM images, these researchers conclude that the ruptured outer shell then slides over the inner tubes, which then in turn bear the load [16,37].

To conclude this section on the measurement of mechanical properties, it was demonstrated that the Young's modulus is very high for individual, and well graphitized carbon nanotubes. Their high strength makes them promising candidates in reinforcement applications. There are many problems remaining to be overcome before composite materials can be fabricated, which reflect the exceptionally good mechanical properties of the individual nanotubes. As well as optimizing the material properties of the individual tubes, the tubes must be bonded to a surrounding matrix in an efficient way to enable load transfer from the matrix to the tubes. To improve their mechanical properties, the nanotube surfaces should be functionalized, and cross-linking should be generated between the SWNTs within and between the ropes.

## 7 Summary

In summary, we again emphasize the very rich physics encountered in the study of multi-walled carbon nanotubes. In this review we gave, at best, a snap-shot of the present status of this rapidly changing field. As an illustration of the enormous potential of the multi-walled carbon nanotubes in basic and applied science, we present the manipulation of MWNTs of the *Zettl* group at Berkeley [166] of which we have become aware during the writing of this chapter. Describing their breakthrough, best summarizes the rapid progress of this field. They managed to peel a MWNT and to sharpen its tip at 'will' by using Joule heating applied via an STM tip in a transmission electron microscope (Fig. 38). The outer, supporting shortened cylinders reinforce the stability of the nanotube. This gives the possibility to improve considerably the resolution of a scanning probe tip. By attaching the STM tip to the innermost nanotube when the outer layers were peeled off, they have created an extremely low-friction nanoscale linear bearing and a constant force nanospring. We believe that in the future such manipulations of MWNTs will revolutionize nanoscale architectures and studies at the nanoscale.

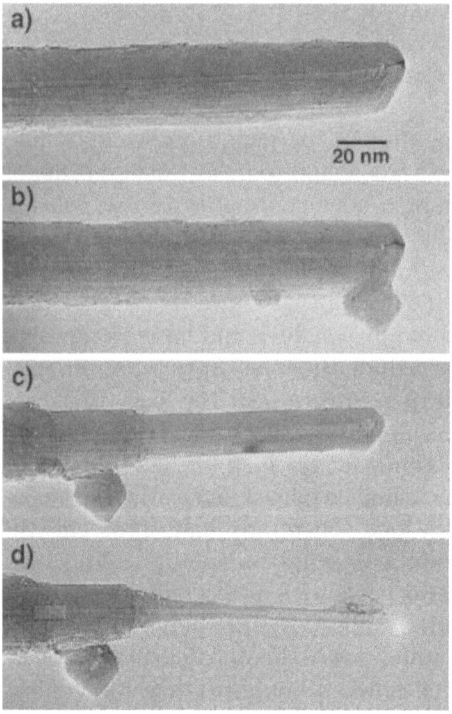


**Fig. 38.** Sequence of transmission electron microscope (TEM) images of a MWNT, showing the sharpening of the MWNT tip by applying current pulses with an STM tip [166]

## Acknowledgments

The authors acknowledge the contributions to the work presented here by all their past and present collaborators, especially J.-M.Bonard, J.-P. Salvetat, A. Kulik, K. Hernadi, S. Pekker, Walt A. de Heer, D. Ugarte, A. Bachtold, and M. Buitelaar. The work in Lausanne and Basel is supported by the Swiss National Science Foundation Program NFP36.

## References


1. S. Iijima, Nature **354**, 56 (1991)
2. D. S. Bethune, C. H. Kiang, M. S. de Vries, G. Gorman, R. Savoy, J. Vazquez, R. Beyers, Nature **363**, 605 (1993)
3. S. Iijima, T. Ichihashi, Nature **363**, 603 (1993)
4. A. Thess, R. Lee, P. Nikolaev, H. Dai, P. Petit, J. Robert, C. Xu, Y. H. Lee, S. G. Kim, A. G. Rinzler, D. T. Colbert, G. E. Scuseria, D. Tománek, J. E. Fischer, R. E. Smalley, Science **273**, 483 (1996)
5. W. Krätschmer, L. D. Lamb, K. Foristopoulos, D. R. Huffman, Nature **347**, 354 (1990)
6. T. W. Ebbesen, P. M. Ajayan, Nature **358**, 220 (1992)
7. L. Forró, J.-P. Salvetat, J.-M. Bonard, R. Bacsa, N. H. Thomson, S. Garaj, L. Thien-Nga, R. Gaál, A. J. Kulik, B. Ruzicka, L. Degiorgi, A. Bachtold, C. Schönenberger, S. Pekker, K. Hernadi, in *Science and Application of Nanotubes*, D. Tománek, R. J. Enbody (Eds.) (Kluwer Academic/Plenum Publishers, New York 1999) p. 297
8. J-M. Bonard, T. Stora, J.-P. Salvetat, F. Meier, T. Stökli, C. Düsch, L. Forró, W. A. de Heer, A. Chatelain, Adv. Mater. **9**, 827 (1997)
9. W. Bacsa, unpublished
10. C. S. Tsang, Y. K. Chen, P. J. F. Harris, M. L. Green, Nature **372**, 159 (1994)
11. H. Hiura, T. W. Ebbesen, K. Tanigaki, Adv. Mater. **7**, 275 (1995)
12. C. Journet, P. Bernier, Appl. Phys. A **67**, 1 (1998)
13. A. Fonseca, K. Hernadi, P. Piedigrosso, J.-F. Colomer, K. Mukhopadhyay, R. Doome, S. Lazarescu, L. P. Biro, P. Lambin, P. A. Thiry, D. Bernaerts, J. B. Nagy, Appl. Phys. A **67**, 11 (1998)
14. M. Endo, K. Takeuchi, K. Kobori, K. Takahashi, H. Kroto, A. Sarkar, Carbon **33**, 873 (1995)
15. M. S. Dresselhaus, G. Dresselhaus, K. Sugihara, I. L. Spain, H. A. Goldberg, in *Graphite Fibers and Filaments.* Springer Ser. in Mater. Sci. **5** (Springer, Berlin, Heidelberg 1988)
16. M. S. Dresselhaus, M. Endo, chapter of this volume
17. X. B. Zhang, X. F. Zhang, D. Bernaerts, G. Van Tendeloo, S. Amelinckx, J. Van Landuyt, V. Ivanov, J. B. Nagy, Ph. Lambin, A. A. Lucas, Europhys. Lett. **27**, 141 (1994)
18. V. Ivanov, J. B. Nagy, Ph. Lambin, A. A. Lucas, X. B. Zhang, X. F. Zhang, D. Bernaerts, G. Van Tendeloo, S. Amelinckx J. Van Lunduyt, J. Chemical Physics Letters **223**, 329 (1994)
19. A. Fonseca, K. Hernadi, J. B. Nagy, P. Lambin, A. Lucas, Synth. Met. **77**, 235 (1996)
20. N. Demoncy, O. Stephan, N. Brun, C. Colliex, A. Loiseau, H. Pascard, Synth. Met. **103**, 2380 (1999)
21. S. Amelinckx, D. Bernaerts, G. van Tendeloo, J. van Landuyt, A. A. Lucas, M. Mathot, P. Lambin, in *Physics and Chemistry of Fullerenes and Derivatives*, Proceedings of the International Winterschool on Electronic Properties of Novel Materials, Kirchberg 1995, H. Kuzmany, J. Fink, M. Mehring, S. Roth (Eds.) (World Scientific, Singapore 1995) p. 515
22. S. Amelinckx, D. Bernaerts, X. B. Zhang, G. Van Tendeloo, J. Van Landuyt, Science **267**, 1334-8 (1995)

23. H. Kind, J.-M. Bonard, C. Emmenegger, L.-O. Nilsson, K. Hernadi, E. Maillard-Schaller, L. Schlapbach, L. Forró, K. Kern, Adv. Mater. **11**, 1285 (1999)
24. N. Yao, V. Lordi, S. X. C. Ma, E. Dujardin, A. Krishnan, M. M. J. Treacy, T. W. Ebbesen, J. Mater. Res. **13**, 2432-2437 (1998)
25. J.-M. Bonard, J.-P. Salvetat, T. Stökli, L. Forró, A. Chatelain, Appl. Phys. A **69**, 245 (1999)
26. M. S. Dresselhaus, G. Dresselhaus, R. Saito, Phys. Rev. B **45**, 6234 (1992)
27. M. S. Dresselhaus, G. Dresselhaus, R. Saito, Carbon **33**, 883 (1995)
28. S. G. Louie, chapter in this volume
29. M. S. Dresselhaus, Ph. Avouris, chapter in this volume
30. N. B. Brandt, Y. G. Chudinov, *Graphite and its Compounds* (North-Holland, Amsterdam 1988)
31. H. W. Kroto, J. R. Heath, S. C. O'Brien, R. F. Curl, R. E. Smalley, Nature **318**, 162 (1985)
32. S. Iijima, Solid State Phys. **27**, 39-45 (1992)
33. V. H. Crespi, N. G. Chopra, M. L. Cohen, A. Zettl, S. G. Louie, Phys. Rev. B **54**, 5927
34. K. L. Lu, R. M. Lago, Y. K. Chen, M. L. H. Green, P. J. F. Harris, S. C. Tsang, Carbon **34**, 814 (1996)
35. J. Liu, A. G. Rinzler, H. Dai, J. H. Hafner, R. K. Bradley, P. J. Boul, A. Lu, T. Iverson, K. Shelimov, C. B. Huffman, F. Rodriguex-Macia, D. T. Colbert, R. E. Smalley, Science **280**, 1253-1256. (1998)
36. C. Guerret-Plecourt, Y. Le Bouar, A. Loiseau, H. Pascard, Nature **372**, 761 (1994)
37. P. Ajayan, O. Zhou, chapter in this volume
38. A. Loiseau, N. Demoncy, O. Stéphan, C. Colliex, H. Pascard, in D. Tománek, R. J. Enbody (Eds.) (Kluwer Academic/Plenum Publishers, New York 1999)
39. P. M. Ajayan, T. W. Ebbesen, T. Ichihashi, S. Iijima, K. Tanigaki, H. Hiura, Nature **362**, 522 (1993)
40. S. C. Tsang, P. J. F. Harris, M. L. H. Green, Nature, **362**, 520 (1993)
41. E. Djuradin, T. W. Ebbesen, H. Hiura, K. Tanigaki, Science **265**, 1850 (1994)
42. Y. K. Chen, M. L. H. Green, S. C. Tsang, Chem. Commun. 2489 (1996)
43. D. Ugarte, A. Chatelain, W. A. de Heer, Science **274**, 1897 (1996)
44. C. Schönenberger, A. Bachtold, C. Strunk, J.-P. Salvetat, L. Forró, Appl. Phys. A **69** 283 (1999)
45. S. Frank, P. Poncharal, Z. L. Wang, W. A. de Heer, Science **280**, 1744 (1998)
46. A. Bachtold, C. Strunk, J. P. Salvetat, J. M. Bonard, L. Forró, T. Nussbaumer, C. Schönenberger, Nature **397**, 673 (1999)
47. R. Saito, G. Dresselhaus, M. S. Dresselhaus, *Physical Properties of Carbon Nanotubes* (Imperial College Press, London 1998)
48. C. Dekker, Phys. Today, May issue, **52**, 22–28 (1999)
49. J. W. Mintmire, B. I. Dunlap, C. T. White, Phys. Rev. Lett. **68**, 631–634 (1992)
50. T. Hamada, M. Furuyama, T. Tomioka, M. Endo, J. Mater. Res. **7**, 1178–1188 (1992). ibid., 2612-2620
51. R. Saito, M. Fujita, G. Dresselhaus, M. S. Dresselhaus, Appl. Phys. Lett. **60**, 2204 (1992)
52. R. Saito, G. Dresselhaus, M. S. Dresselhaus, J. Appl. Phys. **73**, 494 (1993)

53. M. S. Dresselhaus, G. Dresselhaus, P. C. Eklund, *Science of Fullerenes and Carbon Nanotubes* (Academic, New York 1996)
54. P. R. Wallace, Phys. Rev. **71**, 622 (1947)
55. R. Saito, H. Kataura, chapter of this volume
56. R. Saito, G. Dresselhaus, M. S. Dresselhaus, Phys. Rev. B **61**, 2981 (2000)
57. G. S. Painter, D. E. Ellis, Phys. Rev. B **1**, 4747 (1970)
58. D. P. DiVincenzo E. J. Mele, Phys. Rev. **29** 1685 (1984)
59. J. W. Mintmire, D. H. Robertson, C. T. White, J. Phys. Chem. Solids **54**, 1835 (1993)
60. R. Bacon, J. Appl. Phys. **31**, 283–290 (1960)
61. L. Piraux, J. Mater. Res. **5**, 1285 (1990)
62. H. Ajiki, T. Ando, J. Phys. Soc. Jpn. **62**, 2470–2480 (1993). Erratum: ibid p. 4267
63. J. W. G. Wildöer, L. C. Venema, A. G. Rinzler, R. E. Smalley, C. Dekker, Nature (London) **391**, 59–62 (1998)
64. T. W. Odom, J. L. Huang, P. Kim, C. M. Lieber, Nature (London) **391**, 62–64 (1998)
65. Young-Kyun Kwon, D. Tománek, Phys. Rev. B **58**, R16001 (1998)
66. Ph. Lambin, V. Meunier, A. Rubio, in *Science and Application of Nanotubes*, D. Tománek, R. J. Enbody (Eds.) (Kluwer Academic/Plenum Publishers, New York 1999) p. 17
67. C.-H. Kiang, M. Endo, P. M. Ajayan, G. Dresselhaus, M. S. Dresselhaus, Phys. Rev. Lett. **81**, 1869 (1998)
68. S. N. Song, X. K. Wang, R. P. H. Chang, J. B. Ketterson, Phys. Rev. Lett. **72**, 697 (1994)
69. L. Langer, V. Bayot, E. Grivei, J. P. Issi, J. P. Heremans, C. H. Olk, L. Stockman, C. Van Haesendonck, Y. Bruynseraede, Phys. Rev. Lett. **76**, 479 (1996)
70. G. Bergmann, Phys. Rep. **107**, 1 (1984)
71. B. L. Altshuler, A. G. Aharonov, M. E. Gershenson, Y. V. Sharvin, in *Soviet Scientific Reviews, Section A: Physics Reviews*, I. M. Khalatnikov (Ed.) (Harwood Academic, New York 1987)
72. B. L. Altshuler, P. A. Lee, Phys. Today, Dec. issue, 36 (1988)
73. A. Aharonov, Phys. Scr. **T49**, 28 (1993)
74. S. J. Tans, M. H. Devoret, H. Dai, A. Thess, R. E. Smalley, L. J. Geerligs, C. Dekker, Nature **386**, 474 (1997)
75. M. Bockrath, D. H. Cobden, P. L. McEuen, N. G. Chopra, A. Zettl, A. Thess, R. E. Smalley, Science **275**, 1922 (1997)
76. A. Bachtold, C. Strunk, C. Schönenberger, J. P. Salvetat, L. Forró, *Proceedings of the XIIth International Winterschool on Electronic Properties of Novel Materials*, H. Kuzmany, J. Fink, M. Mehring, S. Roth (Eds.) (AIP, New York 1998)
77. A. Yu. Kasumov, R. Deblock, M. Kociak, B. Reulet, H. Bouchiat, I. I. Khodos, Yu. B. Gorbatov, V. T. Volkov, C. Journet, M. Burghard, Science **284**, 1508 (1999)
78. K. Tsukagoshi, B. W. Alphenaar, H. Ago, Nature **401**, 572 (1999)
79. H. Dai, E. W. Wong, C. M. Lieber, Science **272**, 523–526 (1994)
80. A. Thess, R. Lee, P. Nikolaev, H. Dai, P. Petit, J. Robert, C. Xu, Y. H. Lee, S. G. Kim, A. G. Rinzler, D. T. Colbert, G. E. Scuseria, D. Tománek, J. E. Fischer, R. E. Smalley, Science **273**, 483–487 (1996)

81. A. Yu. Kasumov, H. Bouchiat, B. Reulet, O. Stephan, I. I. Khodos, Yu. B. Gorbatov, C. Colliex, Europhys. Lett. **43**, 89 (1998)
82. P. J. de Pablo, E. Graugnard, B. Walsh, R. P. Andres, S. Datta, R. Reifenberger, Appl. Phys. Lett. **74**, 323 (1999)
83. P. J. de Pablo, S. Howell, S. Crittenden, B. Walsh, E. Graugnard, R. Reifenberger, Appl. Phys. Lett. **75**, 3941 (1999)
84. H. R. Shea, R. Martel, T. Hertel, T. Schmidt, Ph. Avouris, Microel. Eng., **46**, 101 (1999)
85. J. Kong, H. T. Soh, A. M. Cassell, C. F. Quate, H. Dai, Nature **395**, 878 (1998)
86. G. Baumgartner, M. Carrard, L. Zuppiroli, W. Basca, W. A. de Heer, L. Forró, Phys. Rev. B **55**, 6704 (1997)
87. J. Hone, I. Ellwood, M. Muno, Ari Mizel, Marvin L. Cohen, A. Zettl, Andrew G. Rinzler, R. E. Smalley, Phys. Rev. Lett. **80**, 1042 (1998)
88. W. Yi, L. Lu, Zhang Dian-lin, Z. W. Pan, S. S. Xie, Phys. Rev. B **59**, R9015 (1999)
89. L. Grigorian, G. U. Sumanasekera, A. L. Loper, S. Fang, J. L. Allen, P. C. Eklund, Phys. Rev. B **59**, R11309 (1999)
90. P. G. Collins, K. Bradley, M. Ishigami, A. Zettl, Science **287**, 1801 (2000)
91. G. U. Sumanasekera, C. Adu, S. Fang, P. C. Eklund, Phys. Rev. Lett. (2000). submitted
92. Li Lu, Wie Yi, Z. Pan, S. S. Xie, Bull. APS **45**, 414 (2000)
93. T. W. Ebbesen, H. Hiura, M. E. Bisher, M. M. J. Treacy, J. L. Shreeve-Keyer, R. C. Haushalter, Adv. Mater. **8**, 155 (1996)
94. Adrian Bachtold, Dissertation, Univ. of Basel (1999)
95. A. Bachtold, M. Henny, C. Terrier, C. Strunk, C. Schönenberger, J.-P. Salvetat, J.-M. Bonard, L. Forró, Appl. Phys. Lett. **73**, 274 (1998)
96. Y. Imry, Physics of mesoscopic systems in *Directions in Condensed Matter Physics*, G. Grinstein, G. Mazenko, (Eds.) (World Scientific, Singapore 1986)
97. A. Bachtold, C. Terrier, M. Kruger, M. Henny, T. Hoss, C. Strunk, R. Huber, H. Birk, U. Staufer, C. Schönenberger, Microel. Engin. **41-42**, 571 1998).
98. M. Buitelaar, A. Bachtold, C. Schönenberger, L. Forró, unpublished
99. H. Grabert, M. H. Devoret, *Single Charge Tunneling: Coulomb Blockade Phenomena in Nanostructures* (Plenum, New York 1992)
100. B. L. Altshuler, Pis'ma Zh. Eksp. Teor. Fiz. **41**, 530 (1985) [JETP Lett. **41**, 648 (1985)]
101. P. A. Lee, A. D. Stone, Phys. Rev. Lett. **55**, 1622 (1985)
102. P. A. Lee, A. D. Stone, H. Fukuyama, Phys. Rev. B **35**, 1039 (1987)
103. P. Delaney, M. Di Ventra, S. T. Pantelides, Appl. Phys. Lett. **75**, 3787 (1999)
104. Walt de Heer et al., unpublished
105. Z. Yao, C. L. Kane, C. Dekker, Phys. Rev. Lett. **84**, 2941 (2000).
106. W. Tian, S. Datta, Phys. Rev. B **49**, 5097 (1994)
107. A. G. Aharonov, Yu. V. Sharvin, Rev. Mod. Phys. **59**, 755 (1987)
108. B. L. Altshuler, A. G. Aharonov, B. Z. Spivak, Pis'ma Zh. Eksp. Teor. Fiz. **33**, 101 (1981) [JETP Lett. **33**, 94 (1981)]
109. M. Baxendale, V. Z. Mordkovich, S. Yoshimura, R. P. H. Chang, Phys. Rev. B **56**, 2161 (1997)
110. A. Fujiwara et al., *Proceedings of the International Winterschool on Electronic Properties of Novel Materials 1997*, H. Kuzmany, J. Fink, M. Mehring, S. Roth (Eds.) (AIP, New York 1997)

111. G. T. Kim, E. S. Choi, D. C. Kim, D. S. Suh, Y. W. Park, K. Liu, G. Duesberg, S. Roth, Phys. Rev. B **58**, 16064 (1998)
112. B. L. Altshuler, A. G. Aharonov, D. E. Khmelnitsky, Solid State Commun. **39**, 619 (1981)
113. P. A. Lee, T. V. Ramakrishnan, Rev. Mod. Phys. **57**, 287 (1985)
114. B. L. Altshuler, A. G. Aharonov, in *Electron-Electron Interactions in Disordered Systems*, M. Pollak, A. L. Efros, (Eds.) (North-Holland, Amsterdam 1984) pp. 1–153
115. H. Fukuyama, in *Electron-Electron Interactions in Disordered Systems*, M. Pollak, A. L. Efros, (Eds.) (North-Holland, Amsterdam 1984) pp. 1–153
116. R. Egger, A. O. Gogolin, Phys. Rev. Lett. **79**, 5082 (1997)
117. C. Kane, L. Balents, M. P. A. Fisher, Phys. Rev. Lett. **79**, 5086 (1997)
118. A. Komnik, R. Egger, Phys. Rev. Lett. **80**, 2881 (1998)
119. A. A. Odintsov, H. Yoshioka, Phys. Rev. Lett. **82** 374 (1999)
120. R. Egger, A. O. Gogolin, Eur. Phys. J. B **3**, 281 (1998)
121. R. Egger, Phys. Rev. Lett. **83**, 5547 (1999)
122. C. T. White, J. W. Mintmire, Nature **394**, 29 (1998)
123. J. W. Mintmire, C. T. White, Phys. Rev. Lett. **81**, 2506 (1998)
124. A. Bachtold, C. Schönenberger, L. Forró, unpublished
125. M. Bockrath, D. H. Cobden, J. Lu, A. G. Rinzler, R. E. Smalley, L. Balents, P. L. McEuen, Nature **397**, 598 (1999)
126. For a review see: M. P. A. Fisher, L. Glazman, in *Mesoscopic Electron Transport*, L. L. Sohn, L. P. Kouwenhoven, G. Schön, (Eds.), NATO ASI Ser. E: Appl. Sci. **345** (Kluwer Academic, Dordrecht 1997)
127. Z. Yao, H. W. Ch. Postma, L. Balents, C. Dekker, Nature **402**, 273 (1999)
128. A. Bachtold, private communication
129. E. Graugnard, B. Walsh, P. J. de Pablo, R. P. Andres, S. Datta, R. Reifenberger, Bull. APS March **45**, 487 (2000)
130. D. Tománek, R. J. Enbody (Kluwer Academic/Plenum Publishers, New York 1999)
131. D. L. Carroll, P. Redlich, P. M. Ajayan, J.-C. Charlier, X. Blase, A. De Vita, R. Car, Phys. Rev. Lett. **78**, 2811 (1997)
132. M. Kociak, A. Yu Kasumov, S. Gu'eron, B. Reulet, L. Vaccarini, I. I. Khodos, Yu, B. Gorbatov, V. T. Volkov, H. Bouchiat, unpublished
133. G. Wagoner, Phys. Rev. **118**, 647 (1960)
134. Y. Yafet, Solid State Phys. **14**, 1 (1963)
135. R. J. Elliott, Phys. Rev. **96**, 266 (1954)
136. O. Chauvet, L. Forró, W. Bacsa, D Chatelain, D. Ugarte, W. De Heer, Phys. Rev. B **52**, R6963 (1995)
137. F. Beuneu, P. Monod, Phys. Rev. B **18**, 2422 (1978)
138. Y. Tomkiewicz, E. M. Engler, T. D. Schultz, Phys. Rev. Lett. **35**, 456 (1975)
139. A. N. Bloch, T. F. Carruthers, T. O. Poehler D. O. Cowan, in *Chemistry and Physics of One-Dimensional Metals*, H. J. Keller (Ed.) (Plenum, New York 1977 ) p. 47
140. L. Forró, J. R. Cooper, G. Sekretarczyk, M. Krupski, K. Kamarás, J. Phys. (Paris) **48**, 413 (1987)
141. J.-P. Salvetat, J.-M. Bonard, L. Forró, unpublished
142. L. S. Singer, G. Wagoner, J. Chem. Phys. **37**, 1812 (1962)
143. G. Baumgartner, M. Carrard, L. Zuppiroli, W. Bacsa, W. A. de Heer, L. Forró, Phys. Rev. B **55** , 6704 (1997)

144. A. G. Rinzler, J. H. Hafner, P. Nikolaev L. Lou, S. G. Kim, D. Tománek, P. Nordlander, D. T. Colbert, R. E. Smalley, Science **269**, 1550 (1995)
145. W. A. de Heer, A. Chatelain, D. Ugarte, Science **270**, 1179 (1995)
146. W. A. de Heer, J.-M. Bonard, Kai Fauth, A. Chatelain, L. Forró, D. Ugarte, Adv. Mater. **9**, 87-9 (1997)
147. J.-M. Bonard, T. Stökli, F. Meier, W. A. de Heer, A. Chatelain, J.-P. Salvetat, L. Forró, Phys. Rev. Lett. **81**, 1441 (1998)
148. J-M. Bonard, F. Meier, T. Stöckli, L. Forró, A. Chatelain, W. de Heer, J.-P. Salvetat, L. Forró, Ultramicroscopy **77**, 7 (1998)
149. P. Kim, T. W. Odom, J.-L. Huang, C. M. Lieber, Phys. Rev. Lett. **82**, 1225 (1999)
150. A. De Vita, J.-C. Charlier, X. Blase, R. Car: Appl. Phys. A **68**, 283 (1999)
151. M. Fransen: Towards high brightness, monochromatic electron sources, Dissertation , Technical University Delft (1999)
152. R. Coratger et al, preprint
153. R. Berndt, R. Gaisch, J. K. Gimzewski, B. Reihl, R. R. Schittler, W. D. Schneider, M. Tschudy, Science **262**, 1425 (1993)
154. S. Frank, P. Poncharal, Z. L. Wang, W. A. de Heer, Science **280**, 1744 (1998)
155. M. M. J. Treacy, T. W. Ebbesen, J. M. Gibson, Nature **381**, 678 (1996)
156. E. W. Wong, P. E. Sheehan, C. M. Lieber, Science **277**, 1971 (1997)
157. A. Krishnan,. E. Dujardin, T. W. Ebbesen, P. N. Yianilos, M. M. J. Treacy, Phys. Rev. B **58** (20) 14013 (1998)
158. J.-P. Salvetat, G. A. D. Briggs, J.-M. Bonard, R. R. Basca, A. J. Kulik, T. Stöckli, N. A. Burnham, L. Forró, Phys. Rev. Lett. **82** 944 (1999)
159. J.-P. Salvetat, A. J. Kulik, J.-M. Bonard, G. A. D. Briggs, T. Stöckli, K. Méténier, S. Bonnamy, F. Béguin, N. A. Burnham, L. Forró, Adv. Mater. **11** (2) 161 (1999)
160. P. Poncharal, Z. L. Wang, D. Ugarte, W. A. de Heer, Science **283**, 1513 (1999)
161. D. A. Walters, L. M. Ericson, M. J. Casavant, J. Liu, D. T. Colbert, K. A. Smith, R. E. Smalley, Appl. Phys. Lett. **74**, 3803 (1999)
162. L. S. Schadler, S. C. Giannaris, P. M. Ajayan, Appl. Phys. Lett. **73**, 3842 (1998)
163. J.-P. Salvetat, J.-M. Bonard, N. H. Thomson, A. J. Kulik, L. Forró, W. Benoit, L. Zuppiroli, Appl. Phys. A **69**, 255 (1999)
164. F. Beuneu, C. L'Huillier, J.-P. Salvetat, J.-M. Bonnard, L. Forró, Phys. Rev. B **59**, 5945 (1999)
165. Min-Feng Yu, O. Lourie, M. J. Dyer, K. Moloni, T. F. Kelly, R. S. Ruoff, Science **287**, 637 (2000)
166. J. Cumings, A. Zettl, Nature **406**, 586 (2000)

# Applications of Carbon Nanotubes

Pulickel M. Ajayan[1] and Otto Z. Zhou[2]

[1] Department of Materials Science and Engineering
Rensselaer Polytechnic Institute, Troy, NY 12180-3590, USA
Ajayan@rpi.edu

[2] Curriculum in Applied and Materials Sciences
Department of Physics and Astronomy
University of North Carolina at Chapel Hill
Chapel Hill, NC 27599-3255, USA
Zhou@physics.unc.edu

**Abstract.** Carbon nanotubes have attracted the fancy of many scientists worldwide. The small dimensions, strength and the remarkable physical properties of these structures make them a very unique material with a whole range of promising applications. In this review we describe some of the important materials science applications of carbon nanotubes. Specifically we discuss the electronic and electrochemical applications of nanotubes, nanotubes as mechanical reinforcements in high performance composites, nanotube-based field emitters, and their use as nanoprobes in metrology and biological and chemical investigations, and as templates for the creation of other nanostructures. Electronic properties and device applications of nanotubes are treated elsewhere in the book. The challenges that ensue in realizing some of these applications are also discussed from the point of view of manufacturing, processing, and cost considerations.

The discovery of fullerenes [1] provided exciting insights into carbon nanostructures and how architectures built from $sp^2$ carbon units based on simple geometrical principles can result in new symmetries and structures that have fascinating and useful properties. Carbon nanotubes represent the most striking example. About a decade after their discovery [2], the new knowledge available in this field indicates that nanotubes may be used in a number of practical applications. There have been great improvements in synthesis techniques, which can now produce reasonably pure nanotubes in gram quantities. Studies of structure–topology–property relations in nanotubes have been strongly supported, and in some cases preceded, by theoretical modeling that has provided insights for experimentalists into new directions and has assisted the rapid expansion of this field [3,4,5,6,7,8].

Quasi-one-dimensional carbon whiskers or nanotubes are perfectly straight tubules with diameters of nanometer size, and properties close to that of an ideal graphite fiber. Carbon nanotubes were discovered accidentally by Sumio *Iijima* in 1991, while studying the surfaces of graphite electrodes used in an electric arc discharge [2]. His observation and analysis of the nanotube structure started a new direction in carbon research, which complemented the excitement and activities then prevalent in fullerene research. These tiny

M. S. Dresselhaus, G. Dresselhaus, Ph. Avouris (Eds.): Carbon Nanotubes,
Topics Appl. Phys. **80**, 391–425 (2001)

carbon tubes with incredible strength and fascinating electronic properties appear to be ready to overtake fullerenes in the race to the technological marketplace. It is the structure, topology and size of nanotubes that make their properties exciting compared to the parent, planar graphite-related structures, such as are for example found in carbon fibers.

The uniqueness of the nanotube arises from its structure and the inherent subtlety in the structure, which is the helicity in the arrangement of the carbon atoms in hexagonal arrays on their surface honeycomb lattices. The helicity (local symmetry), along with the diameter (which determines the size of the repeating structural unit) introduces significant changes in the electronic density of states, and hence provides a unique electronic character for the nanotubes. These novel electronic properties create a range of fascinating electronic device applications and this subject matter is discussed briefly elsewhere in this volume [9], and has been the subject of discussion in earlier reviews [8]. The other factor of importance in what determines the uniqueness in physical properties is topology, or the closed nature of individual nanotube shells; when individual layers are closed on to themselves, certain aspects of the anisotropic properties of graphite disappear, making the structure remarkably different from graphite. The combination of size, structure and topology endows nanotubes with important mechanical properties (e.g., high stability, strength and stiffness, combined with low density and elastic deformability) and with special surface properties (selectivity, surface chemistry), and the applications based on these properties form the central topic of this chapter. In addition to the helical lattice structure and closed topology, topological defects in nanotubes (five member Stone–Wales defects near the tube ends, aiding in their closure) [9,10], akin to those found in the fullerenes structures, result in local perturbations to their electronic structure [11]; for example, the ends or caps of the nanotubes are more metallic than the cylinders, due to the concentration of pentagonal defects [11]. These defects also enhance the reactivity of tube ends, giving the possibility of opening the tubes [12], functionalizing the tube ends [13], and filling the tubes with foreign substances [14,15,16].

The structure of nanotubes remains distinctly different from traditional carbon fibers that have been industrially used for several decades (e.g., as reinforcements in tennis rackets, airplane frame parts and batteries to name a few) [17,18]. Most importantly, nanotubes, for the first time represent the ideal, most perfect and ordered, carbon fiber, the structure of which is entirely known at the atomic level. It is this predictability that mainly distinguishes nanotubes from other carbon fibers and puts them along with molecular fullerene species in a special category of prototype materials. Among the nanotubes, two varieties, which differ in the arrangement of their graphene cylinders, share the limelight. Multi-Walled NanoTubes (MWNT), are collections of several concentric graphene cylinders and are larger structures compared to Single-Walled NanoTubes (SWNTs) which are individual cylinders

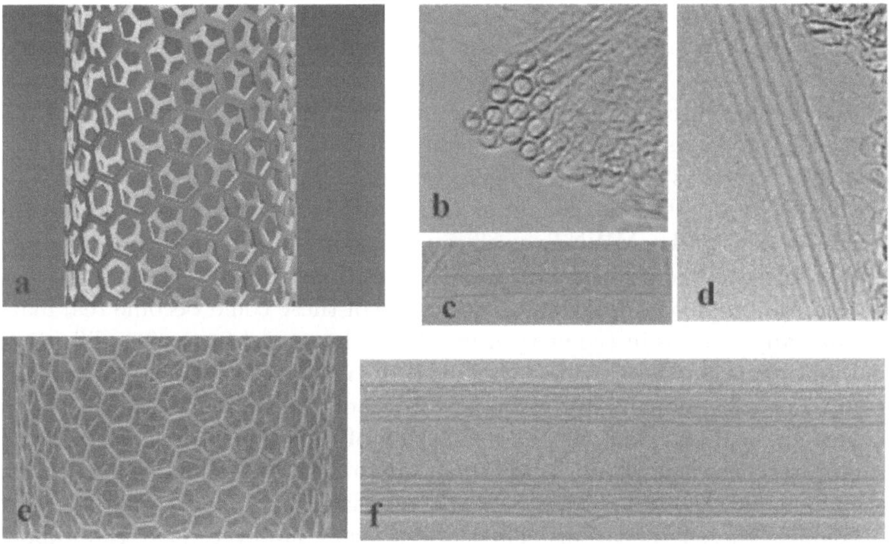


**Fig. 1.** Structure of Single-Walled (SWNT) (**a-d**) and Multi-Walled (MWNT) carbon NanoTubes (**e,f**). (**a**) Shows a schematic of an individual helical SWMT. (**b**) Shows a cross-sectional view (TEM image) of a bundle of SWNTs [transverse view shown in (**d**)]. Each nanotube has a diameter of ~1.4 nm and the tube-tube distance in the bundles is 0.315 nm. (**c**) Shows the high-resolution TEM micrograph of a 1.5 nm diameter SWNT. (**e**) is the schematic of a MWNT and (**f**) shows a high resolution TEM image of an individual MWNT. The distance between horizontal fringes (layers of the tube) in (**f**) is 0.34 nm (close to the interlayer spacing in graphite)

of 1–2 nm diameter (see Fig. 1). The former can be considered as a mesoscale graphite system, whereas the latter is truly a single large molecule. However, SWNTs also show a strong tendency to bundle up into ropes, consisting of aggregates of several tens of individual tubes organized into a one-dimensional triangular lattice. One point to note is that in most applications, although the individual nanotubes should have the most appealing properties, one has to deal with the behavior of the aggregates (MWNT or SWNT ropes), as produced in actual samples. The best presently available methods to produce ideal nanotubes are based on the electric arc [19,20] and laser ablation processes [21]. The material prepared by these techniques has to be purified using chemical and separation methods. None of these techniques are scalable to make the industrial quantities needed for many applications (e.g., in composites), and this has been a bottleneck in nanotube R&D. In recent years, work has focused on developing Chemical Vapor Deposition (CVD) techniques using catalyst particles and hydrocarbon precursors to grow nanotubes [22,23,24,25]; such techniques have been used earlier to produce hollow nanofibers of carbon in large quantities [17,18]. The drawback of the catalytic CVD-based nanotube production is the inferior quality of the structures that contain gross defects (twists, tilt boundaries etc.), particularly because the structures are created at much lower temperatures (600–1000°C) compared to the arc or laser processes (~2000°C).

Since their discovery in 1991, several demonstrations have suggested potential applications of nanotubes. These include the use of nanotubes as electron field emitters for vacuum microelectronic devices, individual MWNTs and SWNTs attached to the end of an Atomic Force Microscope (AFM) tip for use as nanoprobe, MWNTs as efficient supports in heterogeneous catalysis and as microelectrodes in electrochemical reactions, and SWNTs as good media for lithium and hydrogen storage . Some of these could become real marketable applications in the near future, but others need further modification and optimization. Areas where predicted or tested nanotube properties appear to be exceptionally promising are mechanical reinforcing and electronic device applications. The lack of availability of bulk amounts of well-defined samples and the lack of knowledge about organizing and manipulating objects such as nanotubes (due to their sub-micron sizes) have hindered progress in developing these applications. The last few years, however, have seen important breakthroughs that have resulted in the availability of nearly uniform bulk samples. There still remains a strong need for better control in purifying and manipulating nanotubes, especially through generalized approaches such as chemistry. Development of functional devices/structures based on nanotubes will surely have a significant impact on future technology needs. In the following sections we describe the potential of materials science-related applications of nanotubes and the challenges that need to be overcome to reach these hefty goals.

In the following sections we describe several interesting applications of carbon nanotubes based on some of the remarkable materials properties of nanotubes. Electron field emission characteristics of nanotubes and applications based on this, nanotubes as energy storage media, the potential of nanotubes as fillers in high performance polymer and ceramic composites, nanotubes as novel probes and sensors, and the use of nanotubes for template-based synthesis of nanostructures are the major topics that are discussed in the sections that follow.

# 1 Potential Application of CNTs in Vacuum Microelectronics

Field emission is an attractive source for electrons compared to thermionic emission. It is a quantum effect. When subject to a sufficiently high electric field, electrons near the Fermi level can overcome the energy barrier to escape to the vacuum level. The basic physics of electron emission is well developed. The emission current from a metal surface is determined by the Fowler–Nordheim equation: $I = aV^2 \exp(-b\phi^{3/2}/\beta V)$ where $I$, $V$, $\phi$, $\beta$, are the current, applied voltage, work function, and field enhancement factor, respectively [26,27].

Electron field emission materials have been investigated extensively for technological applications, such as flat panel displays, electron guns in electron microscopes, microwave amplifiers [28]. For technological applications, electron emissive materials should have low threshold emission fields and should be stable at high current density. A current density of 1–10 mA/cm$^2$ is required for displays [29] and $> 500\,\mathrm{mA/cm^2}$ for a microwave amplifier [30]. In order to minimize the electron emission threshold field, it is desirable to have emitters with a low work function and a large field enhancement factor. The work function is an intrinsic materials property. The field enhancement factor depends mostly on the geometry of the emitter and can be approximated as: $\beta = 1/5r$ where $r$ is the radius of the emitter tip. Processing techniques have been developed to fabricate emitters such as Spindt-type emitters, with a sub-micron tip radius [28]. However, the process is costly and the emitters have only limited lifetime. Failure is often caused by ion bombardment from the residual gas species that blunt the emission tips. Table 1 lists the threshold electrical field values for a 10 mA/cm$^2$ current density for some typical materials.

**Table 1.** Threshold electrical field values for different materials for a 10 mA/cm$^2$ current density (data taken from [31,32])

| Material | Threshold electrical field (V/μm) |
|---|---|
| Mo tips | 50–100 |
| Si tips | 50–100 |
| $p$-type semiconducting diamond | 130 |
| Undoped, defective CVD diamond | 30–120 |
| Amorphous diamond | 20–40 |
| Cs-coated diamond | 20–30 |
| Graphite powder(<1 mm size) | 17 |
| Nanostructured diamond[a] | 3–5 (unstable >30 mA/cm$^2$) |
| Carbon nanotubes[b] | 1–3 (stable at 1 A/cm$^2$) |

[a]Heat-treated in H plasma.
[b]random SWNT film

Carbon nanotubes have the right combination of properties – nanometer-size diameter, structural integrity, high electrical conductivity, and chemical stability – that make good electron emitters [27]. Electron field emission from carbon nanotubes was first demonstrated in 1995 [33], and has since been studied intensively on various carbon nanotube materials. Compared to conventional emitters, carbon nanotubes exhibit a lower threshold electric field, as illustrated in Table 1. The current-carrying capability and emission stability of the various carbon nanotubes, however, vary considerably depending on the fabrication process and synthesis conditions.

The I–V characteristics of different types of carbon nanotubes have been reported, including individual nanotubes [33,34], MWNTs embedded in epoxy matrices [35,36], MWNT films [37,38], SWNTs [39,40,41,42] and aligned MWNT films [32]. Figure 2 shows typical emission I–V characteristics measured from a random SWNT film at different anode-cathode distances, and the Fowler–Nordheim plot of the same data is shown as the inset. Turn-on and threshold fields are often used to describe the electrical field required for emission. The former is not well-defined and typically refers to the field that is required to yield 1 nA of total emission current, while the latter refers to the field required to yield a given current density, such as 10 mA/cm$^2$. For random SWNT films, the threshold field for 10 mA/cm$^2$ is in the range of 2–3V/μm. Random and aligned MWNTs [fabricated at the University of North Carolina (UNC) and AT&T Bell Labs] were found to have threshold fields slightly larger than that of the SWNT films and are typically in the range of 3–5 V/μm for a 10 mA/cm$^2$ current density [32] (Fig. 3). These values for the threshold field are all significantly better than those from conventional field emitters such as the Mo and Si tips which have a threshold electric field of 50–100 V/μm (Table 1). It is interesting to note that the aligned MWNT films do not perform better than the random films. This is due to the electrical screening effect arising from closely packed nanotubes [43]. The low threshold field for electron emission observed in carbon nanotubes is a direct result of the large field enhancement factor rather than a reduced electron work function. The latter was found to be 4.8 eV for SWNTs, 0.1–0.2 eV larger than that of graphite [44].

SWNTs generally have a higher degree of structural perfection than either MWNTs or CVD-grown materials and have a capability for achieving higher current densities and have a longer lifetime [32]. Stable emission above 20 mA/cm$^2$ has been demonstrated in SWNT films deposited on Si substrates [40]. A current density above 4 A/cm$^2$(measured by a 1 mm local probe) was obtained from SWNTs produced by the laser ablation method [40]. Figure 4 is a CCD (Charge Coupled Device) image of the set-up for electron emission measurement, showing a Mo anode (1 mm diameter) and the edge of the SWNT cathode in a vacuum chamber. The Mo anode is glowing due to bombarding from field emitted electrons, demonstrating the high current capability of the SWNTs. This particular image was taken at a current density of 0.9 A/cm$^2$. The current densities observed from the carbon nanotubes are significantly higher than from conventional emitters, such as nano-diamonds which tend to fail below 30 mA/cm$^2$ current density [31]. Carbon nanotube emitters are particularly attractive for a variety of applications including microwave amplifiers.

Although carbon nanotube emitters show clear advantageous properties over conventional emitters in terms of threshold electrical field and current density, their emission site density (number of functioning emitters per unit area) is still too low for high resolution display applications. Films presently

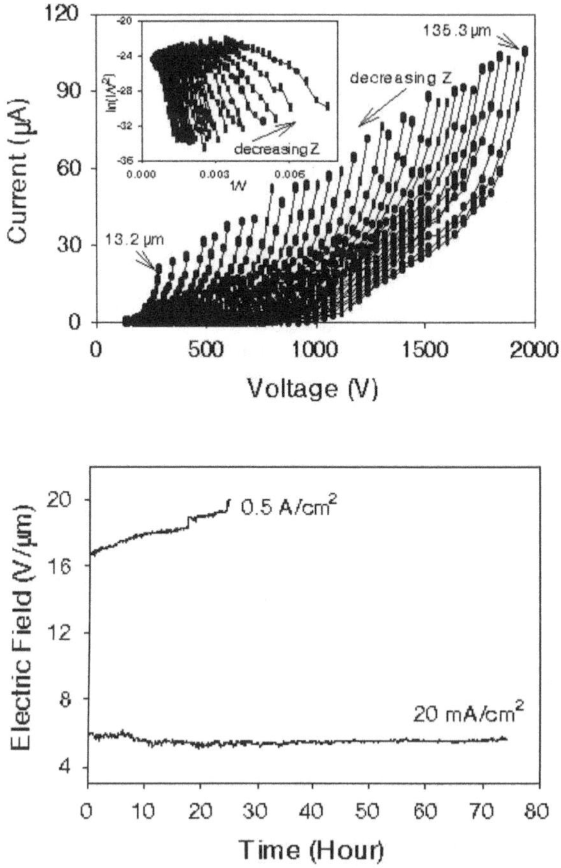


**Fig. 2.** (*Top*): Emission I–V characteristics of a random single-walled carbon nanotube film measured at different anode-cathode distances at $10^{-8}$ torr base pressure. The same data are plotted as [$\ln(I/V^2)$ vs $1/V$] in the *inset*. Deviations from the ideal Fowler–Nordheim behavior are observed at high current. (*Bottom*): Stability test of a random laser-ablation-grown SWNT film showing stable emission at $20\,mA/cm^2$ (from [40])

fabricated [32] have typical emission site densities of $10^3$–$10^4/cm^2$ at the turn-on field, and $\sim 10^6/cm^2$ is typically required for high resolution display devices.

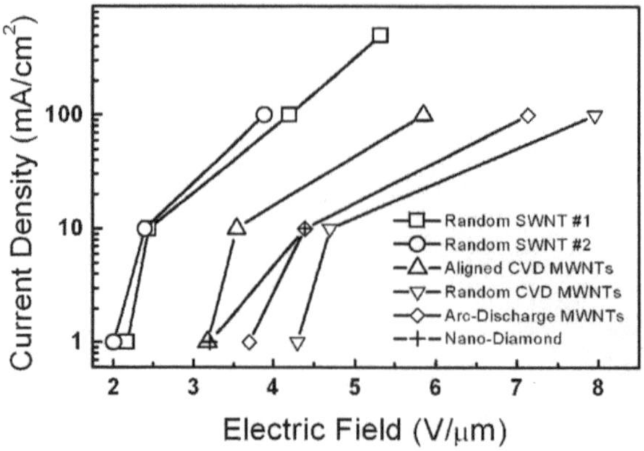


**Fig. 3.** Current density versus electric field measured for various forms of carbon nanotubes (data taken from *Bower* et al. [32])

**Fig. 4.** A CCD image showing a glowing Mo anode (1 mm diameter) at an emission current density of 0.9 A/cm$^2$ from a SWNT cathode. Heating of the anode is due to field emitted electrons bombarding the Mo probe, thereby demonstrating a high current density (image provided by Dr. Wei Zhu of Bell Labs)

## 1.1 Prototype Electron Emission Devices Based on Carbon Nanotubes

### 1.1.1 Cathode-Ray Lighting Elements

Cathode ray lighting elements with carbon nanotube materials as the field emitters have been fabricated by Ise Electronic Co. in Japan [45]. As illustrated in Fig. 5, these nanotube-based lighting elements have a triode-type design. In the early models, cylindrical rods containing MWNTs, formed as a deposit by the arc-discharge method, were cut into thin disks and were glued to stainless steel plates by silver paste. In later models, nanotubes are now screen-printed onto the metal plates. A phosphor screen is printed on the inner surfaces of a glass plate. Different colors are obtained by using different fluorescent materials. The luminance of the phosphor screens measured on the tube axis is $6.4 \times 10^4$ cd/cm$^2$ for green light at an anode current of 200 μA,

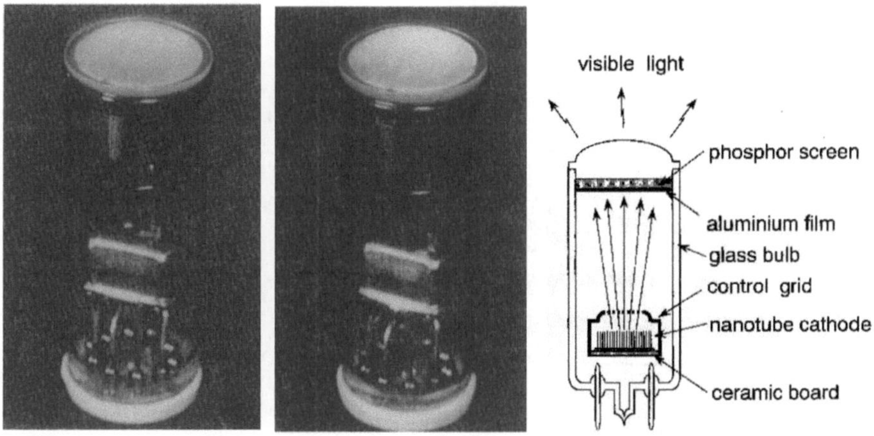


**Fig. 5.** Demonstration field emission light source using carbon nanotubes as the cathodes (fabricated by Ise Electronic Co., Japan) [45]

which is two times more intense than that of conventional thermionic Cathode Ray Tube (CRT) lighting elements operated under similar conditions [45].

### 1.1.2 Flat Panel Display

Prototype matrix-addressable diode flat panel displays have been fabricated using carbon nanotubes as the electron emission source [46]. One demonstration (demo) structure constructed at Northwestern University consists of nanotube-epoxy stripes on the cathode glass plate and phosphor-coated Indium-Tin-Oxide (ITO) stripes on the anode plate [46]. Pixels are formed at the intersection of cathode and anode stripes, as illustrated in Fig. 6. At a cathode-anode gap distance of 30μm, 230 V is required to obtain the emission current density necessary to drive the diode display ($\sim 76\,\mu\text{mA/mm}^2$). The device is operated using the half-voltage off-pixel scheme. Pulses of $\pm 150$ V are switched among anode and cathode stripes, respectively to produce an image.

Recently, a 4.5 inch diode-type field emission display has been fabricated by Samsung (Fig. 6), with SWNT stripes on the cathode and phosphor-coated ITO stripes on the anode running orthogonally to the cathode stripes [47]. SWNTs synthesized by the arc-discharge method were dispersed in isopropyl alcohol and then mixed with an organic mixture of nitro cellulose. The paste was squeezed into sodalime glasses through a metal mesh, 20μm in size, and then heat-treated to remove the organic binder. $Y_2O_2S$:Eu, ZnS:Cu,Al, and ZnS:Ag,Cl, phosphor-coated glass is used as the anode.

### 1.1.3 Gas-Discharge Tubes in Telecom Networks

Gas discharge tube protectors, usually consisting of two electrodes parallel to each other in a sealed ceramic case filled with a mixture of noble gases,

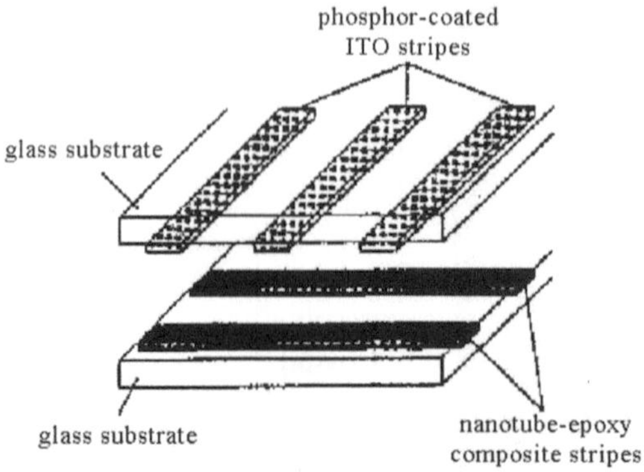

**Fig. 6.** *Left:* Schematic of a prototype field emission display using carbon nanotubes (adapted from [46]). *Right:* A prototype 4.5″ field emission display fabricated by Samsung using carbon nanotubes (image provided by Dr. W. Choi of Samsung Advanced Institute of Technologies)

is one of the oldest methods used to protect against transient over-voltages in a circuit [48]. They are widely used in telecom network interface device boxes and central office switching gear to provide protection from lightning and ac power cross faults on the telecom network. They are designed to be insulating under normal voltage and current flow. Under large transient voltages, such as from lightning, a discharge is formed between the metal electrodes, creating a plasma breakdown of the noble gases inside the tube. In the plasma state, the gas tube becomes a conductor, essentially short-circuiting the system and thus protecting the electrical components from over-voltage damage. These devices are robust, moderately inexpensive, and have a relatively small shunt capacitance, so they do not limit the bandwidth of high-frequency circuits as much as other nonlinear shunt components. Compared to solid state protectors, GDTs can carry much higher currents. However, the current Gas Discharge Tube (GDT) protector units are unreliable from the standpoint of mean turn-on voltage and run-to-run variability.

Prototype GDT devices using carbon nanotube coated electrodes have recently been fabricated and tested by a group from UNC and Raychem Co.[49]. Molybdenum electrodes with various interlayer materials were coated with single-walled carbon nanotubes and analyzed for both electron field emission and discharge properties. A mean dc breakdown voltage of 448.5 V and a standard deviation of 4.8 V over 100 surges were observed in nanotube-based GDTs with 1 mm gap spacing between the electrodes. The breakdown reliability is a factor of 4–20 better and the breakdown voltage is ~30% lower than the two commercial products measured (Fig. 7). The enhanced performance shows that nanotube-based GDTs are attractive over-voltage protection units in advanced telecom networks such as an Asymmetric-Digital-Signal-Line

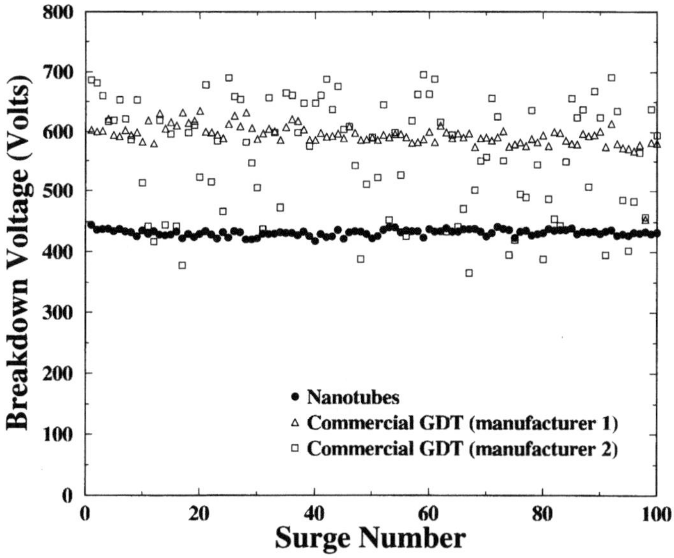


**Fig. 7.** DC breakdown voltage of a gas discharge tube with SWNT coated electrodes, filled with 15 torr argon with neon added and 1 mm distance between the electrodes. The commercial GDTs are off-the-shelf products with unknown filling gas(es), but with the same electrode-electrode gap distances. The breakdown voltage of the GDT with SWNT coated electrodes is 448.5 V, with a standard deviation of 4.8 V over 100 surges. The commercial GDT from one manufacturer has a mean breakdown voltage of 594 V and a standard deviation of 20 V. The GDT from the second manufacturer has a breakdown voltage of 563 V and a standard deviation of 93 V (from *Rosen* et al. in [49])

(ADSL), where the tolerance is narrower than what can be provided by the current commercial GDTs.

## 2 Energy Storage

Carbon nanotubes are being considered for energy production and storage. Graphite, carbonaceous materials and carbon fiber electrodes have been used for decades in fuel cells, battery and several other electrochemical applications [50]. Nanotubes are special because they have small dimensions, a smooth surface topology, and perfect surface specificity, since only the basal graphite planes are exposed in their structure. The rate of electron transfer at carbon electrodes ultimately determines the efficiency of fuel cells and this depends on various factors, such as the structure and morphology of the carbon material used in the electrodes. Several experiments have pointed out that compared to conventional carbon electrodes, the electron transfer kinetics take place fastest on nanotubes, following ideal Nernstian behavior [51]. Nanotube microelectrodes have been constructed using a binder and have been successfully used in bioelectrochemical reactions (e.g., oxidation of dopamine). Their performance has been found to be superior to other carbon electrodes in terms of reaction rates and reversibility [52]. Pure MWNTs and MWNTs deposited with metal catalysts (Pd, Pt, Ag) have been used to electro-catalyze an oxygen reduction reaction, which is important for fuel cells [53,54,55]. It is seen from several studies that nanotubes could be excellent replacements for conventional carbon-based electrodes. Similarly, the improved selectivity of nanotube-based catalysts have been demonstrated in heterogeneous catalysis. Ru-supported nanotubes were found to be superior to the same metal on graphite and on other carbons in the liquid phase hydrogenation reaction of cinnamaldehyde [55]. The properties of catalytically grown carbon nanofibers (which are basically defective nanotubes) have been found to be desirable for high power electrochemical capacitors [56].

### 2.1 Electrochemical Intercalation of Carbon Nanotubes with Lithium

The basic working mechanism of rechargeable lithium batteries is electrochemical intercalation and de-intercalation of lithium between two working electrodes. Current state-of-art lithium batteries use transition metal oxides (i.e., $Li_xCoO_2$ or $Li_xMn_2O_4$) as the cathodes and carbon materials (graphite or disordered carbon) as the anodes [57]. It is desirable to have batteries with a high energy capacity, fast charging time and long cycle time. The energy capacity is determined by the saturation lithium concentration of the electrode materials. For graphite, the thermodynamic equilibrium saturation concentration is $LiC_6$ which is equivalent to 372 mA h/g. Higher Li concentrations have been reported in disordered carbons (hard and soft carbon) [58,59] and metastable compounds formed under pressure [60].

It has been speculated that a higher Li capacity may be obtained in carbon nanotubes if all the interstitial sites (inter-shell van der Waals spaces, inter-tube channels, and inner cores) are accessible for Li intercalation. Electrochemical intercalation of MWNTs [61,62] and SWNTs [63,64] has been investigated by several groups. Figure 8 (top) shows representative electrochemical intercalation data collected from an arc-discharge-grown MWNT sample using an electrochemical cell with a carbon nanotube film and a lithium foil as the two working electrodes [64]. A reversible capacity ($C_{rev}$) of 100–640 mA h/g has been reported, depending on the sample processing and annealing conditions [61,62,64]. In general, well-graphitized MWNTs such as those synthesized by the arc-discharge method have a lower $C_{rev}$ than those prepared by the CVD method. Structural studies [65,66] have shown that alkali metals can be intercalated into the inter-shell spaces within the individual MWNTs through defect sites.

Single-walled nanotubes are shown to have both high reversible and irreversible capacities [63,64]. Two separate groups reported 400–650 mA h/g reversible and $\sim$1000 mA h/g irreversible capacities in SWNTs produced by the

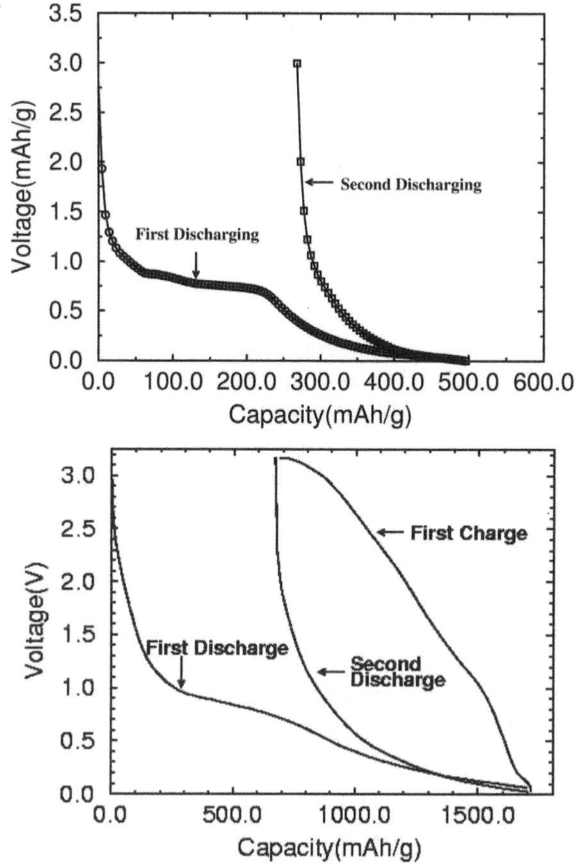


**Fig. 8.** *Top:* Electrochemical intercalation of MWNTs with lithium. Data were collected using 50 mA/h current. The electrolyte was $LiClO_4$ in ethylene carbonate/dimethyl carbonate. *Bottom:* Charge-discharge data of purified and processed SWNTs. The reversible capacity of this material is 1000 mA h/g. (Figures are from *Gao* et al. in [64])

laser ablation method. The exact locations of the Li ions in the intercalated SWNTs are still unknown. Intercalation and *in-situ* TEM and EELS measurements on *individual* SWNT bundles suggested that the intercalants reside in the interstitial sites between the SWNTs [67]. It is shown that the Li/C ratio can be further increased by ball-milling which fractures the SWNTs [68]. A reversible capacity of 1000 mA h/g [64] was reported in processed SWNTs. The large irreversible capacity is related to the large surface area of the SWNT films (~300 m$^2$/g by BET characterization) and the formation of a solid-electrolyte-interface. The SWNTs are also found to perform well under high current rates. For example, 60% of the full capacity can be retained when the charge-discharge rate is increased from 50 mA/h to 500 mA/h [63].

The high capacity and high-rate performance warrant further studies on the potential of utilizing carbon nanotubes as battery electrodes. The large observed voltage hysteresis (Fig. 8) is undesirable for battery application. It is at least partially related to the kinetics of the intercalation reaction and can potentially be reduced/eliminated by processing, i.e., cutting the nanotubes to short segments.

## 2.2 Hydrogen Storage

The area of hydrogen storage in carbon nanotubes remains active and controversial. Extraordinarily high and reversible hydrogen adsorption in SWNT-containing materials [69,70,71,72] and graphite nanofibers (GNFs) [73] has been reported and has attracted considerable interest in both academia and industry. Table 2 summarizes the gravimetric hydrogen storage capacity reported by various groups [74]. However, many of these reports have not been independently verified. There is also a lack of understanding of the basic mechanism(s) of hydrogen storage in these materials.

**Table 2.** Summary of reported gravimetric storage of $H_2$ in various carbon materials (adapted from [74])

| Material | Max. wt% $H_2$ | $T$(K) | $P$ (MPa) |
|---|---|---|---|
| SWNTs(low purity) | 5–10 | 133 | 0.040 |
| SWNTs(high purity) | ~4 | 300 | 0.040 |
| GNFs(tubular) | 11.26 | 298 | 11.35 |
| GNFs(herringbone) | 67.55 | 298 | 11.35 |
| GNS(platelet) | 53.68 | 298 | 11.35 |
| Graphite | 4.52 | 298 | 11.35 |
| GNFs | 0.4 | 298-773 | 0.101 |
| Li-GNFs | 20 | 473-673 | 0.101 |
| Li-Graphites | 14 | 473-674 | 0.101 |
| K-GNFs | 14 | <313 | 0.101 |
| K-Graphite | 5.0 | <313 | 0.101 |
| SWNTs(high purity) | 8.25 | 80 | 7.18 |
| SWNTs(~50% pure) | 4.2 | 300 | 10.1 |

Materials with high hydrogen storage capacities are desirable for energy storage applications. Metal hydrides and cryo-adsorption are the two commonly used means to store hydrogen, typically at high pressure and/or low temperature. In metal hydrides, hydrogen is reversibly stored in the interstitial sites of the host lattice. The electrical energy is produced by direct electrochemical conversion. Hydrogen can also be stored in the gas phase in the metal hydrides. The relatively low gravimetric energy density has limited the application of metal hydride batteries. Because of their cylindrical and hollow geometry, and nanometer-scale diameters, it has been predicted that the carbon nanotubes can store liquid and gas in the inner cores through a capillary effect [76]. A Temperature-Programmed Desorption (TPD) study on SWNT-containing material (0.1–0.2 wt% SWNT) estimates a gravimetric storage density of 5–10 wt% SWNT when $H_2$ exposures were carried out at 300 torr for 10 min at 277 K followed by 3 min at 133 K [69]. If all the hydrogen molecules are assumed to be inside the nanotubes, the reported density would imply a much higher packing density of $H_2$ inside the tubes than expected from the normal $H_2$–$H_2$ distance. The same group recently performed experiments on purified SWNTs and found essentially no $H_2$ absorption at 300 K [75]. Upon cutting (opening) the nanotubes by an oxidation process, the amount of absorbed $H_2$ molecules increased to 4–5 wt%. A separate study on higher purity materials reports ~8 wt% of $H_2$ adsorption at 80 K, but using a much higher pressure of 100 atm [77], suggesting that nanotubes have the highest hydrogen storage capacity of any carbon material. It is believed that hydrogen is first adsorbed on the outer surface of the crystalline ropes.

An even higher hydrogen uptake, up to 14–20 wt%, at 20–400°C under ambient pressure was reported [70] in alkali-metal intercalated carbon nanotubes. It is believed that in the intercalated systems, the alkali metal ions act as a catalytic center for $H_2$ dissociative adsorption. FTIR measurements show strong alkali–H and C–H stretching modes. An electrochemical absorption and desorption of hydrogen experiment performed on SWNT-containing materials (MER Co, containing a few percent of SWNTs) reported a capacity of 110 mA h/g at low discharge currents [72]. The experiment was done in a half-cell configuration in 6 M KOH electrolyte and using a nickel counter electrode. Experiments have also been performed on SWNTs synthesized by a hydrogen arc-discharge method [71]. Measurements performed on relatively large amount materials (~50% purity, 500 mg) showed a hydrogen storage capacity of 4.2 wt% when the samples were exposed to 10 MPa hydrogen at room temperature. About 80% of the absorbed $H_2$ could be released at room temperature [71].

The potential of achieving/exceeding the benchmark of 6.5 wt% $H_2$ to system weight ratio set by the Department of Energy has generated considerable research activities in universities, major automobile companies and national laboratories. At this point it is still not clear whether carbon nanotubes will have real technological applications in the hydrogen storage applications area. The values reported in the literature will need to be verified on well-characterized materials under controlled conditions. What is also lacking is a detailed understanding on the storage mechanism and the effect of materials processing on hydrogen storage. Perhaps the ongoing neutron scattering and proton nuclear magnetic resonance measurements will shed some light in this direction.

In addition to hydrogen, carbon nanotubes readily absorb other gaseous species under ambient conditions which often leads to drastic changes in their electronic properties[78,79,112]. This environmental sensitivity is a double-edged sword. From the technological point of view, it can potentially be utilized for gas detection[112]. On the other hand, it makes very difficult to deduce the intrinsic properties of the nanotubes, as demonstrated by the recent transport[78] and nuclear magnetic resonance[79] measurements. Care must be taken to remove the adsorbed species which typically requires annealing the nanotubes at elevated temperatures under at least $10^{-6}$ torr dynamic vacuum.

## 3 Filled Composites

The mechanical behavior of carbon nanotubes is exciting since nanotubes are seen as the "ultimate" carbon fiber ever made. The traditional carbon fibers [17,18] have about *fifty times* the specific strength (strength/density) of steel and are excellent load-bearing reinforcements in composites. Nanotubes should then be ideal candidates for structural applications. Carbon fibers have been used as reinforcements in high strength, light weight, high performance composites; one can typically find these in a range of products ranging from expensive tennis rackets to spacecraft and aircraft body parts. NASA has recently invested large amounts of money in developing carbon nanotube-based composites for applications such as the futuristic Mars mission.

Early theoretical work and recent experiments on individual nanotubes (mostly MWNTs) have confirmed that nanotubes are one of the stiffest structures ever made [81,82,83,84,85]. Since carbon–carbon covalent bonds are one of the strongest in nature, a structure based on a perfect arrangement of these bonds oriented along the axis of nanotubes would produce an exceedingly strong material. Theoretical studies have suggested that SWNTs could have a Young's modulus as high as 1 TPa [82], which is basically the in-plane value of defect free graphite. For MWNTs, the actual strength in practical situations would be further affected by the sliding of individual graphene cylinders with respect to each other. In fact, very recent experiments have evaluated the tensile strength of individual MWNTs using a nano-stressing stage located within a scanning electron microscope [86]. The nanotubes broke by a sword-in-sheath failure mode [17]. This failure mode corresponds to the sliding of the layers within the concentric MWNT assembly and the breaking of individual cylinders independently. Such failure modes have been observed previously in vapor grown carbon fibers [17]. The observed tensile strength of individual MWNTs corresponded to $<$60 GPa. Experiments on individual SWNT ropes are in progress and although a sword-in-sheath failure mode cannot occur in SWNT ropes, failure could occur in a very similar fashion. The individual tubes in a rope could pull out by shearing along the rope axis, resulting in the final breakup of the rope, at stresses much below the tensile strength of individual nanotubes. Although testing of individual nanotubes is challenging, and requires specially designed stages and nano-size loading devices, some clever experiments have provided valuable insights into the mechanical behavior of nanotubes and have provided values for their modulus and strength. For example, in one of the earlier experiments, nanotubes projecting out onto holes in a TEM specimen grid were assumed to be equivalent to clamped homogeneous cantilevers; the horizontal vibrational amplitudes at the tube ends were measured from the blurring of the images of the nanotube tips and were then related to the Young's modulus [83]. Recent experiments have also used atomic force microscopy to bend nanotubes attached to substrates and thus obtain quantitative information about their mechanical properties [84,87].

Most of the experiments done to date corroborate theoretical predictions suggesting the values of Young's modulus of nanotubes to be around 1 TPa (Fig. 9). Although the theoretical estimate for the tensile strength of individual SWNTs is about 300 GPa, the best experimental values (on MWNTs) are close to ~50 GPa [86], which is still an order of magnitude higher than that of carbon fibers [17,18].

The fracture and deformation behavior of nanotubes is intriguing. Simulations on SWNTs have suggested very interesting deformation behavior; highly deformed nanotubes were seen to switch reversibly into different morphological patterns with abrupt releases of energy. Nanotubes gets flattened, twisted and buckled as they deform (Fig. 9 ). They sustain large strains (40%) in tension without showing signs of fracture [82]. The reversibility of deformations, such as buckling, has been recorded directly for MWNT, under TEM observations [7]. Flexibility of MWNTs depends on the number of layers that make up the nanotube walls; tubes with thinner walls tend to twist and flatten more easily. This flexibility is related to the in-plane flexibility of a planar graphene sheet and the ability for the carbon atoms to rehybridize, with the degree of $sp^2$–$sp^3$ rehybridization depending on the strain. Such flexibility of nanotubes under mechanical loading is important for their potential application as nanoprobes, for example, for use as tips of scanning probe microscopes.

Recently, an interesting mode of plastic behavior has been predicted in nanotubes [88]. It is suggested that pairs of 5–7 (pentagon-heptagon) pair defects, called a Stone–Wales defect [89], in $sp^2$ carbon systems, are created at high strains in the nanotube lattice and that these defect pairs become mobile. This leads to a step-wise diameter reduction (localized necking) of the nanotube. These defect pairs become mobile. The separation of the defects creates local necking of the nanotube in the region where the defects have moved. In addition to localized necking, the region also changes lattice orientation (similar in effect to a dislocation passing through a crystal). This extraordinary behavior initiates necking but also introduces changes in helicity in the region where the defects have moved (similar to a change in lattice orientation when a dislocation passes through a crystal). This extraor-

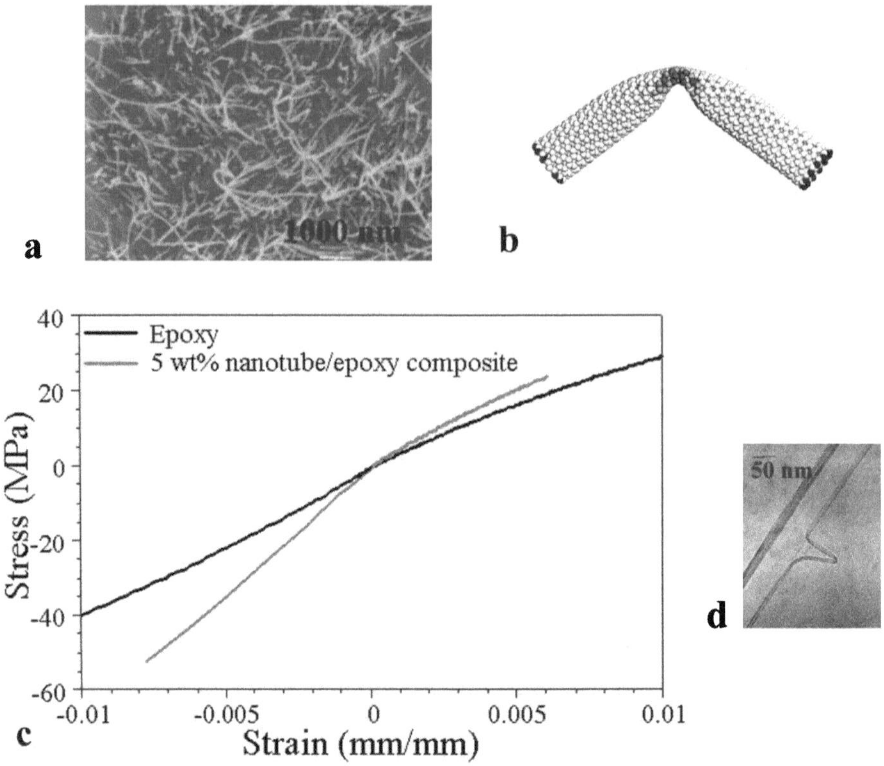


**Fig. 9.** Results of the mechanical properties from MWNT-polymer (epoxy) composites. (**a**) SEM micrograph that shows good dispersion of MWNTs in the polymer matrix. The tubes are, however, elastically bent due to their highly flexible nature. Schematic of an elastically bent nanotube is shown in (**b**) (courtesy Prof. Boris Yakobson). The strain is concentrated locally near the bend. (**c**) Stress-strain relationship observed during the tension/compression testing of the nanotube-epoxy (5 wt% MWNTs) composite (the curve that shows larger slope, both on the tension and compression sides of the stress-strain curve, belongs to the nanotube epoxy composite). It can be seen that the load transfer to the nanotube is higher during the compression cycle (seen from the deviation of the composite curve from that of the pure epoxy), because in tension the individual layers of the nanotubes slide with respect to each other. (**d**) TEM image of a thicker straight MWNT as well as a buckled MWNT in an epoxy matrix after loading. The smaller diameter nanotubes have more tendency to bend and buckle [86]

dinary behavior could lead to a unique nanotube application: a new type of probe, which responds to mechanical stress by changing its electronic character. High temperature fracture of individual nanotubes under tensile loading, has been studied by molecular dynamics simulations [90]. Elastic stretching elongates the hexagons until, at high strain, some bonds are broken. This local defect is then redistributed over the entire surface, by bond saturation and surface reconstruction. The final result of this is that instead of fracturing, the nanotube lattice unravels into a linear chain of carbon atoms. Such behavior is extremely unusual in crystals and could play a role in increasing the toughness (by increasing the energy absorbed during deformation) of nanotube-filled ceramic composites during high temperature loading.

The most important application of nanotubes based on their mechanical properties will be as reinforcements in composite materials. Although nanotube-filled polymer composites are an obvious materials application area, there have not been many successful experiments, which show the advantage of using nanotubes as fillers over traditional carbon fibers. The main problem is in creating a good interface between nanotubes and the polymer matrix and attaining good load transfer from the matrix to the nanotubes, during loading. The reason for this is essentially two-fold. First, nanotubes are atomically smooth and have nearly the same diameters and aspect ratios (length/diameter) as polymer chains. Second, nanotubes are almost always organized into aggregates which behave differently in response to a load, as compared to individual nanotubes. There have been conflicting reports on the interface strength in nanotube-polymer composites [91,92,93,94,95,96]. Depending on the polymer used and processing conditions, the measured strength seems to vary. In some cases, fragmentation of the tubes has been observed, which is an indication of a strong interface bonding. In some cases, the effect of sliding of layers of MWNTs and easy pull-out are seen, suggesting poor interface bonding. Micro-Raman spectroscopy has validated the latter, suggesting that sliding of individual layers in MWNTs and *shearing* of individual tubes in SWNT ropes could be limiting factors for good load transfer, which is essential for making high strength composites. To maximize the advantage of nanotubes as reinforcing structures in high strength composites, the aggregates needs to be broken up and dispersed or cross-linked to prevent slippage [97]. In addition, the surfaces of nanotubes have to be chemically modified (functionalized) to achieve strong interfaces between the surrounding polymer chains (Fig. 10).

There are certain advantages that have been realized in using carbon nanotubes for structural polymer (e.g., epoxy) composites. Nanotube reinforcements will increase the toughness of the composites by absorbing energy during their highly flexible elastic behavior. This will be especially important for nanotube-based ceramic matrix composites. An increase in fracture toughness on the order of 25% has been seen in nano-crystalline alumina nanotube (5% weight fraction) composites, without compromising on hardness [98]. Other interesting applications of nanotube-filled polymer films will be in adhesives where a decoration of nanotubes on the surface of the polymer films could alter the characteristics of the polymer chains due to interactions between the nanotubes and the polymer chains; the high surface area of the nanotube structures and their dimensions being nearly that of the linear di-

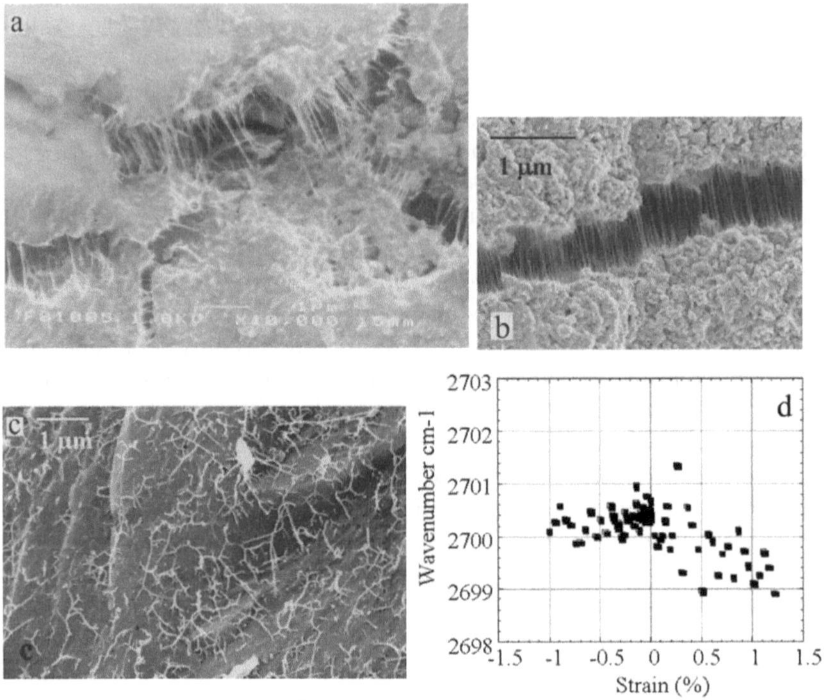


**Fig. 10.** Results of mechanical properties measurements on SWNT-polymer composites. (**a**) SEM micrograph that shows a partially fractured surface of a SWNT-epoxy composite, indicating stretched nanotubes extending across cracks. (**b**) Shows a similar event illustrating the stretching and aligning of SWNT bundles across a long crack in a SWNT-carbon soot composite. (**c**) SEM micrograph that shows the surface of a fractured SWNT-epoxy composite where the nanotube ropes have been completely pulled out and have fallen back on the fractured surface, forming a loose random network of interconnected ropes. (**d**) Shows results of micro-Raman spectroscopy that detects peak-shifts (in wave-numbers) as a function of strain. In both tension and compression of the SWNT-epoxy specimens, the peak shifts are negligible, suggesting no load transfer to the nanotubes during the loading of the composites [97]

mensions of the polymer chains could give such nanocomposites new surface properties. The low density of the nanotubes will clearly be an advantage for nanotube-based polymer composites, in comparison to short carbon fiber reinforced (random) composites. Nanotubes would also offer multifunctionality, such as increased electrical conduction. Nanotubes will also offer better performance during compressive loading in comparison to traditional carbon fibers due to their flexibility and low propensity for carbon nanotubes to fracture under compressive loads.

Other than for structural composite applications, some of the unique properties of carbon nanotubes are being pursued by filling photo-active polymers with nanotubes. Recently, such a scheme has been demonstrated in a conjugated luminescent polymer, poly(m-phenylenevinylene-co-2,5-dioctoxy-p-phenylenevinylene) (PPV), filled with MWNTs and SWNTs [99]. Nanotube/PPV composites have shown large increases in electrical conductivity (by nearly eight orders of magnitude) compared to the pristine polymer, with little loss in photoluminescence/electro-luminescence yield. In addition, the composite is far more robust than the pure polymer regarding mechanical strength and photo-bleaching properties (breakdown of the polymer structure due to thermal effects). Preliminary studies indicate that the host polymer interacts weakly with the embedded nanotubes, but that the nanotubes act as nano-metric heat sinks, which prevent the build up of large local heating effects within the polymer matrix. While experimenting with the composites of conjugated polymers, such as PPV and nanotubes, a very interesting phenomenon has been recently observed [80]; it seems that the coiled morphology of the polymer chains helps to wrap around nanotubes suspended in dilute solutions of the polymer. This effect has been used to separate nanotubes from other carbonaceous material present in impure samples. Use of the non-linear optical and optical limiting properties of nanotubes has been reported for designing nanotube-polymer systems for optical applications, including photo-voltaic applications [100]. Functionalization of nanotubes and the doping of chemically modified nanotubes in low concentrations into photo-active polymers, such as PPV, have been shown to provide a means to alter the hole transport mechanism and hence the optical properties of the polymer. Small loadings of nanotubes are used in these polymer systems to tune the color of emission when used in organic light emitting devices [101]. The interesting optical properties of nanotube-based composite systems arise from the low dimensionality and unique electronic band structure of nanotubes; such applications cannot be realized using larger micron-size carbon fibers (Fig. 11).

There are other less-explored areas where nanotube-polymer composites could be useful. For example, nanotube filled polymers could be useful in ElectroMagnetic Induction (EMI) shielding applications where carbon fibers have been used extensively [17]. Membranes for molecular separations (especially biomolecules) could be built from nanotube-polycarbonate systems, making use of the remarkable small pores sizes that exist in nanotubes. Very recently, work done at RPI suggests that composites made from nanotubes (MWNTs) and a biodegradable polymer (polylactic acid; PLA) act more efficiently than carbon fibers for osteointegration (growth of bone cells), especially under electrical stimulation of the composite.

There are challenges to be overcome when processing nanotube composites. One of the biggest problems is dispersion. It is extremely difficult to separate individual nanotubes during mixing with polymers or ceramic ma-

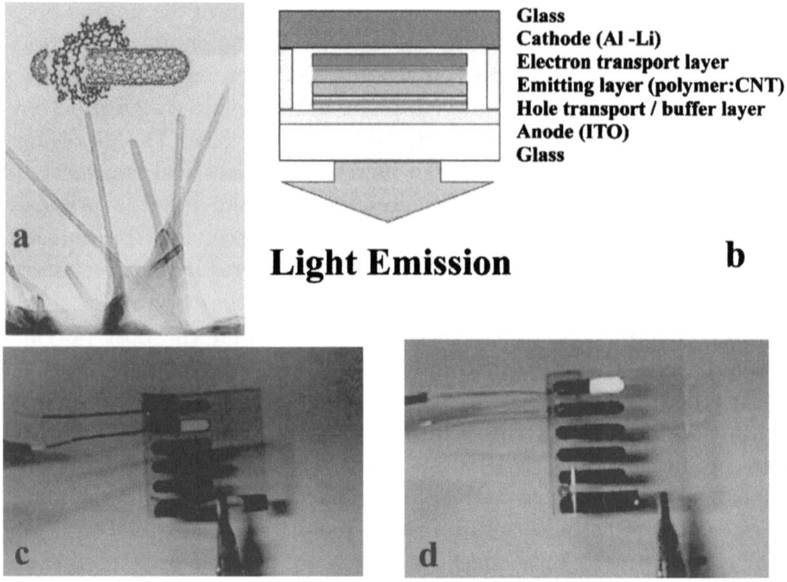


**Fig. 11.** Results from the optical response of nanotube-doped polymers and their use in Organic Light Emitting Diodes (OLED) . The construction of the OLED is shown in the schematic of (*top*). The *bottom* figure shows emission from OLED structures. Nanotube doping tunes the emission color. With SWNTs in the buffer layer, holes are blocked and recombination takes place in the transport layer and the emission color is red [101]. Without nanotubes present in the buffer layer, the emission color is green (not shown in the figure) (figures are courtesy of Prof. David Carroll)

terials and this creates poor dispersion and clumping together of nanotubes, resulting in a drastic decrease in the strength of composites. By using high power ultrasound mixers and using surfactants with nanotubes during processing, good nanotube dispersion may be achieved, although the strengths of nanotube composites reported to date have not seen any drastic improvements over high modulus carbon fiber composites. Another problem is the difficulty in fabricating high weight fraction nanotube composites, considering the high surface area for nanotubes which results in a very high viscosity for nanotube-polymer mixtures. Notwithstanding all these drawbacks, it needs to be said that the presence of nanotubes stiffens the matrix (the role is especially crucial at higher temperatures) and could be very useful as a matrix modifier [102], particularly for fabricating improved matrices useful for carbon fiber composites. The real role of nanotubes as an efficient reinforcing fiber will have to wait until we know how to manipulate the nanotube surfaces chemically to make strong interfaces between individual nanotubes (which are really the strongest material ever made) and the matrix materials. In the meanwhile, novel and unconventional uses of nanotubes will have to take the center stage.

# 4 Nanoprobes and Sensors

The small and uniform dimensions of the nanotubes produce some interesting applications. With extremely small sizes, high conductivity, high mechanical strength and flexibility (ability to easily bend elastically), nanotubes may ultimately become indispensable in their use as nanoprobes. One could think of such probes as being used in a variety of applications, such as high resolution imaging, nano-lithography, nanoelectrodes, drug delivery, sensors and field emitters. The possibility of nanotube-based field emitting devices has been already discussed (see Sect. 1). Use of a single MWNT attached to the end of a scanning probe microscope tip for imaging has already been demonstrated (Fig. 12) [104]. Since MWNT tips are conducting, they can be used in STM, AFM instruments as well as other scanning probe instruments, such as an electrostatic force microscope. The advantage of the nanotube tip is its slenderness and the possibility to image features (such as very small, deep surface cracks), which are almost impossible to probe using the larger, blunter etched Si or metal tips. Biological molecules, such as DNA can also be imaged with higher resolution using nanotube tips, compared to conventional STM tips. MWNT and SWNT tips were used in a tapping mode to image biological molecules such as amyloid-b-protofibrils (related to Alzheimer's disease), with resolution never achieved before [105]. In addition, due to the high elasticity of the nanotubes, the tips do not suffer from crashes on contact with the substrates. Any impact will cause buckling of the nanotube, which generally is reversible on retraction of the tip from the substrate. Attaching

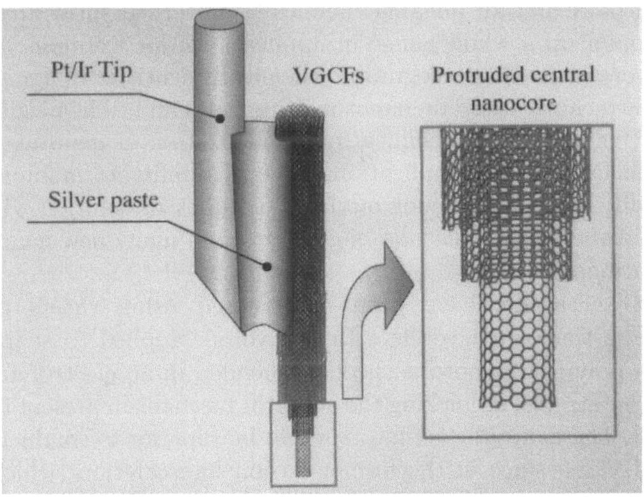


**Fig. 12.** Use of a MWNT as an AFM tip (after *Endo* [103]). At the center of the Vapor Grown Carbon Fiber (VGCF) is a MWNT which forms the tip [18]. The VGCF provides a convenient and robust technique for mounting the MWNT probe for use in a scanning probe instrument

individual nanotubes to the conventional tips of scanning probe microscopes has been the real challenge. Bundles of nanotubes are typically pasted on to AFM tips and the ends are cleaved to expose individual nanotubes (Fig. 12 and also [27]). These tip attachments are not very controllable and will result in vibration problems and in instabilities during imaging, which decrease the image resolution. However, successful attempts have been made to grow individual nanotubes onto Si tips using CVD [106], in which case the nanotubes are firmly anchored to the probe tips. Due to the longitudinal (high aspect) design of nanotubes, nanotube vibration still will remain an issue, unless short segments of nanotubes can be controllably grown (Fig. 12).

In addition to the use of nanotube tips for high resolution imaging, it is also possible to use nanotubes as active tools for surface manipulation. It has been shown that if a pair of nanotubes can be positioned appropriately on an AFM tip, they can be controlled like tweezers to pick up and release nanoscale structures on surfaces; the dual nanotube tip acts as a perfect nano-manipulator in this case [107]. It is also possible to use nanotube tips in AFM nano-lithography. Ten nanometer lines have been written on oxidized silicon substrates using nanotube tips at relatively high speeds [108], a feat that can only be achieved with tips as small as nanotubes.

Since nanotube tips can be selectively modified chemically through the attachment of functional groups [109], nanotubes can also be used as molecular probes, with potential applications in chemistry and biology. Open nanotubes with the attachment of acidic functionalities have been used for chemical and biological discrimination on surfaces [110]. Functionalized nanotubes were used as AFM tips to perform local chemistry, to measure binding forces between protein-ligand pairs and for imaging chemically patterned substrates. These experiments open up a whole range of applications, for example, as probes for drug delivery, molecular recognition, chemically sensitive imaging, and local chemical patterning, based on nanotube tips that can be chemically modified in a variety of ways. The chemical functionalization of nanotubes is a major issue with far-reaching implications. The possibility to manipulate, chemically modify and perhaps polymerize nanotubes in solution will set the stage for nanotube-based molecular engineering and many new nanotechnological applications.

Electromechanical actuators have been constructed using sheets of SWNTs. It was shown that small voltages (a few volts), applied to strips of laminated (with a polymer) nanotube sheets suspended in an electrolyte, bends the sheet to large strains, mimicking the actuator mechanism present in natural muscles [111]. The nanotube actuators would be superior to conducting polymer-based devices, since in the former no ion intercalation (which limits actuator life) is required. This interesting behavior of nanotube sheets in response to an applied voltage suggests several applications, including nanotube-based micro-cantilevers for medical catheter applications and as novel substitutes, especially at higher temperatures, for ferroelectrics.

Recent research has also shown that nanotubes can be used as advanced miniaturized chemical sensors [112]. The electrical resistivities of SWNTs were found to change sensitively on exposure to gaseous ambients containing molecules of $NO_2$, $NH_3$ and $O_2$. By monitoring the change in the conductance of nanotubes, the presence of gases could be precisely monitored. It was seen that the response times of nanotube sensors are at least an order of magnitude faster (a few seconds for a resistance change of one order of magnitude) than those based on presently available solid-state (metal-oxide and polymers) sensors. In addition, the small dimensions and high surface area offer special advantages for nanotube sensors, which could be operated at room temperature or at higher temperatures for sensing applications.

## 5 Templates

Since nanotubes have relatively straight and narrow channels in their cores, it was speculated from the beginning that it might be possible to fill these cavities with foreign materials to fabricate one-dimensional nanowires. Early calculations suggested that strong capillary forces exist in nanotubes, strong enough to hold gases and fluids inside them [113]. The first experimental proof was demonstrated in 1993, by the filling and solidification of molten lead inside the channels of MWNTs [14]. Wires as small as 1.2 nm in diameter were fabricated by this method inside nanotubes. A large body of work now exists in the literature [14,15,16], to cite a few examples, concerning the filling of nanotubes with metallic and ceramic materials. Thus, nanotubes have been used as templates to create nanowires of various compositions and structures (Fig. 13).

The critical issue in the filling of nanotubes is the wetting characteristics of nanotubes, which seem to be quite different from that of planar graphite, because of the curvature of the tubes. Wetting of low melting alloys and solvents occurs quite readily in the internal high curvature pores of MWNTs and SWNTs. In the latter, since the pore sizes are very small, filling is more difficult and can be done only for a selected few compounds. It is intriguing that one could create one-dimensional nanostructures by utilizing the internal one-dimensional cavities of nanotubes. Liquids such as organic solvents wet nanotubes easily and it has been proposed that interesting chemical reactions could be performed inside nanotube cavities [16]. A whole range of experiments remains to be performed inside these constrained one-dimensional spaces, which are accessible once the nanotubes can be opened.

The topology of closed nanotubes provides a fascinating avenue to open them through the simple chemical method of oxidation [114]. As in fullerenes, the pentagonal defects that are concentrated at the tips are more reactive than the hexagonal lattice of the cylindrical parts of the nanotubes. Hence, during oxidation, the caps are removed prior to any damage occurring to the tube body, thus easily creating open nanotubes. The opening of nanotubes

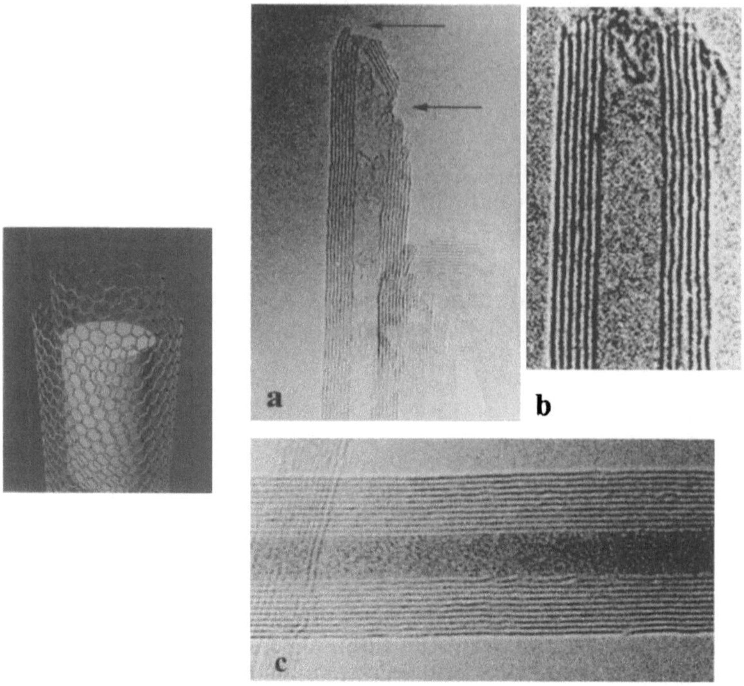


**Fig. 13.** Results that show the use of nanotubes as templates. The *left-hand* figure is a schematic that shows the filling of the empty one-dimensional hollow core of nanotubes with foreign substances. (**a**) Shows a high-resolution TEM image of a tube tip that has been attacked by oxidation; the preferential attack begins at locations where pentagonal defects were originally present (*arrows*) and serves to open the tube. (**b**) TEM image that shows a MWNT that has been completely opened by oxidation. (**c**) TEM image of a MWNT with its cavity filled uniformly with lead oxide. The filling was achieved by capillarity [15]

by oxidation can be achieved by heating nanotubes in air (above 600°C) or in oxidizing solutions (e.g., acids). It is noted here that nanotubes are more stable to oxidation than graphite, as observed in Thermal Gravimetric Analysis (TGA) experiments, because the edge planes of graphite where reaction can initiate are conspicuous by their absence in nanotubes.

After the first set of experiments, reporting the opening and filling of nanotubes in air, simple chemical methods, based on the opening and filling nanotubes in solution, were discovered to develop generalized solution-based strategies to fill nanotubes with a range of materials [15]. In these methods an acid is first used to open the nanotube tip and to act as a low surface tension carrier for solutes (metal-containing salts) to fill the nanotube hollows. Calcination of solvent-treated nanotubes leaves deposits of oxide material (e.g., NiO) inside nanotube cavities. The oxides can then be reduced to metals by annealing in reducing atmospheres. Observation of solidification inside the one-dimensional channels of nanotubes provides a fascinating study of phase stabilization under geometrical constraints. It is experimentally found that when the channel size gets smaller than a certain critical diameter, solidification results in new and oftentimes disordered phases (e.g., $V_2O_5$) [115]. Crystalline bulk phases are formed in larger cavities. Numerous modeling studies are under way to understand the solidification behavior of materials inside nanotubes and the physical properties of these unique, filled nano-composite materials.

Filled nanotubes can also be synthesized *in situ*, during the growth of nanotubes in an electric arc or by laser ablation. During the electric arc formation of carbon species, encapsulated nanotubular structures are created in abundance. This technique generally produces encapsulated nanotubes with carbide nanowires (e.g., transition metal carbides) inside [116,117]. Laser ablation also produces heterostructures containing carbon and metallic species. Multi-element nanotube structures consisting of multiple phases (e.g., coaxial nanotube structures containing SiC, SiO, BN and C) have been successfully synthesized by reactive laser ablation [118]. Similarly, post-fabrication treatments can also be used to create heterojunctions between nanotubes and semiconducting carbides [119,120]. It is hoped that these hybrid nanotube-based structures, which are combinations of metallic, semiconducting and insulating nanostructures, will be useful in future nanoscale electronic device applications.

Nanocomposite structures based on carbon nanotubes can also be built by coating nanotubes uniformly with organic or inorganic structures. These unique composites are expected to have interesting mechanical and electrical properties due to a combination of dimensional effects and interface properties. Finely-coated nanotubes with monolayers of layered oxides have been made and characterized (e.g., vanadium pentoxide films) [115]. The interface formed between nanotubes and the layered oxide is atomically flat due to the absence of covalent bonds across the interface. It has been demonstrated that after the coating is made, the nanotubes can be removed by oxidation leaving behind freely-standing nanotubes made of oxides, with nanoscale wall thickness. These novel ceramic tubules, made using nanotubes as templates, could have interesting applications in catalysis. Recently, researchers have also found that nanotubes can be used as templates for the self-assembly of protein molecules [121]. Dipping MWNTs in a solution containing proteins, results in monolayers of proteins covering nanotubes; what is interesting is that the organization of the protein molecules on nanotubes corresponds directly to the helicity of the nanotubes. It seems that nanotubes with controlled helicities could be used as unique probes for molecular recognition, based on the helicity and dimensions, which are recognized by organic molecules of comparable length scales.

There are other ways in which pristine nanotubes can be modified into composite structures. Chemical functionalization can be used to build macromolecular structures from fullerenes and nanotubes. The attachment of organic functional groups on the surface of nanotubes has been achieved, and with the recent success in breaking up SWNTs into shorter fragments, the possibility of functionalizing and building structures through chemistry has become a reality. Decoration of nanotubes with metal particles has been achieved for different purposes, most importantly for use in heterogeneous catalysis [55]. SWNT bundles have been doped with alkali metals, and with the halogens $Br_2$ and $I_2$, resulting in an order of magnitude increase in electrical conductivity [122]. In some cases, it is observed that the dopants form a linear chain and sit in the one-dimensional interstitial channels of the bundles. Similarly, Li intercalation inside nanotubes has been successfully carried out with possible impact on battery applications, which has already been discussed in a previous section. The intercalation and doping studies suggest that nanotube systems provide an effective host lattice for the creation of a range of carbon-based synthetic metallic structures.

The conversion of nanotubes through vapor chemistry can create unique nanocomposites with nanotubes as a backbone. When volatile gases such as halogenated compounds or $SiO_x$ are reacted with nanotubes, the tubes get converted into carbide nano-rods of similar dimensions [123]. These reactions can be controlled, such that the outer nanotube layers can be converted to carbides, keeping the inner graphite layer structure intact. The carbide rods so produced (e.g., SiC, NbC) should have a wide range of interesting electrical and mechanical properties, which could be exploited for applications as reinforcements and nanoscale electrical devices [124].

## 6 Challenges and Potential for Carbon Nanotube Applications

Carbon nanotubes have come a long way since their discovery in 1991. The structures that were first reported in 1991 were MWNTs with a range of diameters and lengths. These were essentially the distant relatives of the highly defective carbon nanofibers grown via catalytic chemical vapor deposition. The latter types of fibers (e.g., the lower quality carbon nanofibers made commercially by the Hyperion Corporation and more perfect nanotube structures revealed by *Endo* in his 1975 Ph.D. thesis [125]) had existed for more than a decade. The real molecular nanotubes arrived when they were found accidentally while a catalyst (Fe, Co) material was inserted in the anode during electric-arc discharge synthesis. For the first time, there was hope that molecular fibers based purely on carbon could be synthesized and the excitement was tremendous, since many physical properties of such a fiber had already been predicted by theory. It was really the theoretical work proposed on SWNTs and the availability of nanoscale technology (in characterization and measurements) that made the field take off in 1991.

The greatness of a single-walled nanotube is that it is a macro-molecule and a crystal at the same time. The dimensions correspond to extensions of fullerene molecules and the structure can be reduced to a unit-cell picture, as in the case of perfect crystals. A new predictable (in terms of atomic structure–property relations) carbon fiber was born. The last decade of research has shown that indeed the physical properties of nanotubes are remarkable, as elaborated in the various chapters of this book. A carbon nanotube is an extremely versatile material: it is one of the strongest materials, yet highly elastic, highly conducting, small in size, but stable, and quite robust in most chemically harsh environments. It is hard to think of another material that can compete with nanotubes in versatility.

As a novel material, fullerenes failed to make much of an impact in applications. It seems, from the progress made in recent research, that the story of nanotubes is going to be very different. There are already real products based on nanotubes on the market, for example, the nanotube attached AFM tips used in metrology. The United States, Europe and Japan have all invested heavily in developing nanotube applications. Nanotube-based electronics tops this list and it is comforting that the concepts of devices (such as room-temperature field-effect transistors based on individual nanotubes) have already been successfully demonstrated. As in the case of most products, especially in high technology areas, such as nano-electronics, the time lag between concept demonstration and real products could be several years to decades and one will have to wait and see how long it is going to take nanotube electronics to pervade high technology. Other more obvious and direct applications are some of the bulk uses, such as nanotube-based polymer composites and electrochemical devices. These, although very viable applications, face challenges, as detailed in this review. What is also interesting is that new and novel applications are emerging, as for example, nanotubes affecting the transport of carriers and hence luminescence in polymer-based organic light-emitting diodes, and nanotubes used as actuators in artificial muscles. It can very well be said that some of these newly found uses will have a positive impact on the early stages of nanotube product development.

There are also general challenges that face the development of nanotubes into functional devices and structures. First of all, the growth mechanism of nanotubes, similar to that of fullerenes, has remained a mystery [126]. With this handicap, it is not really possible yet to grow these structures in a controlled way. There have been some successes in growing nanotubes of certain diameter (and to a lesser extent, of predetermined helicity) by tuning the growth conditions by trial and error. Especially for electronic applications, which rely on the electronic structure of nanotubes, this inability to select the size and helicity of nanotubes during growth remains a drawback. More so, many predictions of device applicability are based on joining nanotubes via the incorporation of topological defects in their lattices. There is no controllable way, as of yet, of making connections between nanotubes. Some recent reports, however, suggest the possibility of constructing these interconnected structures by electron irradiation and by template mediated growth and manipulation.

For bulk applications, such as fillers in composites, where the atomic structure (helicity) has a much smaller impact on the resulting properties, the quantities of nanotubes that can be manufactured still falls far short of what industry would need. There are no available techniques that can produce nanotubes of reasonable purity and quality in kilogram quantities. The industry would need tonnage quantities of nanotubes for such applications. The market price of nanotubes is also too high presently (~$200 per gram) for any realistic commercial application. But it should be noted that the starting prices for carbon fibers and fullerenes were also prohibitively high during their initial stages of development, but have come down significantly in time. In the last 2–3 years, there have been several companies that were set up in the US to produce and market nanotubes. It is hoped that in the next few years nanotubes will be available to consumers for less than US $100/pound.

Another challenge is in the manipulation of nanotubes. Nano-technology is in its infancy and the revolution that is unfolding in this field relies strongly on the ability to manipulate structures at the atomic scale. This will remain a major challenge in this field, among several others.

## 7 Conclusions

This review has described several possible applications of carbon nanotubes, with emphasis on materials science-based applications. Hints are made to the electronic applications of nanotubes which are discussed elsewhere [9]. The overwhelming message we would like to convey through this chapter is that the unique structure, topology and dimensions of carbon nanotubes have created a superb all-carbon material, which can be considered as the most perfect fiber that has ever been fabricated. The remarkable physical properties of nanotubes create a host of application possibilities, some derived as an extension of traditional carbon fiber applications, but many are new possibilities, based on the novel electronic and mechanical behavior of nanotubes. It needs to be said that the excitement in this field arises due to the versatility of this material and the possibility to predict properties based on its well-defined perfect crystal lattice. Nanotubes truly bridge the gap between the molecular realm and the macro-world, and are destined to be a star in future technology.

## References


1. H. W. Kroto, J. R. Heath, S. C. O'Brien, S. C. Curl, R. E. Smalley, Nature **318**, 162 (1985)
2. S. Iijima, Nature **354**, 56 (1991)
3. M. S. Dresselhaus, G. Dresselhaus, P. C. Eklund, *Science of Fullerenes and Carbon Nanotubes* (Academic, New York 1996)
4. T. W. Ebbesen, *Carbon Nanotubes: Preparation and Properties* (CRC, Boca Raton 1997)
5. R. Saito, G. Dresselhaus, M. S. Dresselhaus, *Physical Properties of Carbon Nanotubes* (Imperial College Press, London 1998)
6. B. I. Yakobson, R. E. Smalley, American Scientist **85**, 324 (1997)
7. P. M. Ajayan, Chem. Rev. **99**, 1787 (1999)
8. C. Dekker, Phys. Today, **22** (May 1999)
9. S. G. Louie, see chapter in this volume
10. P. M. Ajayan, T. Ichihashi, S. Iijima, Chem. Phys. Lett. **202**, 384 (1993)
11. D. L. Carroll, P. Redlich, P. M. Ajayan, J. C. Charlier, X. Blase, A. De Vita, R. Car, Phys. Rev. Lett. **78**, 2811 (1997)
12. T. W. Ebbessen, P. M. Ajayan, H. Hiura, K. Tanigaki, Nature **367**, 519 (1994)
13. J. Liu, A. Rinzler, H. Dai, J. Hafner, R. Bradley, P. Boul, A. Lu, T. Iverson, K. Shelimov, C. Huffman, F. Rodriguez-Macias, Y. Shon, R. Lee, D. Colbert, R. E. Smalley, Science **280**, 1253 (1998)
14. P. M. Ajayan, S. Iijima, Nature **361**, 333 (1993)
15. S. C. Tsang, Y. K. Chen, P. J. F. Harris, M. L. H. Green, Nature **372**, 159 (1994)
16. E. Dujardin, T. W. Ebbesen, T. Hiura, K. Tanigaki, Science **265**, 1850 (1994)
17. M. S. Dresselhaus, G. Dresselhaus, K. Sugihara, I. L. Spain, H. A. Goldberg, *Graphite Fibers and Filaments* (Springer, Berlin, Heidelberg 1988)
18. M. S. Dresselhaus, M. Endo, see chapter in this volume
19. T. W. Ebbesen, P. M. Ajayan, Nature **358**, 220 (1992)
20. C. Journet, W. K. Maser, P. Bernier, A. Loiseau, M. Lamy de la Chapelle, S. Lefrant, P. Deniard, R. Lee, J. E. Fischer, Nature **388**, 756 (1997)
21. A. Thess, R. Lee, P. Nikdaev, H. Dai, P. Petit, J. Robert, C. Xu, Y. H. Lee, S. G. Kim, A. G. Rinzler, D. T. Colbert, G.E . Scuseria, D. Tomanek, J. E. Fischer, R. E. Smalley, Science **273**, 483 (1996)
22. W. Z. Li, S. S. Xie, L. X. Qian, B. H. Chang, B. S. Zou, W. Y. Zhou, R. A. Zhao, G. Wang, Science **274**, 1701 (1996)
23. M. Terrones, N. Grobert, J. Olivares, J. P. Zhang, H. Terrones, K. Kordatos, W. K. Hsu, J. P. Hare, P. D. Townshend, K. Prassides, A. K. Cheetham, H. W. Kroto, Nature **388**, 52 (1997)
24. Z. F. Ren, Z. P. Huang, J. W. Xu, J. H. Wang, P. Bush, M. P. Siegal, P. N. Provencio, Science **282**, 1105 (1998)
25. J. Kong, H. T. Soh, A. M. Cassell, C. F. Quate, H. Dai, Nature **395**, 878 (1998)
26. R. Gomer, *Field Emission and Field Ionization* (Harvard Univ. Press, Cambridge, MA 1961)
27. L. Forró, C. Schönenberger, see chapter in this volume
28. I. Brodie, C. Spindt, Adv. Electron. Electron Phys. **83**, 1 (1992)

29. J. A. Castellano, *Handbook of Display Technology* (Academic Press, San Diego 1992)
30. A. W. Scott, *Understanding Microwaves* (Wiley, New York 1993)
31. W. Zhu, G. Kochanski, S. Jin, Science **282**, 1471 (1998)
32. C. Bower, O. Zhou, W. Zhu, A. G. Ramirez, G. P. Kochanski, S. Jin, in *Amorphous and Nanostructured Carbon*, J. P. Sullivan, J. R. Robertson, B. F. Coll, T. B. Allen, O. Zhou (Eds.) (Mater. Res. Soc.) (in press)
33. A. G. Rinzler, J. H. Hafner, P. Nikolaev, L. Lou, S. G. Kim, D. Tomanek, D. Colbert, R. E. Smalley, Science **269**, 1550 (1995)
34. Y. Saito, K. Hamaguchi, T. Nishino, K. Hata, K. Tohji, A. Kasuya, Y. Nishina, Jpn. J. Appl. Phys. **36**, L1340-1342, (1997)
35. P. Collins, A. Zettl, Appl. Phys. Lett. **69**, 1969 (1996)
36. Q. H. Wang, T. D. Corrigan, J. Y. Dai, R. P. H. Chang, A. R. Krauss, Appl. Phys. Lett. **70**, 3308 (1997)
37. W. de Heer, A. Châtelain, D. Ugarte, Science **270**, 1179 (1995)
38. O. Kuttel, O. Groening, C. Emmenegger, L. Schlapbach, Appl. Phys. Lett. **73**, 2113 (1998)
39. J. M. Bonnard, J. P. Salvetat, T. Stockli, W. A. de Herr, L. Forro, A. Chatelain, Appl. Phys. Lett. **73**, 918 (1998)
40. W. Zhu, C. Bower, O. Zhou, G. P. Kochanski, S. Jin, Appl. Phys. Lett. **75**, 873 (1999)
41. K. A. Dean, B. R. Chalamala, J. Appl. Phys. **85**, 3832 (1999)
42. K. A. Dean, B. R. Chalamala, Appl. Phys. Lett. **76**, 375 (2000)
43. J. Robertson, J. Vac. Sci. Technol. B **17** (1999)
44. S. Suzuki, C. Bower, Y. Watanabe, O. Zhou, Appl. Phys. Lett. (in press)
45. Y. Saito, S. Uemura, K. Hamaguchi, Jpn. J. Appl. Phys. **37**, L346 (1998)
46. Q. H. Wang, A. A. Setlur, J. M. Lauerhaas, J. Y. Dai, E. W. Seelig, R. H. Chang, Appl. Phys. Lett. **72**, 2912 (1998)
47. W. B. Choi, D. S. Chung, J. H. Kang, H. Y. Kim, Y. W. Jin, I. T. Han, Y. H. Lee, J. E. Jung, N. S. Lee, G. S. Park, J. M. Kim, Appl. Phys. Lett. **75**, 20 (1999)
48. R. Standler, *Protection of Electronic Circuits from Over-voltages* (Wiley, New York 1989)
49. R. Rosen, W. Simendinger, C. Debbault, H. Shimoda, L. Fleming, B. Stoner, O. Zhou, Appl. Phys. Lett. **76**, 1197 (2000)
50. R. L. McCreery, Electroanal. Chem., **17**, (ed. A. J. Bard) (Marcel Dekker, New York 1991)
51. J. Nugent, K. S. V. Santhanam, A. Rubio, P. M. Ajayan, J. Phys. Chem. submitted
52. P. J. Britto, K. S. V. Santhanam, P. M. Ajayan, Bioelectrochem. Bioenergetics **41**, 121 (1996)
53. P. J. Britto, K. S. V. Santhanam, A. Rubio, A. Alonso, P. M. Ajayan, Adv. Mater. **11**, 154 (1999)
54. G. Che, B. B. Lakshmi, E. R. Fisher, C. R. Martin, Nature **393**, 346 (1998)
55. J. M. Planeix, N. Coustel, B. Coq, V. Brotons, P. S. Kumbhar, R. Dutartre, P. Geneste, P. Bernier, P. M. Ajayan, J. Am. Chem. Soc. **116**, 7935 (1994)
56. C. Niu, E. K. Sichel, R. Hoch, D. Moy, D. H. Tennet, Appl. Phys. Lett. **7**, 1480 (1997)
57. M. Whittingham (Ed.), *Recent Advances in Rechargeable Li Batteries*, Solid State Ionics **69** (3,4) (1994)

58. M. Winter, J. Besenhard, K. Spahr, P. Novak, Adv. Mater. **10**, 725 (1998)
59. J. R. Dahn, T. Zhang, Y. Liu, J. S. Xue, Science **270**, 590(1995)
60. V. Avdeev, V. Nalimova, K. Semenenko, High Pressure Res. **6**, 11 (1990)
61. E. Frackowiak, S. Gautier, H. Gaucher, S. Bonnamy, F. Beguin, Carbon **37**, 61 (1999)
62. G. T. Wu, C. S. Wang, X. B. Zhang, H. S. Yang, Z. F. Qi, P. M. He, W. Z. Li, J. Electrochem. Soc. **146**(5), 1696-1701 (1999)
63. A. Claye, R. Lee, Z. Benes, J. Fischer, J. Electrochem. Soc. (in press)
64. B. Gao, A. Kelinhammes, X. P. Tang, C. Bower, Y. Wu, O. Zhou, Chem. Phys. Lett. **307**, 153 (1999)
65. O. Zhou, R. M. Fleming, D. W. Murphy, C. T. Chen, R. C. Haddon, A. P. Ramirez, S. H. Glarum, Science **263**, 1744 (1994)
66. S. Suzuki, M. Tomita, J. Appl. Phys. **79**, 3739 (1996)
67. S. Suzuki, C. Bower, O. Zhou, Chem. Phys. Lett. **285**, 230 (1998)
68. B. Gao, C. Bower, O. Zhou (unpublished results)
69. A. C. Dillon, K. M. Jones, T. A. Bekkedahl, C. H. Kiang, D. S. Bethune, M. J. Heben, Nature **386**, 377 (1997)
70. P. Chen, X. Wu, J. Lin, K. Tan, Science **285**, 91 (1999)
71. C. Liu, Y. Y. Fan, M. Liu, H. T. Cong, H. M. Cheng, M. S. Dresselhaus, Science **286**, 1127 (1999)
72. C. Nutenadel, A. Zuttel, D. Chartouni, L. Schlapbach, Solid-State Lett. **2**, 30 (1999)
73. A. Chambers, C. Park, R. T. K. Baker, N. M. Rodriguez, J. Phys. Chem. B **102**, 4253 (1998)
74. M. S. Dresselhaus, K. A. Williams, P. C. Eklund, MRS Bull., **24**, (11), 45 (1999)
75. M. J. Heben, Kirchberg (private communication) (2000)
76. M. Pederson, J. Broughton, Phys. Rev. Lett. **69**, 2689 (1992)
77. Y. Ye, C. C. Ahn, C. Witham, B. Fultz, J. Liu, A. G. Rinzler, D. Colbert, K. A. Smith, R. E. Smalley, Appl. Phys. Lett. **74**, 2307 (1999)
78. P. G. Collins, K. Bradley, M. Ishigami, A. Zettl, Science, **287**, 1801-1804 (2000)
79. X. P.Tang, A. Kleinhammes, H. Shimoda, L. Fleming, K. Y. Bennoune, C. Bower, O. Zhou, Y. Wu, Science, **288**, 492-494 (2000)
80. J. N. Coleman, A. B. Dalton, S. Curran, A. Rubio, A. P. Davey, A. Drury, B. McCarthy, B. Lahr, P. M. Ajayan, S. Roth, R. C. Barklie, W. Blau, Adv. Mater. **12**, 213 (2000)
81. G. Overney, W. Zhong, D. Tomanek, Z. Phys. D **27**, 93 (1993)
82. B. I. Yakobson, C. J. Brabec, J. Bernholc, Phys. Rev. Lett. **76**, 2511 (1996)
83. M. M. J. Treacy, T. W. Ebbesen, J. M. Gibson, Nature **381**, 678 (1996)
84. E. W. Wong, P. E. Sheehan, C. M. Lieber, Science **277**, 1971 (1997)
85. B. Yakobson, Ph. Avouris, see chapter in this volume. (hoped for chapter)
86. M. Yu, O. Lourie, M. J. Dyer, K. Moloni, T. F. Kelly, R. S. Ruoff, Science **287**, 637 (2000)
87. M. R. Falvo, C. J. Clary, R. M. Taylor, V. Chi, F. P. Brooks, S. Washburn, R. Superfine, Nature **389**, 582 (1997)
88. B. I. Yakobson, Appl. Phys. Lett. **72**, 918 (1998)
89. A. J. Stone, D. J. Wales, Chem. Phys. Lett. **128**, 501 (1986)
90. B. I. Yakobson, C. J. Brabec, J. Bernholc, J. Computer Aided Materials Design **3**, 173 (1996)

91. P. M. Ajayan, O. Stephan, C. Colliex, D. Trauth, Science **265**, 1212 (1994)
92. H. D. Wagner, O. Lourie, Y. Feldman, R. Tenne, Appl. Phys. Lett. **72**, 188 (1998)
93. L. Jin, C. Bower, O. Zhou, Appl. Phys. Lett. **73**, 1197 (1998)
94. L. S. Schadler, S. C. Giannaris, P. M. Ajayan, Appl. Phys. Lett. **73**, 26 (1999)
95. C. Bower, R. Rosen, L. Jin, J. Han, O. Zhou, Appl. Phys. Lett. **74**, 3317 (1999)
96. P. Calvert, Nature **399**, 210 (1999)
97. P. M. Ajayan, L. S. Schadler, C. Giannaris, A. Rubio, Adv. Mater. (in press)
98. S. Chang, R. H. Doremus, P. M. Ajayan, R. W. Siegel, unpublished results
99. S. Curran, P. M. Ajayan, W. Blau, D. L. Carroll, J. Coleman, A. B. Dalton, A. P. Davey, B. McCarthy, A. Strevens, Adv. Mater. **10**, 1091 (1998)
100. H. Ago, K. Petritsch, M. S. P. Shaffer, A. H. Windle, R. H. Friend, Adv. Mater. **11**, 1281 (1999)
101. D. L. Carroll, unpublished results
102. M. S. P. Shaffer, A. H. Windle, Adv. Mater. **11**, 937 (1999)
103. M. Endo, unpublished results
104. H. J. Dai, J. H. Hafner, A. G. Rinzler, D. T. Colbert, R. E. Smalley, Nature **384**, 147 (1996)
105. S. S. Wong, J. D. Harper, P. T. Lansbury C. M. Lieber, J. Am. Chem. Soc. **120**, 603 (1998)
106. J. H. Hafner, C. L. Cheung, C. M. Lieber, Nature **398**, 761 (1999)
107. P. Kim, C. M. Lieber, Science **286**, 2148 (1999)
108. S. S. Wong, A. T. Woolley, T. W. Odom, J. L. Huang, P. Kim, D. Vezenov, C. Lieber, Appl. Phys. Lett. **73**, 3465 (1998)
109. J. Chen, M. Hamon, H. Hu, Y. Chen, A. Rao, P. C. Eklund, R. C. Haddon, Science **282**, 95 (1998)
110. S. S. Wong, E. Joselevich, A. T. Woolley, C. L. Cheung, C. M. Lieber, Nature **394** (1998)
111. R. H. Baughman, C. Cui, A. A. Zhakhidov, Z. Iqbal, J. N. Barisci, G. M. Spinks, G. G. Wallace, A. Mazzoldi, D. D. Rossi, A. G. Rinzler, O. Jaschinski, S. Roth, M. Kertesz, Science **284**, 1340 (1999)
112. J. Kong, N. R. Franklin, C. Zhou, M. C. Chapline, S. Peng, K. Cho, H. Dai, Science **287**, 622 (2000)
113. M. R. Pederson, J. Q. Broughton, Phys. Rev. Lett. **69**, 2689 (1992)
114. P. M. Ajayan, T. W. Ebbesen, T. Ichihashi, S. Iijima, K. Tanigaki, H. Hiura, Nature **362**, 522 (1993)
115. P. M. Ajayan, O. Stephan, P. Redlich, C. Colliex, Nature **375**, 564 (1995)
116. R. S. Ruoff, D. C. Lorents, B. Chan, R. Malhotra, S. Subramoney, Science **259**, 346 (1992)
117. C. Guerret-Plecourt, Y. Le Bouar, A. Loiseau, H. Pascard, Nature **372**, 761 (1994)
118. Y. Zhang, K. Suenaga, C. Colliex, S. Iijima, Science **281**, 973 (1998)
119. J. Hu, M. Ouyang, P. Yang, C. M. Lieber, Nature **399**, 48 (1999)
120. Y. Zhang, T. Ichihashi, E. Landree, F. Nihey, S. Iijima, Science **285**, 1719 (1999)
121. F. Balavoine, P. Schultz, C. Richard, V. Mallouh, T. W. Ebbesen, C. Mioskowsk, Angew. Chem. **111**, 2036 (1999)
122. R. S. Lee, H. J. Kim, J. E. Fischer, A. Thess, R. E. Smalley, Nature **388**, 255 (1997)

123. H. J. Dai, E. W. Wong, Y. Z. Lu, S. S. Fan, C. M. Lieber, Nature **375**, 769 (1995)
124. R. Tenne, A. Zettl, see chapter in this volume
125. M. Endo, Ph.D. Thesis (1975)
126. J. C. Charlier, S. Iijima, see chapter in this volume

# Index

# Topics in Applied Physics

63 **Hydrogen in Intermetallic Compounds I**
Electronic, Thermodynamic, and Crystallographic Properties, Preparation
By L. Schlapbach (Ed.) 1988. 118 figs. 12 tabs. XIV, 350 pages

64 **Sputtering by Particle Bombardment III**
Characteristics of Sputtered Particles, Technical Applications
By R. Behrisch and K. Wittmaack (Eds.) 1991. 190 figs. 6 tabs. XV, 410 pages

65 **Laser Spectroscopy of Solids II**
By W. M. Yen 1989. 144 figs. XI, 307 pages

66 **Light Scattering in Solids V**
Superlattices and Other Microstructures
By M. Cardona and G. Güntherodt (Eds.) 1989. 184 figs. XIII, 351 pages

67 **Hydrogen in Intermetallic Compounds II**
Surface and Dynamic Properties, Applications
By L. Schlapbach (Ed.) 1992. 126 figs. XIV, 328 pages

68 **Light Scattering in Solids VI**
Recent Results, Including High-$T_c$ Superconductivity
By M. Cardona and G. Güntherodt (Eds.) 1991. 267 figs., 31 tabs., XIV, 526 pages

69 **Unoccupied Electronic States**
Fundamentals for XANES, EELS, IPS and BIS
By J. C. Fuggle and J. E. Inglesfield (Eds.) 1992. 175 figs. XIV, 359 pages

70 **Dye Lasers: 25 Years**
By M. Stuke (Ed.) 1992. 151 figs. XVI, 247 pages

71 **The Monte Carlo Method in Condensed Matter Physics**
By K. Binder (Ed.) 2nd edn. 1995. 83 figs. XX, 418 pages

72 **Glassy Metals III**
Amorphization Techniques, Catalysis, Electronic and Ionic Structure
By H. Beck and H.-J. Güntherodt (Eds.) 1994. 145 figs. XI, 259 pages

73 **Hydrogen in Metals III**
Properties and Applications
By H. Wipf (Ed.) 1997. 117 figs. XV, 348 pages

74 **Millimeter and Submillimeter Wave Spectroscopy of Solids**
By G. Grüner (Ed.) 1998. 173 figs. XI, 286 pages

75 **Light Scattering in Solids VII**
Christal-Field and Magnetic Excitations
By M. Cardona and G. Güntherodt (Eds.) 1999. 96 figs. X, 310 pages

76 **Light Scattering in Solids VIII**
C60, Semiconductor Surfaces, Coherent Phonons
By M. Cardona and G. Güntherodt (Eds.) 1999. 86 figs. XII, 228 pages

77 **Photomechanics**
By P. K. Rastogi (Ed.) 2000, 314 Figs. XVI, 472 pages

78 **High-Power Diode Lasers**
By R. Diehl (Ed.) 2000, 260 Figs. XIV, 416 pages

79 **Frequency Measurement and Control**
Advanced Techniques and Future Trends
By A. N. Luiten (Ed.) 2001, 169 Figs. XIV, 394 pages

80 **Carbon Nanotubes**
Synthesis, Structure, Properties, and Applications
By M. S. Dresselhaus, G. Dresselhaus, Ph. Avouris (Eds.) 2001, 235 Figs. XVI, 448 pages

MIX
Papier aus verantwortungsvollen Quellen
Paper from responsible sources
FSC® C105338

If you have any concerns about our products,
you can contact us on
**ProductSafety@springernature.com**

In case Publisher is established outside the EU,
the EU authorized representative is:
**Springer Nature Customer Service Center GmbH**
**Europaplatz 3, 69115 Heidelberg, Germany**

Printed by Libri Plureos GmbH
in Hamburg, Germany